Spectroscopy: Processing, Analysis and Interpretation

Spectroscopy: Processing, Analysis and Interpretation

Edited by
Ryan Cooper

MURPHY & MOORE
www.murphy-moorepublishing.com

Published by Murphy & Moore Publishing,
1 Rockefeller Plaza,
New York City, NY 10020, USA

ISBN: 978-1-63987-503-0

Cataloging-in-Publication Data

Spectroscopy : processing, analysis and interpretation / edited by Ryan Cooper.
p. cm.
Includes bibliographical references and index.
ISBN 978-1-63987-503-0
1. Spectrum analysis. 2. Spectroscope. I. Cooper, Ryan.
QD95 .S64 2022
543.085 8--dc23

For information on all Murphy & Moore Publications
visit our website at www.murphy-moorepublishing.com

MURPHY & MOORE

Contents

Preface

The study of the interaction between matter and electromagnetic radiation as a function of the wavelength or frequency of the radiation is known as spectroscopy. It focuses on the precise study of color as generalized from visible light to all bands of the electromagnetic spectrum. It is also involved in the study of the wavelength dependence of the absorption by a gas-phase matter of visible light dispersed by a prism. Spectroscopy is a fundamental exploratory tool in the fields of physics, chemistry, and astronomy within the electromagnetic spectrum. Some of the important applications that arise from biomedical spectroscopy are in the areas of tissue analysis and medical imaging. This book outlines the processes and applications of spectroscopy in detail. It will also provide interesting topics for research which interested readers can take up. As this field is emerging at a rapid pace, the contents of this book will help the readers understand the modern concepts and applications of the subject.

Various studies have approached the subject by analyzing it with a single perspective, but the present book provides diverse methodologies and techniques to address this field. This book contains theories and applications needed for understanding the subject from different perspectives. The aim is to keep the readers informed about the progresses in the field; therefore, the contributions were carefully examined to compile novel researches by specialists from across the globe.

Indeed, the job of the editor is the most crucial and challenging in compiling all chapters into a single book. In the end, I would extend my sincere thanks to the chapter authors for their profound work. I am also thankful for the support provided by my family and colleagues during the compilation of this book.

Editor

1

An Approach to Rapid Determination of Tween-80 for the Quality Control of Traditional Chinese Medicine Injection by Partial Least Squares Regression in Near-Infrared Spectral Modeling

Jin-Fang Ma,[1,2] Tian-Ling Chen,[2,3] Xiang-Dong Zhang,[2] Xue Xiao,[3] and Fa-Huan Ge[1,2]

[1]*School of Pharmaceutical Sciences, Sun Yat-sen University, Guangzhou 510006, China*
[2]*Nansha Research Institute, Sun Yat-sen University, Guangzhou 511458, China*
[3]*Guangdong Pharmaceutical University, Guangzhou 510006, China*

Correspondence should be addressed to Xue Xiao; erxiaohappy@163.com and Fa-Huan Ge; gefahuan@mail.sysu.edu.cn

Academic Editor: Jose S. Camara

This study established an approach to rapidly determine the Tween-80 in traditional Chinese medicine (TCM) injection by using near-infrared (NIR) spectroscopy. Totally 133 standard solutions of Tween-80 were prepared and randomly divided into calibration set and validation set, containing 109 and 24 samples, respectively. Spectral data were preprocessed and then subjected to establish a predictive model using partial least-squares (PLS). The standard error of cross validation (SECV), standard deviation of calibration (SEC), and the determination coefficient (R) of the established model were 0.0561, 0.0525, and 0.9986, respectively. The model was successfully applied to determine Tween-80 contents in 25 *XBJ Injections* and 40 *FFSX Injections*, and it produced satisfactory quantitative analysis results with average relative deviations 0.60% and 0.16%, respectively, for 25 *XBJ Injections* and 40 *FFSX Injections*, and the maximum relative deviations 8.57% and 7.60%, respectively. This work shows that NIR model displayed quite good predictive ability for Tween-80 quantitative analysis, which could potentially be applied to rapid determination of Tween-80 in the production process of TCM injections and other TCM products.

1. Introduction

Traditional Chinese medicine (TCM) injection is a new dosage form developed in the modernization of TCM. It displays exact efficacy in clinical practice but raises a controversy either due to its possible adverse reactions [1–3]. TCM injections are generally produced from medicinal materials with sources of plants or animals, and both the complicated ingredients thereof and the production process may result in uncontrollable and random factors affecting the quality of TCM injections. Moreover, TCM injections are typically added with excipients, such as various solubilizers, which may also affect the quality of the products. Therefore, TCM injection is facing high-quality risks, and accordingly, more perfect means of quality control are urgently required [4–6].

However, the current quality control method of TCM injections is still at infancy. According to Chinese Pharmacopoeia (2015), quality control of TCM injection is mainly conducted via determining some active ingredients, while nonbioactive additives and excipients thereof were often ignored. A representative pharmaceutical excipient is Tween-80 (see its chemical structure in Figure 1).

As a common solubilizer, emulsifier, or stabilizer, Tween-80 is often used in a number of TCM injections, such as *XBJ Injection* and *FFSX Injection*. In general, nearly 2.5% Tween-80 [7] is commonly added to TCM injection, particularly in TCM injection with volatile components as the main active ingredients. Tween-80 is a nonbioactive adjuvant, but recent studies have found that drugs containing Tween-80 exhibit sensitization, hemolysis, hepatotoxicity, or peripheral neurotoxicity activities [8–10] and may even correlate with anaphylactic reaction [1–3]. Therefore, the FDA has issued a security warning for Tween-80 [11]. So far, adverse reactions of TCM injections have been frequently

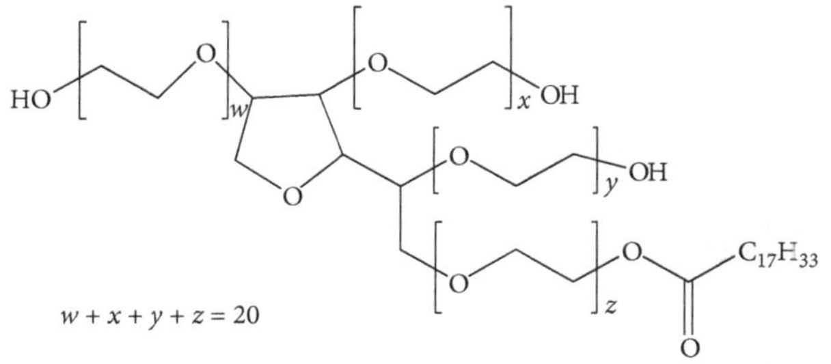

Figure 1: Chemical structural of Tween-80.

reported [1–3], and there are evidences indicating that Tween-80 might be, at least in part, responsible for those issues [12, 13]. So, Tween-80 should attract our more attention in the quality control of TCM injections.

Some routine analytical methods, such as gas chromatography and ultraviolet spectrophotometry, were widely used to determine Tween-80 content [14–17]. However, these methods are mostly cumbersome and time-consuming, and the determination process may suffer from slight deviations; as a result, sample quality cannot be ascertained in a timely manner. Thus, a rapid, efficient, and accurate quality control method is needed.

Near-infrared (NIR) spectroscopy has been ever increasingly accepted as a fast and green process analysis technology in chemical industry, agriculture, food, medicine, tobacco, and some other fields [18–22]. The frequency doubling of the stretching vibration of compounds X-H (X = C, O, N, etc.) causes absorption in NIR spectroscopy. NIR combined with chemometrics algorithms can qualitatively and quantitatively determine unknown samples [23, 24], thereby achieving the purpose of quality control. Researchers have used different concentrations of ethanol solution to quickly determine the alcoholicity of liquor and have found good application prospect [25].

This work intends to use the NIR spectrum of different concentrations of Tween-80 standard solutions to establish an optimized PLS model. By using the proposed model, Tween-80 contents can be predicted in TCM injections (see Figure 2). This work is expected to provide a promising method for the quality control of TCM injection and gain further insight into the preparation of medicines containing Tween-80.

2. Materials and Methods

2.1. Materials. Tween-80 (batch number 20120201) was obtained from Nanjing Well Chemical Co., Ltd. (Nanjing, China). 25.0 g Tween-80 was accurately weighted and completely dissolved in pure water to a final volume of 500 mL. Then, 133 Tween-80 standard solutions were then prepared by stock solution dilution, with concentrations uniformly distributed at the range of 0.1~5% (g/g).

25 *XBJ Injection* samples (Tween-80 labeled amount 0.14%, g/g) were provided by the manufacturer A (Tianjin, China). These samples involve nine batches, with batch numbers as follows: 1212051, 1212151, 1212211, 1212231, 1301061, 1301121, 1301151, 1301161, and 1303171. 40 *FFSX Injection* samples (Tween-80 labeled amount 2.00%, g/g) were provided by the manufacturer B (Shijiazhuang, China). These samples involve eight batches, with batch numbers as follows: 12021031, 12021131, 12021141, 12022041, 12022441, 12022941, 12022942, and 12022942.

2.2. NIR Apparatus and Software. A Bruker spectrometer (Bruker Optics, Germany) was used for spectral acquisition. Transmission spectra of samples were scanned in a 2 mm light path. Each spectrum was acquired as the average of 32 scans on the same sample in the wavelength range of 12,000 to 4000 cm^{-1}. The spectral resolution was kept to 8.0 cm^{-1}.

Processes of spectral wavebands selection, spectral pretreatment, and PLS modeling were performed using THUNIR software (Tsinghua University version 3.0, Beijing) [26–29].

2.3. Spectral Data Pretreatment. In order to obtain reliable, accurate, and stable calibration models, it is necessary to preprocess the spectral data before modeling. Spectral pretreatment was performed using Savitsky-Golay (SG) smoothing, first-order and second-order convolutional derivation, multiplicative scatter correction (MSC), standard normal variant (SNV), and normalization. The effects of different pretreatment methods were compared, and the optimized combination was selected.

2.4. Calibration of Models. The predictive model was constructed based on NIR data of 133 Tween-80 standard solution samples. Partial least squares (PLS) combines data matrix decomposition with regression interaction. The PLS model is robust and reduces the interference caused by light scattering and other components. PLS can eliminate noises to establish a well-stabilized model with prediction ability, thereby solving the model's overfitting and variable collinearity. In NIR-based modeling, PLS is the most commonly used linear regression method [30–32]. In this study, the PLS method was thus used to extract valid information from the NIR spectrum and a quantitative calibration model was established. The performance of the established model was evaluated via indicators of standard error of cross validation (SECV), the standard error of calibration (SEC), the primary factor (LV), and the determination coefficient (R) [28, 33–35].

The established quantitative calibration model was used to predict the Tween-80 content in 25 *XBJ Injection* samples and 40 *FFSX Injection* samples.

3. Results and Discussion

3.1. NIR Spectral Features. The selected spectral scanning range was 12000–4000 cm^{-1}, and the measured characteristics were the fundamental frequencies of molecular vibration, particularly the frequency doubling and combined frequency absorption bands of the stretching and bending vibration [36–39]. The original and pretreated spectra of Tween-80 are shown in Figure 3. In the original spectra, 5500–5800 cm^{-1} and 6500–7300 cm^{-1} had significant

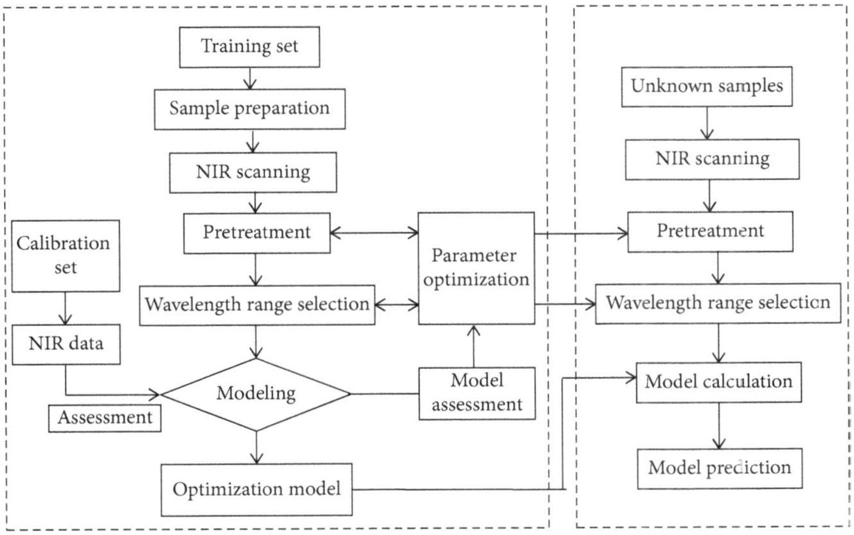

FIGURE 2: Experimental flow chart.

responses, which were related to the second-order frequency doubling of -CH_2 and -CH and the three-stage frequency doubling of C=O [40–43]. Figure 4(a) summarizes the shifts in the NIR spectrum baseline. These three bonds are widespread in the target molecules, so their concentration information can be reflected in the NIR spectra and this can be viewed as the foundation of NIR quantitative analysis of the Tween-80.

3.2. Calibration of the Model

3.2.1. Calibration and Validation Sets. 133 samples were randomly divided into calibration and validation sets, which contained, respectively, 109 and 24 samples. The calibration set was used to construct a predictive model for quantitative determination of Tween-80 in TCM injections, and the validation set was applied as blind samples to investigate the accuracy of the proposed method.

3.2.2. Data Pretreatment. NIR spectroscopy inevitably contains noises that are independent of the nature of the sample. Pretreatment of the spectral data can effectively filter the noise of the NIR spectrum, thereby ensuring a good correlation between the spectral data and the components, reducing the complexity of the quantitative model, and improving the accuracy of the model [22]. We investigated different spectral preprocessing methods and compared their SEC, SECV, and R values, as listed in Table 1. After SG smoothing on spectra, the SEC and SECV values obviously decreased and R values were close to 1, which suggested pretreatment of convolution smoothing allows a superior model performance.

3.2.3. Waveband Selection. The band of near-infrared spectrum was 14286–4000 cm^{-1}, which was divided into short-wave and long-wave near-infrared regions. The short-wave near-infrared region was a high-frequency absorption band, and the long-wave near-infrared region was a first- and second-order frequency doubling absorption band [44]. To obtain more information, the available area was covered up to the entire near-infrared band, which had a wider spectral range than that of the short-wave or long-wave near-infrared alone. As shown in Figure 3, most of the spectral regions were smooth. However, noise was detected in the spectra of 5300–4800 and 420–4000 cm^{-1}, which were discarded during the calibration process. In this research, the selected wavelength range was 5780–5870.7 cm^{-1}. To verify the correctness of the band selection, the relevant coefficients of the spectrum in this band were studied. As shown in Figure 4(b), the variables with higher coefficients (>0.4) were mostly distributed in the selected region. Therefore, the selected band (5780–5870.7 cm^{-1}) was the best modeling band.

3.2.4. Optimum LVs. Selection of the primary factors directly affects the prediction ability of calibration model. For instance, if the selected primary factor is excessively small, the useful information of the original spectrum will be lost and may result in insufficient fitting, while the primary factor is excessively large, it will include excessive measured noise, leading to an overfitting phenomenon and a significant increase in the prediction error of the built model. To avoid "under-fitting" and "overfitting," the optimal LV values must be used to establish the appropriate model [45–48]. In this study, we acquire optimum LVs by comparing the correlation diagram of the SECV and LV values. The SECV value is observed decreasing as the LV increasing, and LV reaches the optimal value when the SECV approached a constant value, as shown in Figure 4(c).

3.2.5. Model Establishment. The PLS method has been most commonly used in NIR spectroscopy-based quantitative

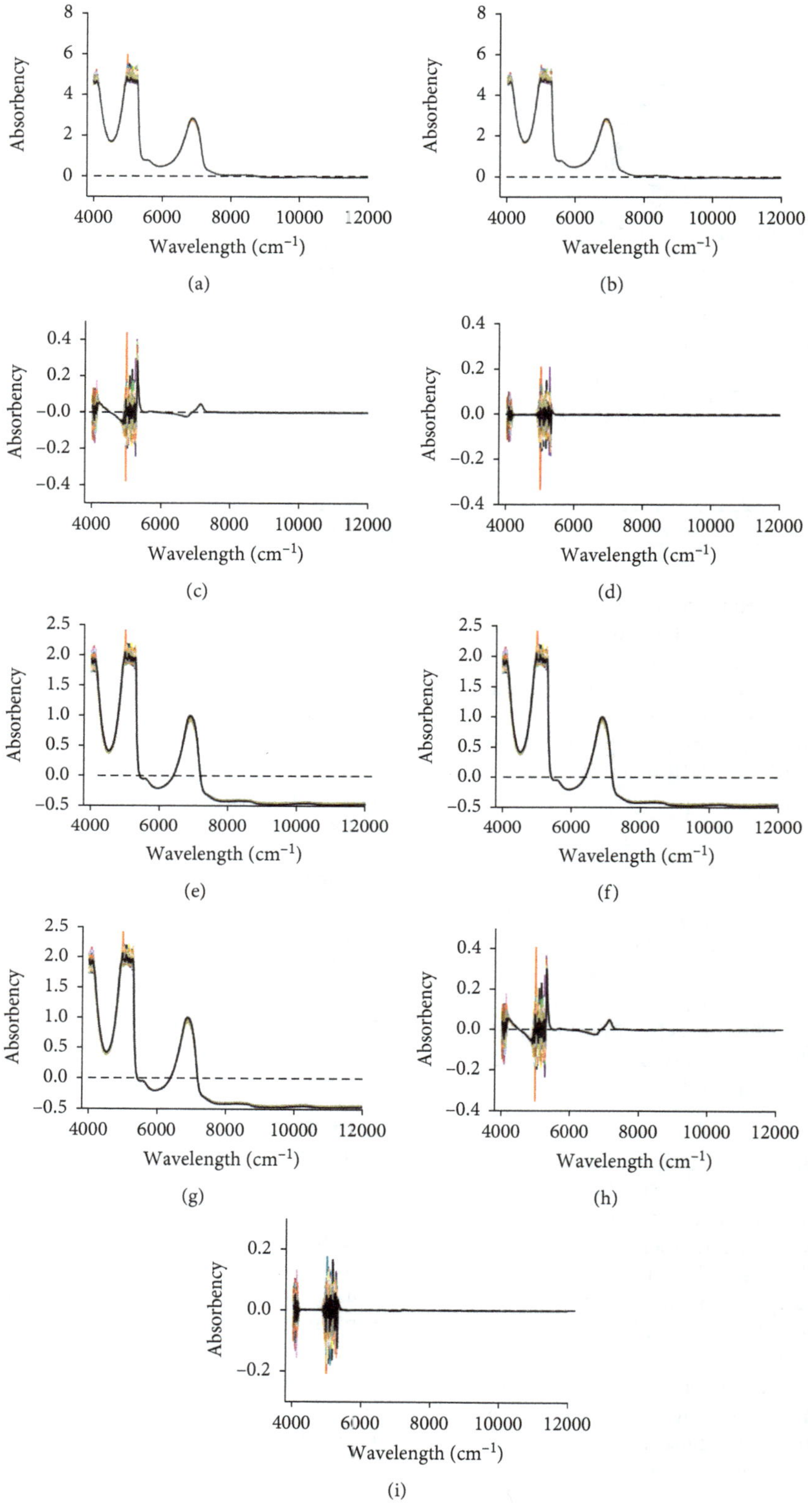

FIGURE 3: Comparison of the spectrum pretreatment results of Tween-80. Raw NIR spectra of full wavelength (a) and spectra preprocessed using SG smoothing (b) and one-dimensional convolution (c), two-dimensional convolution (d), multiplicative scatter correction (e), standard normal variable transformation (f), normalization method (g), one-order derivative convolution + multiplicative scatter correction (h) and two-dimensional convolution + multiplicative scatter correction (i).

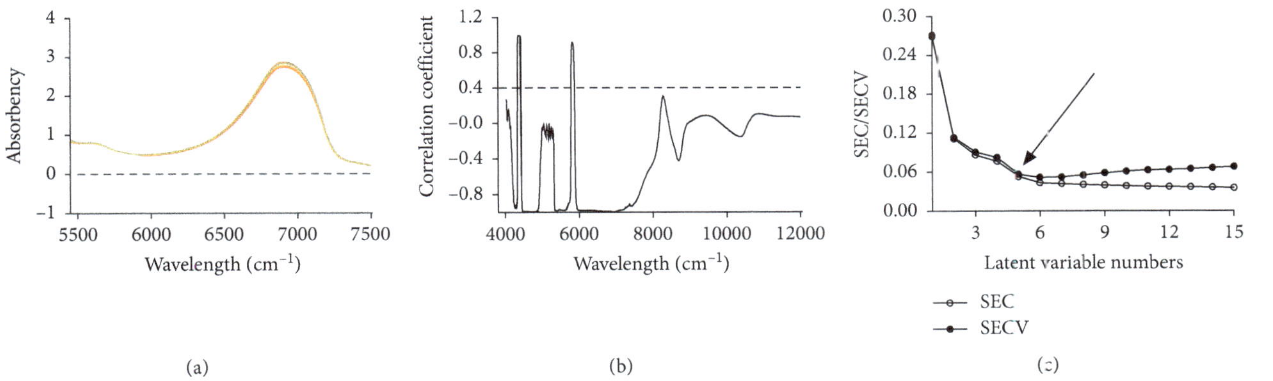

Figure 4: Enlarged view of the shifts in the NIR spectrum baseline (a). Correlogram of the NIR spectra and Tween-80 concentration (b). Relational graph of LV numbers and SECV of Tween-80 (c).

Table 1: Comparison of different calibration models of Tween-80 content developed with different spectral preprocessing methods.

No.	Pretreatment methods	LVs	Calibration		Cross validation	
			R	SEC	R	SECV[c]
B	SG smoothing	5	0.9986	0.0526	0.9984	0.0561
C	One DC	5	0.9985	0.0523	0.9980	0.0635
D	Two DC	5	0.9985	0.0523	0.9984	0.0561
E	MSC	5	0.9983	0.0571	0.9981	0.0615
F	SNV	5	0.9980	0.0614	0.9978	0.0660
G	Normalization	5	0.9985	0.0523	0.9984	0.0561
H	One DC + MSC	4	0.9964	0.0834	0.9958	0.8840
I	Two DC + MSC	4	0.9964	0.834	0.9981	0.8840

LV: latent variable number; SEC: standard error of calibration; SECV: standard error of cross validation; DC: dimension convolution; MSC: multiple scatter correction; SNV: standard normal variate.

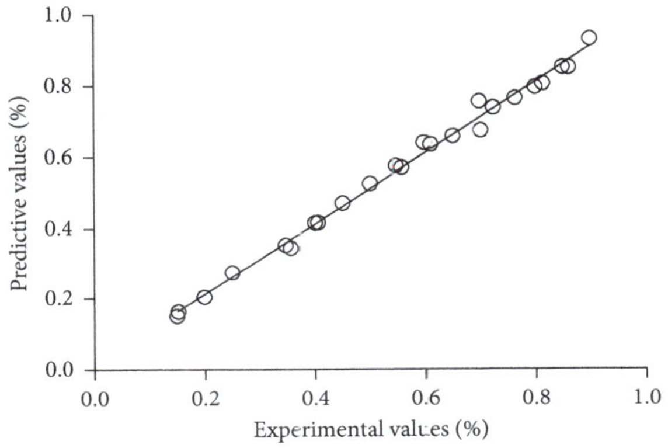

Figure 5: Fitting trend diagram of Tween 80 verification set.

modeling. Data of Tween-80 NIR spectrum were pretreated and then used to establish a calibration model. The SECV, SEC, LV, and R values were investigated to evaluate the model performance, and the result showed that the models had low SEC and SECV values and R value close to 1. The best modeling band (5780–5870.7 cm^{-1}) was selected, which allowed acceptable model performance of the Tween-80 calibration model, with parameters as follows: LVs 5, SECV values 0.0561, SEP values of 0.0526, and the R value 0.9984.

3.2.6. Model Validation. The established calibration model should take an external validation to verify its accuracy. Samples of validation set were tested as blind samples using the established model, and the result indicates that the correlation of the prediction values and factual values R is most close to 1 ($R = 0.9968$). Figure 5 indicates that the calibration model was acceptable.

3.2.7. Model Application. The Tween-80 labeled amounts in *XBJ Injection and FFSX Injection* were 0.14% and 2.00%, respectively. The established Tween-80 calibration model was used to predict the Tween-80 content in 25 samples of *XBJ Injection* and 40 samples of *FFSX Injection*. Table 2 lists the predicted value and the labeled value. The results showed that in samples of both *XBJ Injection* and *FFSX Injection*, the predicted values of the Tween-80 calibration model strongly matched with the true value. For *XBJ Injection* and *FFSX Injection*, the average relative deviations were 0.60% and 0.16%, respectively, and the maximum relative deviations were 8.57% and 7.60%, respectively.

4. Discussion

In the past decade, process analysis technology (PAT) gradually became an important technology for production process controlling in chemical engineering, environmental, agricultural, food, and other fields, along with the acceptance of concept of "*quality by design*." As one of the most common PAT technologies, NIR is now widely applied in these fields. However, either for TCM products or production process, the mainstream quality control method is still an offline analysis approach. In recent years, people come to recognize the advantages of the NIR-based method, i.e., fast, real-time, in situ, nondestructive, and accurate, and NIR shows great expectation in the quality control for TCM injections. Moreover, presently most quality control methods focus on the so-called bioactive components of TCM, while nearly take no consideration on other additive

Table 2: Experimental vs. predictive values of Tween-80 in *XBJ Injection* and *FFSX Injection*.

No.	Batch	EV	PV	Ea	RD	No.	Batch	EV	PV	Ea	RD
1	1212051	0.140	0.151	0.011	7.857	34	12021131	2.000	2.001	0.001	0.050
2		0.140	0.143	0.003	2.143	35		2.000	2.016	0.016	0.800
3	1212151	0.140	0.133	−0.007	−5.000	36		2.000	1.987	−0.013	−0.650
4		0.140	0.140	0.000	0.000	37		2.000	2.009	0.009	0.450
5		0.140	0.152	0.012	8.571	38	12021141	2.000	1.987	−0.013	−0.650
6	1212211	0.140	0.150	0.010	7.143	39		2.000	1.968	−0.032	−1.600
7		0.140	0.149	0.009	6.429	40		2.000	1.965	−0.035	−1.750
8		0.140	0.151	0.011	7.857	41		2.000	1.936	−0.064	−3.200
9	1212231	0.140	0.144	0.004	2.857	42		2.000	1.940	−0.060	−3.000
10		0.140	0.142	0.002	1.429	43	12022041	2.000	1.929	−0.071	−3.550
11		0.140	0.137	−0.003	−2.143	44		2.000	1.943	−0.057	−2.850
12	1301061	0.140	0.138	−0.002	−1.429	45		2.000	1.957	−0.043	−2.150
13		0.140	0.149	0.009	6.429	46		2.000	2.152	0.152	7.600
14		0.140	0.136	−0.004	−2.857	47		2.000	2.130	0.130	6.500
15	1301121	0.140	0.135	−0.005	−3.571	48	12022441	2.000	2.132	0.132	6.600
16		0.140	0.136	−0.004	−2.857	49		2.000	2.126	0.126	6.300
17		0.140	0.134	−0.006	−4.286	50		2.000	2.110	0.110	5.500
18	1301151	0.140	0.139	−0.001	−0.714	51		2.000	2.003	0.003	0.150
19		0.140	0.140	0.000	0.000	52		2.000	2.002	0.002	0.100
20		0.140	0.133	−0.007	−5.000	53	12022941	2.000	1.992	−0.008	−0.400
21	1301161	0.140	0.132	−0.008	−5.714	54		2.000	2.000	0.000	0.000
22		0.140	0.135	−0.005	−3.571	55		2.000	1.984	−0.016	−0.800
23		0.140	0.136	−0.004	−2.857	56		2.000	1.977	−0.023	−1.150
24	1303171	0.140	0.144	0.004	2.857	57		2.000	1.995	−0.005	−0.250
25		0.140	0.142	0.002	1.429	58	12022942	2.000	1.998	−0.002	−0.100
26		2.000	1.982	−0.018	−0.900	59		2.000	1.961	−0.039	−1.950
27		2.000	1.999	−0.001	−0.050	60		2.000	1.974	−0.026	−1.300
28	12021031	2.000	1.984	−0.016	−0.800	61		2.000	1.944	−0.056	−2.800
29		2.000	1.973	−0.027	−1.350	62		2.000	1.949	−0.051	−2.550
30		2.000	1.976	−0.024	−1.200	63	12022943	2.000	1.933	−0.067	−3.350
31		2.000	2.035	0.035	1.750	64		2.000	1.911	−0.089	−4.450
32	12021131	2.000	2.041	0.041	2.050	65		2.000	1.927	−0.073	−3.650
33		2.000	2.041	0.041	2.050	—	—	—	—	—	—
		EV	[a]PV	[b]Ea	RD			EV	PV	Ea	RD
Mean (no. 1–25, *XBJ Injection*)		0.140	0.141	0.001	0.600	Mean (no. 26–65, *FFSX Injection*)		2.000	1.997	−0.003	−0.164
SEP (no. 1–25, *XBJ Injection*)			0.006			SEP (no. 26–65, *FFSX Injection*)			0.059		

PV: predicted values; Ea: absolute error; SEP: standard error of prediction; RD: relative deviation.

nonactive ingredients. For instance, Tween-80, as one of the most commonly used additives in TCM injection, might lead to potential quality risks of products, whereas it has not been paid much attention to the quality control. This work established a NIR-based quality control approach for Tween-80 in TCM injection, which can be used not only for the fast quality detection of Tween-80 in the terminal product of TCM injection but for the real-time detection of Tween-80 in the production process, due to its characteristics of rapid, in situ, and nondestructive. Moreover, the proposed method also can be used PAT technology with good adaptability in other TCM products and produce process.

For NIR modeling, the commonly used algorisms include multivariate linear regression (MLR), principal component regression (PCR), and partial least squares regression (PLSR). MLR often produces unsatisfactory prediction performance due to the problem of collinearity of spectral variables; PCR only decomposes the spectral array to eliminate the noise while PLSR decomposes both the spectral array and the concentration array simultaneously, which not only overcomes the disadvantage of the PCR with only data of spectra decomposed but resolves problems of variable collinearity and model overfitting. Therefore, the PLS modeling method is used in this study to construct calibration model.

This work concerned the quality control of Tween-80 in TCM injection but did not promote the method in the application for other TCM dosage forms containing Tween-80. In our further study, the proposed NIR-based Tween-80 quality control method would be improved and developed to be applied in other TCM, and even in food, cosmetics, and other industries.

5. Conclusions

In this study, a method for rapidly determining the Tween-80 in TCM injection by NIR spectroscopy was established for the first time. The proposed method is simple and accurate, which may be suitable for application in the inspection in production processes and in final product quality assessment for TCM injection, thereby may promote the quality control of TCM injection and even expand the strategy for the quality control of the production process in the pharmaceutical industry.

Additional Points

Highlights. Rapid determination of Tween-80 in TCM injection by using near-infrared spectroscopy was established. This work could promote the quality control of TCM injection and expand the strategies for the quality control of the production process in the pharmaceutical industry.

Disclosure

Jin-Fang Ma and Tian-Ling Chen are the co-first authors.

Conflicts of Interest

The authors declare there are no conflicts of interest.

Acknowledgments

This work was funded by S&T Plan Project of Guangzhou (201803010104), the National Key R&D Program of China (2017YFC1703102), and Pearl River S&T Nova Program of Guangzhou (201610010113).

References

[1] Y. Xu, C. Liu, D. Dou, and Q. Wang, "Evaluation of anaphylactoid constituents *in vitro* and *in vivo*," *International Immunopharmacology*, vol. 43, pp. 79–84, 2017.

[2] Y. B. Xu, T. G. Kang, D. Q. Dou, and H. X. Kuang, "The evaluation and optimization of animal model for anaphylactoid reaction induced by injections," *Asian Pacific Journal of Allergy and Immunology*, vol. 33, no. 4, pp. 330–338, 2015.

[3] E. A. Coors, H. Seybold, H. F. Merk, and V. Mahler, "Polysorbate 80 in medical products and nonimmunologic anaphylactoid reactions," *Annals of Allergy, Asthma & Immunology*, vol. 95, no. 6, pp. 593–599, 2005.

[4] H. Li, J. Deng, L. Deng, X. Ren, and J. Xia, "Safety profile of traditional Chinese herbal injection: an analysis of a spontaneous reporting system in China," *Pharmacoepidemiology and Drug Safety*, vol. 28, no. 7, pp. 1002–1013, 2019.

[5] H. Li, S. Wang, Z. Yue, X. Ren, and J. Xia, "Traditional Chinese herbal injection: current status and future perspectives," *Fitoterapia*, vol. 129, pp. 249–256, 2018.

[6] C. Wang, Q.-P. Shi, F. Ding et al., "Reevaluation of the post-marketing safety of Shuxuening injection based on real-world and evidence-based evaluations," *Drug Design, Development and Therapy*, vol. 12, pp. 757–767, 2018.

[7] H. Y. Zhang, W. K. Wu, Y. C. Wei et al., "Safety status of solubilizing auxiliary materials commonly used for injection," *Shaanxi Journal of Traditional Chinese Medicine*, vol. 32, no. 4, pp. 503–505, 2011.

[8] B. J. Zhu, T. T. Li, Z. L. Zhou, and D. Luo, "The associativity study of polysorbate 80 in shenmai injection and haemolysis rate," *Tech Square Magazine*, vol. 10, pp. 71–73, 2010.

[9] X. Luo, Q. Wang, L. Zhou, Y. Dong, and Y. P. Jiang, "Study on relationship between tween-80 contained in injections of Chinese herbal medicine and anaphylactic reactions," *Australasian Dispute Resolution Journal*, vol. 12, no. 3, pp. 160–166, 2010.

[10] Y. T. Wang, C. Y. Li, Y. Yi et al., "Comparative study on pseudoanaphylactoid reactions induced by medicinal tween 80 and injectable tween80," *China Journal of Chinese Materia Medica*, vol. 37, no. 13, pp. 1890–1898, 2012.

[11] FDA, *Inactive Ingredient Search for Approved Drug Products*, 2018, http://www.accessdata.fda.gov/scripts/cder/iig/index.cfm?event=BasicSearch.page.

[12] S. Q. Chen, *Study on Clinical Application Risk Assessment of Xiyanping Injection and Revision Procedure of Its Specification*, Beijing University of Chinese Medicine, Beijing, China, 2019.

[13] R. Yang, Q.-C. Lao, H.-P. Yu et al., "Tween-80 and impurity induce anaphylactoid reaction in zebrafish," *Journal of Applied Toxicology*, vol. 35, no. 3, pp. 295–301, 2015.

[14] J. Puschmann, D.-H. Evers, C. C. Müller-Goymann, and M. E. Herbig, "Development of a design of experiments optimized method for quantification of polysorbate 80 based on oleic acid using UHPLC-MS," *Journal of Chromatography A*, vol. 1599, pp. 136–143, 2019.

[15] L. X. Nie, Y. Chang, Z. Dai et al., "A new method for determination of polysorbate 80 in shengmai injection based on absorption coefficient," *Spectroscopy and Spectral Analysis*, vol. 39, no. 1, pp. 199–203, 2019.

[16] W. Li, L. Li, and Z. J. Chen, "Determination of CMC of tween-80 in Chinese medicine injection," *Journal of Liaoning University of TCM*, vol. 14, no. 2, pp. 181–183, 2012.

[17] Y. Wu, X. D. Li, R. C. Lin, and S. H. Jin, "Development of a size-exclusion HPLC method with evaporative light-scattering detection for the quantitation of polysorbate 80 in Houttuynia cordata injection," *Journal of AOAC International*, vol. 93, no. 3, pp. 917–921, 2010.

[18] W. Li, H. Han, L. Zhang, Y. Zhang, and H. Qu, "A feasibility study on the non-invasive analysis of bottled Compound E Jiao oral liquid using near infrared spectroscopy," *Sensors and Actuators B: Chemical*, vol. 211, pp. 131–137, 2015.

[19] Z.-S. Wu, L.-W. Zhou, S.-Y. Dai, X.-Y. Shi, and Y.-J. Qiao, "Evaluation of the value of near infrared (NIR) spectromicroscopy for the analysis of glycyrrhizic acid in licorice," *Chinese Journal of Natural Medicines*, vol. 13, no. 4, pp. 316–320, 2015.

[20] C. Y. Hou, L. Y. Duan, Y. L. Li et al., "Study on comprehensive evaluation of *Ligustrum lucidum* based on near infrared spectroscopy," *Journal of Hanan Institute of Science and Technology (Natural Science Edition)*, vol. 47, no. 1, pp. 47–52, 2019.

[21] L. Yin, J. Zhou, D. Chen et al., "A review of the application of near-infrared spectroscopy to rare traditional Chinese medicine," *Spectrochimica Acta Part A: Molecular and Biomolecular Spectroscopy*, vol. 221, no. 5, article 117208, 2019.

[22] R. Peng, Z. He, T. Gou, J. Du, and H. Li, "Detection of parameters in solid state fermentation of Monascus by near infrared spectroscopy," *Infrared Physics & Technology*, vol. 96, pp. 244–250, 2019.

[23] X. Q. Cao, C. Jia, L. F. Huang, X. L. Liu, and D. Wang, "Rapid determination of α-cyperone in cyperi rhizoma stir-fried with

vinegar by near-infrared spectroscopy," *Chinese Journal of Experimental Traditional Medical Formulae*, vol. 24, no. 7, pp. 27–32, 2018.

[24] W. Nie, P. L. Zhu, H. L. Fang, B. Gong, and Y. C. He, "Quality evaluation of Callicarpa kwangtungensis from different origins by HPLC coupled with chemometrics," *Chinese Traditional and Herbal Drugs*, vol. 48, no. 1, pp. 185–191, 2017.

[25] F. Qu, D. Ren, J. Wang, Z. Zhang, N. Lu, and L. Meng, "An ensemble successive project algorithm for liquor detection using near infrared sensor," *Sensors*, vol. 16, no. 1, p. 89, 2016.

[26] W. Pan, J. Ma, X. Xiao et al., "Near-infrared spectroscopy assay of key quality-indicative ingredients of Tongkang tablets," *AAPS PharmSciTech*, vol. 18, no. 3, pp. 913–919, 2016.

[27] Z. L. Zhou, J. Li, S. Q. Huang et al., "Development of chemometric modelling in the application of NIR to the quality control of Chinese herbal medicine: literature review and future perspectives," *Chemical Industry and Engineering Progres*, vol. 35, no. 6, pp. 1627–1645, 2016.

[28] F. Allegrini, J. A. Fernández Pierna, W. D. Fragoso, A. C. Olivieri, V. Baeten, and P. Dardenne, "Regression models based on new local strategies for near infrared spectroscopic data," *Analytica Chimica Acta*, vol. 933, pp. 50–58, 2016.

[29] X. Xiao, J. Ma, F. Ge et al., "Application of near-infrared spectroscopy for the rapid analysis of *Lonicerae japonicae* flos solution extracted by water," *Journal of Innovative Optical Health Sciences*, vol. 7, no. 4, article 1350063, 2014.

[30] J. A. Cayuela and J. F. García, "Sorting olive oil based on alpha-tocopherol and total tocopherol content using near-infra-red spectroscopy (NIRS) analysis," *Journal of Food Engineering*, vol. 202, pp. 79–88, 2017.

[31] J. Hu, X. Ma, L. Liu, Y. Wu, and J. Ouyang, "Rapid evaluation of the quality of chestnuts using near-infrared reflectance spectroscopy," *Food Chemistry*, vol. 231, pp. 141–147, 2017.

[32] F. Patrizia, D. L. Silvia, B. Remo et al., "Near infrared (NIR) spectroscopy-based classification for the authentication of Darjeeling black tea," *Food Control*, vol. 100, pp. 292–299, 2019.

[33] L. F. Liu, J. Liang, J. F. Ma, S. S. Li, X. Xiao, and F. H. Ge, "Rapid determination of calycosin-7-glucoside content in extraction process of Astragalus by near infrared spectroscopy," *Jiangxi Journal of Traditional Chinese Medicine*, vol. 397, no. 47, pp. 68–70, 2016.

[34] C. Schaefer, D. Clicq, C. Lecomte, A. Merschaert, E. Norrant, and F. Fotiadu, "A process analytical technology (PAT) approach to control a new API manufacturing process: development, validation and implementation," *Talanta*, vol. 120, no. 5, pp. 114–125, 2014.

[35] Y. Wu, Y. Jin, Y. Li, D. Sun, X. Liu, and Y. Chen, "NIR spectroscopy as a process analytical technology (PAT) tool for on-line and real-time monitoring of an extraction process," *Vibrational Spectroscopy*, vol. 58, no. 1, pp. 109–118, 2012.

[36] A. Q. Vo, H. He, J. Zhang, S. Martin, R. Chen, and M. A. Repka, "Application of FT-NIR analysis for in-line and real-time monitoring of pharmaceutical hot melt extrusion: a technical note," *AAPS Pharmscitech*, vol. 19, no. 8, pp. 3425–3429, 2018.

[37] L. X. Sun, J. P. Chen, G. M. Zhao, and M. Y. Li, "Research progress in application of chemometrics in food analysis," *Science and Technology of Food Industry*, vol. 33, no. 7, pp. 444–452, 2012.

[38] A. Kuriyama, J. Osuga, Y. Hattori, and M. Otsuka, "In-line monitoring of a high-shear granulation process using the baseline shift of near infrared spectra," *AAPS Pharmscitech*, vol. 19, no. 2, pp. 710–718, 2017.

[39] V. R. Naidu, R. S. Deshpande, M. R. Syed, P. Deoghare, D. Singh, and P. S. Wakte, "PAT-based control of fluid bed coating process using NIR spectroscopy to monitor the cellulose coating on pharmaceutical pellets," *AAPS Pharmscitech*, vol. 18, no. 6, pp. 2045–2054, 2017.

[40] J. Workman and L. Weyer, *Practical Guide to Interpretive Near-Infrared Spectroscopy*, CRC Press, Boca Raton, FL, USA, 2007.

[41] P. M. Santos and L. A. Colnago, "Comparison among MIR, NIR, and LF-NMR techniques for quality control of jam using chemometrics," *Food Analytical Methods*, vol. 11, no. 7, pp. 2029–2034, 2018.

[42] J. Wang, X. Zhang, S. Sun, X. Sun, Q. Li, and Z. Zhang, "Online determination of quality parameters of dried soybean protein-lipid films (Fuzhu) by NIR spectroscopy combined with chemometrics," *Journal of Food Measurement and Characterization*, vol. 12, no. 3, pp. 1473–1484, 2018.

[43] G. F. d. Lima, S. A. C. Andrade, V. H. d. Silva, and F. A. Honorato, "Multivariate classification of UHT milk as to the presence of lactose using benchtop and portable NIR spectrometers," *Food Analytical Methods*, vol. 11, no. 10, pp. 2699–2706, 2018.

[44] W. Li, X. Yan, J. Pan, S. Liu, D. Xue, and H. Qu, "Rapid analysis of the Tanreqing injection by near-infrared spectroscopy combined with least squares support vector machine and Gaussian process modeling techniques," *Spectrochimica Acta Part A: Molecular and Biomolecular Spectroscopy*, vol. 218, pp. 271–280, 2019.

[45] C. Y. Xu, Z. L. Ye, D. S. Li, D. Z. Zho, and J. D. Liu, "Rapid analysis of naringin and neohesperidin by near infrared transmittance spectroscopy in extract of aurantii fructus," *Chinese Journal of Experimental Traditional Medical Formulae*, vol. 11, no. 6, pp. 86–90, 2014.

[46] O. A. Alrezj, K. Patchava, M. Benaissa, and S. A. Alshebeili, "Pre-processing to enhance the quantitative analysis of glucose from NIR and MIR spectra," in *Proceedings of the EMBEC & NBC*, Springer, Tamper, Finland, June 2017.

[47] J. T. Xue, Y. F. Liu, L. M. Ye et al., "Rapid and simultaneous analysis of five alkaloids in four parts of coptidis rhizoma by near-infrared spectroscopy," *Spectrochimica Acta A Molecular Biomolecular Spectroscopy*, vol. 188, pp. 611–618, 2018.

[48] X. Zhuang, L. Wang, Q. Chen, X. Wu, and J. Fang, "Identification of green tea origins by near-infrared (NIR) spectroscopy and different regression tools," *Science China Technological Sciences*, vol. 60, no. 1, pp. 84–90, 2017.

2

Semiempirical Efficiency Calibration in Semiconductor HPGe Gamma-Ray Spectroscopy

Jelena Krneta Nikolić, Milica Rajačić, Dragana Todorović, Marija Janković, Nataša Sarap, Gordana Pantelić, and Ivana Vukanac

Institute of Nuclear Sciences "Vinča", University of Belgrade, Mike Petrovica Alasa 12-16, 11001 Belgrade, Serbia

Correspondence should be addressed to Ivana Vukanac; vukanac@vinca.rs

Academic Editor: Eugen Culea

One of the main problems in quantitative gamma-ray spectroscopy is the determination of detection efficiency, for different energies, source-detector geometries, and composition of samples or sources. There are, in principle, three approaches to this issue: experimental, numerical, and semiempirical. Semiempirical approach is based on the calculation of the efficiency for the measured sample on the basis of an experimental efficiency measured on the same detector, but with a calibration source that can be of different size, geometry, density, or composition—the so-called efficiency transfer. The aim of this paper is to analyze the semiempirical approach, using EFFTRAN and MEFFTRAN software as a typical example. These software were used in the Department of Radiation and Environmental Protection, Vinča Institute of Nuclear Sciences, on three HPGe detectors. The results were compared to the experimentally obtained efficiency, and further validation is performed by measuring reference materials issued within the framework of several interlaboratory intercomparisons. The analysis of the results showed that the efficiency transfer produces good results with the discrepancies within the limits of the measurement uncertainty. Also, for intercomparison measurement, u_{test} criterion for the trueness of the result was applied showing that the majority of the obtained results were acceptable. Some difficulties were identified, and the ways to overcome them were discussed.

1. Introduction

Gamma spectrometry is a widely used method for the measurement of gamma-ray emitting radionuclide content in various materials. It is a method of choice for the measurement of environmental samples conducted, for example, during radiological monitoring of the environment and contamination control. This method is based on the interaction of gamma rays emitted from the source and the active volume of the detector. Semiconductor detectors and among them, high-purity germanium (HPGe) detectors, are proven to be very sensitive and have good energy and time resolution. That is why this type of detector is commonly used for gamma-spectroscopic measurements.

In the absence of background, the result of any gamma-spectroscopic measurement is represented by the spectrum of photons originating from the source that are collected by the multichannel analyzer (MCA) with the number of detected photons in a peak at a specific energy being proportional to the activity of the given radionuclide. One of the main problems in quantitative gamma-ray spectroscopy is the determination of detection efficiency, for different energies, different source-detector geometries, and different compositions of voluminous samples or sources. This task represents the efficiency calibration of the detector.

The quality of the results of gamma-spectrometric measurements is dependent on the knowledge of detector efficiency for different sample geometries, chemical compositions, and sample-detector geometries. In reality, measurement depends on the geometry, structure, and the chemical composition of the sample, and the efficiency calibration for each specific case is not always available. That is why several methods for efficiency calibration were developed. There are, in principle, three approaches to this issue: experimental, numerical, and semiempirical. These methods vary in difficulty, reliability of the obtained results, time and resources required, and so on.

Experimental approach utilizes standardized sources (calibration sources) with composition, density, and geometry that are as close as possible to the measured samples. The direct measurement of different calibration sources containing γ-ray emitters within the energy range of interest, and their subsequent fitting to a parametric function, yields the best results. However, this approach requires a large number of calibration sources, implying a high financial cost, a long counting time, and complicated and time-consuming preparation of the calibration sources. This problem is especially pronounced when environmental samples are of interest due to their diversity in composition and structure [1].

Numerical methods of the efficiency calibration consist of a computer simulation of the processes that contribute to the detection of the emitted photons. Monte Carlo simulation, such as GEANT4, MCNP, EGS, and PENELOPE, can be adapted to the computation of the efficiency of gamma-ray spectrometry detectors [2]. Dedicated codes, such as GESPECOR, LABSOCS, ANGLE, and so on, are specifically tailored to solve most of the problems concerning gamma-spectrometric measurements [3]. Monte Carlo simulation codes are developed to simulate the response of complex particle detectors and for variety of different high energy and nuclear interactions [4]. In case of gamma spectrometry, these codes need defining appropriate processes for interaction of gamma photons with the detector and the corresponding database used in the development of the application for the particular detector, which may be time-consuming and may require proficiency in programming language [5]. Once an application is developed, the use is relatively easy and the results are straightforward. Besides these software packages that simulate detector response to gamma rays, there is also numerical method ("direct matrices multiplication" (DMM) method) based only on general decay scheme developed to determine peak efficiencies as well as the activity of the source [6, 7].

The third approach to the efficiency calibration is the semiempirical approach. It is based on the calculation of the efficiency for the measured sample on the basis of an experimental efficiency curve obtained for the same detector, but with a calibration source that can be of different sizes, geometries, densities, or compositions—the so-called efficiency transfer. The procedure saves time and resources, since sample-specific experimental calibration is avoided. It has been proven especially useful in environmental measurements [8], where an ultimate precision in calibration is usually not required and a variety of different sources might be measured. Many software packages were developed in order to perform efficiency transfer calculations with a known set of parameters. Some are EFFTRAN, MEFFTRAN, ETNA, ANGLE, and so on. The result of the calculation is transfer coefficient, which is the ratio of the efficiency for unknown sample and reference efficiency.

The aim of this paper is to analyze the semiempirical approach to the efficiency calibration of the HPGe detector. A specific example of the efficiency transfer software will be analyzed in more detail, with practical instruction, and advantages and drawbacks pointed out. The results of the testing of this method will also be presented.

1.1. Semiempirical Efficiency Calibration: EFFTRAN and MEFFTRAN. Semiempirical methods for detector efficiency calculation are based on the assumption that the detector efficiency for the measured sample can be calculated knowing the reference efficiency obtained by measuring the calibration source with known composition and activity. This is referred to as "efficiency transfer." The relation between reference efficiency and efficiency for the measured sample is defined in (1), according to the model proposed by Moens et al. [9]:

$$\varepsilon_x = \varepsilon_{\text{ref}} \frac{\overline{\Omega}_x}{\overline{\Omega}_{\text{ref}}}, \quad (1)$$

where ε_x represents the unknown efficiency, ε_{ref} is the reference efficiency, $\overline{\Omega}_x$ is the effective solid angle for the given measurement configuration, and $\overline{\Omega}_{\text{ref}}$ is the effective solid angle for reference configuration. The effective solid angle depends on the geometry of the detector, the source–detector position, and needs to be calculated. Both ε_x and ε_{ref} depend on the gamma photon energy and, through the interaction probabilities, on the chemical composition of the crystal and source.

In order to calculate these effective solid angles, some sort of software is needed for solving the partial differential equations. Often in the calculation, Monte Carlo integration is used as a suitable choice for many-dimensional systems. This calculation requires relatively precise information on geometry of the detector (crystal geometry, housing geometry and composition, active and inactive layers, etc). Also, the information on both calibration source and measured sample has to be provided (diameter, filling height, matrix composition, thickness of the container, etc.). Because the model of the sample, as well as the detector crystal, can be constructed from cylinders only, the only complex operation required in the code is the calculation of the path length traversed through the cylinders defining the counting geometry by a gamma-ray originating at an arbitrary location [10].

The procedure for obtaining the unknown efficiency is as follows. Firstly, a calibration source with known composition and activity is measured, and the detection efficiency ε_{ref} has to be calculated for all energies emitted from the source using the following equation:

$$\varepsilon_{\text{ref}} = \frac{N \cdot C(E)}{t \cdot P_\gamma \cdot A}, \quad (2)$$

where N is the number of detected gamma photons, $C(E)$ is the coincidence summing correction factor, t is the measurement time, P_γ is the emission probability at given energy E, and A is the activity of the radionuclide present in the calibration source.

As with all measurement, also in this case, a measurement uncertainty $u(\varepsilon_{\text{ref}})$ has to be defined. It is expressed as a combined standard uncertainty using the following equation:

$$u_{\text{ref}}(\varepsilon) = \sqrt{(\delta A)^2 + (\delta N)^2 + (\delta M)^2 + (\delta C)^2}, \quad (3)$$

where δA is the relative uncertainty of the activity of the radionuclide present in the calibration source (1-2%), δN is the relative statistical uncertainty, and δM includes all other contributions to the uncertainty such as measurement of the reference material mass, the uncertainty of the positioning of the sample on the detector, measurement time uncertainty, and radioactive decay during measurement, and so on, which are estimated to be 2%. Finally, δC is the uncertainty of the coincidence correction factor (≈1.2%).

Then, the geometrical characteristics of the detector need to be defined. Due to the specific shape of the detector crystal and surrounding material, the geometry of the detector can be defined as a set of cylinders with a certain radius and height, made of known material. These data are usually available from the manufacturer of the detector itself. Furthermore, the reference source and measured sample characteristics are defined, also as a set of cylinders. The material of both calibration source and sample is defined using its chemical formulae, the percent of each compound in the total mass of the source or sample, and the density of the matrix. After all these data are entered, the calculation gives the transfer coefficients which, when multiplied with the reference efficiency obtained by measuring the calibration source, yield the unknown efficiency for the measured sample.

As a typical example of efficiency transfer software, here we will use EFFTRAN and MEFFTRAN software. EFFTRAN, an efficiency transfer software, is dedicated to calculation of efficiency transfer for cylindrical samples, while MEFFTRAN performs the same calculations for the Marinelli beaker [10, 11]. The software is organized as a Microsoft excel file with three modules. There is a module for defining the material of the calibration source and measured sample (Material), a module for calculation of the efficiency transfer (Efficiency Transfer), and a module for calculation of coincidence correction factors (Coincidence Correction). In the module Efficiency Transfer, the geometrical characteristic of the detector crystal has to be defined. This means that also, the material of the crystal and detector window should be defined, as well as the thickness of the dead layers and window to crystal gap.

All the data that are required are not always known and cannot be precisely defined, thus contributing to the measurement uncertainty. However, the lack of precision and uncertainties rising from poor knowledge of the composition of the sample cannot be avoided.

The combined relative uncertainty of the efficiency ε_x obtained by MEFFTRAN or EFFTRAN can estimated according to the following expression:

$$u(\varepsilon_x) = \sqrt{(u(\varepsilon_{\text{ref}}))^2 + (u(\text{C}))^2 + (u_{\text{D}})^2 + (u_{\text{S}})^2}, \quad (4)$$

where $u(\varepsilon_{\text{ref}})$ is the relative uncertainty of the reference efficiency value, defined in (2), $u(\text{C})$ is the uncertainty of the transfer factors calculated by the program as a statistical uncertainty of the Monte Carlo integration (≈1.2%), u_{D} is the uncertainty associated with the geometry of the detector, and u_{S} is the uncertainty associated with the characteristics of the sample. As it can be seen from (4), the parameters contributing to the uncertainty are combined as independent variables.

The uncertainty of the parameters of the detector geometry u_{D} can be estimated by varying each parameter and calculating the amount of change it produces on the transfer factors. In this way, the uncertainty can be estimated to be around 1% for crystal diameter and length, and crystal cavity diameter and length. According to this estimation, the contribution u_{D} can be obtained by combining these uncertainties as a square root of sum of squares [5].

The characteristics of the calibration source and measured samples also contribute to the measurement uncertainty via parameter u_{S}.

In the case of EFFTRAN, the density of the samples was calculated by dividing the measured mass of the sample with the volume of the sample. Therefore, the contribution of the density to the uncertainty parameters is the combined uncertainty of mass and volume (1-2%). The chemical composition of the container can be well defined, but for the sample, the situation is more complicated, and poor knowledge of the chemical composition can be the source of larger uncertainty.

The definition of the chemical composition of the material can significantly contribute to the uncertainty budget. In order to define this contribution, an in-depth analysis has to be performed. This will be the subject of the investigation that is under way. One simple way to estimate this contribution is to vary the composition of the sample by adding or omitting some elements or compounds. Of course, this does not apply to the samples with known composition such as water and charcoal. In case of soil and grass, we have varied the composition in order to see how these changes reflect on the result of the efficiency transfer. It has been determined that omitting Fe from the composition of the soil leads to changes in the result that are around 2% for both EFFTRAN and MEFFTRAN, while changing the content of C leads to the changes in the results that span from 1% in case of EFFTRAN to 8% in case of MEFFTRAN. Much larger changes are observed in case of grass composition, namely, the change of C content with respect to the cellulose content. These changes ranged from 7 to 29%, with the largest influence noticed for the lowest energy. It is noticed that the influence of the composition lowers with the rising energy. Also, in some cases, such as grass, the influence of the chemical composition can be larger than the influence of the sample thickness. How these changes combine with each other (can one simply sum or multiply all changes) is yet to be analyzed. For the sake of simplicity in this paper, the contribution to the uncertainty arising from the chemical composition definition is estimated to be 10% average, for all energies.

In the case of MEFFTRAN, since all samples were Marinelli beakers, the only variable parameters of the sample geometry are sample filling height and mass. The contribution of 0.8%

TABLE 1: The uncertainty budget.

	Uncertainty (%)	
	EFFTRAN	MEFFTRAN
u_D		
Crystal diameter and length	1	1
Crystal cavity diameter and length	1	1
Total u_D	**1.41**	**1.41**
u_S		
Matrix density	1-2	1.64
Chemical composition	1–10	2–10
Total u_S	**10–10.2**	**10.1**
$u(\varepsilon_{ref})$		
Activity (A)	1-2	
Count (N)	<1	
Positioning, time, mass, and so on	2	
Coincidence correction factor (C)	1.2	
Total $u(\varepsilon_{ref})$	**3**	**3**
Total $u(\varepsilon_x)$	**~10.5**	

for the sample filling height was included in the uncertainty budget. This contribution was calculated by conducting the efficiency transfer for 5 different filling heights (105 mm, 100 mm, 95 mm, 90 mm, and 85 mm) and calculating the ratio between the different transfer coefficients. It is then established that the 1% change in height leads to 0.8% change in transfer coefficients, thus contributing to the u_S with 0.8%. The uncertainty of the sample mass is estimated to be 1%. Since the software requires the density of the matrix to be defined, these two were combined to produce the square of the sample density uncertainty equal to 1.64%. The contributions u_D and $u(C)$ are the same as in the case of EFFTRAN.

The uncertainty budget is presented in Table 1.

After all the calculations are done and the efficiency for the measured sample is obtained, it is crucial to perform some sort of validation of the results. This can be done by comparing the calculated efficiency with the experimental one, obtained by measuring a reference sample with the known activity. Of course, this reference sample has to be different from the one used in the calculation as a calibration source. This can be achieved also by taking part in the interlaboratory intercomparisons, where the results obtained using a calculated efficiency are compared to the target value provided by the organizer and the accuracy and the precision of the result is assessed.

2. Results and Discussion

The semiempirical calculation of the detection efficiency is readily performed in the Department of Radiation and Environmental Protection, Vinča Institute of Nuclear Sciences, on three HPGe detectors (two p-type detectors named "Detector 1" and "Detector 3" and one n-type detector named "Detector 2"). For this purpose, EFFTRAN and MEFFTRAN software are used. As a reference source for EFFTRAN, point sources (^{241}Am, ^{109}Cd, ^{137}Cs, ^{54}Mn, and ^{60}Co (LMRI Coffret d'etalons), type gamma ECGS-2, uncertainty about 1%) are used, and for MEFFTRAN, a certified radioactive standard purchased from Czech Meteorological Institute (Marinelli beaker filled with silica resin) is used. The radionuclides present in this standard are ^{214}Am, ^{109}Cd, ^{137}Cs, ^{54}Mn, ^{113}Sn, ^{203}Hg, ^{88}Y, ^{139}Ce, ^{113}Sn, ^{57}Co, and ^{60}Co, with the energies that span from 60 keV to 1836 keV and with the uncertainties from 1.1 to 1.9%. The set of radioactive standards used to validate the accuracy of the transferred efficiency were produced in our laboratory, using certified standard solution 9031-OL-427/12, ERX Prod no. 170712-1374036 and 9031-OL-426/12, ERX Prod no. 170712-1374034. The standard solution used for production of reference materials contained ^{214}Am, ^{109}Cd, ^{137}Cs, ^{54}Mn, ^{113}Sn, ^{203}Hg, ^{88}Y, ^{139}Ce, ^{113}Sn, ^{57}Co, and ^{60}Co, with the energies that span from 60 keV to 1836 keV and with the uncertainties from 0.5 to 1.7%. In Table 2, these reference materials are listed along with their measurement geometries, chemical composition, and packing density, which are all basically the data necessary for the calculation. For EFFTRAN, a total of 11 reference materials were used, while one was used for checking the accuracy of the calculation in MEFFTRAN.

TABLE 2: List of reference materials prepared in the laboratory and used for comparison between the calculated and experimental efficiency.

Matrix material	Chemical composition	Density (g/cm^3)	Measurement geometry
Charcoal	C (100%)	0.486	Cylinder, volume 100 ml
		0.454	Cylinder, volume 200 ml
Soil	SiO_2 (90.00%) K (4.00%) Fe (1%) C (4.00%) Others (1%)	1.381	Cylinder, volume 100 ml
		1.426	Cylinder, volume 200 ml
Sand	SiO_2 (100%)	1.591	Cylinder, volume 100 ml
		1.556	Cylinder, volume 200 ml
		1.9	Marinelli beaker, volume 0.5 l
Mineralized grass	C (99%) Others (1%)	0.300	Cylinder, volume 100 ml
		0.256	Cylinder, volume 100 ml
Water	H_2O (100%)	1.012	Cylinder, volume 125 ml
		1.011	Cylinder, volume 250 ml
Aerosol	SO_4 (15%) NH_4 (15%) NO_3 (15%) CH (50%) Others (5%)	0.572	Vial, volume 50 ml

Since the reference material used for validation of the results contained different radionuclides than point sources, the efficiency calculated from the measurement of point sources was used to obtain a calibration curve. This was achieved by fitting the experimental efficiencies with a function given in (5). This calibration curve was used to calculate the efficiencies for the energies that are emitted by the radionuclides contained in the reference samples:

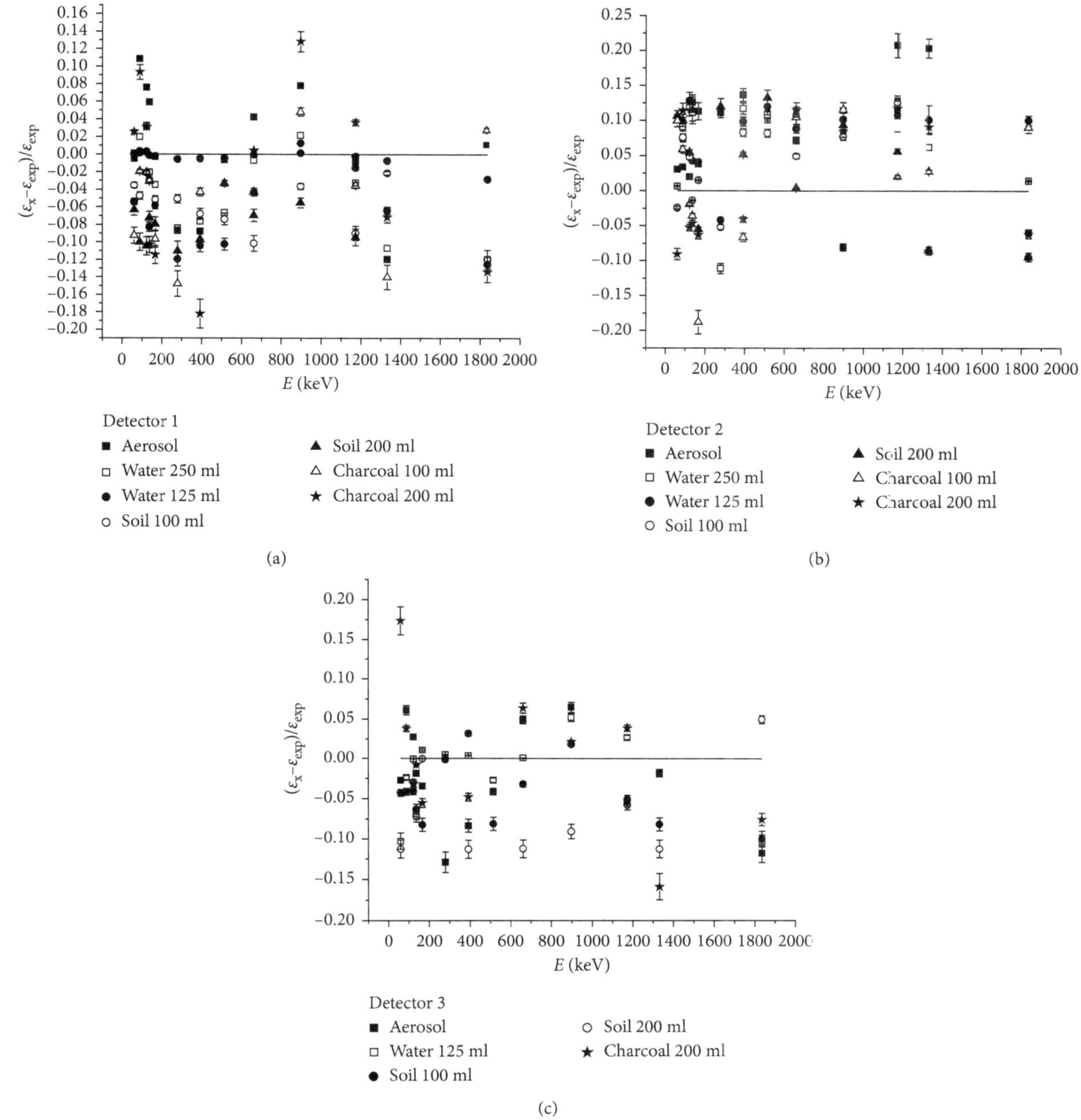

FIGURE 1: The relative discrepancy $(\varepsilon_{EFFTRAN} - \varepsilon_{exp})/\varepsilon_{exp}$ for all three detectors (a–c) and the investigated reference materials.

$$\ln \varepsilon_{ref} = \sum_{i=1}^{5} a_i (\ln E)^{i-1}, \quad (5)$$

where ε_{ref} is the reference efficiency, E is the energy, and a_i are the coefficients of the fit.

The discrepancy between the calculated efficiency, $\varepsilon_{EFFTRAN}$, and the experimental efficiency, ε_{exp}, in case of EFFTRAN is presented in Figure 1, for all three detectors for aerosol, water in the cylinder of 125 ml and 250 ml, charcoal in the cylinder of 100 ml and 200 ml, and soil in the cylinder of 100 ml and 200 ml.

The measurement uncertainty of the results is calculated according to (4).

As it can be seen from Figure 1, relative differences of the efficiency for different matrices placed in different geometries for all three detectors are mainly in the range of ±10% (due to the geometry of the housing and end cap of Detector 3, some geometries were not investigated). As it can be seen, EFFTRAN produces results that are relatively in good accordance with the experimental ones. Majority results are within a 10% margin that falls within the uncertainty limits. This is the case for well defined, low absorbing matrix with simple chemical composition; hence, the accordance between results is expected. It is however noticeable that EFFTRAN produces higher results for Detector 2, which is an n-type detector with Be window. For soil matrix, EFFTRAN produces results that differ from the experimental ones in the mid-energy region for all detectors. The efficiency is mainly underestimated for the p-type and overestimated for the n-type detector. Since the systematic influence of the matrix composition and the density are the largest at low energies and falls off with increasing energy, some discrepancies can be attributed to a lack of point sources that cover some regions of spectrum. Thus, the starting point for EFFTRAN calculation is somewhat incomplete, which can be resolved by measuring more point sources that produce energies in the desired region (e.g., ^{85}Sr and ^{113}Sn, or ^{152}Eu), which are unavailable at present.

Also, since the calculation was performed using point-source efficiency as a starting point, the discrepancies that are of the order of magnitude of the measurement uncertainty are quite acceptable and larger discrepancies should be expected. This can serve as an indication of the limitations of this approach. Using point sources or some other sources that differ significantly from the measurement geometry is not recommendable, better results are achieved using some more similar geometries for the reference efficiency. In that case, the calculations conducted in the laboratory produced results that differ from the experimental ones for 2-3%. [5].

In case of MEFFTRAN, due to its dimensions, Detector 3 was not used for measurements. In the case of Marinelli Sand for Detectors 1 and 2, the discrepancy of the calculated values from the measured values is in the range of −3.9% to +1.5%. These discrepancies are within the limits of measurement uncertainty. The largest discrepancies are noticed for the lowest energy (59 keV), which leads to the conclusion that the reference calibration should be done using calibration standard that contains ^{210}Pb. In that case, the lowest energies in the spectrum would be covered with two points in the reference calibration curve, leading perhaps to a better definition of the curve slope [12].

In order to further check the accuracy of the efficiency transfer, several reference materials issued within the framework of different interlaboratory intercomparisons were measured. The activities were then calculated using the efficiency obtained by EFFTRAN and MEFFTRAN and compared with the reference values provided by the organizer (IAEA). The measured materials were as follows:

(i) Hay sample (IAEA-TEL-2012-03)

(ii) Soil sample (IAEA-TEL-2013-03)

(iii) Seaweed, water, and sediment samples (IAEA-TEL-2014-03)

(iv) Sediment sample 662-1 from the South Transdanubian Inspectorate for Environmental Protection Nature Conservation and Water Management, Hungary.

Not all samples were possible to measure on all detectors, but at least one measurement of each sample was conducted.

The accuracy of the results is tested using u_{test} criterion recommended by the IAEA and defined in the following (5) [13]:

$$u_{\text{test}} = \frac{|A_{\text{IAEA}} - A_{\text{LAB}}|}{\sqrt{u_{\text{IAEA}}^2 + u_{\text{LAB}}^2}}, \tag{6}$$

where A_{IAEA} represents the target value of the activity, A_{LAB} is the activity reported by the participating laboratory, u_{IAEA} is the uncertainty of the target value, and u_{LAB} is the uncertainty of the result reported by the participating laboratory.

According to this criterion, the results are graded as acceptable (A) in case of $u_{\text{test}} \leq 2.5$ and not acceptable (N) in case of $u_{\text{test}} > 2.5.8$.

In Tables 3 and 4, the measurement results of the mentioned reference materials are listed with measurement uncertainty of 1σ level, along with the target value and the u_{test} result. In these tables, the letter N is added to the result which is not acceptable. Also, the target values denoted with the asterisk are obtained as a robust mean value of the results reported by all the participant laboratories with the standard deviation instead of the measurement uncertainty.

In case of EFFTRAN, two results were unacceptable. Since there were 13 radionuclides in total for which the activity was calculated, this stands for good agreement, especially when we notice that the result for the same radionuclide (^{137}Cs) was acceptable in other reference materials.

As it can be seen from Table 4, all but one result are acceptable, proving that semiempirical calculation of the efficiency produces accurate efficiencies for all investigated sources.

Discrepancies between efficiency calibration by EFFTRAN and MEFFTRAN software and experimentally obtained detection efficiencies are mainly due to insufficiently known detector geometry parameters and matrix composition for measured samples and reference materials. In order to achieve the best possible concurrence with experimentally obtained values, the optimization of these parameters was performed. Nevertheless, some of the parameter values like dead layer thickness, bulletization of the germanium crystal, and chemical composition and inhomogeneity of matrix materials could not be ideally optimized and still need improvement. Also, in cases presented in Figure 1, the efficiency transfer was performed based on point-source efficiency calibration. Considering that the initial and transferred efficiency relates to very different geometries and matrices, certain deviations are inevitable. But still, the majority of the results are in agreement within about 10% uncertainties, which are satisfying for environmental sample measurements.

This also shows that semiempirical calibration is not overly sensitive to small variations in sample filling height and mass, while good definition of the matrix chemical composition contributes to the reliability of the results. The good definition of the chemical composition of the material is shown to have the most pronounced influence on the results especially for low energies. Also, it can be recommended that the efficiency transfer should be performed using the most

TABLE 3: The measured activity concentrations of radionuclides obtained with efficiency calculated by EFFTRAN, along with the target values, and evaluation of accuracy using u_{test} parameter.

	A_{LAB} (Bq/kg)		Target value	u_{test}	
	n-Type detector 2	p-Type detectors 1 and 3	(Bq/kg)	n-Type detector 2	p-Type detectors 1 and 3
IAEA-TEL-2014-03 sediment sample					
^{238}U	18 ± 1	21 ± 4	16.0 ± 3.8*	0.51	0.91
^{226}Ra	20.8 ± 0.8	22 ± 1	19.0 ± 4.8*	0.37	0.61
^{137}Cs	14.3 ± 0.5	14.3 ± 0.5	12.0 ± 0.4	**3.59N**	**3.59N**
^{228}Ac	12.5 ± 0.4	12.9 ± 0.7	12.1 ± 1.5*	0.26	0.48
^{40}K	292 ± 9	292 ± 9	270 ± 27*	0.77	0.77
Sediment sample 662-1					
^{137}Cs	18 ± 3	16.9 ± 0.8	14.0 ± 0.8	1.29	2.56
^{228}Ac	38.7 ± 3.1	35 ± 2	37 ± 3	0.39	0.55
^{40}K	555 ± 61	552 ± 20	470 ± 40	1.16	1.83
IAEA-TEL-2012-03 hay sample					
^{137}Cs	760 ± 20	820 ± 30	785 ± 24	0.80	0.91
^{134}Cs	270 ± 20	320 ± 30	306 ± 20	1.27	0.39
IAEA-TEL-2014-03 water					
^{137}Cs	—	12.6 ± 0.4	12.06 ± 0.1	—	1.31
^{134}Cs	—	21.3 ± 0.7	21.4 ± 0.2	—	0.14
^{210}Pb	—	38 ± 5	49.87 ± 1.23	—	2.31

*Robust mean.

TABLE 4: The measured activity concentrations of radionuclides obtained with efficiency calculated by MEFFTRAN, along with the target values, and evaluation of accuracy using u_{test} parameter.

	A_{LAB} (Bq/kg)		Target value	u_{test}	
	p-Type detector 1	n-Type detector 2	(Bq/kg)	p-Type detector 1	n-Type detector 2
Sediment sample IAEA-Tel-2014					
^{210}Pb	—	23 ± 2	19.6 ± 4.5*	—	0.67
^{238}U	23 ± 3	19 ± 2	16 ± 3.8*	1.35	0.59
^{226}Ra	18 ± 1	17.4 ± 1.2	19 ± 4.8*	0.29	0.32
^{137}Cs	12.4 ± 0.7	12.4 ± 0.7	12.0 ± 0.4	0.50	0.50
^{228}Ac	10.9 ± 0.7	11.6 ± 0.8	12.1 ± 1.5*	0.72	0.29
^{40}K	280 ± 20	280 ± 20	270 ± 27*	0.30	0.30
^{212}Pb	9.3 ± 0.6	11 ± 1	12.2 ± 1.5*	1.79	0.67
^{212}Bi	13 ± 1	13 ± 1	12.2 ± 1.2*	0.53	0.62
Seaweed sample IAEA-Tel-2014					
^{137}Cs	21 ± 1	20 ± 1	22.96 ± 0.45	1.79	1.93
^{134}Cs	8.0 ± 0.7	8.0 ± 0.7	8.27 ± 0.2	0.37	0.40
^{228}Ac	2.8 ± 0.4	2.8 ± 0.4	2.5 ± 0.7*	0.37	0.41
^{40}K	1580 ± 70	1580 ± 70	1780 ± 150*	1.21	1.17
^{212}Pb	2.0 ± 0.1	2.0 ± 0.1	2.6 ± 0.6*	0.99	0.94
^{226}Ra	5.7 ± 0.5	—	5.1 ± 3.7*	0.16	—
Hay sample IAEA-TEL-2012					
^{137}Cs	750 ± 40	740 ± 40	785 ± 24	0.64	0.99
^{134}Cs	280 ± 20	300 ± 20	306 ± 20	0.95	0.35
Soil sample IAEA-Tel-2013					
^{210}Pb	480 ± 30	690 ± 40	595 ± 19	**3.24N**	2.15
^{241}Am	—	1.6 ± 0.2	1.8 ± 0.1	—	0.81
^{238}U	—	23 ± 3	23.6 ± 0.7	—	0.19
^{40}K	190 ± 10	190 ± 10	207.7 ± 8.3	1.36	1.36
^{137}Cs	107 ± 6	110 ± 6	118.6 ± 2.9	1.74	1.29
^{208}Tl	10.5 ± 0.8	10.3 ± 0.6	11.5 ± 0.6	1	1.41
^{212}Pb	30 ± 3	27.5 ± 2	31.0 ± 1.2	0.31	1.51
^{228}Ac	30 ± 2	30 ± 2	32.4 ± 1.6	0.94	0.94

*Robust mean.

similar sample geometry and composition as possible; that is, it is better to use almost anything else instead of point source for reference efficiency measurement.

3. Conclusion

One of the main problems in quantitative gamma-ray spectroscopy is the determination of detection efficiency, for different energies, different source-detector geometries, and different compositions of voluminous samples or sources. This task is defined as an efficiency calibration of the detector, and there are several approaches to resolving this problem. Semiempirical approach to efficiency calibration (so-called efficiency transfer) takes advantage of positive attributes of both experimental and numerical approaches, simultaneously minimizing their drawbacks. Application of EFFTRAN or MEFFTRAN software is the most practical for this purpose, giving the values of the transferred efficiencies that agree on a satisfactory level with those obtained experimentally. This approach needs a reference efficiency which should be obtained by measuring the certified calibration source, which is then used for calculation performed by the software. After the unknown efficiency is obtained, it is highly recommended to validate the results. The most practical way to do that is to measure some reference materials with known activity, that is, participate in the interlaboratory intercomparison. Also, an in-depth analysis of uncertainty budget is mandatory in order to assess the contribution of the calculated efficiency to the measurement uncertainty.

Conflicts of Interest

The authors declare that they have no conflicts of interest.

Acknowledgments

This work was financially supported by the Serbian Ministry of Education, Science and Technological Development through National Projects III43009 and OI171018.

References

[1] M. J. Daza, B. Quintana, M. Garcia Tavalera, and F. Fernandez, "Efficiency calibration of a HPGe detector in the 46.54-2000 keV energy range for the measurement of environmental samples," *Nuclear Instruments and Methods in Physics Research Section A*, vol. 470, no. 3, pp. 520–532, 2001.

[2] T. Vidmar, N. Celik, N. Cornejo Dıaz et al., "Testing efficiency transfer codes for equivalence," *Applied Radiation and Isotopes*, vol. 68, no. 2, pp. 355–359, 2010.

[3] O. Sima and D. Arnold, "On the Monte Carlo simulation of HPGe gamma-spectrometry systems," *Applied Radiation and Isotopes*, vol. 67, no. 5, pp. 701–705, 2009.

[4] S. Agostinelli, J. Allison, K. Amako et al., "GEANT4–a simulation toolkit," *Nuclear Instruments and Methods in Physics Research A*, vol. 506, pp. 250–303, 2003.

[5] J. Nikolić, T. Vidmar, D. Joković, M. Rajačić, and D. Todorović, "Calculation of HPGe efficiency for environmental samples: comparison of EFFTRAN and GEANT4," *Nuclear Instruments and Methods in Physics Research A*, vol. 763, pp. 347–353, 2014.

[6] D. Novković, A. Kandić, M. Đurašević, I. Vukanac, Z. Milošević, and L. Nađđerđ, "Coincidence summing of X- and γ-rays in γ-ray spectrometry," *Nuclear Instruments and Methods in Physics Research Section A*, vol. 578, no. 1, pp. 207–217, 2007.

[7] D. Novković, M. Đurašević, A. Kandić, I. Vukanac, B. Šešlak, and Z. Milošević, "Coincidence summing corrections for point and volume ^{152}Eu sources," *Applied Radiation and Isotopes*, vol. 107, pp. p138–p144, 2016.

[8] G. Gilmore, *Practical Gamma-Ray Spectrometry*, Wiley, New York, NY, USA, 2nd edition, 2008.

[9] L. Moens, J. De Donder, X. Lin et al., "Calculation of the absolute peak efficiency of gamma-ray detectors for different counting geometries," *Nuclear Instruments and Methods in Physics Research*, vol. 187, no. 2-3, pp. 451–472, 1981.

[10] T. Vidmar, "EFFTRAN–a Monte Carlo efficiency transfer code for gamma-ray spectrometry," *Nuclear Instruments and Methods in Physics Research A*, vol. 50, no. 3, pp. 603–608, 2005.

[11] MEFFTRAN, http://www.efftran.com/MEFFTRAN.zip.

[12] J. K. Nikolić, M. Rajačić, D. Todorović, and T. Vidmar, "The first experimental test of the MEFFTRAN software on HPGe detector calibration for environmental samples," *Journal of Environmental Radioactivity*, vol. 165, pp. 191–196, 2016.

[13] IAEA, *Analytical Quality in Nuclear Applications Series No. 18*, IAEA, Vienna, Austria, 2010.

3

Structure-Property Relationships for a Series of Ruthenium(II) Polypyridyl Complexes Elucidated through Raman Spectroscopy

Luke O'Neill,[1] Laura Perdisatt,[2] and Christine O'Connor[3]

[1]*FOCAS Research Institute, Dublin Institute of Technology, Kevin St., Dublin 8, Ireland*
[2]*FOCAS Research Institute/School of Chemical and Pharmaceutical Sciences, Dublin Institute of Technology, Kevin St., Dublin 8, Ireland*
[3]*School of Food Science and Environmental Health, Dublin Institute of Technology, Cathal Brugha St., Dublin 1, Ireland*

Correspondence should be addressed to Luke O'Neill; luke.oneill@dit.ie

Academic Editor: Vincenza Crupi

A series of ruthenium polypyridyl complexes were studied using Raman spectroscopy supported by UV/Vis absorption, luminescence spectroscopy, and luminescence lifetime determination by time-correlated single photon counting (TCSPC). The complexes were characterised to determine the influence of the variation of the conjugation across the main polypyridyl ligand. The systematic and sequential variation of the main polypyridyl ligand, 2-(4-formylphenyl)imidazo[4,5-f][1,10]phenanthroline (FPIP), 2-(4-cyanophenyl)imidazo[4,5-f][1,10]phenanthroline (CPIP), 2-(4-bromophenyl)imidazo[4,5-f][1,10]phenanthroline (BPIP), and 2-(4-nitrophenyl)imidazo[4,5-f][1,10]phenanthroline (NPIP) ligands, allowed the monitoring of very small changes in the ligands electronic nature. Complexes containing a systematic variation of the position (*para*, *meta*, and *ortho*) of the nitrile terminal group on the ligand (the *para* being 2-(4-cyanophenyl)imidazo[4,5-f][1,10]phenanthroline (*p*-CPIP), the meta 2-(3-cyanophenyl)imidazo[4,5-f][1,10]phenanthroline (*m*-CPIP) and 2-(2-cyanophenyl)imidazo[4,5-f][1,10]phenanthroline (*o*-CPIP)) were also characterised. Absorption, emission characteristics, and luminescence yields were calculated and correlated with structural variation. It was found that both the electronic changes in the aforementioned ligands showed very small spectral changes with an accompanying complex relationship when examined with traditional electronic methods. Stokes shift and Raman spectroscopy were then employed as a means to directly gauge the effect of polypyridyl ligand change on the conjugation and vibrational characteristics of the complexes. Vibrational coherence as measured as a function of the shifted frequency of the imizodale bridge was shown to accurately describe the electronic coherence and hence vibrational cooperation from the ruthenium centre to the main polypyridyl ligand. The well-defined trends established and elucidated though Raman spectroscopy show that the variation of the polypyridyl ligand can be monitored and tailored. This allows for a greater understanding of the electronic and excited state characteristics of the ruthenium systems when traditional electronic spectroscopy lacks the sensitivity.

1. Introduction

A significant amount of research has been invested into the development of potential novel inorganic therapeutics [1–3]. Since the platinum-based drug *cis*-platins serendipidous discovery, the development of primary and secondary structural analogues has increased exponentially; however, only some of have received FDA worldwide approval, i.e., carboplatin and oxaliplatin. Although the potential is undeniable, several shortcomings within these platinum drugs regarding their clinical efficacy, limited activity against common types of cancers, the drug resistance phenomena which has lowered the impact of these drugs, and a range of deplorable side effects, has limited their potential usage [4]. To this end, ruthenium-based drugs have been proposed as an alternative. To date, several different ruthenium(II) and (III) complexes have exhibited good antitumour and anti-metastatic properties [5–7]. Both ruthenium compounds NAMI-A and KP1019 have successfully completed phase I clinical trials and have commenced to phase II [8].

In 2009, an article by Gianferrara et al. categorised metal anticancer compounds based on their mode of action, which could be divided into five different classes: [9] (1) the metal has a functional role, (2) the metal has a structural role, (3) the metal is a carrier for main ligands that are delivered *in vivo*, (4) the metal compound behaves as a catalyst *in vivo*, and (5) the metal compound is photoactive.

Fundamental studies into the mode of action of anticancer agents are essential. Hence, this requires structure-property relationships to be formulated, starting with the explanation of small changes, which can then be applied to the larger more complex systems for the advantageous properties of ruthenium complexes as novel inorganic therapeutics to fully realise the processes, which govern the reactivity of the complexes, must be clearly established.

Many research teams have embraced the phenanthroline (phen) structure into the synthesis of a variety of polypyridyl ligands with different end group positions in the phenyl ring, developed on from the planar ligand; 2-phenyl-imidazo[4,5-f]1,10-phenanthroline (PIP) consists of a basic phenyl ring. Increasing the surface area of the phen polypyridyl ligand has seen to increase the DNA binding affinity of the complex [8]. Promising activity has been observed for these phen functionalised ligands as ruthenium-based complexes, and Liu et al. synthesised a similar series of these style ligands, *p*-MOPIP with the OMe group (2-(4-methoxyphenyl)imidazo [4,5-f][1,10]phenanthroline) (R_1 (*para*) position on the phenyl ring), *p*-HPIP with the OH group 2-(4-hydroxyphenyl)imidazo[4,5-f][1,10]phenanthroline, and *p*-NPIP with the NO_2 group (2-(4-nitrophenyl)imidazo[4,5-f] [1,10]phenanthroline) [10]. In the last decade, trends in medicinal chemistry are moving away from high-throughput approaches to drug discovery (i.e., those where vast databases of molecules are screened against a biological target) towards structure-based drug discovery (i.e., those where the drug design is based on specific structural information about a biological target) [4].

Establishing the structure-property relationships is a crucial aspect in several other potential applications for which ruthenium-based complexes are at the forefront of research too, e.g., concentrators for solar cells, oxygen sensors and photoactive drugs [11–13].

Research into optimisation of electronic and emissive characteristics in ruthenium complexes, in which the auxiliary ligand is systematically changed, has shown a well-defined relationship between the absorption and emission wavelengths and the electronic effect of the auxiliary ligand as measured by the electronic redox potential [14]. The relationship between the luminescence yield and lifetime establishes that the dominant effect of the auxiliary ligands being systematically changed is a manipulation in the nonradiative decay rate. These structure-property relationships have also been previously shown to exist in organic polymeric and oligomeric systems.

Well-defined relationships have been established showing a well-defined relationship with the vibrational coherence and the luminescence yield with Stokes shift, integrated Raman intensity, and electronic gap [15–18].

To this end, coordination compounds of Ru(II) were synthesised with the following nitrogen donor ligands, 2, 2′-bipyridine (bpy), 1,10-phenanthroline (phen), and 2, 2′-biquinoline (biq). The step-by-step substitution of the functional group on the polypyridyl ligand allows analysis with essentially a single variable, changing the electron affinity of the terminal group on the polypyridyl ligand. The effect of the systematic change on the complexes properties is investigated using a series of ruthenium complexes with the end-group terminations that varied from 2-(4-formylphenyl)imidazo[4,5-f][1,10]phenanthroline [*p*-FPIP], 2-(4-cyanophenyl)imidazo[4,5-f][1,10]phenanthroline [*p*-CPIP], and 2-(4-bromophenyl) imidazo[4,5-f][1,10]phenanthroline [*p*-BPIP] to 2-(4-nitrophenyl)imidazo[4,5-f][1,10]phenanthroline [*p*-NPIP]. The nature and influence of the end-group position was also investigated.

This article presents a study originating in the electronic band gap transitions but extended to the Stokes shift, and thus, the vibrational nature of the complexes is introduced. The study shows that systematic structural variations made to the polypyridyl ligand can be described in terms of the effect on the vibrational coherence across the complex. This variation in the vibrational properties of the complexes can be directly measured via Raman spectroscopy. The vibrational decoupling will allow for quantitative structure-property relationships to be formed and optimisation of the preferred avenue of decay leading to maximising the radiative decay, i.e., luminescence. Hence, the study is aimed at establishing structure-property relationships for both electronic and vibrational characteristics in these systems. Structure-property relationships will aid the smart design of complexes which absorb and emit in required spectral windows. The potential for control and optimisation of luminescence yield and lifetime is also extremely desirable for applications such as photoactive therapeutics, solar cell dyes, and oxygen sensor research where the excited state characteristics are extremely important, and elongation of the luminescence lifetime without compromising the yield is paramount.

2. Experimental

The ruthenium complexes investigated, shown in Figure 1, were $[Ru(bpy)_2L]^{2+}$, $[Ru(phen)_2L]^{2+}$, and $[Ru(biq)_2L]^{2+}$, where L is the main ligand with differing functional group terminations (L = 2-(4-formylphenyl)imidazo[4,5-f][1,10] phenanthroline (*p*-FPIP), 2-(4-nitrophenyl)imidazo[4,5-f] [1,10]phenanthroline (*p*-NPIP), 2-(4-bromophenyl)imidazo [4,5-f][1,10]phenanthroline (*p*-BPIP), 2-(4-cyanophenyl)imidazo [4,5-f][1,10]phenanthroline (*p*-CPIP), 2-(3-cyanophenyl) imidazo[4,5-f][1,10]phenanthroline (*m*-CPIP), and 2-(2-cyanophenyl)imidazo[4,5-f][1,10] phenanthroline (*o*-CPIP).

Synthesis of the ruthenium complexes shown in Table 1 have been described elsewhere [19, 20]. The complexes were prepared in an acetonitrile solution. Concentration-dependent studies were undertaken to ensure the samples were unaffected by aggregation. Absorption spectroscopy was carried out using a PerkinElmer Lambda 900 UV/VIS/NIR absorption spectrometer. The luminescence measurements were performed using a PerkinElmer LS55 luminescence spectrometer. These measurements were used to calculate luminescence yields.

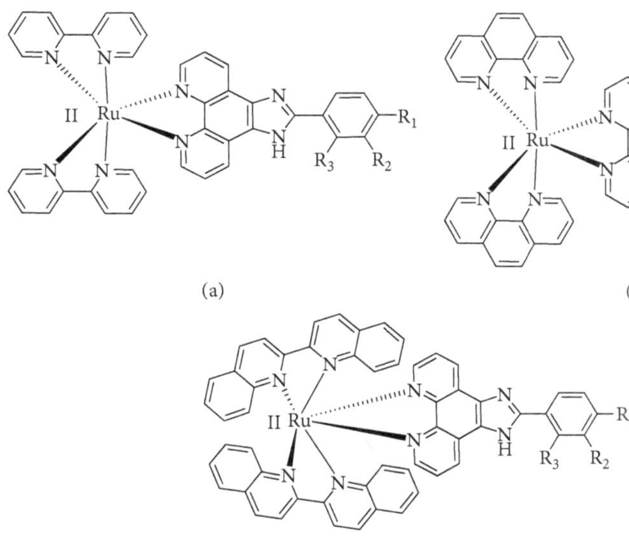

(a) (b) (c)

FIGURE 1: Ruthenium complexes: (a) $[Ru(bpy)_2L]^{2+}$, (b) $[Ru(phen)_2L]^{2+}$, and (c) $[Ru(biq)_2L]^{2+}$, where L is FPIP, NPIP, BPIP, and CPIP with R groups being CHO, NO_2, Br, and CN for the ligands, respectively.

TABLE 1: R groups for each main polypyridyl ligand.

Ligand names	R_1	R_2	R_3
p-FPIP	CHO	H	H
p-BPIP	Br	H	H
p-NPIP	NO_2	H	H
p-CPIP	CN	H	H
m-CPIP	H	CN	H
o-CPIP	H	H	CN

The electron affinity for the terminal group is taken from the literature and used to gauge the electronic state of the main ligand as the only parameter being varied. The electron affinity can be defined as the tendency of a chemical species to acquire electrons and thereby be reduced. Hence, it is a measure of the tenancy of the main ligands to contribute electrons across the Ru centre.

Luminescence lifetimes of aerated samples were measured using a computer-controlled time-correlated single photon counting (TCSPC) Spectrometer FL900 from Edinburgh Instruments. A nanosecond nF900 flashlamp excitation source using deuterium gas at a pressure of ~0.40 bar provided the fluorescence excitation pulses at 300 nm. A Peltier-cooled Hamamatsu R955 side-window photomultiplier tube (PMT) was used in an orthogonal geometry. All decay curves were corrected using a deconvolution with the instrument response function obtained using a scattering solution. The profile of the instrument response pulse had a FWHM of ~1 ns which was the detection limit of the system. All samples were run using a variation in concentrations from 10^{-4} M to 10^{-9} M at 300 K in HPLC-grade acetonitrile, and in all cases, the lifetime was found to be concentration independent within 0.1 ns.

Raman spectroscopy was performed using a Horiba Jobin-Yvon Labram HR800 UV confocal Raman imaging microscope system. A solid-state (785.1 nm/3 mW) laser source was used. The light was imaged to a diffraction-limited spot via the objective lens of an Olympus BX40 microscope. All experiments were carried out at room temperature (300 K).

2.1. Results: Part I—Effect of Changing Electron Affinity of the Terminal Group. The normalised absorption and emission spectra of $[Ru(phen)_2L]^{2+}$ in acetonitrile can be seen in Figures 2(a) and 2(b), respectively. The principle absorption of ruthenium complexes is made up of two main features at ~300 nm and at ~460 nm corresponding to the ligand centred (LC) and a metal to ligand charge transfer (MLCT) absorption, respectively [21–23]. It can be seen as expected that there is a gradual systematic bathochromic shift of the important MLCT transition as the electron affinity of the functional group located on the polypyridyl ligand is increased. The luminescence spectra for the ruthenium complexes in Figure 2(b) exhibit the same trend.

It would be expected that the systematic introduction of more electron withdrawing groups attached to the polypyridyl ligand follows a well-defined trend, starting with the lowest wavelength BPIP to the longest wavelength NPIP via the electronically intermediate aldehyde (FPIP) and nitrile (CPIP) groups [24]. Changes in excited state geometry, vibrational coherence, and electronic delocalisation can play a major role in determining the excited state condition [25, 26].

The electron withdrawing from the functional group at the termination site will lead to the polypyridyl ligand attracting electrons from the metal centre and hence acting to delocalise the electron density. It can be clearly seen form Table 2 that the emission maxima behaves in a similarly well-defined manner as the absorption maxima.

Unlike in the absorption however magnitude of the wavelength shift in emission maxima with the introduction

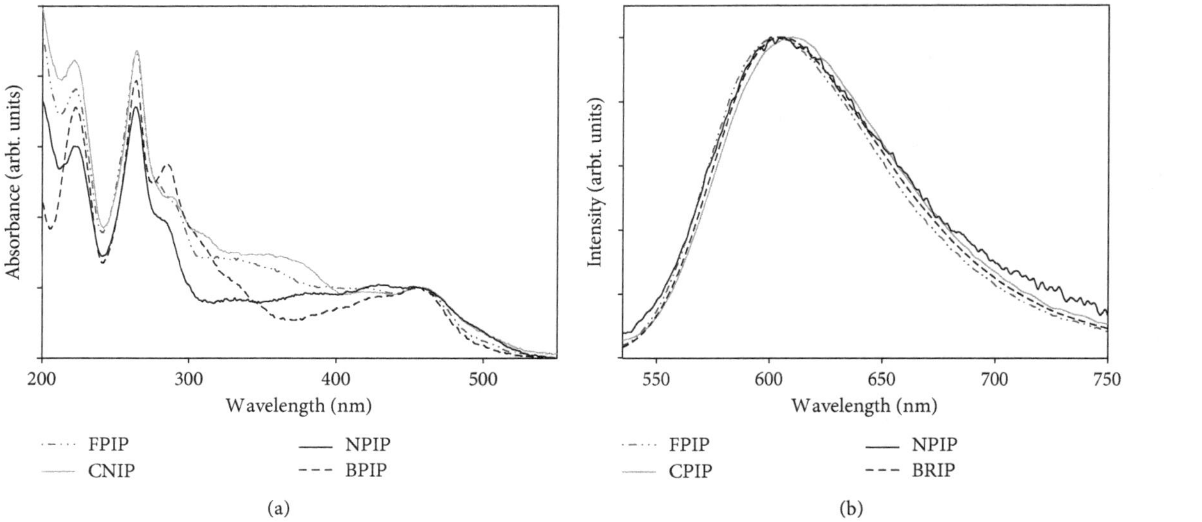

FIGURE 2: UV/Vis absorption (a) and luminescence (b) spectra of ruthenium complexes in acetonitrile with incremented termination group electronegativity for $[Ru(phen)_2L]^{2+}$.

TABLE 2: Electronic data for ruthenium complexes aerated in acetonitrile at 275 K.

	UV/Vis λ_{max} (nm)				Luminescence λ_{max} (nm)			
	p-NPIP 21.5	*p*-CPIP 15.9	*p*-FPIP 14.9	*p*-BPIP 6.2	*p*-NPIP 21.5	*p*-CPIP 15.9	*p*-FPIP 14.9	*p*-BPIP 6.2
EA (eV)								
Bpy	466	460	460	458	628	622	614	612
Phen	463	458	456	453	607	606	605	603
Biq	553	551	550	549	755	752	750	749

of the more electronegative functional groups, from BPIP to NPIP, cannot be explained solely as linear combination of the functional group and auxillary ligand electronic effects. From complexes BPIP to NPIP in the phenanthroline series, a shift of 4 nm is observed; however, in the bipyridine, it increases to 16 nm and then back to a relatively small shift of 6 nm for the biquinoline complexes. It is evident from this behaviour that a more complex relationship is at work here, than the first evidenced by the absorption spectra. It has been previously shown that the luminescence is greatly affected by the polypyridyl ligand [25, 27] and as such direct substitution to the terminal ring causing planarity effects and polarity alternation/interference [28]. This delocalisation will lead to a larger effective conjugation length and hence a bathochromic shift for the absorption and luminescence features [29]. As such, it is expected that the luminescence would exhibit a differing and more complex effect than the absorption with functional group change.

In summation, a reduction in the absorption and emission energy is observed with the changing of the functional group on the polypyridyl ligand from the bromo (BPIP), to the aldehyde (FPIP) to the nitrile (CPIP) and then the nitro- (NPIP-) based ligands. However, of particular interest is the variation in the magnitude of the wavelength shift with the systematic changing of the electronic withdrawing properties of the functional group for differing auxillary ligands (phenanthroline, bipyridine, and biquinoline). The systematic tuning of the band gap with the auxiliary ligands was most effective in the complexes with the greater electronegativity of the terminal functional group of the polypyridyl ligand, [14] and tuning was well defined approaching a linear dependence on the reduction potential of the auxiliary ligand. However, it is apparent when considering the functional group variation on the polypyridyl ligand that this is not a simple one-parameter effect. The effect of changing the electronic susceptibility of the functional group is highly influenced by the available electron density at the Ru centre afforded by the attached auxiliary ligands.

This magnitude of the effect on luminescence maxima of varying the functional group from BPIP to NPIP cannot be reconciled by treatment of the functional group change alone, and the interaction of the auxiliary ligand and the functional group is paramount in determining the extent of the electronic change.

With the Stokes shift, a manifestation of both electronic and vibrational effects, it was an ideal entry point to explore the potential for vibrational spectroscopy as a direct tool to correlate structural and vibrational effects with electronic changes. The stokes shift was calculated, taken as the difference between the absorption maximum and the emission maximum [30–32]. The $[Ru(phen)_2L]^{2+}$ complexes show a variation in Stokes shift with changing of the terminal group from BPIP to NPIP of 4 nm, the $[Ru(bpy)_2L]^{2+}$ showing the largest variation with systematic functional group change (8 nm), and $[Ru(biq)_2L]^{2+}$ the smallest variation (3 nm).

In Figure 3, the Stokes shift is shown as a function of the electronic nature of the terminal group on the polypyridyl ligand. The Stokes shift combines aspects of both the electronic and vibrational nature of the complexes. It can be readily observed that the Stokes does indeed vary in a well-defined manner with electron affinity, but not all complexes behave in a similar manner. The $[Ru(phen)_2L]^{2+}$ series shows a decrease in the Stokes shift as the electron affinity is increased, and conversely the $[Ru(bpy)_2L]^{2+}$ shows a concomitant increase with electron affinity of the terminal group. The $[Ru(biq)_2L]^{2+}$ series is least affected by electronic change on the polypyridyl ligand only showing a minimal increase in Stokes shift with increasing electron affinity.

From analysis of the Stokes shift, with origins in both electronic and vibrational coherence, it can be seen that systematic changing of the polypyridyl ligand is effective in the tuning of the electronic band gap, but the Stokes shift relationship again relays a more complex dependence. The effects, both vibrational and electronic, cannot be wholly described by the electron density changes induced by the functional group alone. The differing behaviour, albeit still well defined, of the Stokes shift with increasing electronegativity of the functional group leads to the possibility that the vibrational changes incurred with the systematic variation of the end-group polarity are hugely different to that previously illustrated for changing auxiliary ligands wherein the end-group contribution became negligible [14].

For the systematic changing of the auxiliary ligand, it was shown that the Stokes shift was inversely proportional to the nonradiative rate for all complexes. As seen earlier, the variation of the functional group does not show the same magnitude of effect across all ruthenium complexes studied. This would lend itself to the inference that the electronic nature and by extension the electronic delocalisation can be considered in terms of a modified push-pull effect where the ability of the ruthenium centre to donate electrons and the ability of the polypyridyl ligand to accept electrons must be accounted for. When working at extremes of the electron donated (the auxiliary ligand) or electron-accepting ability (the polypyridyl ligand), there will always be a saturation point where an increase in either will cease to have a noticeable effect on the electronic delocalisation. As such, linear behaviour as a function of either the auxiliary ligand or the functional group alone cannot be expected to reconcile the spectral changes. Both extremes, fully delocalised and fully localised electron densities situated on the complex would negate any influence on functional group change.

It has been previously shown for auxiliary ligand change that the Stokes shift can be seen as a crude measure of the nonradiative decay [14, 16–18]. A well-defined relationship would reinforce the idea that the vibrational coherence is a dominant effect and suggests the potential for tailoring the electronic and the vibrational (nonradiative decay) characteristics of the complexes. To further investigate the potential for such tailoring even with the complex nature of the spectral variations, the luminescence yield and lifetimes have

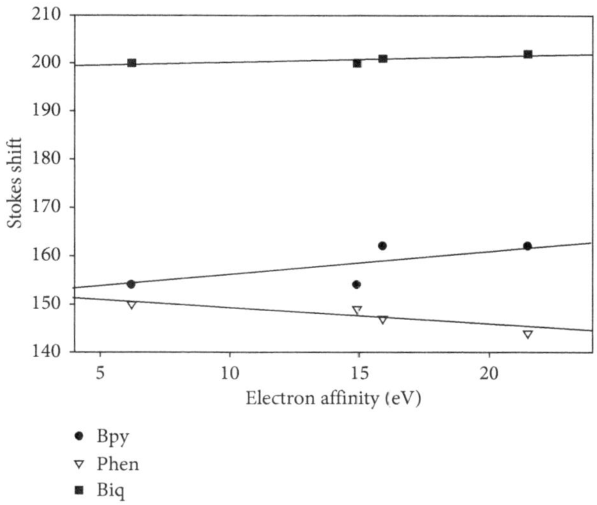

FIGURE 3: Stokes shift plotted as function of the electron density of the polypyridyl ligand terminal group.

been calculated as a means to elucidate the effect on the excited state lifetimes and yield, shown in Table 3.

A well-defined relationship is observed between the Stokes shift and luminescence yield, and a graph of the $[Ru(bpy)_2L]^{2+}$, $[Ru(phen)_2L]^{2+}$, and $[Ru(biq)_2L]^{2+}$ series is plotted as shown in Figure 4. A linear fit to the data points is presented as a guide for the eye; the slope indicates the effect of the changing of the polypyridyl ligand (L) from BPIP-FPIP-CPIP-NPIP.

The variation of luminescence yield with Stokes shift approaches a linear behaviour; however, the point of interest is the variation of the slope. With the Stokes shift taken as a crude measurement of the nonradiative rate, it would therefore be expected that all complexes would have a similar relationship as only the functional group is changing; however, the variation of slope indicates that the effect for the substitution of more electron-withdrawing functional ligands has a differing effect dependent on the attached auxiliary ligand. This variation of slope cannot be explained solely by considering the variation of the electronic nature of the dominant auxiliary ligands as determined by the reduction potential [7]. The slope varies from positive for the $[Ru(phen)_2L]^{2+}$ to slightly negative for the $[Ru(biq)_2L]^{2+}$ to a furthermore negative slope for the $[Ru(bpy)_2L]^{2+}$ with the reduction potential of the auxiliary ligand given a variation which should be in the order of phenanthroline, bipyridine, and biquinoline, respectively.

The variation in the slope further displays characteristics that are not in keeping with a simple linear solution and mirror those shown earlier. These variances can be reconciled by considering the complexes as a function of polarity interference/alternation as mentioned earlier or can be viewed as a modified version of the donor-acceptor push-pull system [33, 34]. The interaction along the conjugated backbone can be separated into two distinct and at times opposing factors. It has been shown that a series of "push-pull" oligomers can be systematically examined with

TABLE 3: Luminescence data for ruthenium complexes aerated in acetonitrile.

	Luminescence lifetimes (ns)				Luminescence yield* × 10^{-2}			
	p-NPIP 21.5	*p*-CPIP 15.9	*p*-FPIP 14.9	*p*-BPIP 6.2	*p*-NPIP 21.5	*p*-CPIP 15.9	*p*-FPIP 14.9	*p*-BPIP 6.2
EA (eV)								
Bpy	72	125	138	156	0.12	3.9	3.7	4.5
Phen	33	107	103	120	0.78	3.2	3.0	3.9
Biq	2.7	3.2	4.6	10.0	0.30	1.8	1.6	2.1

*Relative to 0.095 measured for $[Ru(bpy)_3]^{2+}$.

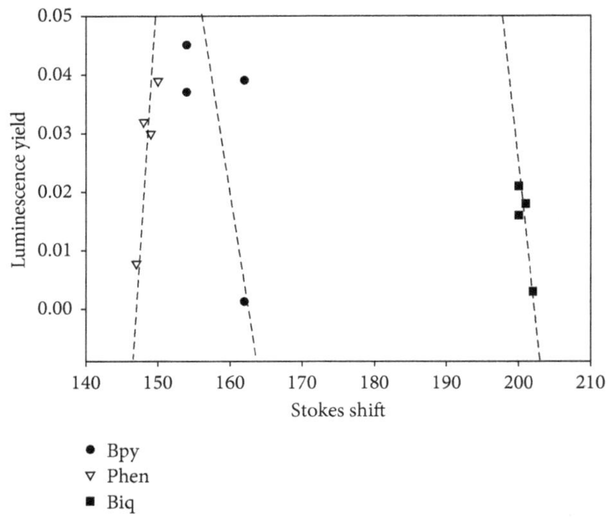

FIGURE 4: Luminescence yield versus peak Stokes shift for Ru(II) complexes (dotted lines are a linear guide for the eye).

a particular reference to alternate combinations of donor and acceptor end terminations.

The proposed effect of the functional group substitution on the polypyridyl ligand in these ruthenium complexes can be considered from a similar viewpoint as shown Figure 5.

Figure 5 shows a graphical representation for the proposed push-pull dynamic. The phenanthroline ligands (Figure 5(a)) contribute the least electron density to the central ruthenium moiety. This electron deficiency allows the changing functional group from Br (push) to NO_2 (pull) to mediate the conjugation length. The biquinoline ligand on the other hand has a huge associated electron density and as such contributes most to the central ruthenium moiety diminishing any effect the changing of the functional group may have. In essence, this push-pull effect determines whether the complexes behaves polymeric in nature as is the case with the biquinoline complexes where the terminal group does not influence the effective conjugation. On the other extreme, the phen complexes, due to a lesser electron density, act akin to oligomeric systems as the functional groups control the effective conjugative and hence the vibrational coherence.

The bipyridine and biquinoline auxiliary ligand with their greater associated electron densities than phen allowing the establishment of a great degree of electronic coherence across the polypyridyl ligand and as such variation of the polarity of the end group results in a pseudopolymeric regime. It has been shown that the bipyridine and biquinoline when attached as auxiliary ligands work to strengthen and planarise the polypyridyl ligand, hence restricting torsional effects facilitating pi electron delocalisation. This planarisation and delocalisation of pi electrons can be evidenced in the switching from a pseudo-oligomeric to a pseudopolymeric behaviour. The slope of the linear fits shows that the electronic nature of the functional group has a major impact on the relationship between the Stokes shift and luminescence yield, and a small variation in Stokes shift results in a large change in luminescence yield. The sensitivity of the luminescence yield to changes in the Stokes shift would point towards the changing of the functional group having a larger contribution to vibrational rather than a purely electronic effect.

In ascertaining that the functional group change has a great influence on the vibrational characteristics rather than the electronic, it was decided to directly examine the vibrational changes in the complexes via Raman spectroscopy. The Raman spectra of a series of complexes can be seen in Figure 6. The imidazole ring stretch of the polypyridyl ligand was examined for changes in vibrational nature with the systematic change of polypyridyl ligand end group as this provides the co-operative bridge for vibrational coherence.

This vibrational mode was chosen as it is unique to the polypyridyl ligand, and conjugation and vibrational coherence across this bridging ring will determine the extent of the effective conjugation. The imidazole ring also provides a large contribution to rotational freedom, and as such, the planarization of this ring structure will have a large effect on the polypyridyl ligand on a whole. This allows variations in both the planarity and electron delocalisation to be manifested in changing stretching frequency of this unique ring structure. The electronic nature of both the auxiliary ligands and polypyridyl functional group will considerably affect the electron density across the imidazole ring and therefore the vibrational frequency.

It is anticipated that direct measurement of relative conjugation across the imidazole bridge via Raman spectroscopy will elucidate both the complex relationship shown earlier and shortcoming of the energy gap law [35–37] when considering monomer to polymeric structures [38, 39].

Figure 7 shows the linear variation of the luminescence yield as a function of the variation of the frequency of the imidazole ring stretch of the polypyridyl ligand. As seen previously for the electronic properties, the bipyridine- and biquinoline-terminated complexes exhibit a similar activity

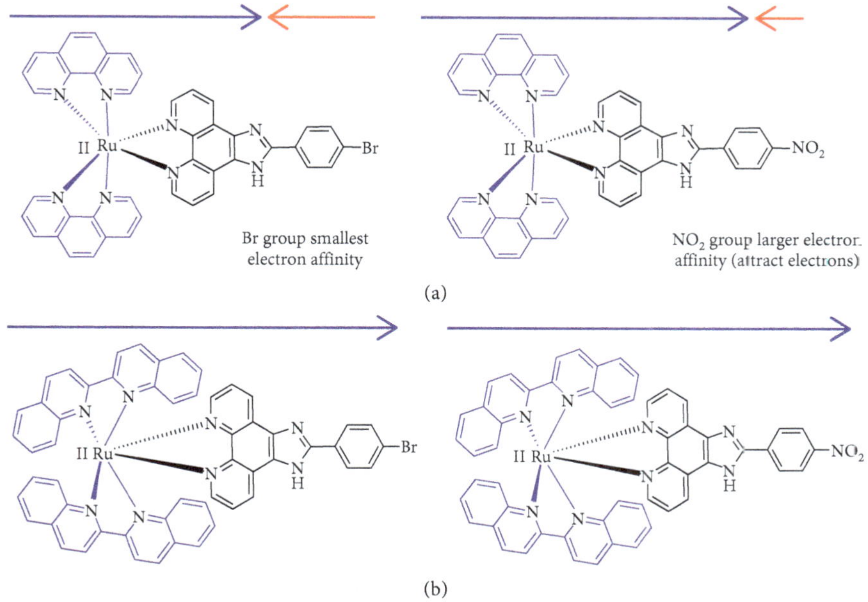

FIGURE 5: Graphical representation of the "donor-acceptor" description of electron density with end-group variation, showing Br to NO_2 for phen (a) and biq (b). Blue arrows indicate the electron delocalisation, and red arrows indicate the ability of the functional group to push back against the delocalisation.

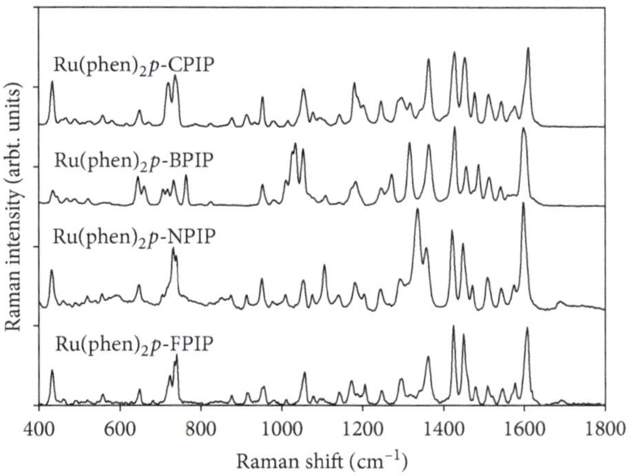

FIGURE 6: Solid-state Raman spectra of $[Ru(phen)_2L]^{2+}$ recorded at 273 K and 785 nm.

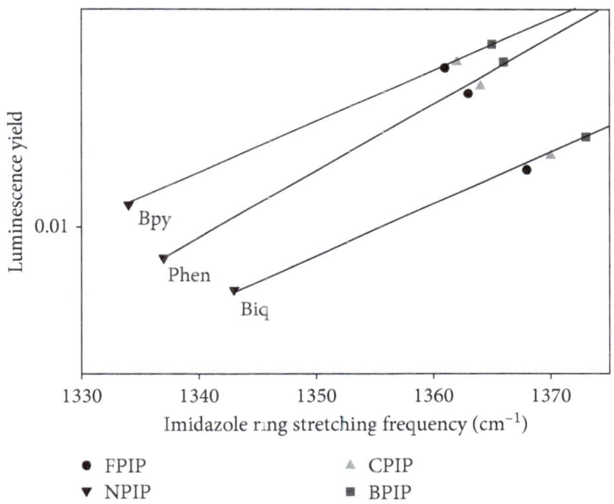

FIGURE 7: Imidazole stretching frequency against the luminescence yield. Linear fit is a guide for the eye.

profile. The phen series shows a linear behaviour but with significantly increased slope.

It is evident that there is a concomitant increase in luminescence yield with increasing ring stretching frequency. This behaviour is diametrical opposed to the behaviour seen when the ruthenium series are examined as a function of auxiliary ligand change where a reduction in the stretching frequency resulted in an increase in luminescence yield [14].

This behaviour can be reconciled by considering that the imidazole ring frequency is a conduit for electronic delocalisation to the functional group. As stated earlier, the substituted benzene ring at the end of the polypyridyl ligand has a large effect in determining the electronic and luminescence properties of the complex due to the lowest unoccupied molecular orbital (LUMO) location being centred on the polypyridyl ligand.

In the variation of the auxiliary ligand with the unchanging polypyridyl functional group, any increase in frequency of the imidazole ring stretch can be attributed to an increase of electron density available across the conduit and thus resulting in a concomitant increase in electronic density/conjugation of the terminal substituted benzene ring. Thus, an increase in frequency results in a decrease in yield, and the increased electron density facilitated greater vibrational coherence. However, when considering the variation of the functional group termination, a different scenario is presented, and the auxiliary ligand (an electron donor) remains constant with only the electron-accepting ability of the terminal group to influence the imidazole conduit. The effect of increasing the electron-accepting ability of the functional group draws electron density from the imidazole bridge, hence resulting in a lower vibrational frequency but greater electron density situated benzene-substituted ring. This effect is mediated as the electron-accepting ability of the functional group is changed from BPIP to NPIP.

2.2. Results: Part II—Effect of Changing the Terminal Group Position. Having fully explored the systematic variation of functional group change on the polypyridyl ligand, it was decided to examine the relationships in the more subtle structural changes of systematic regioisomerism. To this end, the *para*-, *meta*-, and *ortho*-positioned nitrile (CPIP) ligands were synthesised to create the [Ru(phen)$_2$L] [2]+, [Ru (bpy)$_2$L]$^{2+}$, and [Ru(biq)$_2$L]$^{2+}$ variants. Figure 8 show the absorption spectra of [Ru(phen)$_2$L]$^{2+}$-based complexes, and it can be readily seen that the positional variation of the end group has little effect on the spectral profile albeit inducing a minor shifting of the λ_{max} from 455 nm to 458 nm to 458 nm for the *meta*, *para*, and *ortho* positions, respectively.

The luminescence spectra show little or no variation in the electronic band gap when considering positional changes with biquinoline as auxiliary ligands, and a more significant effect is observed when moving to bipyridine and phenanthroline, respectively. This variation is consistent with the redox potential of the auxiliary ligands, bipyridine −0.8 V, biquinoline −1.4 V, and phenanthroline −1.6 V and clearly shows the influence of the auxiliary ligands even when considering the direct substitution on the polypyridyl terminal ring structure.

To further explore the changes to the excited state, as the position of the terminal group is systematically varied, the luminescence yield and lifetimes were calculated and measured, respectively. These values can be found in Table 4.

Due to the minute changes in the spectral profile brought about by varying the position of the substituent from *para*, *ortho*, and *meta* position, it is difficult to correlate the Stokes shift with any electronic parameter. This should be possible as well-defined correlations were seen for the auxiliary ligand changes [7] and functional group change, so it would suggest that the relative dominance of the nonradiative rate which was seen in the earlier trend may be absent. Furthermore, no obvious correlation of Stokes shift with the luminescence yield is observed which further suggested that although the Stokes shift can be seen as a gauge of nonradiative decay, this would only present a linear relationship with yield if the radiative rate was to remain unchanged. The calculated photophysical rates can be found in Table 5.

The radiative and nonradiative rates calculated are shown in Table 5. It is evident that there is no dominant effect; the variations in both rates are small and follow no well-defined trend relative to the electronic parameters. The certainty is that in no way can the nonradiative rate be considered to be dominant, and this is to be expected as the positional change on the terminal benzene ring would not be expected to have a major effect on the vibrational coherence on the ring structure or extend to the bridging imidazole ring.

Figure 9 shows the Raman spectra of all three possible regioisomers. It can be clearly seen in the range from 1330 cm^{-1}–1400 cm^{-1} that there is a shifting of the imidazole ring stretch as the terminal group position is changed.

In the previous section, the Raman stretching frequency of the bridging imidazole was used to reconcile the opposing changes in electronic density across the imidazole ring, and by association, the effect of the functional group changes on the benzene ring. However, in the case of the terminal group change, significant electronic and vibrational effects were observed such as the correlation of stokes shift with luminescence yield which have been absent for the regioisomers.

The position of the imidazole stretch frequency was plotted as a function of luminescence yield. This was undertaken to show that the sensitivity of Raman spectroscopy to the pi conjugated backbone make it a far better tool to gauge electronic conjugation and by extension potential avenue from nonradiative decay. A well-defined relationship with the direct Raman measurement of the complex and the luminescence yield would confirm the validity of this experimental study.

In Figure 10, it can be again seen that the luminescence yield shows a good correlation with the imidazole stretch frequency. Also evident is the positive slope linear trend as seen for the direct substitution on the terminal ring structure. From this well-defined relationship, it can be concluded that the direct measurement of the vibrational activity can provide an accurate gauge of the state of the system with regards to increasing luminescence yield. The ability of Raman spectroscopy to be extremely sensitive to delocalisation of pi electrons across the polypyridyl backbone allows it to account for subtle electronic such as positional changes of the substituent group. Although the study of the positional changes results in neither the radiative or nonradiative process being changed preferentially, the vibrational study seems to be able to account for the changes induced at both a vibrational and electronic level in the complexes.

In summary, the effect of changing the electron withdrawing effects of the functional group works, in two opposing ways, by varying the electron density at both the imidazole ring and the substituted benzene ring. This ability to reduce the electron density at the substituted benzene ring by adding a less electron-accepting functional group, such as Br, limits the contribution across the imidazole bridge. This has been shown to increase the luminescence and hence reducing the vibrational coherence on the benzene ring and thus the potential for nonradiative decay.

It is evident that the variation of the auxiliary ligand plays the major role in the maintenance of vibrational coherence and hence the nonradiative decay. Once the polypyridyl ligand is constant the effect of auxiliary ligand change is relatively easy to reconcile; however, the changing of both polypyridyl and auxiliary ligands will introduce

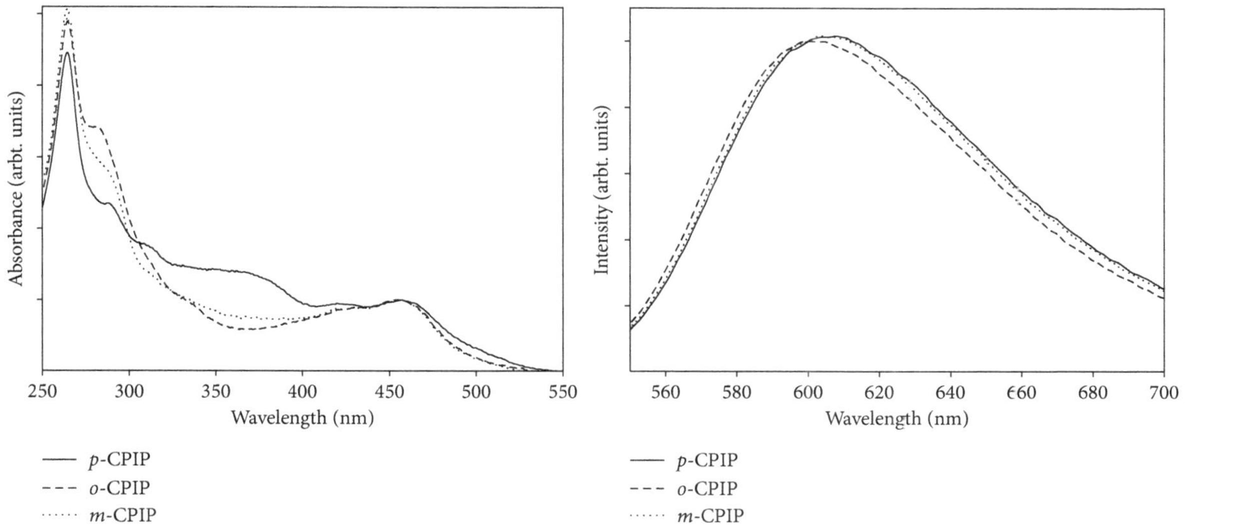

FIGURE 8: Normalised UV/Vis (a) and luminescence (b) spectra of $[Ru(phen)_2L]^{2+}$, where L is *m*-CPIP, *p*-CPIP, and *o*-CPIP, respectively.

TABLE 4: Luminescence data for ruthenium complexes aerated in acetonitrile.

	Luminescence lifetimes (ns)			Relative luminescence yield*		
	p-CPIP	*o*-CPIP	*m*-CPIP	*p*-CPIP	*o*-CPIP	*m*-CPIP
Bpy	128	135	130	0.032	0.034	0.029
Phen	107	107	102	0.039	0.039	0.043
Biq	4	4	4	0.018	0.015	0.019

*Relative to 0.095 as measured for $[Ru(bpy)_3]^{2+}$.

TABLE 5: Aerated photophysical rates for ruthenium polypyridyl complexes.

	k_{rad} (s × 10^5)			k_{nrad} (s × 10^6)		
	p-CPIP	*o*-CPIP	*m*-CPIP	*p*-CPIP	*o*-CPIP	*m*-CPIP
Bpy	2.5	2.52	2.23	7.56	7.16	7.47
Phen	3.64	3.64	4.22	8.98	9.98	9.938
Biq	4.5	3.75	4.75	245.5	246.3	245.2

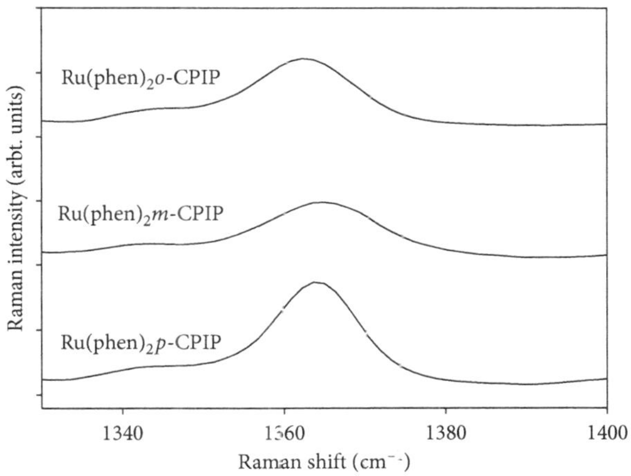

FIGURE 9: Raman spectra of imidazole ring stretch for the positionally substituted CN.

competing effects. These effects make a unified expression of the effect of the auxiliary/polypyridyl ligand change difficult to quantify, but it is apparent that there are both vibrational and electronic components to this effect. Analysis of the variation of Raman bands has shown to be effective in reconciling the changes induced in vibrational and electronic structures. The one drawback being the inability to monitor the terminal benzene substituted ring directly is influential in the excited state characteristics. The indirect measurements using the bridging imidazole ring are effective albeit requiring a great deal of more understanding of the relative variation of the electronic density across the complex.

3. Conclusions

This study focuses on the effect of terminal group change of the main polypyridyl ligand and the manifestation of these structural changes in both the electronic and vibrational spectra of a series of systematically varied ruthenium complexes. It establishes structure-property relationships for the series presented through investigation of the electronic absorption and emission characteristics. A well-defined relationship is found to exist between the absorption and emission wavelengths and the electronic nature of polypyridyl end-group ligand as measured by the electron affinity for the end group.

The relative electron withdrawing effects of the polypyridyl ligands are shown to manifest themselves in a complex push-pull nature due to the electronically dominant auxiliary ligand attached. In tailoring the electronic characteristics of the complexes, it is shown that the effect of systematically changing the terminal group works mediate the effective electron delocalisation originated for the auxiliary ligand on the ruthenium centre. This effect is not linear in nature but rather consisting of two plateaus

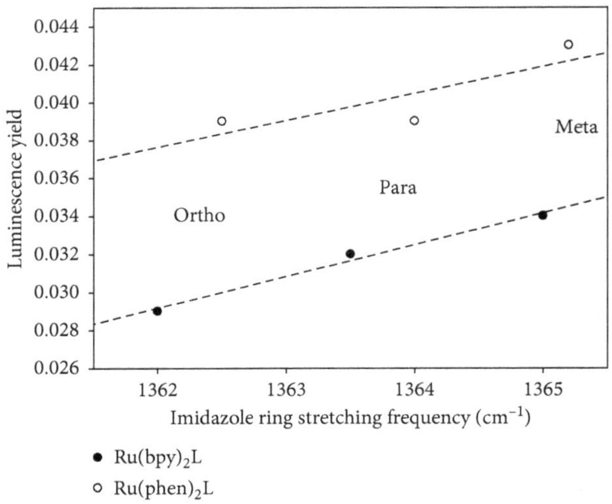

FIGURE 10: Luminescence yield against imidazole ring stretch for the positionally substituted CN.

representing a fully delocalised regime and a fully localised density of electronic states.

The relationship between the luminescence yield and lifetime establishes that the dominant effect of the of end-group variation on the polypyridyl ligands is a manipulation in the nonradiative decay rate. However, the positional change shows a negligible change in both photophysical rates.

The inability of the Stokes shift to accurately account for the variation in luminescence yield led to the examination of the vibrational spectrum in an effort to reconcile both Stokes and the energy gap law as a measure of the luminescence yield for low band gap material and proxyoligomeric structures.

Raman spectroscopy of the complexes leads to the assignment of a vibrational mode unique to the polypyridyl ligand, the frequency of which is shown to vary in a well-defined manner with the luminescence yield, indicating it can be used as an empirical guide to the push-pull effect of polypyridyl ligand change for both positional and end-group changes. This empirical relationship was also shown to be evident in regioisomeric changes where no significant changes in absorption or emission were evidenced. This study highlights how the polypyridal ligand and the molecule inherent vibrational coherence can be examined via Raman spectroscopy. This method can be used a direct measurement of both the electronic and vibrational changes and steric effects caused by end-group variation, where absorption, emission, and photophysical rates proved nonconclusive.

Conflicts of Interest

The authors declare that they have no conflicts of interest.

Acknowledgments

The FOCAS Institute is funded under the National Development Plan 2000–2006 with assistance from the European Regional Development Fund. L. Perdisatt acknowledges the Abbest Scholarship support.

References

[1] A. M. Florea and D. Büsselberg, "Cisplatin as an anti-tumor drug: cellular mechanisms of activity, drug resistance and induced side effects," *Cancer*, vol. 3, no. 1, pp. 1351–1371, 2011.

[2] M. S. Novak, G. E. Buchel, B. K. Keppler, and M. A. Jacupec, "Biological properties of novel ruthenium- and osmium-nitrosyl complexes with azole heterocycles," *Journal of Biological Inorganic Chemistry*, vol. 21, no. 3, pp. 347–356, 2016.

[3] S. T. Haque and E. H. Chowdhury, "Recent progress in delivery of therapeutic and imaging agents utilizing organic-inorganic hybrid nanoparticles," *Current Drug Delivery*, vol. 15, no. 4, pp. 485–496, 2018.

[4] S. Page, "Ruthenium compounds as anticancer agents," *Education in Chemistry*, vol. 26–29, 2012.

[5] E. S. Antonarakis and A. Emadi, "Ruthenium-based chemotherapeutics: are they ready for prime time?," *Cancer Chemotherapy and Pharmacology*, vol. 66, no. 1, pp. 1–9, 2010.

[6] M. P. M. Marques, "Platinum and palladium polyamine complexes as anticancer agents: the structural factor," *ISRN Spectroscopy*, vol. 2013, Article ID 287353, 29 pages, 2013.

[7] G. Mestroni, E. Alessio, G. Sava, S. Pacor, and M. Coluccia, *Metal Complexes in Cancer Chemotherapy*, B. K. Keppler, Ed., Vol. 157, VCH, Weinheim, Germany, 1993.

[8] G. K. Gransbury, P. Kappen, C. J. Glover et al., "Comparison of KP1019 and NAMI-A in tumour-mimetic environments," *Metallomics*, vol. 8, no. 8, pp. 762–773, 2016.

[9] T. Gianferrara, I. Bratsos, and E. Alessio, "A categorization of metal anticancer compounds based on their mode of action," *Dalton Transactions*, vol. 37, pp. 7588–7598, 2009.

[10] J. Liu, W. J. Mei, L. J. Lin et al., "Electronic effects on the interactions of complexes $[Ru(phen)_2(p\text{-}L)]^{2+}$ (L=MOPIP, HPIP, and NPIP) with DNA," *Inorganica Chimica Acta*, vol. 357, no. 1, pp. 285–293, 2004.

[11] H. M. Nguyen, D. N. Nguyen, and N. Kim, "Improved performance of dye-sensitized solar cells by tuning the properties of ruthenium complexes containing conjugated bipyridine ligands," *Advances in Natural Sciences: Nanoscience and Nanotechnology*, vol. 1, article 025001, 2010.

[12] S. J. Payne, G. L. Fiore, C. L. Fraser, and J. N. Demas, "Luminescence oxygen sensor based on a ruthenium(II) star polymer complex," *Analytical Chemistry*, vol. 82, no. 3, pp. 917–921, 2010.

[13] M. J. Rose and P. K. Mascharak, "Photoactive ruthenium nitrosyls: effects of light and potential application as NO donors," *Coordination Chemistry Reviews*, vol. 252, no. 18–20, pp. 2093–2114, 2008.

[14] L. O'Neill, L. Perdisatt, and C. O'Connor, "Correction to: "The influence of auxiliary ligands on the photophysical characteristics of a series of ruthenium (II) polypyridyl complexes"," *Journal of Physical Chemistry A*, vol. 116, no. 48, p. 11987, 2012.

[15] L. O'Neill, P. Lynch, M. McNamara, and H. J. Byrne, "Correlation of vibrational intensity with fluorescence lifetimes in π conjugated polymers," *Polymer*, vol. 49, no. 19, pp. 4109–4114, 2008.

[16] L. Luke O'Neill, *Photophysics of organic conjugated systems*, Ph.D. dissertation, Dublin Institute of Technology, Dublin, Ireland, 2006.

[17] L. O'Neill and H. J. Byrne, "Structure–property relationships for electron–vibrational coupling in conjugated organic

oligomeric systems," *Journal of Physical Chemistry B*, vol. 109, no. 26, pp. 12685–12690, 2005.

[18] L. O'Neill, P. Lynch, M. McNamara, and H. J. Byrne, "Vibrational characterization and fluorescence optimization of polycyclic polymers," *Journal of Physical Chemistry B*, vol. 111, no. 28, pp. 7999–8005, 2007.

[19] L. Perdisatt, S. Moqadasi, L. O'Neill et al., "Synthesis characterisation and DNA intercalation studies of regioisomers of ruthenium (II) polypyridyl complexes," *Journal of Inorganic Biochemistry*, vol. 182, pp. 71–82, 2018.

[20] H. Chao, R.-H. Li, C. W. Jiang, H. Li, L.-N. Ji, and X.-Y. Li, "Mono-, di- and tetra-nuclear ruthenium(II) complexes containing 2,2′-*p*-phenylenebis(imidazo[4,5-*f*]phenanthroline): synthesis, characterization and third-order non-linear optical properties," *Journal of the Chemical Society, Dalton Transactions*, no. 12, pp. 1920–1926, 2001.

[21] G. Whittaker, A. R. Mount, and M. R. Heal, *Physical Chemistry*, BIOS, Oxford, UK, 2000.

[22] L. Messori, F. Abbate, G. Marcon et al., "Gold(III) complexes as potential antitumor agents: solution chemistry and cytotoxic properties of some selected gold(III) compounds," *Journal of Medicinal Chemistry*, vol. 43, no. 19, pp. 3541–3548, 2000.

[23] A. A. Adeniyi and P. A. Ajibade, "The spectroscopic and conductive properties of Ru(II) complexes with potential anticancer properties," *Journal of Spectroscopy*, vol. 2014, Article ID 656830, 14 pages, 2014.

[24] S. Chowdhury, H. Kishi, G. W. Dillow, and P. Kebarle, "Electron affinities of substituted nitrobenzenes," *Canadian Journal of Chemistry*, vol. 67, no. 4, pp. 603–610, 1989.

[25] Q. X. Zhen, Q.-L. Zhang, J.-G. Liu, B.-H. Ye, L.-N. Ji, and L. Wang, "Synthesis, characterization and DNA binding of ruthenim(II) complexes containing the atatp ligand," *Journal of Inorganic Biochemistry*, vol. 78, no. 4, pp. 293–298, 2000.

[26] L. Zang, Y. Che, and J. S. Moore, "One-dimensional self-assembly of planar pi-conjugated molecules: adaptable building blocks for organic nanodevices," *Accounts of Chemical Research*, vol. 41, no. 12, pp. 1596–1608, 2008.

[27] H. Deng, H. Xu, Y. Yang et al., "Synthesis, characterization, DNA-binding and cleavage studies of [Ru(bpy)2(actatp)]2+ and [Ru(phen)2(actatp)]2+ (actatp=acenaphthereno[1,2-b]-1,4,8,9-tetraazariphenylence)," *Journal of Inorganic Biochemistry*, vol. 97, no. 2, pp. 207–214, 2003.

[28] S. Shi, J. Liu, J. Li et al., "Electronic effect of different positions of the $-NO_2$ group on the DNA-intercalator of chiral complexes $[Ru(bpy)_2L]^{2+}$ (L = *o*-npip, *m*-npip and *p*-npip)," *Dalton Transactions*, no. 11, p. 2038, 2005.

[29] C. Sauteret, J. P. Hermann, R. Frey et al., "Optical nonlinearities in one-dimensional-conjugated polymer crystals," *Physical Review Letters*, vol. 36, no. 16, pp. 956–959, 1976.

[30] J. B. Birks, *Organic Molecular Photophysics*, John Wiley & Sons, Hoboken, NJ, USA, 1973.

[31] J. F. Endicott, B. Schlegel, M. D. J. Uddin, and D. Seniveratne, "MLCT excited states and charge delocalization in some ruthenium/ammine/polypyridyl complexes," *Coordination Chemistry Reviews*, vol. 229, no. 1-2, pp. 95–106, 2002.

[32] O. Bräm, F. Messina, A. M. El-Zohry, A. Cannizzo, and M. Chergui, "Polychromatic femtosecond fluorescence studies of metal–polypyridine complexes in solution," *Chemical Physics*, vol. 393, no. 1, pp. 51–57, 2012.

[33] E. Kleinpeter, "Push-pull alkenes: structure and -electron distribution," *Journal of the Serbian Chemical Society*, vol. 71, no. 1, pp. 1–17, 2006.

[34] H. Meier, "Conjugated oligomers with terminal donor–acceptor substitution," *Angewandte Chemie International Edition*, vol. 44, no. 17, pp. 2482–2506, 2005.

[35] J. R. Lakowicz, *Principles of Flourescence Spectroscopy*, Vol. 1, Springer Science &Business Media, Berlin, Germany, 2006.

[36] M. Arias, J. Concepcion, I. Crivelli et al., "Influence of ligand structure and molecular geometry on the properties of d6 polypyridinic transition metal complexes," *Chemical Physics*, vol. 326, no. 1, pp. 54–70, 2006.

[37] C. Doffek, J. Wahsner, E. Kreidt, and M. Seitz, "Breakdown of the energy gap law in molecular lanthanoid luminescence: the smallest energy gap is not universally relevant for non-radiative deactivation," *Inorganic Chemistry*, vol. 53, no. 7, pp. 3263–3265, 2014.

[38] H. L. Chen, Y. F. Huang, C. P. Hsu et al., "Direct measurements of intersystem crossing rates and triplet decays of luminescent conjugated oligomers in solutions," *Journal of Physical Chemistry A*, vol. 111, no. 38, pp. 9424–9430, 2007.

[39] C. Latouche, P.-H. Lanoë, J. A. Gareth Williams, V. Guerchais, A. Boucekkine, and J. L. Fillaut, "Switching of excited states in cyclometalated platinum complexes incorporating pyridyl-acetylide ligands (Pt–C≡C–py): a combined experimental and theoretical study," *New Journal of Chemistry*, vol. 35, no. 10, p. 2196, 2011.

4

Time-Resolved Emission Spectroscopy Measurements during the Restrike Period in Arc Plasma Torch

Alan Mašláni,[1] Peter Ondáč,[1,2] Viktor Sember,[1] and Milan Hrabovský[1]

[1]Institute of Plasma Physics, Za Slovankou 3, 18200 Prague, Czech Republic
[2]Department of Surface and Plasma Science, Faculty of Mathematics and Physics, Charles University, V Holešovičkách 2, 18200 Prague, Czech Republic

Correspondence should be addressed to Alan Mašláni; maslani@ipp.cas.cz

Academic Editor: Arnaud Cuisset

Simultaneous optical, spectroscopic, and electrical measurements in the region of the arc anode attachment of the water-argon plasma torch are presented. A movement of the arc attachment along the anode surface together with its restrike mode is monitored. Temporal evolution of temperature during one cycle of the restrike mode is obtained in three different axial positions in the plasma column. Resulting temperature profiles show how the position of the arc attachment influences the plasma properties.

1. Introduction

Measurement of plasma characteristics with high temporal and spatial resolution is of great importance in order to understand physics of the individual discharge as well as for various applications in plasma processing technologies. Time-resolved measurements are typical for discharges with well-defined temporal appearance, that is, different types of AC and pulsed discharges. It is possible to find such cases also for arc discharges, which are of special interest in relation to the present work [1–3]. As for DC discharges, measurements with high temporal resolution are less frequent because it is often not expected any specific temporal behaviour of such discharges in the timescales of interest.

In case of optical diagnostic, one of the issues is a type of detector used for collection of radiation. CCD sensors give a possibility to reach high spatial and/or spectral resolution; however, they have limits in studying of temporal behaviour. Although CCD-based high-speed cameras are still more frequently used and also their speed quickly increases, they are still relatively expensive. Nevertheless, attempts to combine high-speed CCD and spectroscopy measurements have been done exactly on the arc discharges [4, 5]. On the other hand, photomultipliers offer usually well temporal but lower spatial and spectral resolution. Photomultipliers are also much more sensitive than CCDs; therefore, they are suitable also for very low light measurements. In this work, we encounter limits of used CCD sensor, namely, the temporal resolution of spectroscopic measurements.

In case of DC arc discharges, the well-known time-resolved pattern described in the literature is the so-called restrike mode of the anode attachment (also called arc root) movement in plasma torches [6, 7]. Restrike mode is characterized by sudden appearance of arc attachment in the upstream position of the anode. Then, the attachment moves along the anode downstream of the plasma flow until the next restrike occurs. Typically, the anode has a shape of a hollow cylinder; attachment moves in an inner surface of this cylinder in three dimensions, and the period of this movement is not well defined. However, in this work, we study the plasma torch in which the anode is disc shaped; thus, the attachment moves along the outer surface of this disc in one dimension only. As a result, the periodicity of the restrike mode is much better than in the case mentioned above. Then, we are able to use this period in order to obtain the time-resolved spectroscopic characterization of one cycle of the restrike. Combination of these spectroscopic measurements with high-speed camera movies and time-resolved

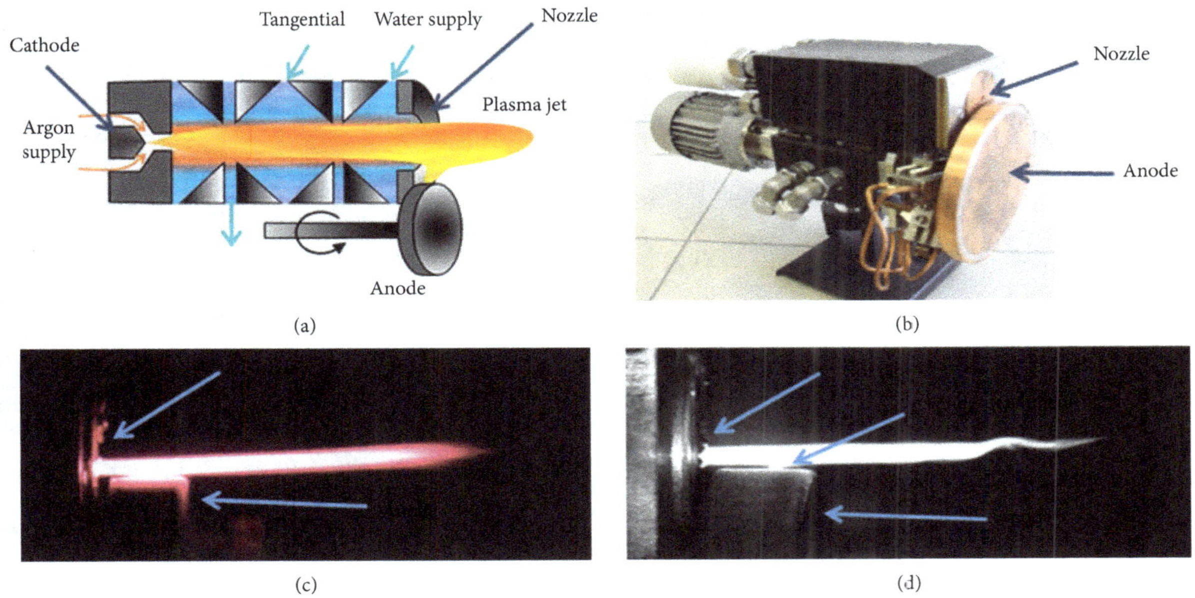

FIGURE 1: Water-argon plasma torch and jet: (a) schematic view of the torch, (b) photograph of the torch, (c) image of the plasma jet (camera (Canon EOS 450D), exposure time in the order of milliseconds), and (d) image of the plasma jet (high-speed camera (Photron), exposure time in the order of microseconds). Images were taken at the arc current 500 A and argon flow rate 12 slm.

measurement of the cathode-anode voltage brings new insight into the arc phenomena in the thermal plasma torches.

The last topic relevant to this work is an analysis of optical emission spectra and consideration of local thermodynamic equilibrium (LTE). Generally saying, LTE conditions are supposed to be valid in DC arc plasmas with good accuracy; therefore, these plasmas are often given as examples of thermal plasmas [8]. The departures from LTE are reported only in outer parts of the arcs, in recombining jets, and in electrode regions [7, 9, 10]. In this work, spectroscopic measurement in the anode region of the arc is performed and LTE validity is assumed. Therefore, we should be aware of the correct interpretation of the results especially from the point of view of electrode phenomena.

First, preliminary results dealing with time-resolved emission spectroscopy in the anode area of the water-argon plasma torch were already published [11]. In comparison with those results, this paper brings new and more detailed data together with their proper interpretation and analysis.

2. Experimental Setup

Plasma source used in this work is the so-called hybrid DC water-argon torch, which was developed on the base of the previously studied water-stabilized torch. The hybrid torch has been successfully applied for plasma spraying [12] as well as for gasification of various organic substances [13, 14]. Details about this torch can also be found elsewhere [15, 16]. Schematic view of the hybrid torch and its photo are shown in Figures 1(a) and 1(b). The arc is stabilized by the argon in the cathode region and by the water vortex surrounding substantial part of the arc column. The arc current can be varied between 200 A and 600 A and the argon flow rate between 8 slm and 40 slm. The rate of water evaporation into the plasma is about 0.3 g/s. This work presents measurements for 500 A and for 12 slm of argon. The cathode-anode voltage reaches about 280 V, which gives the arc power 140 kW.

A cathode, made of thoriated tungsten, is protected by the argon flow; therefore, its erosion is negligible and its lifetime is long. The anode is a copper disc with the thickness 16 mm which is rotating with the frequency 50 Hz in order to assure uniform erosion. Moreover, the anode is cooled by water circulating in its partly hollow body. The anode is located outside of the torch, in the distance 2 mm from the exit of the nozzle in the horizontal direction.

The gap of the nozzle connecting arc chamber with the surrounding environment has 6 mm diameter. From this description and from Figure 1, it is evident that we are able to observe part of the arc between nozzle and anode, and also the arc attachment which is moving along the anode surface with the period in the order of tens of microseconds. Typical photographs of the plasma jet taken by a common camera and by a high-speed camera are shown in Figures 1(c) and 1(d). It is possible to see that the common camera with exposure times in milliseconds does not reveal any details in the anode region. On the other hand, high-speed camera with exposure times in microseconds clearly shows the anode attachment and is able to follow its movement.

The measurement system consists of high-speed camera (Photron), voltage probe (Rigol), PC oscilloscope (PicoScope), and spectrometer (Jobin-Yvon Triax 552). Videos recorded by the high-speed camera are synchronized with the cathode-anode voltage measurements and with spectroscopic measurements.

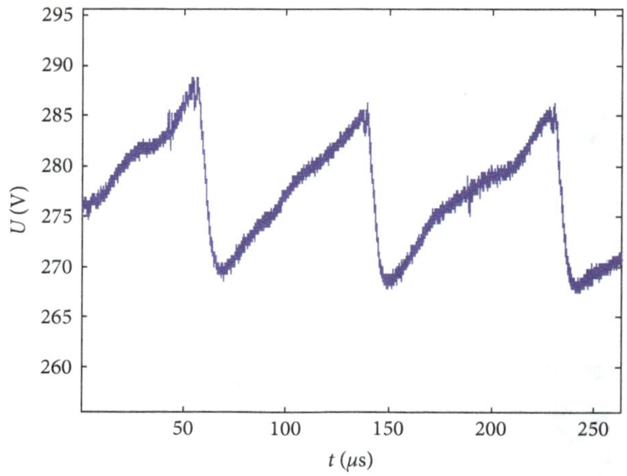

Figure 2: Example of the temporal evolution of the cathode-anode voltage.

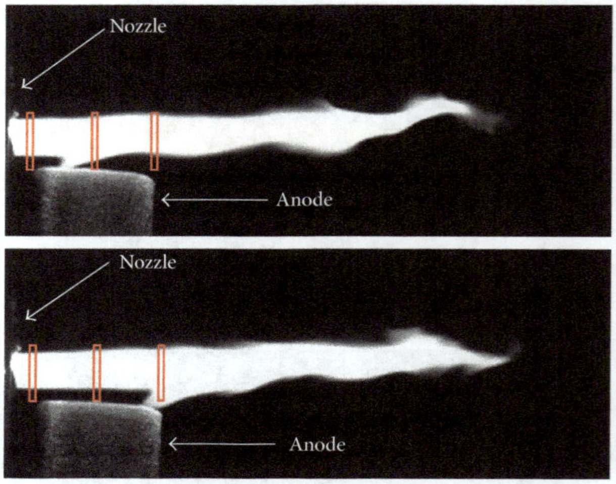

Figure 3: High-speed camera images of the plasma arc and jet for two cases: minimum voltage (a) and maximum voltage (b). Images come from a high-speed movie with 20,000 fps; exposure time is 1 μs. Red rectangles show positions of the entrance slit for spectroscopic measurements.

3. Description of Measurement

We present measurements for the arc current 500 A and argon flow rate 12 slm. Figure 2 shows a sawtooth pattern of cathode-anode voltage, which is typical for restrike mode of the arc attachment movement along the anode surface. Figure 3 shows images of the plasma obtained by the high-speed camera corresponding approximately to a minimum (Figure 3(a)) and maximum (Figure 3(b)) cathode-anode voltage. The voltage monotonically increases from minimum to maximum as the arc attachment moves along the anode downstream of the plasma flow. When the attachment reaches the end of the anode, it disappears, conducting connection between the plasma flow and anode is interrupted, and new attachment is formed upstream. This event is accompanied by a fast drop of the voltage. Thus, the new cycle of arc attachment movement begins and the process repeats. The full sequence of images characterizing this process is shown in Figure 4. In this case, images come from the high-speed movie with the frame rate 300,000 fps. For such high temporal resolution, only small window with anode region of the plasma column is visible (anode is in the bottom part of images).

Figure 3 shows also positions in which we performed spectroscopic measurements; they are marked by the red rectangles and by numbers 1, 2, and 3. In position 1, light is collected from the arc plasma close to the exit of the torch nozzle; it means that measurement is not directly affected by the transition of the arc attachment through the field of view of the slit. Position 2 is above the central part of the anode, and in position 3, we measure region above the downstream edge of the anode surface. In all measurements, the integration time of the spectra is 10 μs; as a typical period of the restrike is about 80 μs, we are able to monitor temporal behaviour during this period spectroscopically. In fact, integration time of 10 μs is the shortest possible time, which gives reasonable results. For shorter times, the signal-to-noise ratio of obtained spectra is too low. One more feature of these measurements is that we collect light from the whole plasma column in a given axial distance. The summary is that, with given experimental setup and using CCD detector, we are able to measure spectra with temporal resolution given by minimum integration time of 10 μs and with spatial resolution in one dimension given by the distance from the exit of the torch nozzle. On the other hand, in our previous studies, we investigated spectra with high spatial resolution but integrated during timescales of tens of milliseconds [17, 18]. In such a way, we obtained local values of temperatures in different parts of the arc column, but temporal resolution did not allow following the arc attachment movement.

Figure 5 shows typically measured emission spectra in a spectral window with the centre near Hβ line (486.1 nm). In spite of the fact that, using given integration time 10 μs, the spectra are noisy, it is possible to see well-recognized and broad Hβ, as well as four emission lines of singly ionized argon (472.7, 473.6, 476.5, and 480.6 nm). All these lines were used for temperature determination. We supposed that the plasma is optically thin for all these lines. Such assumption is usually valid for arc plasmas of these dimensions in this spectral region. The method is based on the ratio of experimental and theoretical emission coefficients of Hβ and ArII lines; a detailed description of this method can be seen in [19]; then, the method was further used for the characterization of the plasma produced by the same torch as in this work [18]. The method uses the fact that the plasma composition in the arc is known—water and argon. Then, equilibrium composition of plasma as a function of temperature can be calculated. From the composition, emission coefficients corresponding to individual transitions can be obtained theoretically, and from measured emission spectral lines, we also have them experimentally. Comparison of experimental and theoretical emission coefficients (and their ratios) gives the temperature. More specifically, the most suitable quantity for temperature determination is the ratio

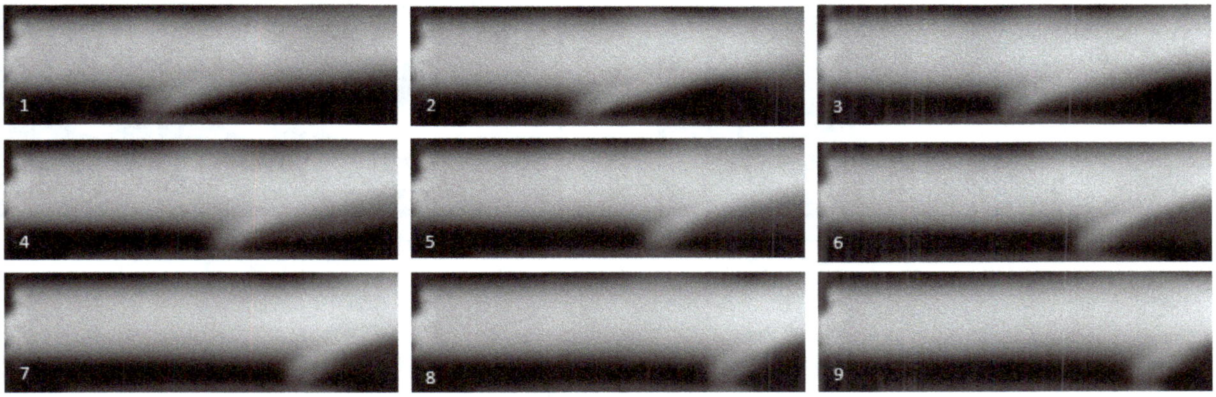

FIGURE 4: A sequence of images covering one period of the restrike process. Images come from a high-speed movie with 300,000 fps; exposure time is 0.29 μs.

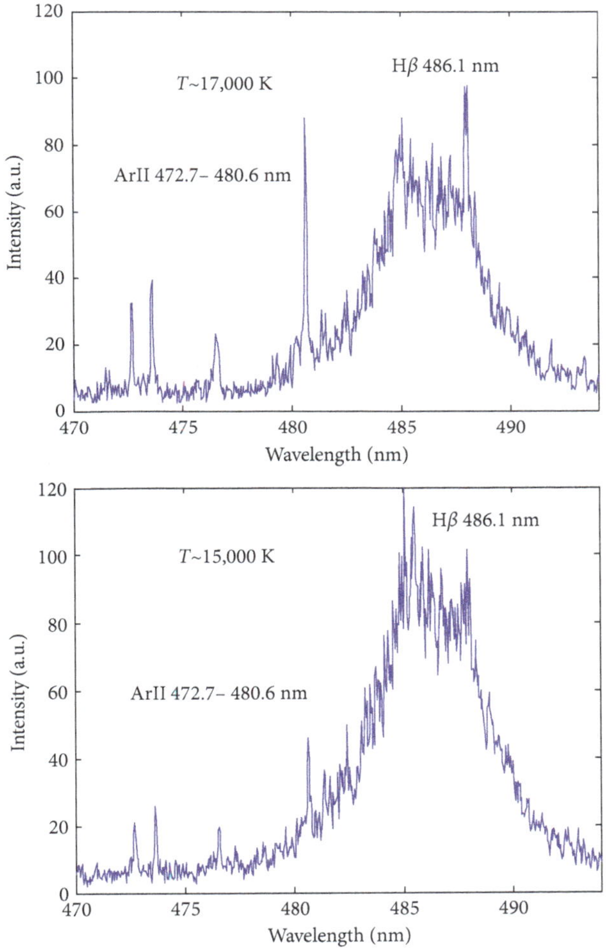

FIGURE 5: Typical emission spectra with the central wavelength near Hβ line. The ratio of intensities of Hβ to ArII lines is apparently different when the temperature is 17,000 K (a) and 15,000 K (b).

of the maximum of Hβ line and total emission coefficient of four ArII lines. This ratio is very sensitive to temperature and requires only relative calibration of spectral response. The advantage of this method is that one spectral window from 470 nm to 500 nm gives enough information to obtain temperature reliably. Comparison of two spectra in Figure 5 shows that the ratio of intensities of Hβ and ArII lines is suitable for temperature determination in the approximate range from 15,000 K to 17,000 K, which is reached in these experiments.

Spectroscopic measurements were performed as follows: for given position of the measurement, we captured a large number of spectra together with high-speed camera movies and temporal dependence of voltage. From these data, we chose only those, which did not include the restrike process itself. Comparison of movies and voltage dependence specifies in which moment during restrike period the spectrum was measured. Camera movies and voltage dependence have temporal resolution with the frame rate 300,000 fps and the sampling rate 80 MHz, respectively. Therefore, identification of moment of capture of each spectrum is well exact. Finally, we excluded those data that were measured during periods, which were much longer or shorter than the average one (80 μs). As a result, we obtained the temporal profile of temperature during the average restrike period.

4. Results and Discussion

Results are shown in Figures 6–8. Each point in these temporal profiles corresponds to the centre of the time interval during which spectra were acquired. As was mentioned above, the integration time of each measurement is 10 μs. It means, for instance, that point with time coordinate 40 μs was measured in the time interval from 35 μs to 45 μs. Time zero corresponds to the moment just after restrike process; that is, it is the moment with minimum voltage when anode attachment is closest to the exit nozzle of the torch (Figure 3(a)). Similarly, at the time 80 μs, there is a maximum voltage (Figure 3(b)).

Temperatures measured in position 1 are shown in Figure 6. It means that we measure in the arc column close to the exit of the nozzle of the torch. We can observe a monotonic increase in temperature during the period of

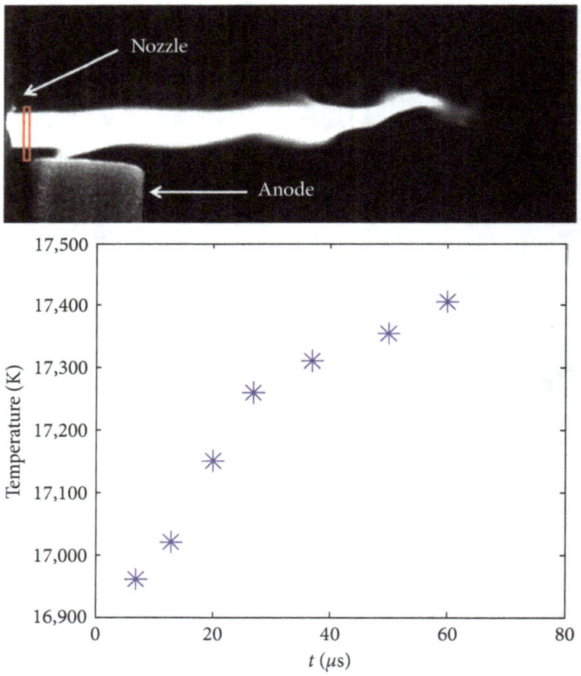

FIGURE 6: Temporal evolution of temperature during one period of restrike mode (measured in position 1 as shown by the red rectangle in high-speed camera image above the graph).

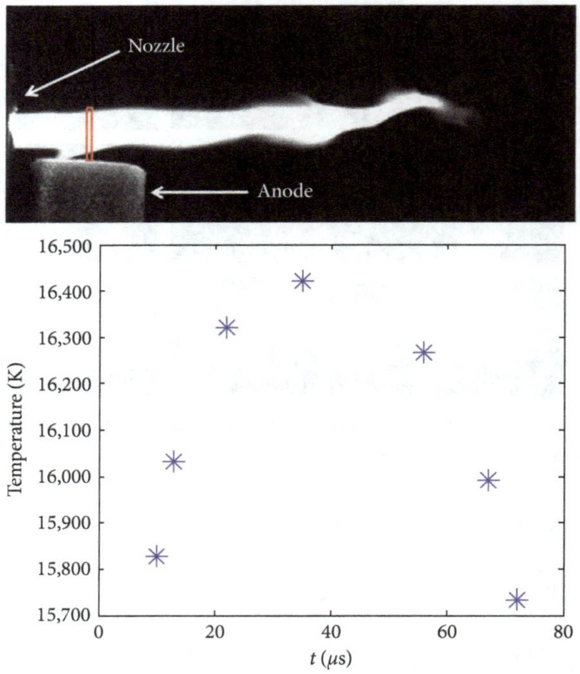

FIGURE 7: Temporal evolution of temperature during one period of restrike mode (measured in position 2 as shown by the red rectangle in high-speed camera image above the graph).

arc attachment movement. Figure 7 presents measurement in position 2 (above the centre of the anode). In this case, the temperature reaches its maximum value in the middle of the time interval at about 40 μs. Finally, Figure 8 shows the results of temperature measurement in position 3, where temperature again monotonically increases up to the maximum value at the end of the interval.

Interpretation of these results is not straightforward. As for position 1, we measure in the arc column without the direct influence of anode processes. However, the length of the arc and therefore also the distance of the measurement from the arc anode attachment change during the restrike period. It is known that temperature distribution in the arc column is generally such that the maximum temperature is in the region between the electrodes from which temperature decreases with approaching to electrodes [20–22]. For this torch, we do not have experimental data of this type, but according to numerical models, such dependence along the axis of the torch is also present [23, 24]. Such temperature distribution can simply explain the dependence as shown in Figure 6; as the arc attachment moves downstream, the position of measurement gets closer to the centre of the arc where the temperature is higher. The assumption is that LTE approximation is valid, which is highly probable for given conditions. Another assumption, which is less clear, is connected with the shape of the temperature distribution function along the arc column for different arc lengths. We must suppose that this function is approximately the same for different lengths of the arc during the restrike period, and moreover, gradient of temperature in a relatively short distance (10–15 mm) must be measurable by means of emission spectra. Then, the abovementioned explanation of temperature profile can be valid. It is evident that detailed knowledge of this function is necessary to give higher credibility to this explanation.

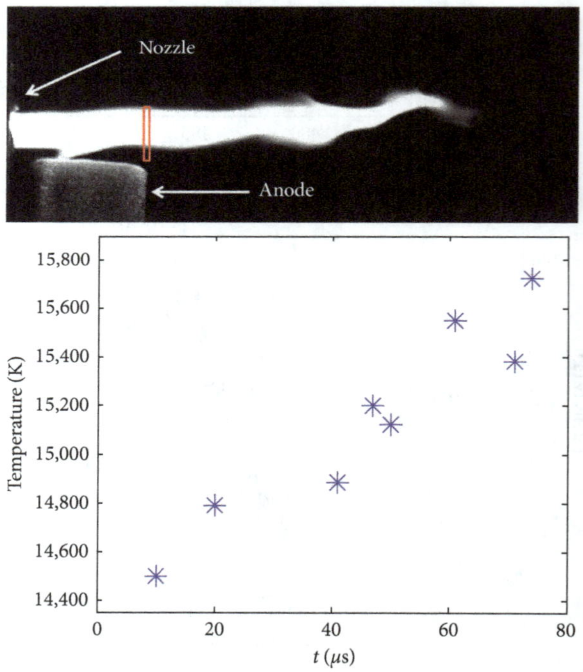

FIGURE 8: Temporal evolution of temperature during one period of restrike mode (measured in position 3 as shown by the red rectangle in high-speed camera image above the graph).

For measurements in positions 2 and 3, it seems that maximum value is connected with the presence of the arc attachment. As it moves along the centre of the anode surface, the maximum temperature is measured (Figure 7). Similarly, the highest temperature at the edge of the anode is obtained at the moment when arc attachment gets there (Figure 8). The region of anode attachment is characterized by the interaction of plasma with the solid electrode. This phenomenon brings strong nonequilibrium effects into the thermal arc plasma behaviour [7]. We can thus simply claim that increase of the temperature in the moments with arc attachment in the field of view of spectroscopic diagnostic is connected with violation of LTE, and then, this temperature does not represent kinetic temperature.

About the measurement in position 2, we can also say that it takes place on both sides of the arc attachment, that is, downstream and upstream of the attachment. In principle, it is expected that the temperature should decrease downstream of the arc attachment in the direction of the freely expanding jet, and on the other hand, temperature should increase in the upstream direction inside the arc. However, Figure 7 does not prove this assumption, which can be explained by the fact that measurement takes place in close vicinity (few millimeters) of the attachment, and thus, nonequilibrium effects dominate in this case. On the other hand, in position 1, we get far enough from the attachment in order to register temperature profile inside the arc. In any case, only higher temporal and spatial resolution of measurement can bring more details into the temperature distributions around the arc attachment.

Regarding the measurement in position 3, its interpretation is probably not so complicated. At the outer edge of the anode, we are most of the time downstream of the arc where recombining jet begins. The temperature in this region is naturally lower than that in the arc. As the arc attachment approaches, the temperature grows. Then, both effects, approaching of the hot arc region and influence of the nonequilibrium near the attachment, can cause the increase in temperature during the restrike period.

We can also say something about the range of measured temperatures in Figures 6–8, which is from approximately 14,500 K to 17,500 K. This range is relatively narrow and can bring questions about the accuracy of the method. The fact is that temperature gradients in arcs are quite strong. However, our measurements are averaged through the whole plasma column in a given axial distance. Therefore, measured averaged temperatures cannot be easily compared with local values of temperatures, which show these strong gradients. Anyway, rough estimation can be done: temperature on the axis of this torch in the arc region for given conditions is above 20,000 K, while in the 3 mm radial distance, which is the edge of the luminous arc, the temperature decreases to about 12,000 K [16]. Then, the average temperatures measured in this work may be correct and also obtained temporal profiles may be reliable. As was mentioned in Section 3, the sensitivity of the method is very high; the variations of temperature in the order of ~100 K have a clear effect on the $H\beta$/ArII ratio (in the temperature range of interest). Systematic error arising from not exactly known concentrations of species can only influence absolute value of temperature (we suppose uncertainty of this absolute value about 10%). However, under the assumption that these concentrations are constant in all three measured positions and in all measured time instants, the changes in spectra can be attributed to temperature changes.

Finally, some comments on thermodynamic equilibrium validity can be added. As was mentioned earlier, the LTE conditions are usually well valid in central parts of high-intensity arc plasmas. This is also the case of the plasma torch used in this research; it was shown that necessary conditions for LTE validity, such as Griem's or McWhirter's criteria, are well fulfilled in the arc region as well as downstream in recombining jet [18]. However, these criteria are not sufficient; in spite of their validity, the departure from LTE can be additionally caused, simply saying, by inhomogeneity and/or transient nature of the plasmas, as summarized in the work of Cristoforetti et al. [25]. The region of arc attachment is characterized not only by high electron density but also by strong inhomogeneity of the plasma. This can be the reason of overestimation of the temperature in this region. One more observation deduced from Figures 6–8 is that temperatures monotonically decrease from position 1 to position 3. For example, if we choose time 10 μs, we have almost a linear decrease of these temperatures in this sequence of positions (~17,000 K, ~15,800 K, and ~14,500 K). Presence of anode attachment does not change the monotonic character of this dependence, but it makes this decrease nonlinear (the most extreme example is ~17,400 K, ~15,700 K, and ~15,600 K for time 70 μs). It means that the most important factor for the temperature value and its variations is still the presence of arc current path, while the non-LTE behaviour only modulates these changes.

5. Conclusion

Time-resolved measurements of arc discharge in the anode region are presented. We were able to follow the movement of the arc anode attachment using high-speed camera movies and arc voltage time evolution as well as by emission spectra. Results include the temporal evolution of temperature during one cycle of the restrike mode. It is shown that temperature in a given axial distance depends on the position of the arc attachment.

Presented measurement can be considered as one of the first measurements of this type. It seems to be natural that higher temporal and spatial resolution at the same moment will bring better insight into the studied phenomena. Therefore, enhancement of the diagnostic apparatus in order to reach this goal is planned as a next step.

Conflicts of Interest

The authors declare that they have no conflicts of interest.

Acknowledgments

This work was supported by the Czech Science Foundation (GACR) under Project GC17-10246J.

References

[1] M. Tanaka, T. Ikeba, Y. Liu, S. Choi, and T. Watanabe, "Investigation of electrode erosion mechanism of multi-phase AC arc by high-speed video camera," *Journal of Physics: Conference Series*, vol. 441, article 012015, 2013.

[2] M. Kühn-Kauffeldt, J.-L. Marqués, and J. Schein, "Thomson scattering diagnostics of steady state and pulsed welding processes without and with metal vapor," *Journal of Physics D: Applied Physics*, vol. 48, no. 1, article 012001, 2015.

[3] J. Hlína, J. Šonský, J. Gruber, and Y. Cressault, "Fast tomographic measurements of temperature in an air plasma cutting torch," *Journal of Physics D: Applied Physics*, vol. 49, no. 10, article 105202, 2016.

[4] B. Bachmann, R. Kozakov, G. Gött et al., "High-speed three-dimensional plasma temperature determination of axially symmetric free-burning arcs," *Journal of Physics D: Applied Physics*, vol. 46, no. 12, article 125203, 2013.

[5] D. Uhrlandt, R. Methling, S. St. Franke Gorchakov, M. Baeva, A. Khakpour, and V. Brüser, "Extended methods of emission spectroscopy for the analysis of arc dynamics and arc interaction with walls," *Plasma Physics and Technology*, vol. 2, no. 3, pp. 280–289, 2015.

[6] S. Paik, P. C. Huang, J. Heberlein, and E. Pfender, "Determination of the arc-root position in a DC plasma torch," *Plasma Chemistry and Plasma Processing*, vol. 13, no. 3, pp. 379–397, 1993.

[7] J. Heberlein, J. Mentel, and E. Pfender, "The anode region of electric arcs: a survey," *Journal of Physics D: Applied Physics*, vol. 43, no. 2, article 023001, 2010.

[8] V. Helbig, "Diagnostics of thermal plasma," *Pure and Applied Chemistry*, vol. 60, no. 5, pp. 675–684, 1988.

[9] L. E. Cram, L. Poladian, and G. Roumeliotis, "Departures from equilibrium in a free-burning argon arc," *Journal of Physics D: Applied Physics*, vol. 21, no. 3, pp. 418–425, 1988.

[10] J. Haidar, "Departures from local thermodynamic equilibrium in high-current free burning arcs in argon," *Journal of Physics D: Applied Physics*, vol. 30, no. 19, pp. 2737–2743, 1997.

[11] A. Mašláni, P. Ondáč, V. Sember, and M. Hrabovský, "Time-resolved optical and spectroscopic study of the restrike mode in arc plasma torch," in *Proceedings of the ISPC 2017: 23rd International Symposium on Plasma Chemistry, Book of Extended Abstracts*, pp. 640–642, International Plasma Chemistry Society, Montréal, Canada, 2017.

[12] T. Kavka, J. Matějíček, P. Ctibor, A. Mašláni, and M. Hrabovský, "Plasma spraying of copper by hybrid water-gas DC arc plasma torch," *Journal of Thermal Spray Technology*, vol. 20, no. 4, pp. 760–774, 2011.

[13] M. Hrabovský, M. Hlína, M. Konrád et al., "Thermal plasma gasification of biomass for fuel gas production," *High Temperature Material Processes*, vol. 13, no. 3-4, pp. 299–313, 2009.

[14] M. Hrabovský, M. Hlína, V. Kopecký et al., "Steam plasma treatment of organic substances for hydrogen and syngas production," *Plasma Chemistry and Plasma Processing*, vol. 37, no. 3, pp. 379–397, 2017.

[15] M. Hrabovský, "Generation of thermal plasmas in liquid-stabilized and hybrid dc-arc torches," *Pure and Applied Chemistry*, vol. 74, no. 3, pp. 429–433, 2002.

[16] M. Hrabovský, V. Kopecký, V. Sember, T. Kavka, O. Chumak, and M. Konrád, "Properties of hybrid water/gas DC arc plasma torch," *IEEE Transactions on Plasma Science*, vol. 34, no. 4, pp. 1566–1575, 2006.

[17] A. Mašláni and V. Sember, "Emission spectroscopy of OH radical in water-argon arc plasma jet," *Journal of Spectroscopy*, vol. 2014, Article ID 952138, 6 pages, 2014.

[18] A. Mašláni, V. Sember, and M. Hrabovský, "Spectroscopic determination of temperatures in plasmas generated by arc torches," *Spectrochimica Acta Part B*, vol. 133, pp. 14–20, 2017.

[19] V. Sember and A. Mašláni, "A simple spectroscopic method for determining the temperature in a H_2O-Ar thermal plasma jet," *High Temperature Material Processes*, vol. 13, no. 2, pp. 217–228, 2009.

[20] A. J. D. Farmer, G. N. Haddad, and L. E. Cram, "Temperature determinations in a free-burning arc: III. Measurements with molten anodes," *Journal of Physics D: Applied Physics*, vol. 19, no. 9, pp. 1723–1730, 1986.

[21] M. Razafinimanana, L. El Hamidi, A. Gleizes, and S. Vacquié, "Experimental study of the influence of anode ablation on the characteristics of an argon transferred arc," *Plasma Sources Science and Technology*, vol. 4, no. 3, pp. 501–510, 1995.

[22] P. Kah, H. Latifi, R. Suoranta, J. Martikainen, and M. Pirinen, "Usability of arc types in industrial welding," *International Journal of Mechanical and Materials Engineering*, vol. 9, no. 1, p. 15, 2014.

[23] J. Jeništa, H. Takana, H. Nishiyama et al., "Integrated parametric study of a hybrid-stabilized argon-water arc under subsonic, transonic and supersonic plasma flow regimes," *Journal of Physics D: Applied Physics*, vol. 44, article 435204, 2011.

[24] J. Jeništa, H. Takana, S. Uehara et al., "Modeling of inhomogeneous mixing of plasma species in argon-steam arc discharge," *Journal of Physics D: Applied Physics*, vol. 51, article 045202, 2018.

[25] G. Cristoforetti, A. De Giacomo, M. Dell'Aglio et al., "Local thermodynamic equilibrium in laser-induced breakdown spectroscopy: beyond the McWhirter criterion," *Spectrochimica Acta Part B*, vol. 65, no. 1, pp. 86–95, 2010.

5

Dynamic Localized SNV, Peak SNV, and Partial Peak SNV: Novel Standardization Methods for Preprocessing of Spectroscopic Data Used in Predictive Modeling

Emily Grisanti,[1,2] Maria Totska,[2] Stefan Huber,[2] Christina Krick Calderon,[2] Monika Hohmann,[2] Dominic Lingenfelser,[2] and Matthias Otto[1]

[1]Institute of Analytical Chemistry, TU Bergakademie Freiberg, Leipziger Str. 29, 09599 Freiberg, Germany
[2]Robert Bosch GmbH, Renningen, 70465 Stuttgart, Germany

Correspondence should be addressed to Emily Grisanti; emily.grisanti@de.bosch.com

Academic Editor: K. S. V. Krishna Rao

An essential part of multivariate analysis in spectroscopic context is preprocessing. The aim of preprocessing is to remove scattering phenomena or disturbances in the spectra due to measurement geometry in order to improve subsequent predictive models. Especially in vibrational spectroscopy, the Standard Normal Variate (SNV) transformation has become very popular and is widely used in many practical applications, but standardization is not always ideal when performed across the full spectrum. Herein, three different new standardization techniques are presented that apply SNV to defined regions rather than to the full spectrum: Dynamic Localized SNV (DLSNV), Peak SNV (PSNV) and Partial Peak SNV (PPSNV). DLSNV is an extension of the Localized SNV (LSNV), which allows a dynamic starting point of the localized windows on which the SNV is executed individually. Peak and Partial Peak SNV are based on picking regions from the spectra with a high correlation to the target value and perform SNV on these essential regions to ensure optimal scatter correction. All proposed methods are able to significantly improve the model performance in cross validation and robustness tests compared to SNV. The prediction errors could be reduced by up to 16% and 29% compared with LSNV for two regression models.

1. Introduction

Chemometric approaches are becoming increasingly popular as they enable more comprehensive extraction of relevant information out of complex data provided by modern instrumental analytics. At the same time, advances in data analysis make it possible to reduce the size of the instrument hardware by compensating for the missing measurement quality of miniaturized instruments. In combination with multivariate calibration, the development of models based on low-cost analytics, such as vibrational spectroscopy, allows the development of models that predict parameters usually determined with cost-intensive measuring instruments or complex methods. Monitoring the alcoholic fermentation [1] and determining the viscosity of engine oil [2, 3] or proteins in milk [4] by spectroscopic means become thus feasible. It has also been possible to determine specific viscosity modifiers and pour point depressant additive compounds in engine oils [5] by FTIR, which is due to the fact that the concentration of a component follows, according to the Lambert–Beer Law, a linear dependency on the light absorbance of the medium [6, 7].

Preprocessing methods play a decisive role for the performance of these models, as spectra can be influenced by various disturbing factors that interfere with the significance of the measurement [8–11]. The main influence comes from the measuring geometry, which includes the sample thickness, the distance from the detector to sample, the contact pressure, and the angle from the light source to sample [12, 13]. The elimination of scattering effects by particles of different size and distribution also plays a major role in preprocessing.

Different spectroscopic measurement techniques suffer from different major disturbing factors. In near-infrared spectroscopy, it is usually a constant or linear baseline offset due to scattering light, Raman spectra often show polynomial fluorescence background, and for mid-infrared spectra, the sample thickness and thus the spectroscopic response plays a crucial role [14, 15]. The information about the sample is present in the shape of the spectrum and independent of the offset (additive effect) and the scaling of the absolute signal intensity (multiplicative effect). The task of preprocessing is to remove these interfering factors from the informative part of the spectrum, and there are different approaches for this.

A method for eliminating constant offset terms is to calculate the first derivative [9]. This procedure can be extended to higher-order derivatives also eliminating offset terms with linear or quadratic baseline curves. The disadvantage of calculating the deviation of a spectrum is that noise effects are amplified.

Multiplicative signal correction (MSC) is another tool which can deal with the two major effects. A reference spectrum, in most cases represented by the mean spectrum of the calibration data set, is defined, and the spectra are corrected for the baseline and the multiplicative amplification effects [16, 17]. The approach is associated with the Kubelka–Munk theory, which takes optical phenomena caused by light scattering into account [18, 19]. For each spectrum, the two correction parameters are estimated via a least squares regression calculation.

Standard normal variate (SNV) removes a constant offset term by subtracting the mean value of the full spectrum and brings all spectra to the same scale by subsequent division by the standard deviation of the full spectrum [20]. Due to its simplicity, SNV is a popular preprocessing method [21]. SNV and MSC usually yield similar results and are often regarded as exchangeable [22]. Since no extra regression step is needed for the SNV transformation to estimate the correction parameters, in the following, the focus lies on SNV as the models should be kept as simple as possible.

Some efforts have been made to optimize standardization techniques. A piecewise MSC (PMSC) method has been proposed by Isaksson and Kowalski [23], which significantly improved the predictive power of several regression models based on near-infrared transmittance spectra. A Localized SNV (LSNV) approach has been introduced by Bi et al. performing the SNV not on the full spectrum but on subsequent sequences [24]. This strategy also yielded very promising results in several regression cases based on benchmark NIR data sets. In the following, a dynamic version of the LSNV algorithm, called DLSNV, is presented. By allowing for a dynamic starting point of the first and subsequent SNV windows, it is more flexible to align the SNV to important vibrational bands in the spectra. PSNV and PPSNV are based on the idea that the standardization can be optimized when performed on distinct wavenumber windows across highly specific regions of the spectrum.

2. Experimental

As a sample set, data originated from an investigation about aging and interaction phenomena in Automatic Transmission Fluids (ATF) were used. Many ATF samples have been stored for different periods at several temperatures to produce artificially aged samples.

The aim of the presented study was to transfer information coming from a highly specific, costly, and complex measurement method (High-Performance Liquid Chromatography coupled with Quadrupole Time-of-Flight-Mass Spectrometry (HPLC-QToF-MS)) to data measured with a low-cost, flexible tabletop instrument (Fourier-Transform Infrared (FTIR) spectrometer). This was achieved by analyzing each sample coming from the storage experiment and determining the additive response signals in these samples by HPLC-QToF-MS. By using these additive responses as reference values, a calibration model was created in order to be able to predict the concentration of the additive compounds in the samples by evaluating the FTIR spectra. The new standardization techniques proposed here are being tested for the regression models.

2.1. Additive Compounds. Two additive compounds from two different ATF oils were analyzed:

Within ATF A: an unsaturated ethoxylated amine known as friction modifier

Within ATF B: a bis-*tert*-butyl-hydroxytoluene (BHT) derivate known as phenolic antioxidant

2.2. Samples and Experiments. For the investigation of degradation phenomena in ATFs, a comprehensive storage experiment had been set up. The effects of different materials on ATFs and the impact of temperature on oil aging should be analyzed. Therefore, the ATFs were stored under various conditions in an oven. Three parameters had been varied: the storage temperature, the storage time, and added materials. The storage times had been adjusted to the temperatures so that a comparable load, according to Arrhenius Law, could be expected. The parameters are listed in Table 1.

For all time/temperature combinations, three interaction experiments have been conducted:

(i) storage with pure oil
(ii) storage with oil plus copper alloy chips
(iii) storage with oil plus chips from copper alloy, iron, and PA66

The samples were prepared by storing 100 ml fresh oil in a glass jar with a screw cap. The lid had been manipulated with a central hole that allowed air exchange.

2.3. Sample Measurements

2.3.1. FTIR. The FTIR spectra were collected in transmission with a Bruker Alpha instrument in combination with the QuickSnapTM transmission sample compartment in the wavenumber region ranging from 4000 to 600 cm^{-1} with a spectral resolution of 4 cm^{-1}.

The samples were measured without any special sample preparation with two different setups: (1) a droplet of ATF between two potassium bromide (KBr) discs separated by

TABLE 1

Temperature (°C)	Storage time (h)			
120	500	1000	2000	3000
140	105	210	415	625
160	25	50	105	165

a teflon spacer with the thickness of about 50 μm, and (2) fixed KBr cuvette of 100 μm thickness filled with ATF.

After each sample measurement, the KBr discs and the cuvette were rinsed several times with petroleum ether in order to prevent cross contamination. The cuvette was dried with N_2 gas after rinsing, and the KBr discs were dried under ambient air. For the measurement type (1), 4 spectra per sample were recorded, and for type (2), one spectrum per sample was recorded.

Due to the sample layer thickness, the hydrocarbon bands are saturated, and therefore, the spectra had to be cut in the wavenumber regions between 3000 and 2815 cm^{-1} (C-H stretching mode) and between 1491 and 1424 cm^{-1} (C-H bending and rocking mode). Additionally, the CO_2 bands were eliminated by cutting out the region from 2387 to 2285 cm^{-1} as well. The spectra of ATF A are shown in Figure 1 in transmission without any preprocessing as measured, in Figure 1(b) after truncation and SNV transformation, and in Figure 1(c), SNV transformed after calculating the absorbance spectra by using $A = -\log(T)$. In Figure 2, the same diagrams are shown for ATF B. In both cases, two series of curves can be discriminated from the raw spectra by the eye. The blue series comes from measurement type (1), and the red set comes from the cuvette measurements (2). To combine the two data sets from the measurement setups (1) and (2) are challenging tasks for a predictive model as the main variance is due to the thickness variation. The data set demonstrates the importance of suitable and sophisticated preprocessing methods in order to eliminate the difference in the spectra induced by the varying sample thickness. The standardization techniques presented here are able to meet this need.

2.3.2. Liquid Chromatography Coupled with Mass Spectrometry. The measurements for the determination of the additive compound signals were performed with an Agilent liquid chromatograph 1260 coupled with a high-resolution QToF 6540 mass spectrometer with methanol/water/ammonium acetate and isopropanol as an eluent. Ionization was carried out by means of electrospray (ESI). The final compound peak area data set was created using the Agilent MassHunter Qualitative Analysis B. 06.00 analysis software.

The response signals of the additive compounds are standardized by subtracting mean and dividing by standard deviation in order to bring all signal values on the same scale. The standardized signals are depicted in Figure 3.

3. Methods

3.1. Implementation. The proposed novel standardization methods and respective optimization processes were implemented via Python scripts.

3.2. Regression Algorithm—Ridge. For the prediction, the ridge regression estimator implemented in the Python scikit-learn framework for machine learning applications was used [25]. It is a linear model which solves a regression task via the least squares loss function $J(w)$ with L_2 regularization [26]. Regularization is an approach to minimize the issue of overfitting, which is particularly important for high-dimensional data such as FTIR spectra, by controlling the quadratic sum of the model coefficient w. This is done by adding the penalizing term L_2 weighted by the hyper parameter λ.

$$\lambda\|w\|^2 = \lambda \sum_{j=1}^{m} w_j^2. \tag{1}$$

Thus, the loss function is defined as

$$J(w) = \sum_{i=1}^{n} \left(y_i - y_{i,\text{pred}}\right)^2 + \lambda\|w\|^2, \tag{2}$$

where y_i stands for the reference value of the ith sample and $y_{i,\text{pred}}$ for the prediction of this sample. Since the performance of the preprocessing methods has to be assessed independently from the actually used predictive regression model, the same regression model with identical hyperparameter λ was applied to the various preprocessed data sets. For the regression of the friction modifier compound of ATF A, $\lambda = 5$, and for the antioxidant of ATF B, $\lambda = 3$ was used. These parameters turned out to be the best choices regarding cross validation and robustness for the SNV transformed data set in a previously conducted internal study.

3.3. Model Performance Evaluation. To assess the performance of our models, two different approaches were chosen, namely, the predictive power under cross validation and noise addition.

3.3.1. Cross Validation. For cross validation, the mean from the different measurements of one sample was calculated. The sample set was randomly divided 50 times into a calibration and validation set by taking 70% of the data as training samples and 30% as test samples in each validation iteration with different combinations. Each separation run was provided with a unique random seed to ensure that the data set was split into the same training and test sets for each model, enabling better comparability of results between the different models.

3.3.2. Robustness against Noise. In order to assess the model performance under noisy input spectra, the model was calibrated by the full original data set. Random Gaussian-distributed white noise was added to each data point. These perturbed samples were predicted by the model and the prediction error was monitored. This was done for different noise levels. The random numbers added to each data point were generated by a standard normal distributed (mean: $\mu = 0$ and standard deviation: $\sigma = 1$) random

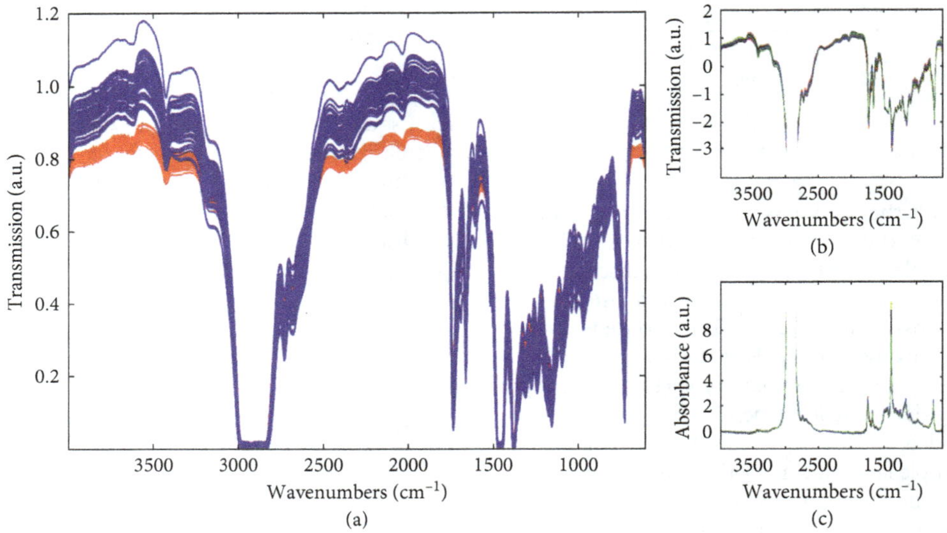

FIGURE 1: FTIR spectra of ATF. (a) Raw full transmission spectra without any preprocessing. The two data sets with different measurement setups can be discriminated by eye. The blue spectra originate from measurement type (1) with two KBr discs separated by a Teflon spacer, and the red set of curves originates from the cuvette measurement (2). (b) SNV-transformed transmission spectra after truncation of the saturated C-H vibrational regions and CO_2 areas and (c) SNV-transformed absorbance spectra after truncation.

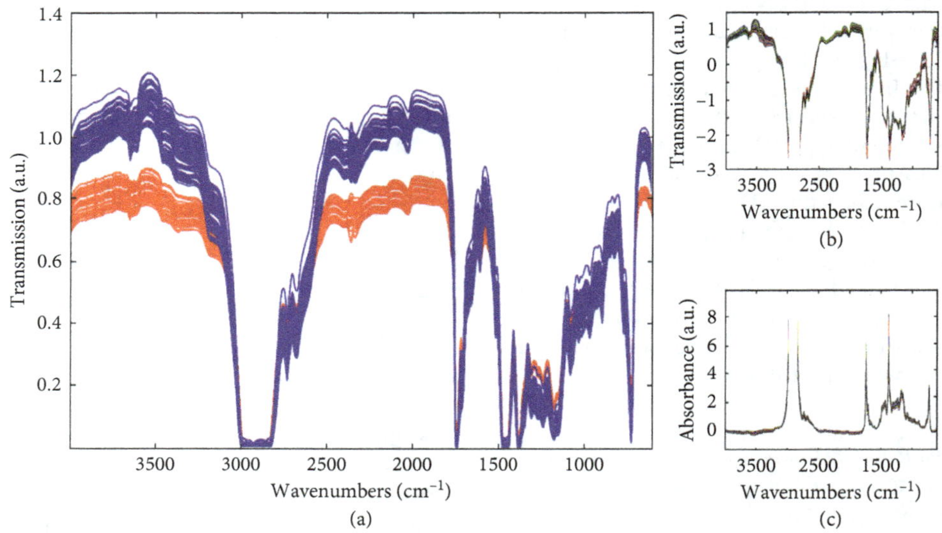

FIGURE 2: FTIR spectra of ATF B. (a) Raw full transmission spectra without any preprocessing. (b) SNV-transformed transmission spectra after truncation of the saturated C-H vibrational regions and CO_2 areas and (c) SNV-transformed absorbance spectra after truncation.

number generator. The noise levels were defined by the factors (0.05, 0.10, 0.15, 0.20, 0.25, 0.30, 0.35, 0.40, and 0.45), which were multiplied with the output of the random number generator. For each noise level, 50 simulated noisy data sets were generated and predicted by the pretrained model in order to be able to make well-founded statements about the model performance under noise perturbation.

The noise robustness workflow is a very helpful tool to investigate whether a good calibration error is a real advantage or if the model ran into overfitting. Using the same regression algorithm twice with different regularization parameters λ, the lower regularized model will generate a lower initial calibration error than the more stringent regularized model. But if the models are tested for robustness, the latter tends to have a lower error slope when the noise level increases.

3.4. Evaluation Metrics. The built-in functions R^2 score and mean squared error (MSE) of the scikit-learn framework were used as performance metrics.

3.4.1. Mean Squared Error (MSE). The mean squared error (MSE) of a prediction is calculated by the squared differences between the predicted value $y_{i,\text{pred}}$ and the reference value y_i

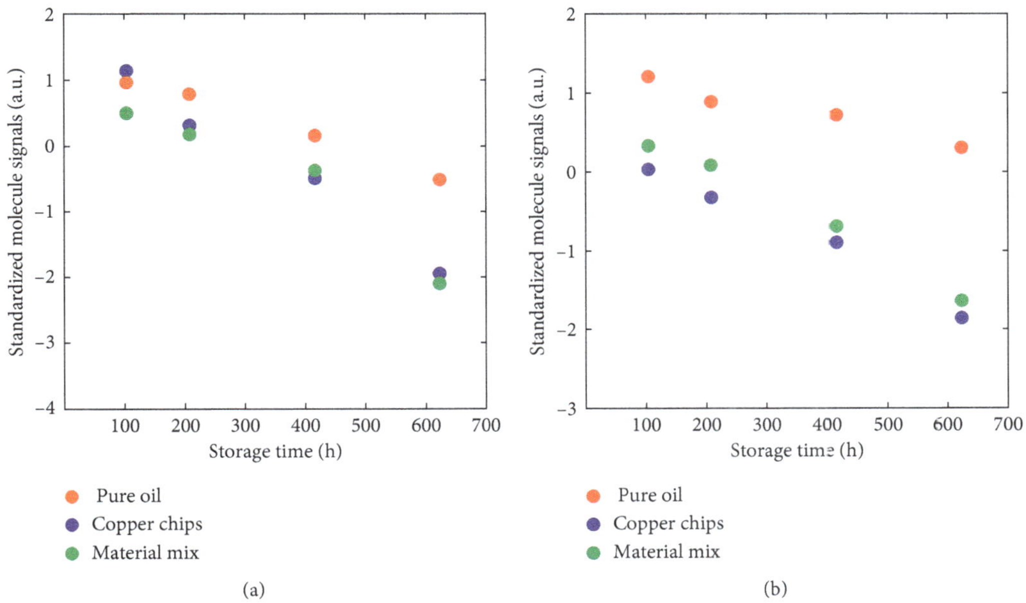

FIGURE 3: Standardized additive responses used as target value for the FTIR regression model for (a) the friction modifier compound and (b) the antioxidant plotted against time for storage temperature 140°C for all three storage experiments measured by HPLC-QToF.

of the ith sample. For a given data set with n samples, the MSE is the average value over all samples. It follows the following formula [27]:

$$\mathrm{MSE}\left(y, y_{\mathrm{pred}}\right) = \frac{1}{n}\sum_{i=1}^{n}\left(y_i - y_{i,\mathrm{pred}}\right)^2. \tag{3}$$

The best possible MSE value is 0, and small values are desirable as the deviation from the correct prediction is low. From MSE, the root-mean-squared error (RMSE) was calculated by taking the square root. The RMSE value has the same dimension as the original reference target values.

3.4.2. R^2 Coefficient of Determination.

R^2 describes the portion of the variance in the target values (dependent variables) that can be predicted from the spectra (independent variables) by the model [28]. The best possible score for R^2 is 1.0. R^2 gets 0.0 for a constant model which predicts a constant value disregarding of the input features. For linear regression modeling with intercept, R^2 is equal to the square of Pearson correlation coefficient between predicted and reference target values [29]. For a data set comprising n samples, the R^2 score is given as

$$R^2\left(y, y_{\mathrm{pred}}\right) = 1 - \frac{\sum_{i=1}^{n}\left(y_i - y_{i,\mathrm{pred}}\right)^2}{\sum_{i=1}^{n}\left(y_i - \overline{y}_{\mathrm{pred}}\right)^2}, \tag{4}$$

where $y_{i,\mathrm{pred}}$ is the model prediction of the ith sample which has a reference value y_i, and $\overline{y}_{\mathrm{pred}}$ is the mean value of all predictions.

$$\overline{y}_{\mathrm{pred}} = \frac{1}{n}\sum_{i=1}^{n} y_{i,\mathrm{pred}}. \tag{5}$$

3.5. Standard Normal Variate.

Each spectrum $x = (x_1, x_2, \ldots, x_k)$ with k measured data points is transformed to the standardized form $z = (z_1, z_2, \ldots, z_k)$ by bringing the spectra to zero mean and unit variance. For this purpose, the mean spectrum $\overline{x}$ is subtracted from each data point x_i and divided by the standard deviation.

$$z_i = \frac{x_i - \overline{x}}{\sqrt{\sum_j^k \left(x_i - \overline{x}\right)^2 / k}}, \tag{6}$$

with

$$\overline{x} = \frac{1}{k}\sum_j^k x_j. \tag{7}$$

3.5.1. Dynamic Localized SNV (DLSNV).

The DLSNV workflow is based on the SNV-transformed spectra data set (Figure 4(a)). To calculate the DLSNV data, the spectra are divided into multiple regions. On each of these regions, standardization is performed. To adjust the windows to important areas in the spectrum, a starting point can be defined. In Figure 4(b), the DLSNV spectra are shown, with a starting point of 100 and a window size of 300 pixels.

DLSNV algorithm

(i) Perform SNV on a window of the spectrum ranging from first data point to the sth one

(ii) Subdivide spectra from s^{th} data point into windows of all the same size *ws*

To optimize the two parameters, *window size ws* and *starting point s*, a three-step approach is performed. In each step, the predictive power of the model is assessed via the

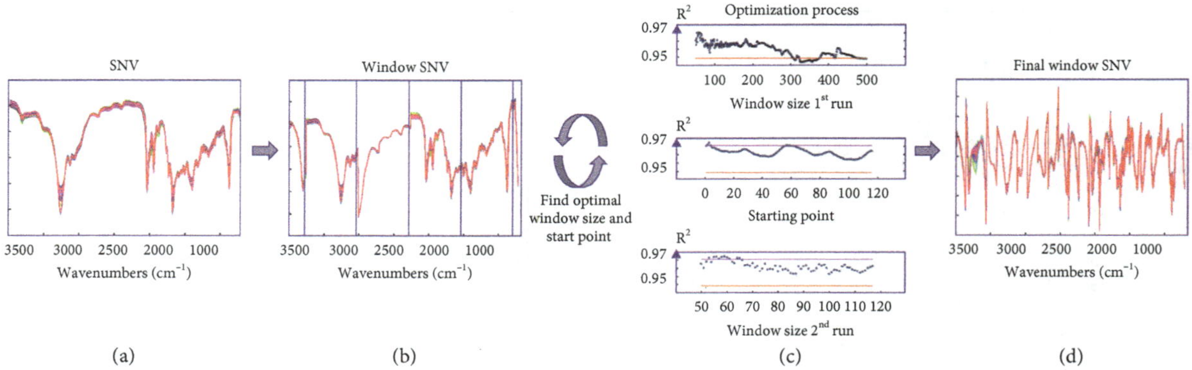

FIGURE 4: Demonstration of the workflow and optimization process for Dynamic Localized SNV. (a) Single SNV, (b) Dynamic Localized SNV with starting point 100 and window 300 for visualization, (c) three-stage optimization process for window, starting point and final window optimization, and (d) optimized DLSNV.

coefficient of determination R^2. The prediction performance of the chosen *window size* in combination with the regression model is benchmarked by fitting the same model to the single SNV data, indicated by a red line in Figure 4(c). The optimization steps can be summarized as follows:

(1) Perform LSNV with window sizes from 50 to 500 pixels, and determine R^2 for all window sizes. Find the optimal window size ws_{opt1}.

(2) Perform LSNV with optimal window size of step 1 $ws_{\text{opt 1}}$, vary the starting point from 0 to $2 \cdot ws_{\text{opt 1}}$, and select the optimal starting point s_{opt}.

(3) Perform LSNV with optimal starting point s_{opt} with window sizes from 50 to $2 \cdot ws_{\text{opt 1}}$ in order to find the best combination of window size $ws_{\text{opt 2}}$ and starting point s_{opt}.

In Figure 4(d), the final DLSNV spectra after optimization are shown. Note that jumps can occur between the individual standardization windows since the mean value of this current window is subtracted for each window. However, this does not affect the regression model.

In Figure 5(a), the ATF A samples are shown with SNV performed on the entire spectral region, and in Figure 5(b), the same spectra are depicted after DLSNV optimization. Figure 5(c) shows a zoom-in view of the highlighted region of Figure 5(a), and in Figure 5(d), the same region is depicted after DLSNV optimization. The baseline is removed for the exact spectra sequence, and thus, peaks are aligned in a way that the different aging levels of the samples can already be recognized by eye. The shown snipped spectrum is the phenolic antioxidant region. Thus, the decrease of this band can be associated with the aging level. Magenta indicates (relatively) fresh samples, whereas red indicates a strong degradation level.

3.5.2. Peak SNV. The idea behind the Peak SNV method is to standardize the important areas of the spectrum independently of each other. The optimization workflow for PSNV is shown in Figure 6, starting from the single SNV transformed data set. Data points with a high correlation with the target values (points of interest, POI) are selected (Figure 6(a)), and the SNV transformation is performed on windows around the centroids. Once the POIs are identified, the PSNV transformation is conducted as follows:

PSNV algorithm

(i) Subdivide spectra into sequences ranging from half the distance from the previous POI to half the distance to the next one (Figure 6(b)). SNV is performed across these windows.

To find the POI, an initial regression model is fitted to the data. In order to identify important regions of the spectra, the model coefficients are assessed. The normalized absolute values of the coefficient vector are fed into a peak-picking algorithm. Since it may occur that POIs are in close proximity, an agglomeration of the POIs is conducted in order to prevent from very narrow standardization windows. Peak centroids are calculated via the mean value of the combined POIs. The task for the optimization process is to find the best window for POI agglomeration, agg_{opt}, which is done by analyzing the calibration R^2 for each agglomeration window and picking the window size with maximal correlation between the predicted and reference target (Figure 6(c)). The steps are summarized as follows:

(1) Fit the data set to the target values (only calibration)

(2) Pick peaks from the normalized model coefficient vector $(|w|/\max(|w|))$, threshold for peaks = 0.1

(3) Combine peaks which are within a certain window agg, and calculate the centroid of the agglomerated POIs

(4) Perform PSNV across the centroid of the POIs

(5) Evaluate performance via R^2 for agg between 10 and 50 data points, and choose agg_{opt} according to maximal R^2

After optimization, each window has an individual window size and range over the peak centroid of important signals in the spectrum. On these windows, SNV transformation provides an optimal baseline and scatter effect removal. The optimized spectrum is shown in Figure 6(d).

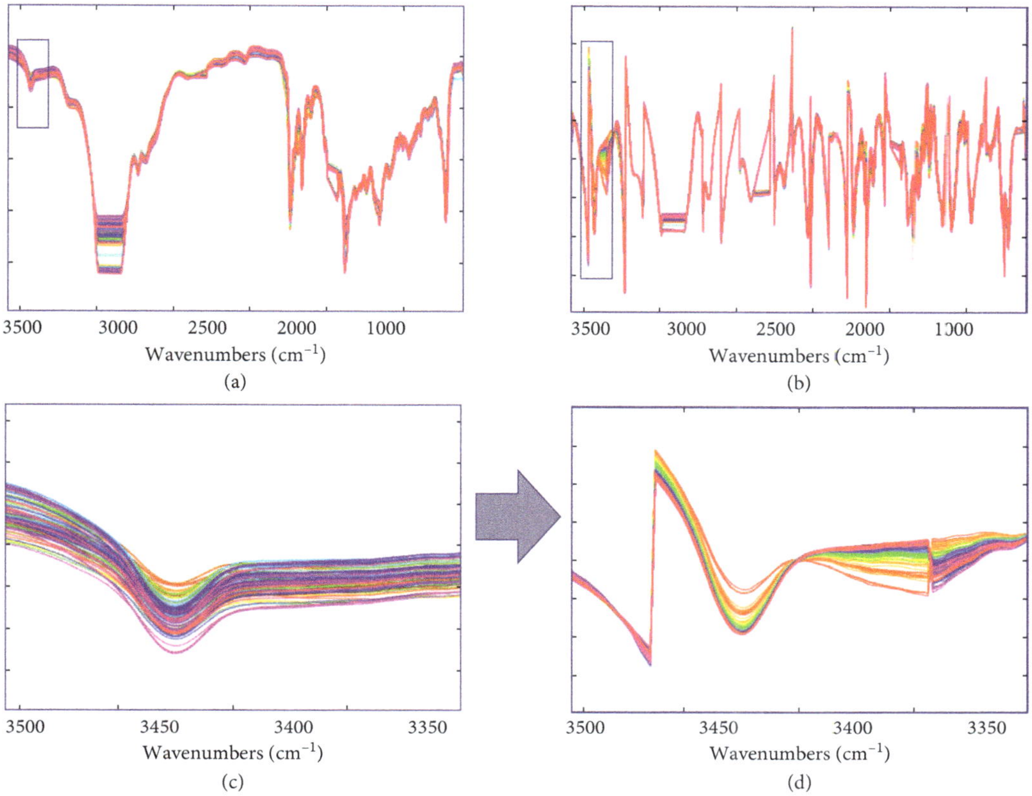

FIGURE 5: Demonstration of the improvement of peak alignment for DLSNV. (a) SNV-transformed transmission spectra with marked zoom level of (c). (b) Optimized DLSV spectra with marked zoom area of (d). Magenta indicates (relatively) fresh samples, whereas red indicates a strong degradation level.

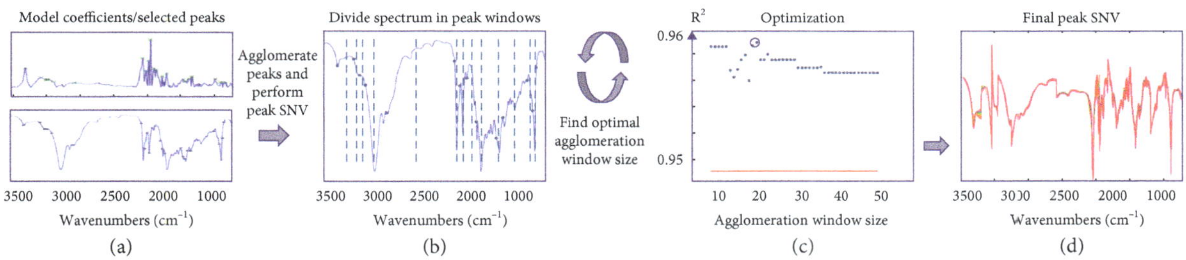

FIGURE 6: Demonstration of the workflow and optimization process for Dynamic Localized SNV. (a) Picked peaks of normalized absolute coefficient vector and indication of the POIs in one spectrum, (b) spectrum separation according to agglomerated peaks, (c) optimization process in order to find optimal agglomeration window size, and (d) optimized PSNV.

3.5.3. Partial Peak SNV. The idea behind Partial Peak SNV is similar to PSNV: picking the regions of the spectrum which show a high correlation with the target values, agglomerating POIs in close proximity, and standardizing these important spectral features (Figure 7(a)). But unlike for PSNV, not only the whole spectrum is finally taken into account but also a small window around the POI. It may occur that the same data point appears several times in different standardizations (see overlapping regions in Figures 7(b) and 7(d)). Due to this workflow, the PPSNV spectrum may have more data points (due to overlapping) or less (because not the entire spectrum is taken into account) than those of the original spectrum. A PPSNV spectrum is calculated as follows:

PPSNV algorithm

(i) Perform SNV across the POIs with a left and right margin of pw

The optimization focuses on the adjustment of the window size pw around the POIs in which the SNV is applied for maximal predictive power in calibration (Figure 7(c)). The optimization process is divided into the following steps:

(1) Fit the data set to the target values (only calibration)

(2) Pick peaks from the normalized model coefficient vector ($|w|/\max(|w|)$), threshold for peaks = 0.1

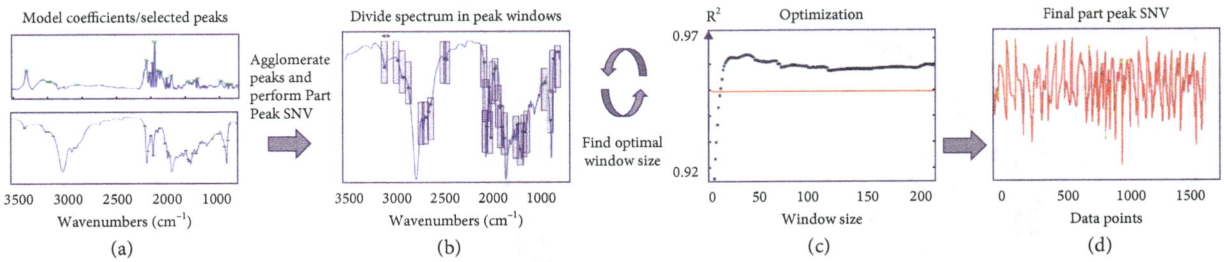

FIGURE 7: Demonstration of the workflow and optimization process for Dynamic Localized SNV. (a) Picked peaks of normalized absolute coefficient vector and indication of the POIs in one spectrum, (b) spectrum separation according to agglomerated peaks, (c) optimization process in order to find optimal window size around the POIs, and (d) optimized PPSNV.

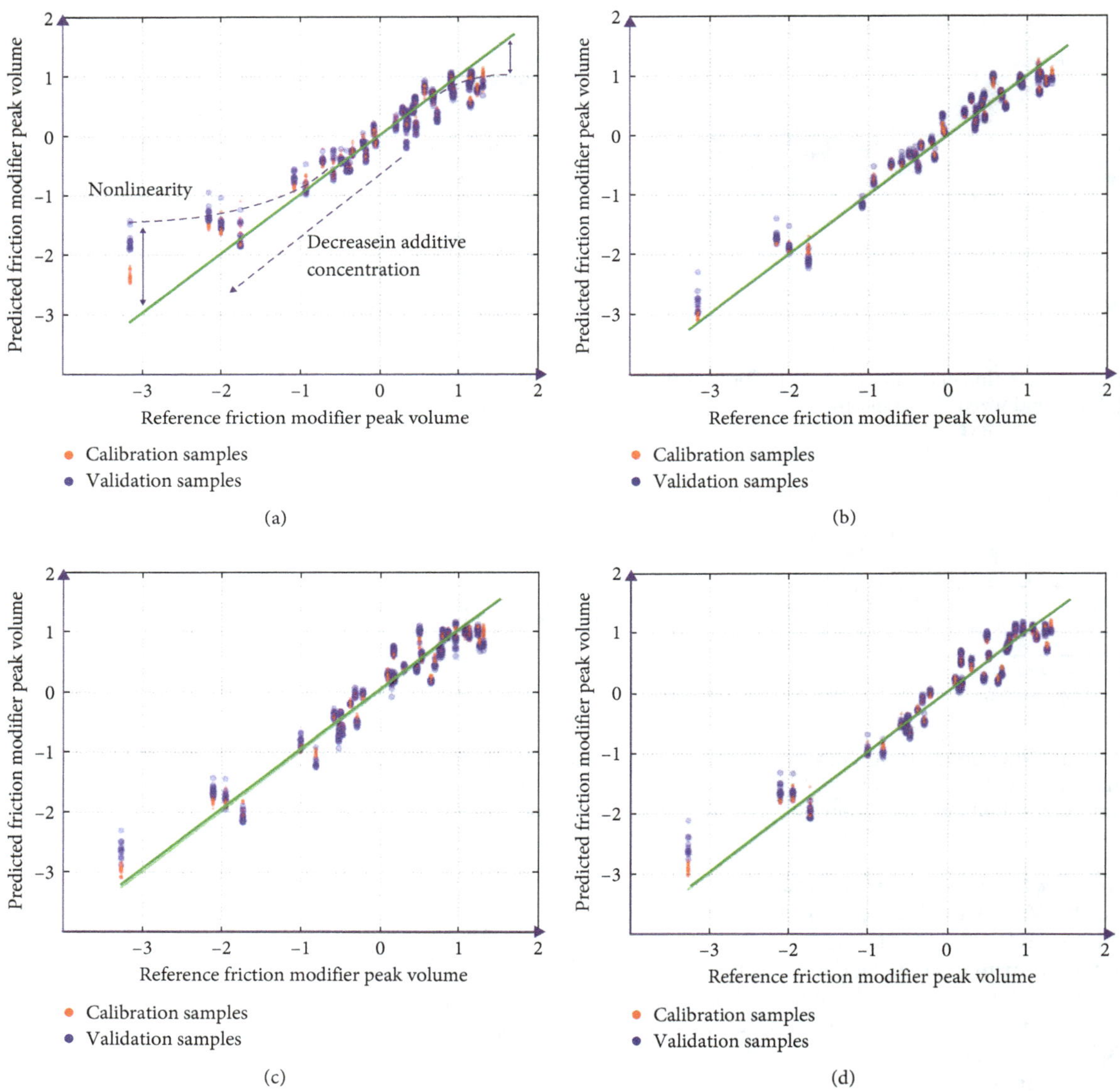

FIGURE 8: Cross validation results for (a) SNV, (b) Dynamic Localized SNV, (c) Peak SNV, and (d) Partial Peak SNV preprocessed spectra in the regression case of the friction modifier additive. The reference values were measured by HPLC-QToF. Red dots indicate prediction of calibration samples, and blue dots represent prediction of the hold-out validation samples. The dashed arrow in (a) indicates the increasing degradation of the samples with increasing time as the friction modifier additive concentration gets lower.

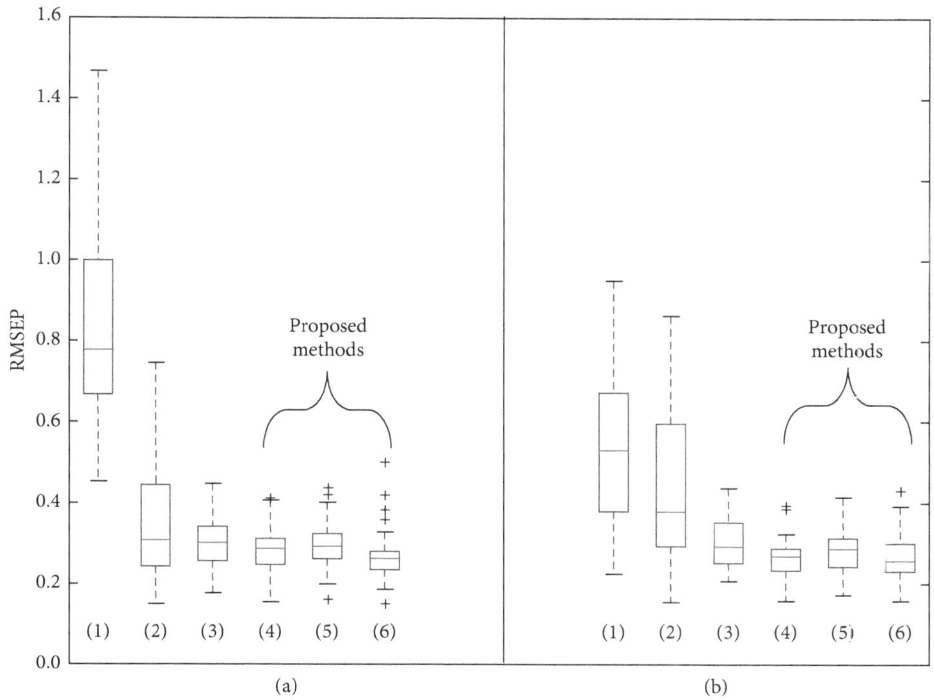

FIGURE 9: Box-and-whisker plot representation of the root-mean-squared error of prediction of the cross validation strategy (50 folds, random train test split of 70/30% of the data) for the friction modifier compound of ATF A. In (a), the RMSEP values for the transmission spectra are shown, and in (b), the RMSEP values for the absorbance spectra are shown. Boxplot (1) is without standardization, (2) is with a single SNV transformation on the full spectrum, (3) is with optimized LSNV, (4) is with optimized Dynamic DLSNV, (5) is with optimized PSNV, and (6) is with optimized PPSNV.

(3) Perform PPSNV across the peaks with the window size *pw*

(4) Evaluate the performance via R^2 for *pw* between 1 and 200 data points, and choose pw_{opt} according to maximal R^2

4. Results and Discussion

4.1. Cross Validation. In Figure 8(a), the cross validation recovery function for predictions of the SNV preprocessed spectra of the regression on the friction modifier compound is shown. A 50-fold cross validation strategy with a calibration/validation splitting of 70%/30% was used. Red dots represent the prediction of calibration, and blue dots represent validation samples. It is obvious that the linear model struggles to predict the high and low compound intensity regions correctly. The nonlinearity is visualized by an arrow and a dashed line to guide the eye.

In Figure 8 also, the cross validation recovery function for predictions after Dynamic Localized SNV (Figure 8(b)), Peak SNV (Figure 8(c)), and Partial Peak SNV (Figure 8(d)) optimization are shown. The saturation effect in the low intensity area of the compound response is almost completely removed in the latter three cases. It is also notable that the scattering around the green bisecting line is significantly reduced. Thus, the confidence interval for the predictions is improved.

The RMSEP values during the cross validation of the regression of the friction modifier component are summarized in Figure 9 in a box-and-whisker plot representation. The red line indicates the median, within the boxes, the interquartile range (IQR) (contains 50% of the data) is depicted, and the margins of the whiskers represent $Q_1 - 1.5 \cdot \text{IQR}$ and $Q_3 + 1.5 \cdot \text{IQR}$ for the lower and upper bound, respectively (Q_1 means the smallest 25% of the data set are smaller than this value and Q_3 means the smallest 75% are smaller than this value). Subplot Figure 9(a) refers to the transmission spectra and Figure 9(b) refers to the absorbance spectra. The labels are associated with (1) without standardization, (2) single SNV transformation on the full spectral range, (3) Localized SNV, (4) Dynamic Localized SNV, (5) Peak SNV, and (6) Partial Peak SNV.

It is noticeable that the RMSEP is very poor in case of the crude transmission spectra and that SNV has a very useful impact on them, whereas the improvement after SNV is low for absorbance spectra.

For all sophisticated optimized standardization approaches DLSNV, PSNV, and PPSNV, the median and the scattering around the median of RMSEP decreases drastically with respect to the SNV-transformed full spectra but also LSNV seems to be a reasonable choice. DLSNV on absorbance spectra is characterized by the lowest median and the smallest scattering confirmed by Table 2, summarizing the mean values and standard deviation of RMSEP.

TABLE 2: Summary of the model performances described by the mean value and the standard deviation of the RMSEP values during cross validation. The relative improvements and respective p-values compared with LSNV are also listed.

Method	Friction modifier ATF A		Antioxidant ATF B	
	Transmission	Absorbance	Transmission	Absorbance
Raw	0.83 ± 0.22	0.53 ± 0.16	0.90 ± 0.14	0.70 ± 0.11
Single SNV	0.34 ± 0.14	0.42 ± 0.17	0.61 ± 0.11	0.70 ± 0.13
LSNV	0.30 ± 0.07	0.31 ± 0.07	0.24 ± 0.04	0.24 ± 0.04
DLSNV	0.28 ± 0.06	**0.26 ± 0.05**	0.21 ± 0.04	0.21 ± 0.04
Rel. improvement	9% ($p < 0.05$)	**16%** ($p < 0.001$)	13% ($p < 0.001$)	13% ($p < 0.001$)
PSNV	0.28 ± 0.08	0.28 ± 0.06	0.33 ± 0.06	0.31 ± 0.07
Rel. improvement	8% ($p > 0.05$)	9% ($p < 0.05$)	−41% ($p < 0.001$)	−33% ($p < 0.001$)
PPSNV	0.26 ± 0.06	0.26 ± 0.06	**0.17 ± 0.04**	0.24 ± 0.05
Rel. improvement	13% ($p < 0.01$)	15% ($p < 0.001$)	**29%** ($p < 0.001$)	−3% ($p > 0.05$)

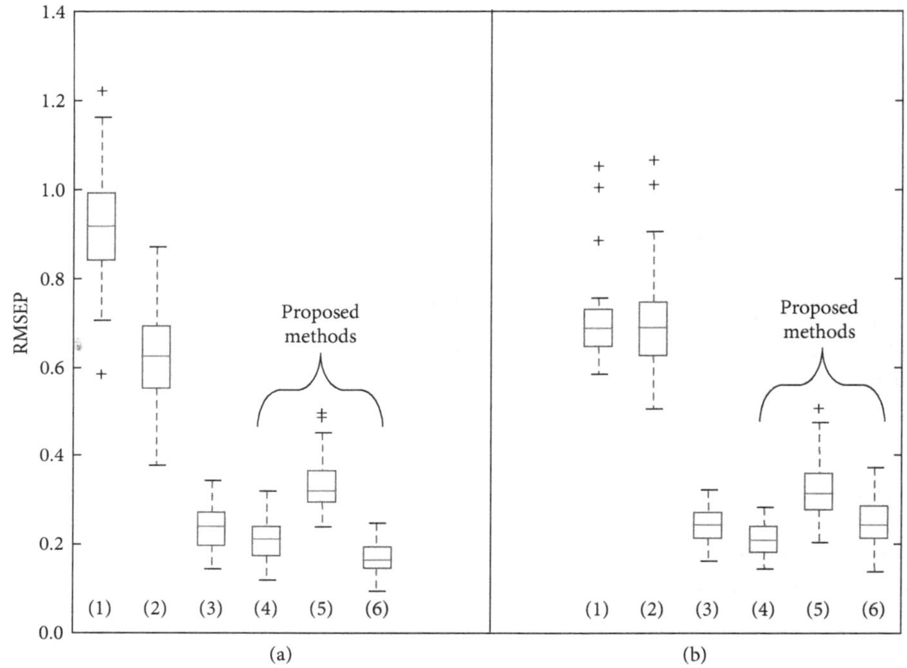

FIGURE 10: Box-and-whisker plot representation of the root-mean-squared error of prediction of the cross validation strategy for the phenolic antioxidant compound of ATF B. In (a), the RMSEP values for the transmission spectra are shown, and in (b), the RMSEP values for the absorbance spectra are shown. Boxplot (1) is without standardization, (2) is with a single SNV transformation on the full spectrum, (3) is with optimized LSNV, (4) is with optimized Dynamic DLSNV, (5) is with optimized PSNV, and (6) is with optimized PPSNV.

The summarized RMSEPs of the regression to the antioxidant additive of ATF B are shown in Figure 10 in a box-and-whisker plot representation where Figure 10(a) refers to the transmission spectra and Figure 10(b) to the absorbance spectra. In this case, DLSNV, PSNV, and PPSNV reduce both the median and the scattering around the median enormously when compared with SNV on full spectra. The best performance is achieved by PPSNV conducted on the transmission spectra confirmed by Table 2. On transmission spectra, DLSNV and PPSNV perform better than LSNV, but PSNV only has a positive effect when compared to SNV. In relation to LSNV, using PSNV, the predictive power is reduced. The fact that PPSNV is the best choice for this regression use case suggests that it is beneficial to only use spectral regions with high correlation with the target value and drop regions without or low correlation.

4.1.1. Noise Robustness. In Figure 11, the performance of the regression model for the prediction of noisy spectra is shown for the friction modifier. In subplot Figure 11(a), the curves for all preprocessings, are depicted and in Figure 11(b), a zoomed view is shown. Without any preprocessing, the initial calibration error for both transmission and absorbance spectra is very poor and rises very fast with the increasing noise level factor. Although the sophisticated preprocessing methods LSNV, DLSNV, PNSV, and PPSNV show a lower initial calibration error, the slope of the error is lower than for SNV. In Figure 11(b), the trend of the transmission spectra having low error steepness is visible. One may say that the three proposed standardization techniques show a very similar noise robustness behavior and are significantly better than none or

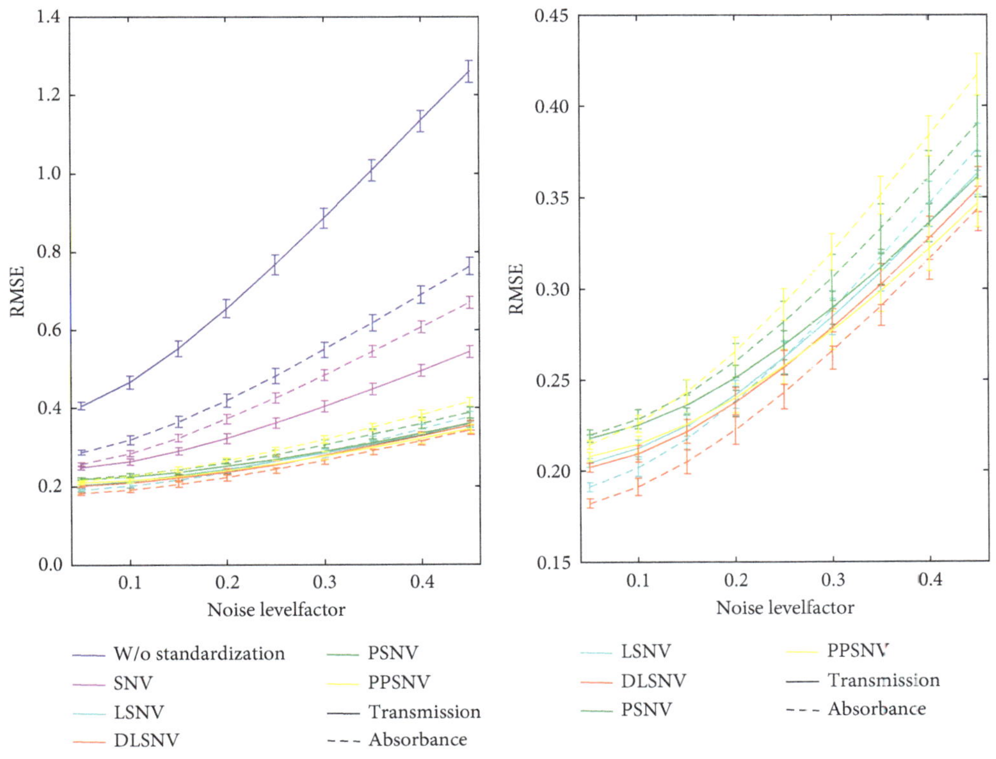

FIGURE 11: RMSE as a function of the noise level factor for the regression of the friction modifier compound of ATF A calibrated by the unperturbed full data set. The error bars represent the standard deviation of the prediction error calculated from the statistics of 50 repetitions of noise addition. In subplot (a), all curves are shown, and in (b), the sophisticated standardizations are shown.

SNV preprocessing, indicating that the model did not overfit the data as mentioned in Section 3.3.2.

In Figure 12, the performance of the regression model of the antioxidant for the prediction of noisy spectra is shown. The absorbance spectra with or without preprocessing show a similar noise trend as the SNV-transformed spectra. In Figure 12(b), the localized versions are shown in a zoomed view. The PPSNV preprocessing on the transmission spectra is characterized by the flattest noise dependency. These resuts demonstrate the superiority of the PPSNV method in this use case. As mentioned above, PSNV is not advantageous in this application and shows the lowest noise immunity, but it is preferable to the SNV across the whole spectrum.

4.2. Summary. The optimized parameters for the preprocessing methods are summarized in Table 3. The LSNV optimization process selects the same window size as DLSNV. Thus, the second window size run has no influence on the final result in these two cases but the starting point produces an improvement.

As already mentioned in Table 2, the cross validation performances of the tested methods are summarized as mean values and standard deviation for all cross validation runs. For the friction modifier, the performances of DLSNV, PSNV, and PPSNV are very similar. Table 2 also lists the relative improvements against the benchmark preprocessing, LSNV, accompanied by corresponding p values from a two-sided t-test, which tests the significance of the mean values being different (the deviation for relative improvements when the same mean value is given due to the fact that the improvements were calculated from exact values rather than rounded values).

The best mean RMSEP value for the regression model for the friction modifier of 0.26 is produced by DLSNV based on absorbance spectra. The antioxidant compound is modeled best by PPSNV preprocessing of the transmission spectra and yields a very low prediction error of 0.17.

To summarize, one may say that all proposed methods performed very well reducing both mean and standard deviation of the cross validation error compared with SNV. PSNV is not reasonable for the antioxidant additive as the performance is poor compared with the benchmark preprocessing method LSNV.

Which preprocessing method is the best depends on the actual regression use case, but in general, it is shown that PPSNV outperforms PSNV. This suggests that it is beneficial to drop spectral regions showing low or no dependency on the target value and to only consider highly correlated peaks.

For the antioxidant compound, PPSNV yielded an enormous improvement. This could be explained as the

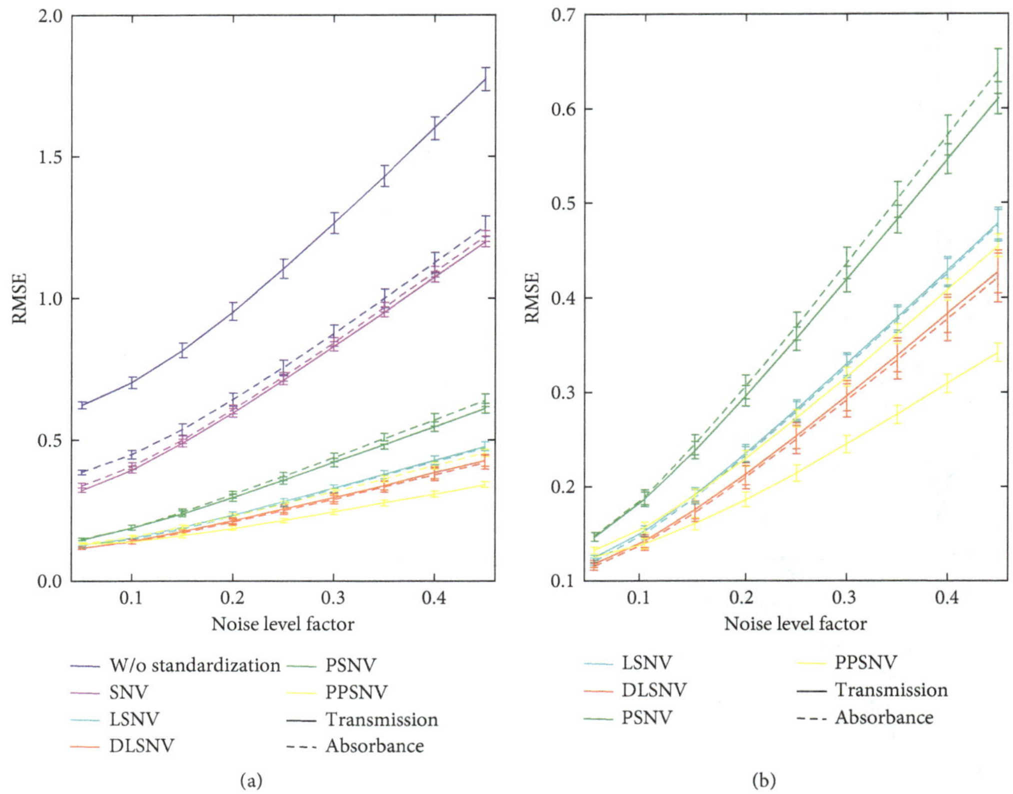

Figure 12: RMSE as a function of the noise level factor for the regression of the antioxidant compound of ATF B calibrated by the original full data set. The error bars represent the standard deviation of the prediction error calculated from the statistics of 50 repetitions of noise addition. In subplot (a) all curves are shown, and in (b), the sophisticated standardizations are shown.

phenolic aging inhibitor is a compound with very narrow vibrational band in the ATF B and thus does not have a great impact when the SNV is carried out across the entire spectrum. This may lead to a suboptimal alignment of this band. In case of novel standardizations, the SNV is optimized to the high correlative bands, and scatter effects can be compensated for these exact regions.

The fact that PSNV is unsuccessful for the antioxidant may be because the POIs are not centered to the middle of the SNV window and may have large left and right margins if they are far away from other POIs. As a result, they may not be optimally standardized. This is shown in Figure 6(b), where the single POI at about 2700 cm^{-1} has a large single SNV window.

The study provides an overview of model performances when using transmission or absorbance spectra suggesting that both cases can lead to valid regression models. However, for quantitative models built on transmission spectra, the SNV is vital, whereas in the absorbance case, the predictive power does not depend on SNV transformation. In absorbance spectra, the influence of the baseline constant is reduced because high transmission values are converted into low absorbance values.

To conclude, DLSNV, PSNV, and PPSNV were able to improve both transmission and absorbance predictive models. The scattering around the mean values are also drastically reduced because the model does not have to learn how to compensate for the baseline shift in each cross validation step leading to more reproducible results. Each vibrational band is optimally aligned so that the additive depletion trend is encoded in the absolute signal intensity, and the model does not have to weigh a data point as background correction.

Table 3: Summary of the preprocessing parameters. For DLSNV window size and starting point, for PSN agglomeration window, and for PPSNV, window width around POI is shown.

Method	ATF A		ATF B	
	Transmission	Absorbance	Transmission	Absorbance
LSNV	$w_{\text{opt}} = 52$	$w_{\text{opt}} = 52$	$w_{\text{opt}} = 51$	$w_{\text{opt}2} = 51$
DLSNV	$w_{\text{opt}2} = 52$ $s_{\text{opt}} = 5$	$w_{\text{opt}2} = 52$ $s_{\text{opt}} = 7$	$w_{\text{opt}2} = 51$ $s_{\text{opt}} = 48$	$w_{\text{opt}2} = 51$ $s_{\text{opt}} = 48$
PSNV	$\text{agg}_{\text{opt}} = 5$	$\text{agg}_{\text{opt}} = 14$	$\text{agg}_{\text{opt}} = 6$	$\text{agg}_{\text{opt}} = 10$
PPSNV	$pw_{\text{opt}} = 17$	$pw_{\text{opt}} = 25$	$pw_{\text{opt}} = 15$	$pw_{\text{opt}} = 38$

5. Conclusion

The results presented in this study demonstrate the outperformance of the proposed novel standardization strategies Dynamic Localized SNV, Peak SNV, and Partial Peak SNV to improve both the mean and scatter of RMSEP values in cross validation and the robustness against noise drastically with respect to SNV transformation executed on the entire spectrum. Against the benchmark LSNV, an enhancement of the predictive power of a ridge regression model by up to 16% and 29% could be achieved for the friction modifier and the antioxidant compound, respectively. The demonstrated optimization workflows for performing SNV on specific regions of the spectrum have been introduced here for the first time. Therefore, the standardization methods used in this paper are capable of eliminating nonlinearities by flexible rescaling in defined areas. To our knowledge, such standardization techniques have not been presented elsewhere.

Conflicts of Interest

The authors declare that they have no conflicts of interest.

References

[1] M. Zeaiter, J. M. Roger, and V. Bellon-Maurel, "Dynamic orthogonal projection. a new method to maintain the on-line robustness of multivariate calibrations. application to nir-based monitoring of wine fermentations," *Chemometrics and Intelligent Laboratory Systems*, vol. 80, no. 2, pp. 227–235, 2006.

[2] R. M. Balabin and R. Z. Safieva, "Motor oil classification by base stock and viscosity based on near infrared (nir) spectroscopy data," *Fuel*, vol. 87, no. 12, pp. 2745–2752, 2008.

[3] A. Borin and R. J. Poppi, "Multivariate quality control of lubricating oils using fourier transform infrared spectroscopy," *Journal of the Brazilian Chemical Society*, vol. 15, no. 4, pp. 570–576, 2004.

[4] J. Huang, D. Brennan, L. Sattler, J. Alderman, B. Lane, and C. O'Mathuna, "A comparison of calibration methods based on calibration data size and robustness," *Chemometrics and Intelligent Laboratory Systems*, vol. 62, no. 1, pp. 25–35, 2002.

[5] M. A. Al-Ghouti and L. Al-Atoum, "Virgin and recycled engine oil differentiation: a spectroscopic study," *Journal of Environmental Management*, vol. 90, no. 1, pp. 187–195, 2009.

[6] R. Kellner, J. M. Mermet, M. Otto, M. Valcárcel, and H. M. Widmer, *Analytical Chemistry*, John Wiley & Sons Australia Limited, Milton, Australia, 2004.

[7] B. G. Osborne, T. Fearn, P. T. Hindle, and B. G. Osborne, *Practical nir Spectroscopy with Applications in Food and Beverage Analysis*, Longman Scientific & Technical, Wiley, Harlow, Essex, UK, 1993.

[8] J. D. Donald and D. D. Kevin, *Interpreting Diffuse Reflectance and Transmittance*, NIR, Chichester, UK, 2007.

[9] A. Rinnan, F. van den Berg, and S. B. Engelsen, "Review of the most common pre-processing techniques for near-infrared spectra," *TrAC Trends in Analytical Chemistry*, vol. 28, no. 10, pp. 1201–1222, 2009.

[10] E. Arendse, O. Amos Fawole, L. Samukelo Magwaza, N. Helene, and U. Linus Opara, "Comparing the analytical performance of near and mid infrared spectrometers for evaluating pomegranate juice quality," *LWT*, vol. 91, pp. 180–190, 2018.

[11] J. C. Machado, M. A. Faria, I. M. P. L. V. O. Ferreira, R. N. M. J. Pascoa, and J. A. Lopes, "Varietal discrimination of hop pellets by near and mid infrared spectroscopy," *Talanta*, vol. 180, pp. 69–75, 2018.

[12] H. W. Siesler, "Vibrational spectroscopy," in *Reference Module in Materials Science and Materials Engineering*, Elsevier, New York, NY, USA, 2016.

[13] A. Paula Craig, B. G. Botelho, L. S. Oliveira, and A. S. Franca, "Mid infrared spectroscopy and chemometrics as tools for the classification of roasted coffees by cup quality," *Food Chemistry*, vol. 245, pp. 1052–1061, 2018.

[14] S. R. Khandasammy, M. A. Fikiet, E. Mistek et al., "Bloodstains, paintings, and drugs: raman spectroscopy applications in forensic science," *Forensic Chemistry*, vol. 8, pp. 111–133, 2018.

[15] J. Engel, G. Jan, E. Szymanska et al., "Breaking with trends in pre-processing?," *TrAC Trends in Analytical Chemistry*, vol. 50, pp. 96–106, 2013.

[16] H. Martens, S. Jensen, and P. Geladi, "Multivariate linearity transformation for near-infrared reflectance spectrometry," in *Proceedings of Nordic symposium on Applied Statistics*, pp. 205–234, Stavanger, Norway, June 1983.

[17] P. Geladi, D. MacDougall, and H. Martens, "Linearization and scatter-correction for near-infrared reflectance spectra of meat," *Applied Spectroscopy*, vol. 39, no. 3, pp. 491–500, 1985.

[18] P. Kubelka and F. Munk, "Ein beitrag zur optik der farbanstriche," *Zeitschrift fur Technische Physik*, vol. 12, pp. 593–601, 1931.

[19] A. Claus Andersson, "Direct orthogonalization," *Chemometrics and Intelligent Laboratory Systems*, vol. 47, no. 1, pp. 51–63, 1999.

[20] R. J. Barnes, M. S. Dhanoa, and S. J. Lister, "Standard normal variate transformation and de-trending of near-infrared diffuse reflectance spectra," *Applied Spectroscopy*, vol. 43, no. 5, pp. 772–777, 1989.

[21] M. Zeaiter, J.-M. Roger, and V. Bellon-Maurel, "Robustness of models developed by multivariate calibration. part ii: the influence of pre-processing methods," *TrAC Trends in Analytical Chemistry*, vol. 24, no. 5, pp. 437–445, 2005.

[22] T. Fearn, C. Riccioli, A. Garrido-Varo, and J. Emilio Guerrero-Ginel, "On the geometry of SNV and msc," *Chemometrics and Intelligent Laboratory Systems*, vol. 96, no. 1, pp. 22–26, 2009.

[23] T. Isaksson and B. Kowalski, "Piece-wise multiplicative scatter correction applied to near-infrared diffuse transmittance data from meat products," *Applied Spectroscopy*, vol. 47, no. 6, pp. 702–709, 1993.

[24] Y. Bi, K. Yuan, W. Xiao et al., "A local pre-processing method for near-infrared spectra, combined with spectral segmentation and standard normal variate transformation," *Analytica Chimica Acta*, vol. 909, pp. 30–40, 2016.

[25] F. Pedregosa, G. Varoquaux, A. Gramfort et al., "Scikit-learn: machine learning in python," *Journal of Machine Learning Research*, vol. 12, pp. 2825–2830, 2011.

[26] S. Raschka, *Python Machine Learning*, Packt Publishing, Birmingham, UK, 2015.

[27] E. L. Lehmann and G. Casella, *Theory of Point Estimation*, Springer Texts in Statistics Springer, New York, NY, USA, 2003.

[28] O. Heinisch, "Steel, R. G. D., and J. H. Torrie: principles and procedures of statistics. (with special reference to the biological sciences) mcgraw-hill book company, New York, Toronto, London 1960, 481 s., 15 abb.; 81 s 6 d," *Biometrische Zeitschrift*, vol. 4, no. 3, pp. 207-208, 1962.

6

Estimating Salt Concentrations Based on Optimized Spectral Indices in Soils with Regional Heterogeneity

Yasenjiang Kahaer [1,2] and Nigara Tashpolat [1,2]

[1] *College of Resources and Environment Sciences, Xinjiang University, Urumqi 830046, China*
[2] *Ministry of Education Key Laboratory of Oasis Ecology, Xinjiang University, Urumqi 830046, China*

Correspondence should be addressed to Nigara Tashpolat; 952544768@qq.com

Academic Editor: Nikša Krstulović

Soil salinity is one of the most damaging environmental problems worldwide, especially in arid and semiarid regions. The objectives of this study were to improve the inversion accuracy of soil salt content (SSC) in soils with spectral heterogeneity by using optimized spectral indices. Soil samples at a 0–20 cm depth were taken from Keriya Oasis (98 soil samples), Ugan-Kuqa Oasis (49 soil samples), and Ebinur Lake Basin (57 soil samples). SSC and spectral reflectance (SR) of all the 204 soil samples were determined. To comprehensively analyze the field-collected hyperspectral data, various band combinations were used to calculate a normalized difference spectral index (NDSI) and ratio spectral index (RSI). Then, the relationships between the indices and SSC were examined, and the most robust relationships were demonstrated. The partial least squares regression (PLSR) method was utilized to develop a predictive model of SSC, and the variable importance in the projection (VIP) method was used during modeling. The results revealed that (i) the salinized soils in different regions had apparent differences in both reflectance and spectral curve morphology, but the optimized spectral indices method effectively overcame the regional heterogeneity of salinized soil hyperspectral characteristics, and the correlation with SSC was always kind, with correlation coefficients up to 0.748 at 0.001 level of significance; (ii) the VIP filtering method effectively selected the optimal independent model, and the modeling accuracy was better than the single optimization index ($R^2_{\text{Pre}} = 0.83$ and $\text{RMSE}_{\text{Pre}} = 2.31\,\text{g}\cdot\text{kg}^{-1}$) by using the combination of two optimal indices; (iii) although the global modeling accuracy was significantly lower than the local modeling accuracy due to the inconsistent salt sensitivity bands of salinized soils in different regions, combined with cross-validation analysis, the global model had the ability to predict soil salinization accurately ($R^2_{\text{Pre}} = 0.69$ and $\text{RMSE}_{\text{Pre}} = 8.45\,\text{g}\cdot\text{kg}^{-1}$). The methods developed in this study can be applied in other arid and semiarid areas. Besides, the study also provides examples for aerospace hyperspectral remote sensing of cross-regional soil salinization.

1. Introduction

Soil salinization can lead to decline in soil fertility, decrease in crop productivity, and deterioration of the ecological environment, which are the main factors restricting the sustainable development of agricultural production and the ecological environment [1–3]. The rational development, improvement, and treatment of salinized soil require rapid, accurate, and dynamic information acquisition about salinized soil. Traditional soil salinization monitoring adopts the field fixed-point survey method, which is time-consuming, laborious, and destructive, with few measuring points and poor representativeness [4, 5]. It is difficult to obtain salt information across a large salinization area quickly and dynamically. Hyperspectral remote sensing has overcome the shortcomings of traditional artificial ground monitoring methods and has become an advanced method in the field of soil salinization monitoring due to its substantial advantages of multiple bands and continuous, abundant information and high quantitative inversion accuracy.

Construction of a soil salinity hyperspectral quantitative inversion model is one of the essential aspects of soil salinization hyperspectral remote sensing monitoring. A large number of studies have carried out the salt hyperspectral quantitative inversion of salinized soils in different areas. Weng et al. [6, 7] used the Yellow River Basin in Shandong Province of China as the research area. The hyperspectral quantitative inversion model for soil salinity based on a laboratory spectrum and Hyperion image was established using the univariate regression method, and the determination coefficients of the model were 0.873 and 0.627, respectively. Zhang et al. [8] also studied salinized soil in the Yellow River Basin by constructing a partial least squares model based on hyperspectral vegetation indices, but the coefficient of determination for the model was only 0.58. Kumar et al. [9] used the Indo-Gangetic plains of India as the study area and constructed a linear regression model with Hyperion data and salinity parameters of electrical conductivity (EC), sodium adsorption ratio (SAR), and exchangeable sodium percentage (ESP), with determination coefficients of 0.777, 0.801, and 0.804, respectively. Zeng et al. [10] established partial least square regression (PLS) and orthogonal projection to latent structure (OPLS) models for soil salinity from field-measured hyperspectral data collected in the Hetao Irrigation District of Inner Mongolia. The soil salt estimation model had a coefficient of determination up to 0.755. An et al. [11] studied a hyperspectral field estimation stepwise multiple linear regression model for salt content in coastal saline soils of the Yellow River delta with a coefficient of determination of 0.867. Rocha Neto et al. [12] used a variety of regression methods (extreme learning machine (ELM), ordinary least square regression (OLS), multilayer perceptron (MLP), and partial least squares regression (PLSR)), based on laboratory spectroscopy, to construct a hyperspectral prediction model for conductivity of saline soil in the Morada Nova Irrigation District. The highest accuracy of the model reached an adjusted $R^2 = 0.953$. Wang et al. [13] established a bootstrap-BP neural network SSC prediction model using field-measured hyperspectral data over the Ebinur Lake Wetland National Nature Reserve, and the prediction accuracy reached $R^2 = 0.95$ based on the optimal spectral indices. These studies can be used as references for follow-up research that seeks to use hyperspectral remote sensing to monitor the status of soil salinization.

Soil salinization has spatiotemporal variability, and the main factors affecting spatial and temporal heterogeneity are as follows: (i) The effect of soil formation on soil water and salt: the soil parent material desalinates the soluble salts during the erosion process under climatic conditions, and the salt gradually accumulates to form salinized land. The human cultivation and maturation process, the irrigation siltation process, and the salinization process are the main soil-forming processes. (ii) Local small terrain plays a crucial role in the accumulation of soil salinity. Generally, in low-lying areas, the soil particles are finely distributed, the water permeability is weak, and the groundwater is shallow. Under hot climatic conditions, the water volatilizes, leaving more salt and salt accumulates in this process, eventually resulting in a slice of salinized land. (iii) The influence of groundwater depth and water quality on soil water and salt. The main reason for the change in the balance of soil salinity is the change in water resources and expenditure. In the arid area, the evaporation is much larger than the precipitation, the surface runoff is basically absent, the groundwater becomes the primary source of water, and the groundwater quality and depth have spatial and temporal changes. The change in space is characterized by deeper groundwater depth in the upstream of the alluvial fan and lower salinity, while the downstream groundwater has a shallower depth and a higher degree of mineralization. The water level in the groundwater depth gradually increases, and the salinity concentration of the groundwater gradually increases, accelerating the accumulation of soil salinity in the upper layer of the soil. The change in time is due to the intercontinental variation of groundwater depth. Generally, it is affected by agricultural irrigation from April to October, and the water level is high. From March to February, due to winter soil freezing and phreatic evaporation, groundwater depth increased and the water level showed a minimum. (iv) The impact of land use on soil water and salt: the land use pattern will directly affect the nature of the soil and has a significant impact on the soil water cycle and the change of salt content.

Hyperspectral prediction models established for soil salinity, at present, are based on data from a specific area. However, the sensitive bands of hyperspectral remote sensing inversion of soil salinization in different regions are affected by regional heterogeneity. Soil salinization results using hyperspectral remote sensing for soil salinization monitoring can vary based on differences in soil organic matter, soil salinity, salt ion composition, and other factors in a particular region. Therefore, it is challenging to apply salt hyperspectral prediction models to other regions. Reduction in soil spectral heterogeneity and improvement of prediction accuracy and universality of the model have always been a hot issue for remote sensing researchers. Mashimbye et al. [14], in terms of the hyperspectral inversion of the conductivity of different types of salinized soils in South Africa, showed that first-order differential treatment could effectively eliminate the influence of soil types and other factors, thus highlighting the correlation between spectral reflectance and electrical conductivity. Shi et al. [15] conducted first-order differential processing and spectral classification modeling on 1581 soil samples from 16 soil types collected from Tibet, Xinjiang, Heilongjiang, and Hainan, China, and found the prediction accuracy of the first-order differential-spectral classification-local modeling is significantly higher than that of the first-order differential-global modeling. The authors also indicated that the first-order differential treatment is better than the original data. Consequently, to improve the universality and accuracy of the model, the first-order differential spectroscopy preprocessing method may be an effective method to improve further the global model presented in this paper.

Remote sensing monitoring of soil salinization can use the macromonitoring advantage of remote sensing technology to establish a unified quantitative inversion model for soil salinization at the national or even global scale to meet the needs of large-scale monitoring. Nevertheless, at the large-scale range, there are apparent differences in the hyperspectral characteristics of soils with different parent materials and soil-forming processes, which may be critical factors that reduce the accuracy of large-scale inversion models relative to regional inversion models. To identify and solve this problem, it is necessary to successfully establish a unified quantitative model for soil salinization at the national and global scales. Therefore, based on the differences in hyperspectral characteristics of soil salinization in three typical salinization areas in Xinjiang, China, this study proposed constructing the soil salinization hyperspectral quantitative inversion model in Xinjiang to provide a reference for future monitoring and application of soil salinization on a large scale by space hyperspectral remote sensing.

2. Materials and Methods

2.1. Study Area. Xinjiang, China (Figure 1), is located in central Eurasia and the northwestern border of China, with a total area of 1,664,900 km^2, accounting for one-sixth of China's total land area [16]. The topography is characterized by "three mountains and two basins." Xinjiang is inland and far from the sea, with a bright temperate continental climate, with Tianshan as the boundary. The temperature in southern Xinjiang is higher than in northern Xinjiang, and the precipitation in northern Xinjiang is higher than in southern Xinjiang. The oases are distributed in the basin margins and river basins. The total oasis area accounts for 5% of the total area with typical oasis ecological characteristics. According to the data from China's Second Soil Census, the soil types distributed in Xinjiang plain oases are mainly anthrosols, gleysol, fluvisol, phaeozems, and solonchaks.

The area of saline soil in Xinjiang accounts for 22% of the total area of saline soil in China, and it is also the largest saline soil distribution area in the country. Xinjiang is known as the salinized soil museum in arid areas with a wide range of distribution and variety. As the most significant agricultural land reserve area in northwestern China, the total area of salinized land in the Xinjiang irrigation area accounts for 32.07% of the total area. Soil salinization has become a significant obstacle to the efficient utilization, sustainable development, and construction of modern agriculture in Xinjiang. The formation of salinized soil in Xinjiang has gone through a long process. Southern Xinjiang began at least from the Cretaceous. The subsequent Quaternary river-lake sediments, mainly the modern sediments, played a critical role in the redistribution of soil water and salt. The deep saline water in soil participated in the regional water-salt cycle in the form of fissure water and confined water, which has a history of millions of years. In arid desert conditions, due to the meager precipitation and extreme evaporation, the salt in the crust continues to accumulate on the surface to form a typical saline soil with high salt content and thick salt layer. The unique geological and geomorphological conditions of Xinjiang have formed a broad closed-flow area, which makes the vast basins and low plains of Xinjiang an enrichment area for salt migration. Xinjiang's natural geographical environment, coupled with unsustainable utilization of water and land resources, will inevitably cause more land to face soil salinization problems. Accurate and efficient determinations of local soil salinity will be beneficial to local water resources management, agricultural land-use planning, and prevention of soil salinization expansion.

Ebinur Lake Basin is the lowest depression and water-salt pool in the southwestern margin of Junggar Basin; the range was 44°19′~45°11′N and 82°02′~83°42′E. The climate type is a temperate semiarid continental monsoon climate. Frequent dust and salt dust storms on the particular terrain are remarkable characteristics of this region. In recent years, Ebinur Lake has shrunk dramatically in volume, owing to climate change and human influence, leaving a saline playa greater than 500 km^2 [17, 18]. The delta oasis of the Ugan and Kuqa rivers is located to the south of Tianshan Mountains and the north of the Tarim Basin; it is a typical and complete piedmont alluvial-pluvial fan plain. The actual study area is bounded by 82°05′E~83°42′E longitude and 40°55′N~41°50′N latitude. The climate exhibits extreme aridity. The type of soil in the study area is mostly light-soil land and sandy loam soil, and the primary component of the salinized soil is chloride. Approximately 26% of the non-agricultural land is heavily salinized [19]. The Keriya Oasis (36°43′~37°13′N, 81°08′~81°47′E) is located on the southern edge of the Taklamakan Desert, at the northern foot of the Kunlun Mountains. This region has a temperate continental arid climate characterized by hyperaridity. The oasis is situated on a fluvial plain with relatively flat terrain, loose soil, high salt concentrations, and low soil fertility [20].

2.2. Field Sampling and Laboratory Analysis. The study of Xinjiang is focused mainly on the oases that address the survival needs for human beings and to establish a hyperspectral model of soil salinity retrieval in the oases. The soil sample collection area covers three typical and large-scale samples.

Soil samples from Ugan-Kuqa Oasis were obtained during the 11 days from July 7, 2016, to July 17, 2016, with 49 samples collected. Then, we took the collected soil samples back to the laboratory for drying and did not process the soil samples before going to the Keriya Oasis. The reason why we choose to collect soil samples in the field in July is that the sunshine in July is abundant and intense, and the evaporation is high, so the phenomenon of soil salinization is distinct, and because of the low precipitation, the field soil sampling can be guaranteed to proceed normally. Because the groundwater table is very high in September and October, which is the month of serious soil salinization in a year, it is more effective to carry out field sampling at this time. Keriya Oasis soil samples were obtained from September 19, 2016, to September 28, 2016, with 98 samples collected during ten days. After the sampling work, we took the soil samples back to the laboratory for drying (together with the soil samples from the Ugan-Kuqa Oasis), and then we went to Ebinur Lake Basin in October. The soil sampling time in the Ebinur Lake Basin was from October 16, 2016, to October 25, 2016, during which 57 samples were collected over ten days. Similarly, the soil samples were taken back to the laboratory and placed together with the samples from the

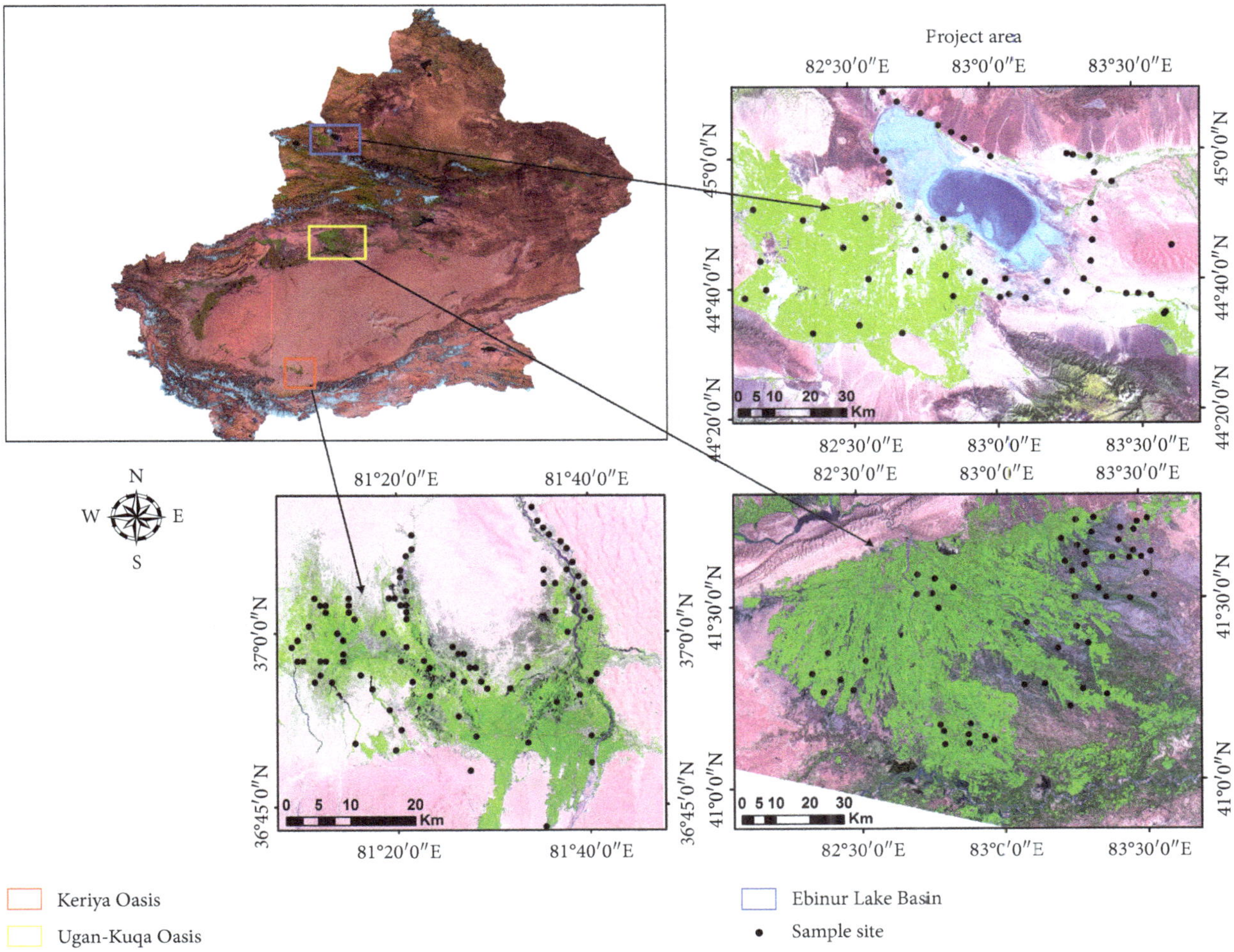

FIGURE 1: Study area with the sample sites.

other two places for drying. In January 2017, after all the soil samples were dried, we concentrated on the indoor laboratory work.

A total of 204 sites were chosen for soil sampling based on the visual determination of land cover conditions and salinity. Quadrats were set (30 m × 30 m) using GPS record position for the use of the 5-point hybrid method for sample collection. Soil samples were collected from the soil surfaces (0–20 cm), bagged, and brought to the laboratory. The collected soil samples were air-dried to eliminate the influence of water content; sieved using a 1 mm sieve to remove large debris, stones, and vegetation; and then split into two groups (one for chemical analysis and the other for spectral measurement). After grinding and filtering all 204 soil samples, the salt content and indoor hyperspectral data of the soil samples were measured. The salt content data of each point are in one-to-one correspondence with the hyperspectral data. Soil salt content was determined by preparing soil solutions with a soil-water ratio of 1 : 5, and the salt content and electrical conductivity (EC 1 : 5) of the solution were measured using Orion 115A^{+} [21, 22].

2.3. Soil Sample Spectrometry. The hyperspectral data of all sampling points were measured centrally on January 15, 2017. Spectral reflectance data for soil samples were obtained using an Analytical Spectral Devices FieldSpec-3 portable spectroradiometer ASD (Analytical Spectral Devices, Inc., USA) with a wavelength range of 350–2500 nm. Spectral measurements were carried out in a dark room with spectral resolutions of 1.4 nm (350–1000 nm) and 2 nm (1000–2500 nm) for different bands, with a sampling interval of 1 nm. The light source was a 50 W halogen lamp, 50 cm from the surface of the soil sample and 15 degrees at the zenith angle. The ground sample was placed in a black container with a diameter of 20 cm and a depth of 2 cm. After filling, the surface was flattened. The sensor probe used a 25° field angle probe, which was 15 cm above the soil sample. The area where the probe receives the soil spectrum was much smaller than the area of the black container so that the reflection spectrum of the soil sample was guaranteed. Before each sampling, a dark current correction and whiteboard calibration were performed. Ten spectral curves were collected for each sample, and the arithmetic average was calculated by ViewSpec-Pro (V6.0.11) software that was used as the reflectance of the soil sample. The ten spectra for each soil sample were averaged to minimize instrument noise. A Savitzky–Golay filter was applied to reduce the noise effect and eliminate the low SNR edge bands (350~399 nm and 2401~2500 nm), smoothing the denoising data of the 400~2400 nm for 204 soil samples. These hyperspectral data were later used to calculate the optimized spectral indices, and then the correlation between the measured salt content data and the optimized spectral indices was obtained.

2.4. Selection of Spectral Soil Indices. One of the most effective approaches to explore the significant relationships between the soil physiological index and hyperspectral data is conducting a comparative analysis of ratio spectral indices (RSIs) and simple normalized difference spectral indices (NDSIs), which are calculated from narrowband reflectance spectra. We identified that the wavelengths or spectral indices (equations (1) and (2)) could estimate soil salt content. NDSI and RSI are applied to identify optimal spectral indices. Spectral indices are defined as

$$\mathrm{NDSI} = \frac{R_i - R_j}{R_i + R_j}, \tag{1}$$

$$\mathrm{RSI} = \frac{R_i}{R_j}, \tag{2}$$

where R is the spectral reflectance value and the subscripts (i_{nm} and j_{nm}) are wavelengths in nanometers (nm).

Spectral indices (RSIs and NDSIs) were calculated for the smoothed soil hyperspectral reflectance using all possible combinations of available bands (i_{nm} and j_{nm}) in the 400–2400 nm spectral region. The two-band combination of optimized indices software V1.0 (No: 2018R11S177501, developed based on the JAVA) was used during the calculation of the indices [23]. Here, we examined the correlation between in situ soil salt content and spectral indices (NDSIs, RSIs) and 2-dimensional maps of correlation coefficients (r). The r maps, made by Matlab (Version 2012), allow the evaluation of the different band combinations and selection of sensitive spectral indices (RSIs and NDSIs) to study soil salt content. The most effective indices (NDSIs and RSIs) were determined using the significance level of 0.001 and the highest soil salt concentration correlation coefficient (r).

2.5. Variable Importance in the Projection Method. The VIP selection method (equation (3)) was first published by Wold and coauthors. VIP scores summarize the influence of individual variables on the PLS model. VIP scores give a measure useful to select the **x** variables which contribute the most to the **y** variance explanation. For a given model and data set, there will always be only one VIP scores vector summarizing all components and **y** variables [24]. The VIP score for the j variable is given as

$$\mathrm{VIP}_j = \sqrt{\frac{\sum_{f=1}^{F} W_{jf}^2 \cdot \mathrm{SSY}_f \cdot J}{\mathrm{SSY}_{\mathrm{total}} \cdot F}}, \tag{3}$$

where W_{jf} is the weight value for j variable and f component and SSY_f is the sum of squares of the explained variance for the f component and J number of variables. $\mathrm{SSY}_{\mathrm{total}}$ is the total sum of squares explained by the dependent variable, and F is the total number of components. The W_{jf2} gives the importance of the j variable in each f component, and VIP_j is a measure of the global contribution of j variable in the complete PLS model. In the case of one-dimensional space **y**, holds $\mathrm{SSY}_f = b_f^2 t_f' t_f$ and $\mathrm{SSY}_{\mathrm{total}} = b^2 T' T$, where T is the variable scores matrix and b is the PLSR inner relation vector of coefficients.

2.6. Partial Least-Squares Regression Method. Partial least-squares regression (PLSR) is a multivariate regression method that specifies a linear relationship between a set of dependent response variables, Y, and a set of predictor variables, X [25]. It is a popular modeling technique applied in chemometrics and commonly used for quantitative spectral analysis. To select the optimum number of factors and avoid overfitting, we calibrated the model by an iterative leave-one-out cross-validation (LOOCV) criterion called the minimum predicted residual sum of squares (PRESS) RMSE. RMSE is minimized by iteratively leaving one sample out of the calibration dataset and calibrating the model from the remaining dataset [26]. Wold et al. [27] described the PLSR method in detail. The PLSR process was performed using DPS® (Version 16.05). The predictive ability of the best selected PLSR model was assessed using the R^2 and RMSE for the independent validation dataset.

2.7. Model Validation. Validation of the models is a crucial step to ensure model quality [28]. Once all the developed models were tested, models with (a) a high R^2, indicating a robust linear relationship, and (b) low root mean square errors for the model variables. The two quantitative criteria between measured and predicted values were calculated by the equation listed in Table 1. As a result, the best-implemented PLSR model that met all the model selection and validation criteria was selected and used to predict the SSC.

3. Results

3.1. Statistical Analysis of SSC. For modeling, the samples were sorted from low to high according to the SSC, and based on the equidistance method, the calibration set and validation set were chosen. The characteristics of the descriptive statistics (Table 2) show that the calibration and validation sets of Ebinur Lake Basin corresponded to the following: the maximum values of SSC were 56.4 $\mathrm{g{\cdot}kg^{-1}}$ and 27.7 $\mathrm{g{\cdot}kg^{-1}}$, respectively; the minimum values were 0 $\mathrm{g{\cdot}kg^{-1}}$ and 1.1 $\mathrm{g{\cdot}kg^{-1}}$, respectively; and the mean values were 8.33 $\mathrm{g{\cdot}kg^{-1}}$ and 8.21 $\mathrm{g{\cdot}kg^{-1}}$, respectively. The coefficients of

Table 1: Cross-validation metrics.

Function name	Equation
Coefficient of determination	$R^2 = ((\sum_{i=1}^{N}(\chi_i - \overline{\chi})(\gamma_i - \overline{\gamma}))/(\sqrt{\sum_{i=1}^{N}(\chi_i - \overline{\chi})^2 + \sum_{i=1}^{N}(\gamma_i - \overline{\gamma})^2}))^2$
Root mean square error	$\text{RMSE} = \sqrt{\sum_{i=1}^{N}(\gamma_i - \chi_i)^2/n}$

Note. χ_i and γ_i are measured and predicted values, respectively; $\overline{\chi}$ and $\overline{\gamma}$ represent the means measured and predicted values, respectively; n is the number of samples.

variation for the calibration and validation sets were 157% and 110%, respectively. The calibration and validation sets of the Ugan-Kuqa Oasis corresponded to the following: the maximum values of SSC were 69.8 g·kg^{-1} and 65.1 g·kg^{-1}, respectively; the minimum values were 0 g·kg^{-1} and 3.9 g·kg^{-1}, respectively; and the mean values were 16.78 g·kg^{-1} and 24.06 g·kg^{-1}, respectively. The coefficients of variation for the calibration and validation sets were 111% and 92%, respectively. The calibration and validation sets of Keriya Oasis corresponded to the following: the maximum values of SSC were 17.8 g·kg^{-1} and 12.9 g·kg^{-1}, respectively; the minimum values were 0 g·kg^{-1} and 0.5 g·kg^{-1}, respectively; and the mean values were 5.36 g·kg^{-1} and 4.225 g·kg^{-1}, respectively. The coefficients of variation for the calibration and validation sets were 81% and 104%. The variation coefficient (CV) reflects the distinctiveness of soil salt content, suggesting the degree of human activity effects on soil salt content. A greater coefficient value indicates that soil is more disturbed by human activities [29]. Variation coefficient ranges are the following: CV < 15% is mild variation; 15% < C.V < 36% is medium variation; and C.V > 36% is highly variable. In this study, SSC had variation coefficients greater than 36% in the three regions. The results showed that SSC had the greatest dispersion in the study area, which was largely impacted by human activities.

A statistical table of measured values for other soil properties is also now incorporated by reference. The unit of soil electrical conductivity (EC 1 : 5) in Table 3 is mS·cm^{-1}, the unit of soil organic matter (SOM) is g·kg^{-1}, and the unit of soil compaction (SC) is Kpa.

Table 4 is a statistical analysis table of the eight major ions in the soil. The measurement unit of the eight major ions in the soil is g·kg^{-1}.

3.2. Hyperspectral Analysis of Salinized Soil in Different Areas. Figure 2 summarizes the reflectance spectra curves of soil in the different study areas (Ebinur Lake Basin, Ugan-Kuqa Oasis, and Keriya Oasis). Figure 2(a) shows the three spectral curves have approximately different characteristics. In the visible near-infrared ranges (400–1300 nm), the soil reflectance in the Ugan-Kuqa Oasis was relatively high and the soil reflectance in the Ebinur Lake Basin and Keriya Oasis was relatively low. In the near-infrared part (1300–2400 nm), Ugan-Kuqa Oasis soil reflectance was highest, followed by the Keriya Oasis soil, with the lowest soil reflectance in the Ebinur Lake Basin. This result may be due to differences in soil texture in different areas.

The spectral curve of soil salinity (SSC: 0 g·mg^{-1}) was selected for analysis to eliminate the influence of salt on the differences between soil spectral curve characteristics in the different regions. Figure 2(b) shows the slope of the soil reflectivity curve of Keriya Oasis is the maximum in the band of 400–600 nm and approximately the same in the Ugan-Kuqa Oasis and the Ebinur Lake Basin. In the water absorption zone, the absorption depth and absorption area near the soil of 1400 and 1900 nm in Ugan-Kuqa Oasis were significantly higher than those in the Ebinur Lake Basin and Keriya Oasis. The reflectance of the soil in Ugan-Kuqa Oasis was smooth after a falling trend at 2100–2300 nm, and the Keriya Oasis and Ebinur Lake Basin showed a downward trend in volatility. Besides, the soil in the Ebinur Lake Basin and the Keriya Oasis had a broad weak absorption peak in the 800–1300 nm, but this was not obvious in the Ugan-Kuqa Oasis. Throughout the entire spectrum, the spectral curves of soils in the two areas (Ebinur Lake Basin and Keriya Oasis) were close to each other, whereas there was a large difference between these and the spectral curve of the soil in Ugan-Kuqa Oasis.

3.3. Correlation between Soil Salt Content and Reflectivity Data in Different Regions. In this paper, the spectral indices (NDSIs and RSIs) were optimized by hyperspectral data from different study areas, and the two-dimensional maps of correlation (r) between optimized spectral indices and SSC are shown in Figure 3. From the analysis, the NDSIs and RSIs with the highest correlation ($|r| \geq 0.7$) were found in the visible and near-infrared (VNIR) spectral range in the Ebinur Lake Basin. Accordingly, the NDSI ($R2086$, $R1880$) and RSI ($R2013$, $R1892$) within this region had a maximum correlation of $|r| = 0.746$ and $|r| = 0.748$, respectively. Furthermore, a spectral region (2000–2200 nm) with the combination of green and red over the NIR region showed a relatively good correlation with SSC ($0.6 \leq |r| < 0.7$). In Ugan-Kuqa Oasis, the spectral ranges (800–900 nm and 1000–1400 nm) correlated relatively well with SSC when only combined with the NIR region ($|r| \geq 0.7$). The NDSI ($R1246$, $R1237$) and RSI ($R1246$, $R1237$) within this region had a maximum correlation of $|r| = 0.730$ and $|r| = 0.720$, respectively. Additionally, the spectral region formed with 1500–1800 nm and 1000–1700 nm regions showed good correlation with SSC ($0.6 \leq |r| < 0.7$). In Keriya Oasis, the strength of the relationship between optimized spectral indices (NDSIs and RSIs) were moderate for SSC (Figures 3(e) and 3(f)). SSC was best correlated with NDSI (($R2006$, $R1896$), ($R2007$, $R1896$), and ($R2008$, $R1896$)) and RSI (($R2001$, $R1896$), ($R2003$, $R1896$), ($R2005$, $R1896$), ($R2006$, $R1896$), ($R2007$, $R1896$), ($R2008$, $R1896$), ($R2001$, $R1897$), ($R2003$, $R1897$), and ($R2001$, $R1898$)) ($|r| = 0.600$), and the most significant region was broad between 2000 and 2200 nm.

According to the soil spectral data for the three research areas, the spectral indices (NDSIs and RSIs) were optimized. As seen, the spectral region formed with 800–1000 nm and 1300–1400 nm, the 1500–1700 nm, 1800–1900 nm, and 2000–2100 nm regions showed good correlation with SSC

Table 2: Statistical characteristics of SSC in sampling area soil ($g \cdot kg^{-1}$).

Field sites	Modeling set	Sample sites	Max value	Min value	Mean value	SD	CV (%)
Ebinur Lake Basin	Whole set	57	56.4	0	8.29	11.80	142
	Calibration set	38	56.4	0	8.33	13.08	157
	Validation set	19	27.7	1.1	8.21	9.04	110
Ugan-Kuqa Oasis	Whole set	49	69.8	0	19.30	19.99	104
	Calibration set	33	69.8	0	16.78	18.65	111
	Validation set	16	65.1	3.9	24.06	22.09	92
Keriya Oasis	Whole set	98	17.8	0	4.99	4.33	87
	Calibration set	65	17.8	0	5.36	4.32	81
	Validation set	33	12.9	0.5	4.225	4.39	104

Table 3: Statistical analysis of chemical composition measurement in sampling area soil.

Field sites	Detection category	Sample sites	Max value	Min value	Mean value	SD
Ebinur Lake Basin	EC 1 : 5	57	81.30	0.07	13.38	17.76
	SOM		45.47	0.22	12.48	6.64
	PH		9.97	7.65	8.26	0.37
	SC		2115	25	514.25	598.28
Ugan-Kuqa Oasis	EC 1 : 5	49	125.50	0.34	36.60	38.16
	SOM		31.70	0.95	12.32	7.05
	PH		9.04	6.35	7.73	0.66
	SC		2352	35	545.51	626.83
Keriya Oasis	EC 1 : 5	98	30.80	0.10	8.28	6.81
	SOM		9.36	0.92	3.38	2.25
	PH		10.56	7.47	8.75	0.63
	SC		1526	15	352.11	425.33

Table 4: Statistical characteristics of the eight major ions in sampling area soil ($g \cdot kg^{-1}$).

Field sites	Statistics	K^+	Na^+	Ca^{2+}	Mg^{2+}	Cl^-	SO_4^{2-}	HCO_3^-	CO_3^{2-}
Ebinur Lake Basin	Max	0.852	158.661	16.248	1.236	182.232	60.519	2.814	0.723
	Min	0.005	0.678	0.048	0.010	0.496	0.284	0.281	0.000
	Mean	0.230	10.089	3.124	0.276	8.303	12.841	1.699	0.014
	SD	0.170	21.328	3.629	0.252	24.768	12.957	0.661	0.100
Ugan-Kuqa Oasis	Max	3.776	161.776	16.012	16.566	205.900	127.770	4.784	1.000
	Min	0.002	0.040	0.028	0.023	0.000	0.182	0.844	0.000
	Mean	0.563	48.294	4.948	1.803	46.365	23.043	2.357	0.018
	SD	0.684	55.599	4.133	3.619	57.746	24.353	0.880	0.134
Keriya Oasis	Max	1.390	72.581	11.124	2.152	82.833	36.429	4.784	0.000
	Min	0.002	0.004	0.040	2.152	0.000	0.072	0.281	0.000
	Mean	0.284	6.341	3.366	0.385	6.916	12.505	2.039	0.000
	SD	0.260	11.114	2.936	0.462	11.971	10.209	0.862	0.000

($0.5 \leq |r| \leq 0.6$). NDSI ($R1332$, $R1287$) and RSI ($R1329$, $R1287$) within this region had a maximum correlation ($|r| = 0.529$ and $|r| = 0.530$). The optimized spectral indices revealed a good correlation with soil salt content in the study areas and illustrated the potential predictive ability for estimating the SSC.

3.4. Inversion Accuracy Analysis of Local Modeling and Global Modeling

3.4.1. Establishment of Local and Global Models. In this paper, the PLSR models with predictive variables were selected based on the rule (VIP) that the addition of another variable should increase the accuracy of the models. The influence of each spectral index in the PLSR model is illustrated in Figure 4 with corresponding VIP values. The VIP method revealed the importance of optimized indices for SSC. In Ebinur Lake Basin, the local VIP values (VIP > 1.0) were found with NDSI (($R2013$, $R1892$), ($R2074$, $R1880$), ($R2073$, $R1880$), ($R2075$, $R1880$), ($R2086$, $R1880$), ($R2082$, $R1880$), ($R2083$, $R1880$), ($R2087$, $R1880$), ($R2089$, $R1880$), ($R2088$, $R1880$)) and RSI (($R2013$, $R1892$), ($R2074$, $R1880$), ($R2073$, $R1880$), ($R2075$, $R1880$), and ($R2086$, $R1880$)) in optimized spectral indices. In Ugan-Kuqa Oasis, the optimized spectral indices NDSI (($R1246$, $R1237$), ($R2227$, $R2219$), and ($R819$, $R811$)) and RSI (($R1246$, $R1237$), ($R2227$, $R2219$), ($R836$, $R825$), ($R1247$, $R1237$), ($R1176$, $R1169$), ($R1245$, $R1237$), and ($R836$, $R826$)) revealed the maximum VIP values (VIP > 1.0). In the Keriya Oasis, the

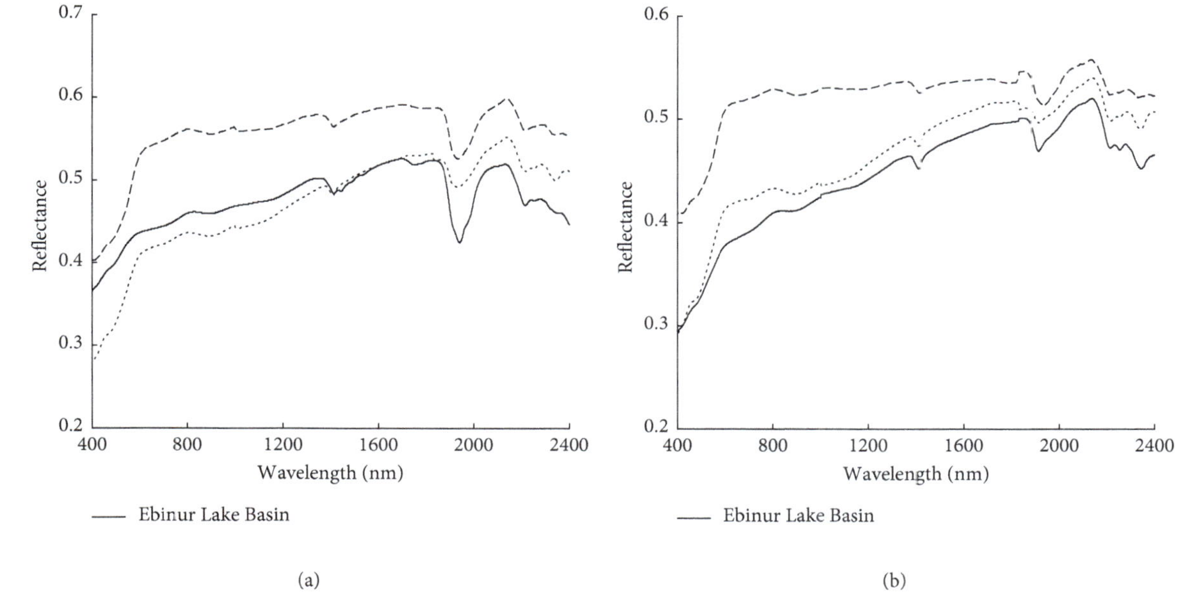

Figure 2: Diffuse reflectance spectra of soil in different areas. (a) Mean reflectance spectra curve of the three regions. (b) Mean reflectance spectra curve of the three regions with SSC ($0\,g{\cdot}kg^{-1}$).

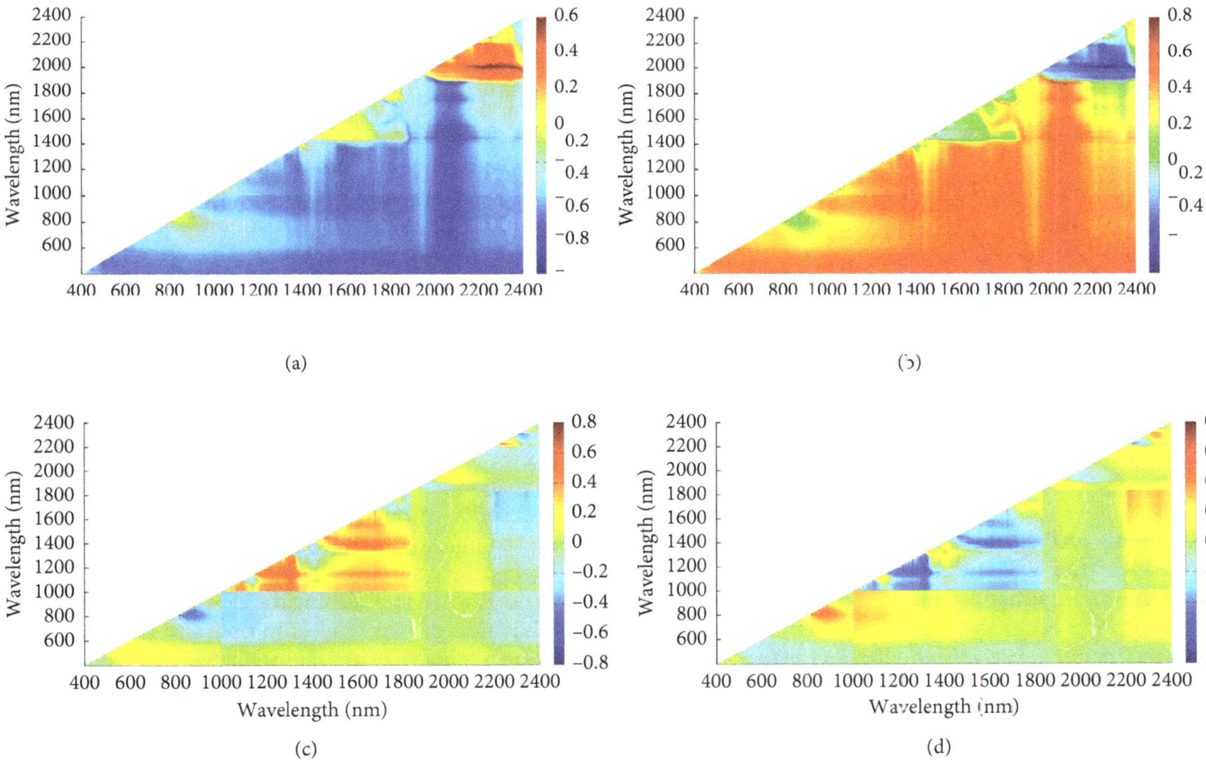

Figure 3: Continued.

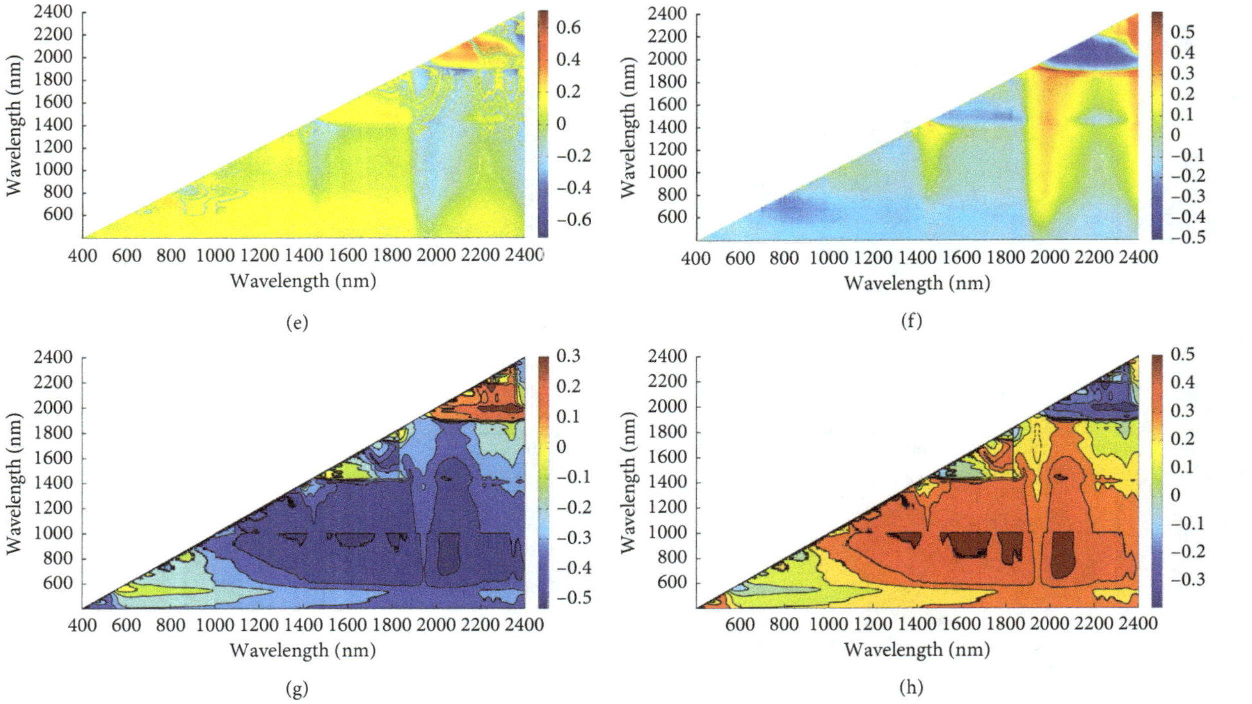

Figure 3: Two-dimensional maps of the correlation between optimized spectral indices and SSC. (a) NDSIs (Ebinur Lake Basin). (b) RSIs (Ebinur Lake Basin). (c) NDSIs (Ugan-Kuqa Oasis). (d) RSIs (Ugan-Kuqa Oasis). (e) NDSIs (Keriya Oasis). (f) RSIs (Keriya Oasis). (g) NDSIs (All data). (h) RSIs (All data).

optimized spectral indices NDSI ((*R*2006, *R*1896), (*R*2008, *R*1896), (*R*2007, *R*1896), and (*R*2005, *R*1897)) and RSI ((*R*2003, *R*1896), (*R*2005, *R*1896), (*R*2006, *R*1896), (*R*2001, *R*1896), (*R*2001, *R*1897), and (*R*2001, *R*1898)) revealed the maximum VIP values (VIP > 1.0).

The best predictive variables for the PLSR model were selected, and ten models were developed for estimating SSC in the research areas; the selected variables and details of PLSR model accuracy are described in Figure 4 and Table 3. The scatter plot (Figure 5) of predicted and measured SCC for the partial least squares regression (PLSR) predictive models was established and illustrated the predictive ability of the PLSR models in SSC estimation using the R^2_{Pre} and $\mathrm{RMSE_{Pre}}$ for the independent validation dataset.

3.4.2. Analysis of Predictive Model Accuracy. Through the application of the VIP method in the previous step, we selected the optimal band combination from the two indices. Next, we used the selected band combination as the model's independent variable and the SSC as the dependent variable to establish a model that reflects the relationship between the two using PLSR. Here, we have a total of ten models, nine partial models, and one global model. Three models were set up for each study area, which are based on the combination of NDSI optimal bands, the combination of RSI optimal bands, and the combination of the optimal band of NDSI and RSI. As seen from Figure 5, the accuracy of the models established by the NDSI in Ebinur Lake Basin was $R^2_{\mathrm{Pre}} = 0.77$ and $\mathrm{RMSE_{Pre}} = 3.81\,\mathrm{g{\cdot}kg^{-1}}$. The accuracy of the model based on the RSI was $R^2_{\mathrm{Pre}} = 0.76$ and $\mathrm{RMSE_{Pre}} = 3.84\,\mathrm{g{\cdot}kg^{-1}}$. The total model accuracy of the two exponential combinations is $R^2_{\mathrm{Pre}} = 0.83$ and $\mathrm{RMSE_{Pre}} = 2.31\,\mathrm{g{\cdot}kg^{-1}}$. The accuracy of the models established by the NDSI in Ugan-Kuqa Oasis is $R^2_{\mathrm{Pre}} = 0.68$ and $\mathrm{RMSE_{Pre}} = 7.56\,\mathrm{g{\cdot}kg^{-1}}$. The accuracy of the model based on the RSI is $R^2_{\mathrm{Pre}} = 0.62$ and $\mathrm{RMSE_{Pre}} = 9.76\,\mathrm{g{\cdot}kg^{-1}}$. The total model accuracy of the two exponential combinations is $R^2_{\mathrm{Pre}} = 0.71$ and $\mathrm{RMSE_{Pre}} = 5.46\,\mathrm{g{\cdot}kg^{-1}}$. The accuracy of the models established by the NDSI in Keriya Oasis is $R^2_{\mathrm{Pre}} = 0.64$ and $\mathrm{RMSE_{Pre}} = 4.86\,\mathrm{g{\cdot}kg^{-1}}$. The accuracy of the model based on the RSI is $R^2_{\mathrm{Pre}} = 0.63$ and $\mathrm{RMSE_{Pre}} = 4.31\,\mathrm{g{\cdot}kg^{-1}}$. The total model accuracy of the two exponential combinations is $R^2_{\mathrm{Pre}} = 0.68$ and $\mathrm{RMSE_{Pre}} = 3.31\,\mathrm{g{\cdot}kg^{-1}}$. Through the evaluation of the model, it was found that the accuracy of the model combined with two kinds of optimized spectral indices in each sample area was better than that of the single index. Therefore, when establishing the global model, we only used a combination of two models to optimize the spectral index, and the accuracy of the global model was $R^2_{\mathrm{Pre}} = 0.63$ and $\mathrm{RMSE_{Pre}} = 8.36\,\mathrm{g{\cdot}kg^{-1}}$. In summary, the prediction accuracy of the combined model in Ebinur Lake Basin was the highest, with R^2_{Pre} reaching 0.83, which was the best for all the models. The combined band for this model was NDSI ((*R*2013, *R*1892), (*R*2074, *R*1880), (*R*2073, *R*1880), (*R*2075, *R*1880), (*R*2086, *R*1880), (*R*2082, *R*1880), (*R*2083, *R*1880), (*R*2087, *R*1880), (*R*2089, *R*1880), and (*R*2088, *R*1880)) and RSI ((*R*2013, *R*1892), (*R*2074, *R*1880), (*R*2073, *R*1880), (*R*2075,

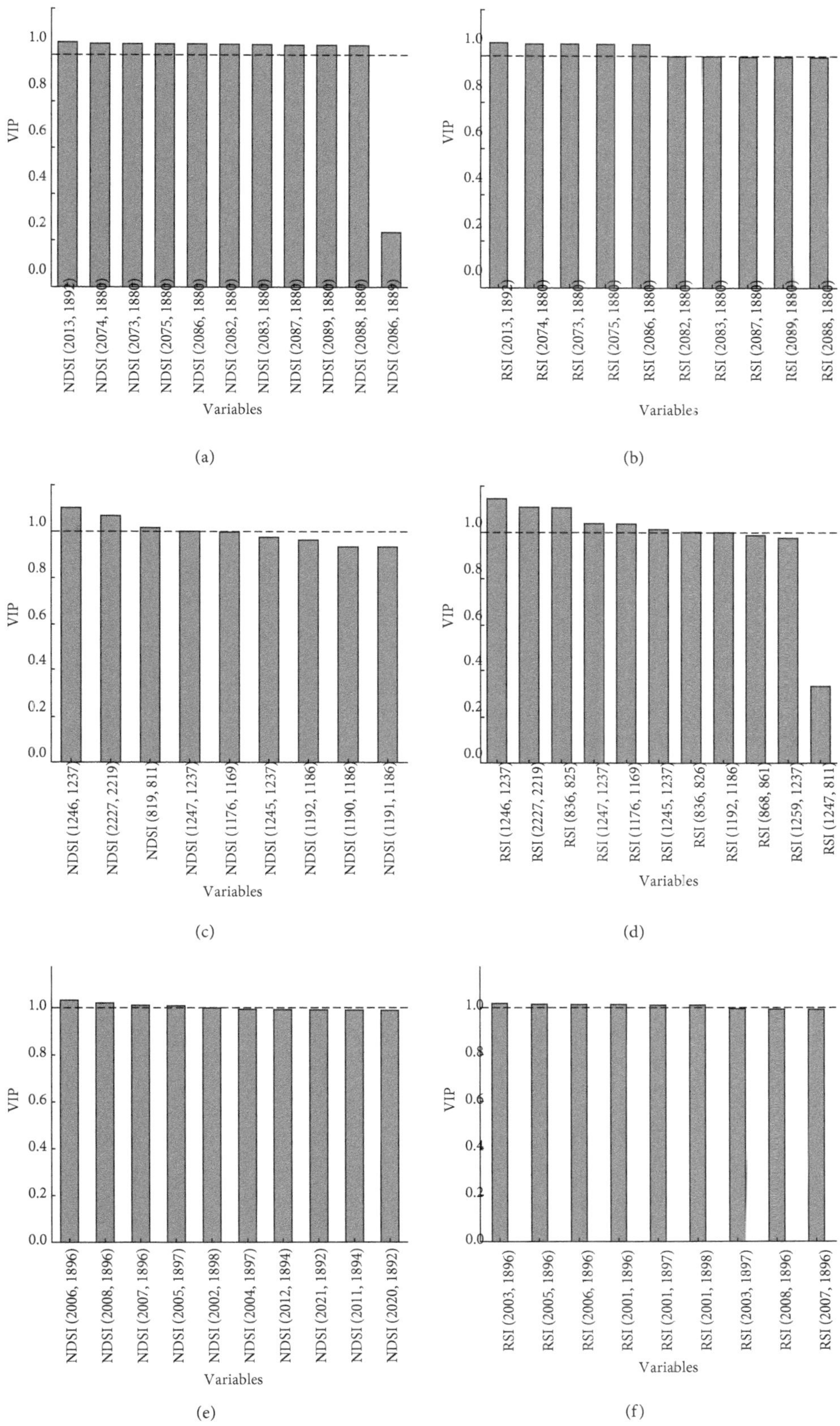

FIGURE 4: Bar chart of the model parameters (VIPs). (a) NDSI-Ebinur Lake Basin. (b) RSI-Ebinur Lake Basin. (c) NDSI-Ugan-Kuqa Oasis. (d) RSI-Ugan-Kuqa Oasis. (e) NDSI-Keriya Oasis. (f) RSI-Keriya Oasis.

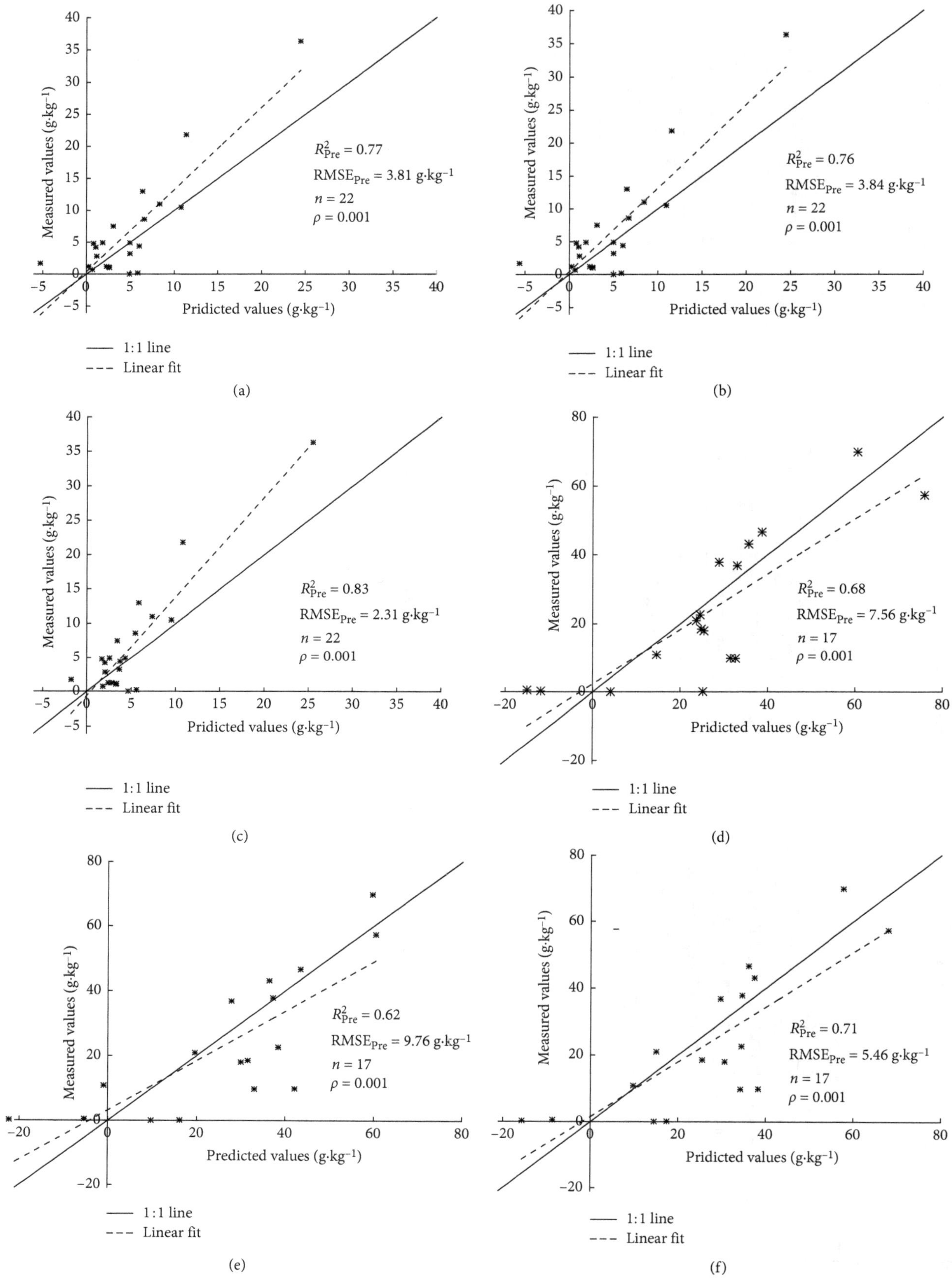

FIGURE 5: Continued.

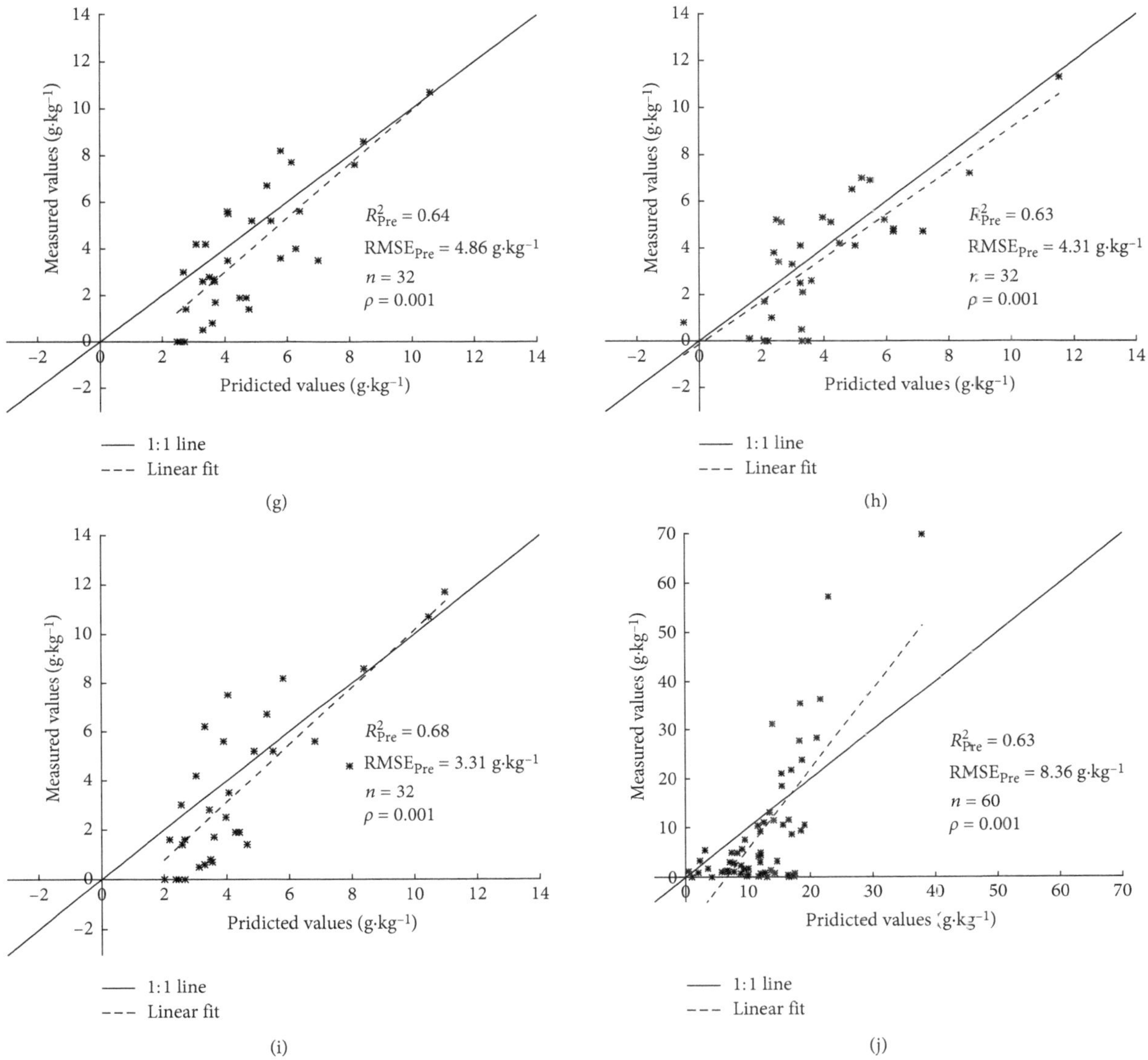

FIGURE 5: Scatter plots of predicted vs. measured SSC. (a) NDSI-Ebinur Lake Basin. (b) RSI-Ebinur Lake Basin. (c) Combination-Ebinur Lake Basin. (d) NDSI-Ugan-Kuqa Oasis. (e) RSI-Ugan-Kuqa Oasis. (f) Combination-Ugan-Kuqa Oasis. (g) NDSI-Keriya Oasis. (h) RSI-Keriya Oasis. (i) Combination-Keriya Oasis. (j) Global model.

*R*1880), and (*R*2086, *R*1880)). Thus, we can think of whether the local model can be applied to the global model and compare it with the prediction of the global model.

In order to further verify the accuracy of the prediction model, a comparative interpolation map of the measured and predicted values of soil salinity in each region was prepared. It can be seen from Figure 6 that the prediction effects of the three regions are sound, and the predicted interpolation map is consistent with the measured map.

3.4.3. Analysis of Cross-Validation Model Accuracy. According to the accuracy analysis and verification of the above model, it can be seen that the local model established by combining two optimized spectral indices has better predictive ability. To select the best stability model, the verification set of 15 samples was used uniformly, and the robustness of the model was evaluated by cross validation of three study areas. As a result, a total of 18 verification schemes were obtained. The results are shown in Table 5. The three models derived from the Ebinur Lake Basin were applied to the Ugan-Kuqa Oasis and the Keriya Oasis. The combined model shows good predictability in Ugan-Kuqa Oasis and Keriya Oasis, and the model accuracies were $R^2_{\text{Pre}} = 0.43$ and $\text{RMSE}_{\text{Pre}} = 9.68\,\text{g}\cdot\text{kg}^{-1}$ and $R^2_{\text{Pre}} = 0.41$ and $\text{RMSE}_{\text{Pre}} = 10.89\,\text{g}\cdot\text{kg}^{-1}$, respectively. The three models derived from the Ugan-Kuqa Oasis were applied to the Ebinur Lake Basin and the Keriya Oasis. The combined model shows good predictability in Keriya Oasis and Ebinur

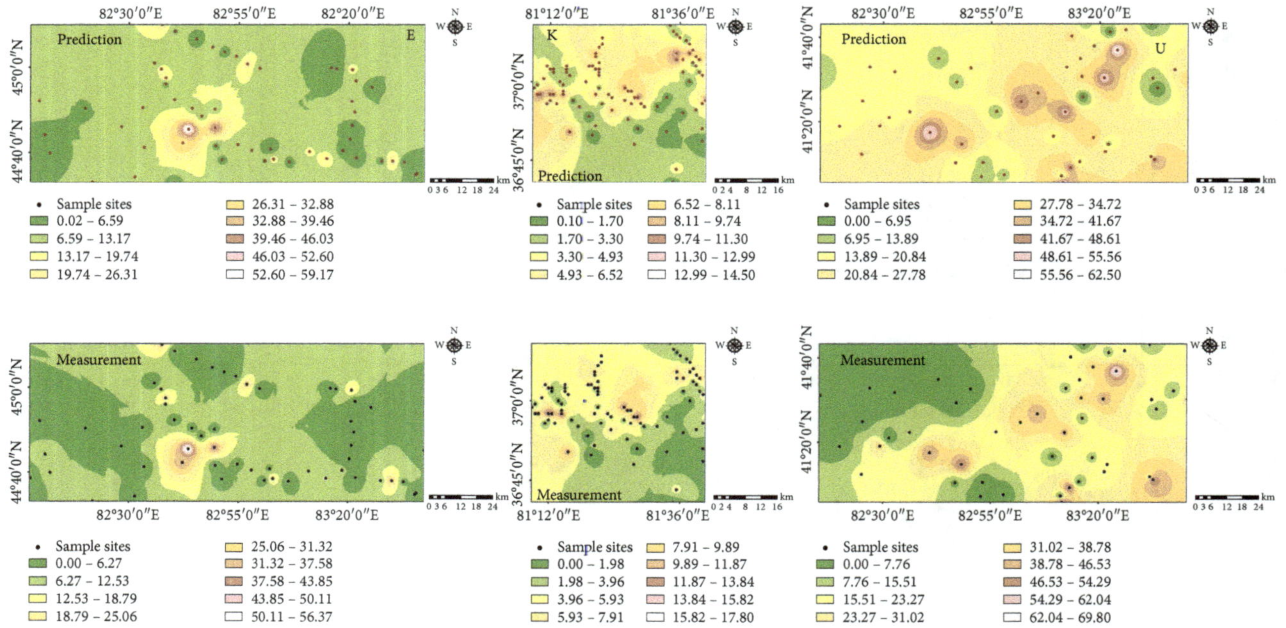

FIGURE 6: Comparative interpolation map of the measured and predicted values of SSC.

TABLE 5: Accuracy of cross-validation dataset.

Predictive models	Calibration/validation dataset	n	R^2_{Cal}	$RMSE_{Cal}$ (g·kg^{-1})	R^2_{Pre}	$RMSE_{Pre}$ (g·kg^{-1})
Cross validation of Ebinur Lake Basin						
E-model (NDSIs)	Ebinur/Ugan-Kuqa	15	0.82	3.82	0.40	10.23
E-model (RSIs)	Ebinur/Ugan-Kuqa	15	0.82	3.84	0.41	10.12
E-model$_{(Com)}$	Ebinur/Ugan-Kuqa	15	0.83	3.79	0.43	9.68
E-model (NDSIs)	Ebinur/Keriya	15	0.82	3.82	0.37	11.96
E-model (RSIs)	Ebinur/Keriya	15	0.82	3.84	0.39	11.54
E-model$_{(Com)}$	Ebinur/Keriya	15	0.83	3.79	0.41	10.89
Cross validation of Ugan-Kuqa Oasis						
U-model (NDSIs)	Ugan-Kuqa/Ebinur	15	0.74	7.38	0.33	12.35
U-model (RSIs)	Ugan-Kuqa/Ebinur	15	0.77	10.50	0.34	12.78
U-model$_{(Com)}$	Ugan-Kuqa/Ebinur	15	0.78	4.96	0.38	11.97
U-model (NDSIs)	Ugan-Kuqa/Keriya	15	0.74	7.38	0.30	11.45
U-model (RSIs)	Ugan-Kuqa/Keriya	15	0.77	10.50	0.40	11.35
U-model$_{(Com)}$	Ugan-Kuqa/Keriya	15	0.78	4.96	0.42	11.14
Cross validation of Keriya Oasis						
K-model (NDSIs)	Keriya/Ebinur	15	0.62	4.53	0.27	11.96
K-model (RSIs)	Keriya/Ebinur	15	0.63	4.22	0.29	11.84
K-model$_{(Com)}$	Keriya/Ebinur	15	0.67	3.18	0.31	11.02
K-model (NDSIs)	Keriya/Ugan-Kuqa	15	0.62	4.53	0.28	12.57
K-model (RSIs)	Keriya/Ugan-Kuqa	15	0.63	4.22	0.29	12.58
K-model$_{(Com)}$	Keriya/Ugan-Kuqa	15	0.67	3.18	0.29	12.31
Revalidation of global model						
Global model	Ebinur/Ugan-Kuqa/Keriya	45	0.77	5.32	0.69	8.45

Lake Basin, and the model accuracies were $R^2_{Pre} = 0.42$ and $RMSE_{Pre} = 11.14$ g·kg^{-1} and $R^2_{Pre} = 0.38$ and $RMSE_{Pre} = 11.97$ g·kg^{-1}. The three models of the Keriya Oasis were validated in the Ebinur Lake Basin and Ugan-Kuqa Oasis. The combined model shows good predictability in the Ebinur Lake Basin and Ugan-Kuqa Oasis, and the model accuracies were $R^2_{Pre} = 0.31$ and $RMSE_{Pre} = 11.02$ g·kg^{-1} and $R^2_{Pre} = 0.29$ and $RMSE_{Pre} = 12.31$ g·kg^{-1}, respectively. It can be seen that each local model was not very useful in the prediction of other sample plots, and the combined model always yielded better prediction accuracy than the single model. The best model was the combination model of Ebinur Lake Basin in Ugan-Kuqa Oasis. However, the model prediction accuracy was only $R^2_{Pre} = 0.43$ and $RMSE_{Pre} = 9.68$ g·kg^{-1}, far less than that of the global model ($R^2_{Pre} = 0.69$, and $RMSE_{Pre} = 8.45$ g·kg^{-1}), indicating that the local model does not have excellent stability and universality and is only applicable to the respective study area. The results of Table 5 show the existence of regional heterogeneity. The local model established in one region cannot be applied to other regions, while the global model can obtain better R^2. This shows that the optimization spectral indices can effectively reduce the heterogeneity to some extent.

4. Discussion

This study showed that the spectral characteristics of salinized soils in different regions were significantly different. The reasons for the spectral heterogeneity of salinized soils in different regions may lie in the following aspects: (i) First, due to regional differences, the soil formation conditions, soil formation processes, and soil parent materials are different, resulting in substantial differences in soil minerals. Minerals, as the soil components with the highest quality, have the most significant impact on soil spectral characteristics. (ii) Second, the soil salinity composition was different in different regions. The salt in the soil is a mixture of many single salts. Different salts have their unique spectral characteristics. Therefore, the soil salinity formed by mixing different proportions of salts also has different spectral characteristics. (iii) Third, the different parts of the salt and soil particles combined in different ways, and the soil salinity content in Ugan-Kuqa Oasis was higher than that in Ebinur Lake Basin and Keriya Oasis. However, the salt content in the soil needs to reach a particular content to exist in crystalline form, which shows as visible salt particles. However, soil with lower content is in the form of salt molecules, so the spectral characteristics of soils in Ugan-Kuqa Oasis are different from those in Ebinur Lake Basin and Keriya Oasis. (iv) Fourth, the composition of organic matter in soils in different areas is different; that is, the content ratio of humic acid, humin, and fulvic acid is different. Climate is one of the essential factors affecting the composition of soil organic matter. The organic matter of different composition states also has its spectral characteristics and will be reflected in the fact that Ebinur Lake Basin, Ugan-Kuqa Oasis, and Keriya Oasis are in different climatic zones, and there will be some differences in the organic matter composition of the soil.

The methods developed in this paper have the following advantages: (i) The continuous narrow-band spectrum contains more salt information, and the two-band optimization algorithm of massive spectral data can fully extract the relevant bands with the most considerable correlation with soil salinity on the two-dimensional level and achieve rapid optimization in complex hyperspectral parameters [30]. The effect of deep mining hyperspectral data further improves the accuracy of hyperspectral estimation of soil salinity, reducing the impact of environmental factors on modeling, and has better sensitivity than traditional single band. (ii) VIP is a useful method for selecting essential bands, and the VIP method is used to further filter the independent variables in this paper, which is beneficial to improve the practicability of the model. (iii) Compared with the independent model, the combined model can achieve the best prediction effect. This aspect may be because the combination of multiple spectral indices can make full use of the characteristics of each spectral index to achieve complementary advantages. On the contrary, the modeling algorithm used in the study may be related because it is more efficient to use partial least squares regression when the variables are highly linearly correlated internally. However, whether further optimization of the spectral index method can be successfully applied to the research of hyperspectral remote sensing monitoring of soil salinization at a larger scale and across a full area must be further studied and verified.

5. Conclusions

By analyzing the differences in the hyperspectral characteristics of salinized soils in three regions (Ebinur Lake Basin, Ugan-Kuqa Oasis, and Keriya Oasis) and using the optimized spectral indices approach to improve the modeling accuracy, the hyperspectral quantitative inversion of salinity in salinized soil with spectral heterogeneity was studied. The major conclusions include the following: (i) The salinized soils in different regions had apparent differences in both reflectance and spectral curve morphology, but the optimized spectral index method effectively overcame the regional heterogeneity of salinized soil hyperspectral characteristics, and the correlation with SSC was always right (correlation coefficient was up to 0.748). (ii) The regional heterogeneity of salinized soil hyperspectral characteristics caused the sensitivity of salinized soils in different regions to be inconsistent. At the same time, the global modeling accuracy was significantly lower than the local modeling accuracy. (iii) The VIP filtering method can effectively select the optimal independent model, and the modeling effect of combined models was always optimal. (iv) The cross-validation results showed that the local model had definite regional limitations, and the global model had the best prediction ability, with $R^2_{Pre} = 0.69$ and $RMSE_{Pre} = 8.45\ g{\cdot}kg^{-1}$. This research idea provides a valid reference for hyperspectral remote sensing monitoring for large-scale soil salinization across regions.

Conflicts of Interest

The authors declare that there are no conflicts of interest regarding the publication of this paper.

Acknowledgments

This work was jointly supported by multiple grants from the National Science Foundation of China (nos. 41761077 and U1603241), College of Resources and Environment Science, and the Key Lab of Oasis Ecology at Xinjiang University. The authors thank research partners who assisted long and strenuous hours in collecting field data.

References

[1] R. L. Dehaan and G. R. Taylor, "Field-derived spectra of salinized soils and vegetation as indicators of irrigation-induced soil salinization," *Remote Sensing of Environment*, vol. 80, no. 3, pp. 406–417, 2002.

[2] R. Dehaan and G. R. Taylor, "Image-derived spectral end-members as indicators of salinisation," *International Journal of Remote Sensing*, vol. 24, no. 4, pp. 775–794, 2003.

[3] G. I. Metternicht and J. A. Zinck, "Remote sensing of soil salinity: potentials and constraints," *Remote Sensing of Environment*, vol. 85, no. 1, pp. 1–20, 2003.

[4] C.-W. Chang and D. A. Laird, "Near-infrared reflectance spectroscopic analysis of soil C and N," *Soil Science*, vol. 167, no. 2, pp. 110–116, 2002.

[5] K. D. Shepherd and M. G. Walsh, "Development of reflectance spectral libraries for characterization of soil properties," *Soil Science Society of America Journal*, vol. 66, no. 3, pp. 988–998, 2002.

[6] Y. Weng, P. Gong, and Z. Zhu, "Reflectance spectroscopy for the assessment of soil salt content in soils of the Yellow River Delta of China," *International Journal of Remote Sensing*, vol. 29, no. 19, pp. 5511–5531, 2008.

[7] Y.-L. Weng, P. Gong, and Z.-L. Zhu, "A spectral index for estimating soil salinity in the Yellow River Delta region of China using EO-1 hyperion data," *Pedosphere*, vol. 20, no. 3, pp. 378–388, 2010.

[8] T.-T. Zhang, S.-L. Zeng, Y. Gao et al., "Using hyperspectral vegetation indices as a proxy to monitor soil salinity," *Ecological Indicators*, vol. 11, no. 6, pp. 1552–1562, 2011.

[9] S. Kumar, G. Gautam, and S. K. Saha, "Hyperspectral remote sensing data derived spectral indices in characterizing salt-affected soils: a case study of Indo-Gangetic plains of India," *Environmental Earth Sciences*, vol. 73, no. 7, pp. 3299–3308, 2015.

[10] W.-Z. Zeng, J.-S. Huang, C. Xu, T. Ma, and J.-W. Wu, "Hyperspectral reflectance models for soil salt content by filtering methods and waveband selection," *Ecological Chemistry and Engineering S*, vol. 23, no. 1, pp. 117–130, 2016.

[11] D. An, G. Zhao, C. Chang et al., "Hyperspectral field estimation and remote-sensing inversion of salt content in coastal saline soils of the Yellow River Delta," *International Journal of Remote Sensing*, vol. 37, no. 2, pp. 455–470, 2016.

[12] O. Rocha Neto, A. Teixeira, R. Leão, L. Moreira, and L. Galvão, "Hyperspectral remote sensing for detecting soil salinization using ProSpecTIR-VS aerial imagery and sensor simulation," *Remote Sensing*, vol. 9, no. 9, p. 42, 2017.

[13] X. Wang, F. Zhang, J. Ding, H.-t. Kung, A. Latif, and V. C. Johnson, "Estimation of soil salt content (SSC) in the Ebinur Lake Wetland National Nature Reserve (ELWNNR), Northwest China, based on a Bootstrap-BP neural network model and optimal spectral indices," *Science of the Total Environment*, vol. 615, no. 12, pp. 918–930, 2018.

[14] Z. E. Mashimbye, M. A. Cho, J. P. Nell, W. P. De Clercq, A. Van Niekerk, and D. P. Turner, "Model-based integrated methods for quantitative estimation of soil salinity from hyperspectral remote sensing data: a case study of selected South African soils," *Pedosphere*, vol. 22, no. 5, pp. 640–649, 2012.

[15] Z. Shi, Q. Wang, J. Peng et al., "Development of a national VNIR soil-spectral library for soil classification and prediction of organic matter concentrations," *Science China: Earth Sciences*, vol. 57, no. 7, pp. 1671–1680, 2014.

[16] P. J. Currie and X. J. Zhao, "A new carnosaur (Dinosauria, Theropoda) from the Jurassic of Xinjiang, people's Republic of China," *Canadian Journal of Earth Sciences*, vol. 30, no. 10, pp. 2037–2081, 1993.

[17] L. Wang, Z. Li, F. Wang, and R. Edwards, "Glacier shrinkage in the Ebinur lake basin, Tien Shan, China, during the past 40 years," *Journal of Glaciology*, vol. 60, no. 220, pp. 245–254, 2014.

[18] W. Dong and Z. H. Liu, "Comprehensive evaluation of water resources carrying capacity in Ebinur Lake Basin," *Arid Land Geography*, vol. 33, no. 2, pp. 217–223, 2010.

[19] J. Ding, W. Chen, and Y. Chen, "Soil salinization disaster warning in arid zones: a case study in the ugan-kuqa oasis," *Journal of Desert Research*, vol. 36, no. 4, pp. 1079–1086, 2016.

[20] M. Eziz, H. Yimit, A. Mohammad, and H. Zhifang, "Oasis land-use change and its effects on the oasis eco-environment in Keriya Oasis, China," *International Journal of Sustainable Development & World Ecology*, vol. 17, no. 3, pp. 244–252, 2010.

[21] R. Li, F. Zhang, Y. Gao et al., "Surface hydrochemistry characteristics and controlling factors in the Ebinur Lake region during dry and wet seasons," *Journal of Glaciology and Geocryology*, vol. 38, no. 5, pp. 1394–1403, 2016.

[22] H. Zhang, F. Zhang, and Z. Li, "Differentiation of soil water and salt characteristics under background of different environmental variation in Ebinur Lake region," *Arid Land Geography*, vol. 40, no. 3, pp. 606–613, 2017.

[23] S. Rukeya, K. Nijat, A. Abdugheni et al., "Possibility of optimized indices for the assessment of heavy metal contents in soil around an open pit coal mine area," *International Journal of Applied Earth Observation and Geoinformation*, vol. 73, pp. 14–25, 2018.

[24] S. Wold, "PLS for multivariate linear modeling," *Chemometric Methods in Molecular Design*, vol. 2, no. 1, pp. 195–218, 1995.

[25] D. M. Haaland and E. V. Thomas, "Partial least-squares methods for spectral analyses. 2. Application to simulated and glass spectral data," *Analytical Chemistry*, vol. 60, no. 11, pp. 1202–1208, 1988.

[26] A. Sidike, S. Zhao, and Y. Wen, "Estimating soil salinity in Pingluo County of China using QuickBird data and soil reflectance spectra," *International Journal of Applied Earth Observation and Geoinformation*, vol. 26, no. 2, pp. 156–175, 2014.

[27] S. Wold, M. Sjöström, and L. Eriksson, "PLS-regression: a basic tool of chemometrics," *Chemometrics and Intelligent Laboratory Systems*, vol. 58, no. 2, pp. 109–130, 2001.

[28] S. B. Karunaratne, T. F. A. Bishop, J. A. Baldock, and I. O. A. Odeh, "Catchment scale mapping of measureable soil organic carbon fractions," *Geoderma*, vol. 219-220, pp. 14–23, 2014.

[29] A Allbed, L. Kumar, and P. Sinha, "Mapping and modelling spatial variation in soil salinity in the Al Hassa oasis based on remote sensing indicators and regression techniques," *Remote Sensing*, vol. 6, no. 2, pp. 1137–1157, 2014.

[30] F. Li, B. Mistele, Y. Hu, X. Chen, and U. Schmidhalter, "Optimising three-band spectral indices to assess aerial N concentration, N uptake and aboveground biomass of winter wheat remotely in China and Germany," *ISPRS Journal of Photogrammetry and Remote Sensing*, vol. 92, no. 2, pp. 112–123, 2014.

7

Prototype of the Near-Infrared Spectroscopy Expert System for Particleboard Identification

Anna Sandak,[1,2] Jakub Sandak,[1,2,3] Dominika Janiszewska,[4] Salim Hiziroglu,[5] Marta Petrillo,[1] and Paolo Grossi[1]

[1]*CNR-IVALSA Trees and Timber Institute, 38010 San Michele all'Adige, Italy*
[2]*InnoRenew CoE, 6310 Izola, Slovenia*
[3]*University of Primorska, Faculty of Mathematics, Natural Sciences and Information Technology, 6000 Koper, Slovenia*
[4]*Composite Wood Products Department, Wood Technology Institute, Poznan, Poland*
[5]*Natural Resource Ecology and Management, Oklahoma State University, Stillwater, Oklahoma, USA*

Correspondence should be addressed to Jakub Sandak; jakub.sandak@innorenew.eu

Academic Editor: Jose S. Camara

The overall goal of this work was to develop a prototype expert system assisting quality control and traceability of particleboard panels on the production floor. Four different types of particleboards manufactured at the laboratory scale and in industrial plants were evaluated. The material differed in terms of panel type, composition, and adhesive system. NIR spectroscopy was employed as a pioneer tool for the development of a two-level expert system suitable for classification and traceability of investigated samples. A portable, commercially available NIR spectrometer was used for nondestructive measurements of particleboard panels. Twenty-five batches of particleboards, each containing at least three independent replicas, was used for the original system development and assessment of its performance. Four alternative chemometric methods (PLS-DA, kNN, SIMCA, and SVM) were used for spectroscopic data classification. The models were developed for panel recognition at two levels differing in terms of their generality. In the first stage, four among twenty-four tested combinations resulted in 100% correct classification. Discrimination precision with PLS-DA and SVMC was high (>99%), even without any spectra preprocessing. SNV preprocessed spectra and SVMC algorithm were used at the second stage for panel batch classification. Panels manufactured by two producers were 100% correctly classified, industrial panels produced by different manufacturing plants were classified with 98.9% success, and the experimental panels manufactured in the laboratory were classified with 63.7% success. Implementation of NIR spectroscopy for wood-based product traceability and quality control may have a great impact due to the high versatility of the production and wide range of particleboards utilization.

1. Introduction

Particleboard is a panel product manufactured from lignocellulosic materials, combined with an adhesive system and bonded together under heat and pressure. Particleboards are easy to process and flexible in application. They are mainly used for furniture production, but in combination with other materials, they might be used for parquet, insulation materials, sheathing boards for timber framed walls, packaging, or "do-it-yourself products." The major types of particles used to manufacture particleboard include wood shavings, flakes, wafers, chips, and sawdust [1].

A new strategy of EU aims at increasing the share of bioenergy in the EU's total energy production. It is expected that 20% of the energy produced in Europe after 2020 will be from renewable resources, where 80% of the above will be related to lignocellulosic feedstock. Simultaneously, a significant increase in the production volumes for other wood-based products (pulp, boards, and furniture) is expected. Therefore, even if supply and demand of wood on the market is balanced at present, it will most probably not be in equilibrium within the coming years [2]. As a consequence, several alternative resources, mainly agricultural and industrial residues, fast-growing shrubs and plantations, and postconsumer wood, have received considerable attention in recent years [3]. Effective use of bagasse [4], oil palm waste [5], bark [6] paper sludge, [3], kenaf stalks [7] wheat straw [8], needle litter [9], vine pruning's [10] waste tissue paper, and corn peel [11] among others was previously reported. Such materials are ecological, functional, and environmental-friendly; therefore, they serve as sustainable raw materials for manufacturing particleboards. The type of raw resources used for panels manufacturing, its quality, size of particles, moisture content as well as type and amount of bonding system have significant effect on particleboard properties [12]. The most important quality assessment aspects of manufactured panels are emissions, which mainly depends on the type and amount of resin [1]. The recent trend to reduce the formaldehyde release from manufactured wood products has led to the substitution of urea-formaldehyde (UF) resin with several alternatives [12]. Liquefied wood was previously reported as an interesting replacement for commonly used adhesives [13–15]. The influence of liquefied wood on particleboard properties was previously investigated by the authors [16, 17]. However, alternative quality control tools with potential application for online process control are still desired.

Portable spectroscopic equipment operating in the NIR range is a highly interesting technology for the wood-based sector [18]. Meder et al. [19] reported successful implementation of FT-NIR spectroscopy for at-line measurement for quality control of melamine-urea-formaldehyde resin in composite wood-panel production. Taylor and Via [20] used visible and near-infrared spectroscopy to quantify phenol formaldehyde resin content in oriented strandboard. Campos et al. [21] used FT-NIR to evaluate composition of agro-based particleboards. Janiszewska et al. [22] reported the application of FT-NIR spectroscopy for clustering raw materials used for liquefaction and their transformation products. Even though several examples of NIR use for product and process quality monitoring have been reported, the majority of them rely on FT-NIR instruments that are relatively costly (both time and investment wise) and with limited applicability for inline measurement.

Numerous chemometric techniques for multivariate classification and discrimination were reported as suitable for processing of NIR spectra. These are usually divided into discriminant or class-modelling methods. In discriminant analysis, an unknown sample is always assigned and can be allocated only to one of the classes given in the training set. The class-modelling approach is more flexible. The sample can be accepted by more than one class model and therefore be recognized as confused. The most common classification methods are linear discriminant analysis (LDA), partial least-squares discriminant analysis (PLS-DA), k-nearest neighbours (kNN), SIMCA, and SVM [23].

LDA is a probabilistic method, which assumes that each sample belonging to a particular class follows a multivariate Gaussian distribution. It requires the explicit calculation of this probability for the formulation of the classification rule. The main disadvantage of this method is the fact that it requires a significantly higher number of training samples than the number of variables, the variables themselves cannot be correlated, and all categories have the same within-class scatter, which is hard to assure in many experimental cases [23].

PLS-DA is a multivariate inverse least-squares analysis method used to classify samples. It decomposes the spectra as linear combinations of principal components (PCs) that express the majority of the information (variability) contained in the global dataset. The predictor variables or latent variables (LVs) are generated from the input variables to maximize the variance between sample classes in the model [24].

KNN is nonparametric and instance-based algorithm. This means that it does not make any assumption on the underlying data distribution and does not use the training data to do generalization. The kNN makes decision based on the entire training data set. The object is classified by a majority vote of neighbours. According to Adeniyi et al. [25], the main advantages of this method are capability of handling training data that are too large to fit in memory, use of simple Euclidean distance to measure the similarities between training and test data, providing a faster and more accurate recommendation as a result of straightforward application.

Soft independent modelling of class analogy (SIMCA) is the most common supervised modelling method, representing class-modelling approach. It requires a training data set of samples with a set of attributes and their class membership. In SIMCA, a principal component analysis (PCA) is performed on each class in the data set; thus, a principal component model is used to represent each class. In SIMCA, there is no restriction on the number of measurement variables, and few samples per class are enough to run the model.

Data classification is a common task in machine learning. The support vector machine (SVM) is a very flexible method that makes no assumption regarding data. It is a nonlinear classification method that constructs a set of hyperplanes in a high- or infinite-dimensional space. Good separation is achieved by the hyperplane that has the largest distance to the nearest training data point of any class [24]. It works by obtaining the optimal boundary of two groups in a vector space independent of the probabilistic arrangements of vectors in training set. When the linear boundary in low-dimension input space is not enough to separate two classes, SVM can create a hyperplane that allows linear separation in the higher dimension feature space [26]. SVM has been successfully used for data mining, pattern recognition, and artificial intelligence fields [18].

Practical implementation of discrimination methods in the wood-based industry is a challenging task. Wood is a heterogeneous and anisotropic biological material. The correct implementation of NIR spectroscopy for bio-based materials requires a complex approach [27]. A two-level expert system is proposed here for identification of particleboard panels with a portable, commercially available NIR spectrometer. The overall goal of this work was to develop a prototype system that might assist quality control and traceability of particleboard panels on the production floor.

2. Materials and Methods

Four different types of particleboards manufactured in the laboratory and in industrial plants were evaluated. The materials differed in terms of panel type (single-layer or three-layer), composition (industrial particles, recycled wood, or alternative lignocellulosic materials including fast-growing species), and the adhesive system (urea-formaldehyde (UF) resin, UF resin modified with liquefied wood (LW), or UF resin modified with starch). A summary of investigated panels is presented in Table 1. In total, 170 samples representing 25 variants were investigated. A colour image of representative samples surface for each variant is presented in Figure 1.

2.1. Particleboards Manufactured with Recycled Wood (D). Industrial wood particles obtained from a local wood-processing sawmill, containing 25% recycled wood content, were used as a raw material for particleboard production. The particles were sorted using an Allgaier vibration screening machine with screens of mesh diameters 8, 2, 1, and 0.5 mm. Particle fractions ≤8 mm and ≥1 mm were selected for particleboard production.

The adhesive was prepared as a mixture of an industrial urea-formaldehyde resin and liquefied wood (10–20% relative to the dry weight of the resin). The liquefied wood was prepared from four types of wood-processing industry wastes: mixed hardwood-softwood powder (LWP), pine (LP), beech sawdust (LB), and bark (LB) [16].

The industrial urea-formaldehyde glue resin characteristics were as follows: gel time of 75 seconds, viscosity 336 mPa·s, total solids content 69.4%, and pH 7.3. Urea-ammonium nitrate solution (46%) was used as a curing agent, constituting 1% of the resin dry mass.

Single-layered particleboards of 12 mm thickness with 0, 10, 15, and 20% liquefied wood content in the adhesive resin were produced in the laboratory. The nominal density of the panels was 650 kg·m^{-3}. All boards were conditioned after pressing at 20°C and 65% relative humidity. Nine types of single-layer panels were prepared with different adhesives system, as listed in Table 1. Each panel type was produced in two independent batches. Six replicates for each panel were analysed; in total, 108 samples were investigated.

2.2. Particleboards Manufactured with Fast-Growing Wood Species (S). Single- and three-layer particleboard panels were manufactured from Eastern red cedar (*Juniperus virginiana* L.) using 9% urea-formaldehyde, or a combination of 15% modified corn starch and 2% urea-formaldehyde adhesive. Three types of samples were prepared: type A: single-layer board with 9% UF; type B: three-layer board with 9% UF; and type C: three-layer board with a combination of 15% starch and 2% UF. All boards were produced in a laboratory press using a pressure of 5 MPa, at a temperature of 165°C for 5 and 10 minutes in the case of modified corn starch bonded samples [28]. Six independent replicates for each panel were analysed with an NIR spectrometer.

2.3. Particleboards Manufactured with Alternative to Wood Lignocellulosic Plants (G). Three-layer particleboards with the core made from different biomasses were produced, including Black locust (*Robinia pseudoacacia* L.), miscant (*Miscanthus sinensis giganteus*), willow (*Salix viminalis*), and rapeseed (*Brassica napus*). The stalks of lignocellulosic materials were reduced in size with a Pallmann's chipper. Fractions smaller than 10 mm and bigger than 1 mm were used for the core of the panels. The chips were used for manufacturing four types of three-layer boards of 16 mm thickness, with the raw density of 680 kg·m^{-3} [29]. Six independent replicates for each panel were analysed with an NIR spectrometer.

2.4. Industrially Manufactured Particleboards (P). Three-layer urea-formaldehyde resin-bonded particleboards, type P2, suitable for non-load-bearing purposes in dry areas, manufactured by six diverse manufacturing plants of a corporation were used as reference industrial samples. In addition, panels with different thicknesses (38 mm, 28 mm, 18 mm, and 8 mm) prepared by a single producer were investigated. Three replicates for each panel type were analysed using NIR spectroscopy.

2.5. Spectroscopic Measurement and Data Mining. The MicroNIR 1700 compact sensor produced by Viavi Solutions (Santa Monica, CA, USA) was used for spectroscopic measurements. Each spectrum was measured as an average of 10 consecutive scans. The scanning frequency was 50 Hz, corresponding to 20 ms of integration time. The spectral range was from 950 to 1650 nm (10526–6060 cm^{-1}). 128 spectral points were defined for each spectrum and corresponded to pixels of the CCD detector of the instrument. Ten independent measurements were done on each panel assuring measurements at different locations over its surface. Each set of 10 spectra was preprocessed with extended multiplicative scatter correction (EMSC) and after that averaged to homogenize the spectral fingerprint of heterogeneous surfaces of panel. The MicroNIR instrument has the proven potential for in-field and inline applications due to its rigid construction and integration of all optical, electronic, and mechanical components.

PLS_Toolbox 8.0 (Eigenvector Research, Manson, WA, USA) and LabView 13 (National Instruments, Austin TX, USA) were used for data processing and mining. Four discriminant analysis methods: PLS-DA, SVM, KNN, and SIMCA, were used for spectra classification. Models were calculated considering use of the raw spectra and spectra preprocessed with normalization, standard normal variate (SNV), EMSC, and 1st and 2nd derivatives (Savitzky–Golay algorithm, 2nd polynomial order, and 15 smoothing points). The order of data was randomized and then divided into calibration (66%) and independent validation (34%) data sets. The samples used for model validation were all different from those used for calibration.

TABLE 1: Manufactured panels investigated in this research.

Source	Sample code	Manufacturing	Panel type	Raw material	Adhesive system
Project LIDER	D1	Laboratory	Single-layer	Industrial particles containing 25% of recycled wood	0% LW + 100% UF1
	D2	Laboratory	Single-layer	Industrial particles containing 25% of recycled wood	10% LWP +90% UF1
	D3	Laboratory	Single-layer	Industrial particles containing 25% of recycled wood	15% LWP + 85% UF1
	D4	Laboratory	Single-layer	Industrial particles containing 25% of recycled wood	20% LWP + 80% UF1
	D5	Laboratory	Single-layer	Industrial particles containing 25% of recycled wood	20% LP + 80% UF1
	D6	Laboratory	Single-layer	Industrial particles containing 25% of recycled wood	20% LB + 80% UF1
	D7	Laboratory	Single-layer	Industrial particles containing 25% of recycled wood	20% LBK + 80% UF1
	D8	Laboratory	Single-layer	Industrial particles containing 25% of recycled wood	0% LW + 100% UF2
	D9	Laboratory	Single-layer	Industrial particles containing 25% of recycled wood	10% LWP + 90% UF2
OSU	S1	Laboratory	Single-layer	100% red cedar particles	9% UF
	S2	Laboratory	Three-layer	100% red cedar particles	9% UF
	S3	Laboratory	Three-layer	100% red cedar particles	15% starch and 2% UF
ITD	G1	Laboratory	Three-layer	Industrial pine particles (75%) + black locust particles (25%)	Core 8% UF, face layers 12%
	G2	Laboratory	Three-layer	Industrial pine particles (75%) + *Miscanthus* particles (25%)	Core 8% UF, face layers 12%
	G3	Laboratory	Three-layer	Industrial pine particles (75%) + willow particles (25%)	Core 8% UF, face layers 12%
	G4	Laboratory	Three-layer	Industrial pine particles (75%) + rapeseed particles (25%)	Core 8% UF, face layers 12%
Chipboard panel producer	P1	Industry	Three-layer	100% industrial particles	UF
	P2	Industry	Three-layer	100% industrial particles	UF
	P3	Industry	Three-layer	100% industrial particles	UF
	P4	Industry	Three-layer	100% industrial particles	UF
	P5	Industry	Three-layer	100% industrial particles	UF
	P6a	Industry	Three-layer	100% industrial particles	UF
	P6b	Industry	Three-layer	100% industrial particles	UF
	P6c	Industry	Three-layer	100% industrial particles	UF
	P6d	Industry	Three-layer	100% industrial particles	UF

3. Results and Discussion

3.1. General Concept of the Expert System Architecture. Expert systems are innovative process tools that enhance the user's productivity in accomplishing a task or solving a problem. These vary in at least two important dimensions, including knowledge and technological complexities [30]. Knowledge complexity is mainly determined by the degree of depth and specialization of the internalized knowledge, the scope of the decision, and the level of expertise required to solve the problem. The technological complexity includes diversity of hardware and software, the complexity of the user's environment, the scale of the software design effort, the complexity of required database accesses, and special user interfaces among others [30].

The main goal of this work was to develop an expert system capable of detecting and tracing different particleboards. The general concept of such a system implemented in the furniture factory is presented in Figure 2. The system relies on the spectroscopic measurements acquired during arrival of the new batch to the factory floor with the easy-to-handle and portable instrument. The particular characteristics of the instrument investigated within this project allow its direct implementation in the real-life applications. It is recommended to average several NIR spectra collected from the characterized sample in order to minimize the heterogeneity effect of the complex surfaces of particleboards. A spectrum acquired in that way is considered as a fingerprint of the distinct batch and is recorded in the database for further analysis. The usage of spectra is twofold. Firstly, it is applied for the further improvement of the chemometric models by providing new case data. Secondly, it is used for future tracking and identification of elements during production as well as in final products.

The NIR spectrum of an unknown sample is confronted at the first level of the expert system with broad classes representing different particleboard manufacturers or panel types.

Figure 1: Appearance of particleboard surfaces representing 25 investigated categories.

In the case of successful identification, the same spectrum is an input for the second-level chemometric model, where affinity to the specific batches recorded previously is assessed. Such a numerical tool might be highly useful for the quality control of panels arriving to the production floor, as well as for authentication of their origin and composition. The demonstrative implementation of the proposed expert system was performed within this research project. The statistical evaluation of the discriminations success rate, performed at both system levels, is presented in the following section.

3.2. Prototype Expert System for Particleboard Identification. The concept of the abovementioned expert system implemented for demonstration of its feasibility is presented in Figure 3. It consists of five chemometric models implemented at two discrimination levels. Model #1 is used to screen the unknown spectrum and determine the most probable class corresponding to the particleboard type. Four classes were therefore defined at the first level of the expert system corresponding to "sample sources" as summarized in the first column of Table 1. The second and more specific discrimination is executed afterward by implementing one of the second-level models (#2 to #5). The number of classes in each of these models corresponds to the number of previously defined batches. It has to be mentioned that it is critical to continuously feed the expert system with the most recent spectra of new batches in order to assure constant improvement of the discrimination

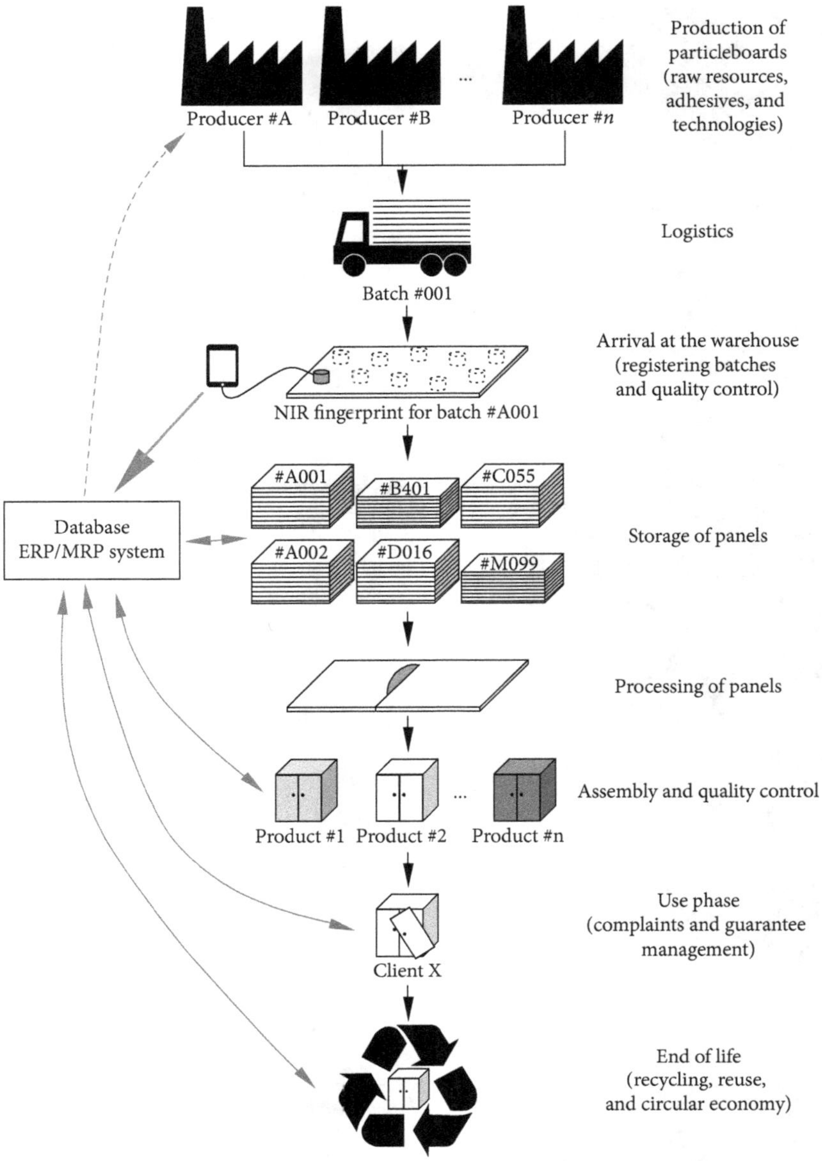

Figure 2: Concept for the traceability of particleboards in the furniture industry.

system and proper identification of all historically used particleboard batches.

An expert system approach to quality control at the industrial scale was proposed by Paladini [31]. He described three-step quality control approach that included precontrol activities, inspection, and a decision stage. Liukkonen et al. [32] described the intelligent optimization and modelling system for electronics production. They proposed three modules consisting of appropriate mathematical tools specifically tailored to each task: preprocessing, variable selection, and optimization. A data-driven approach was proposed to achieve proactive quality improvement of the production process. Hobballah et al. [33] proposed a casual map to show what variables should be included in the design optimization and how the components interact causally. This approach was successfully used for the preliminary design of an insulating composite mat based on wood fibers.

3.3. Discriminant Analysis. Four alternative classification methods: KNN, PLS-DA, SVM, and SIMCA, were tested for the optimal NIR spectral data classification. All these methods are supervised techniques, as they use predefined information about the class membership for all samples selected for model calibration. In that way, the model can be tuned to classify new unknown samples in one of the known classes on the basis of its individual pattern [26]. Discriminating techniques usually build models based on all the categories concerned for the discrimination. Therefore, samples can be classified into one of the predefined categories, even if actually they do not belong to

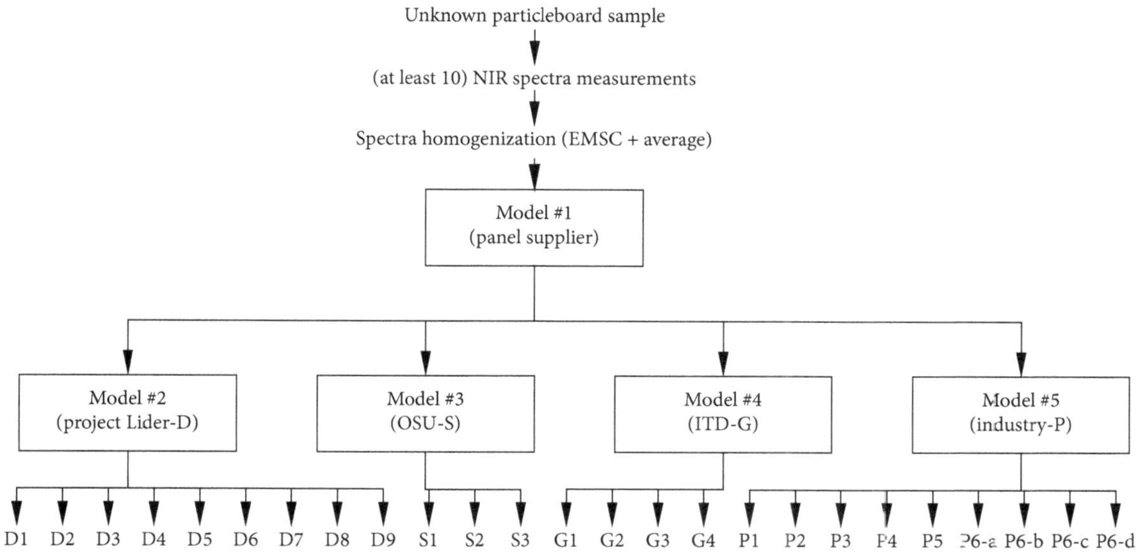

FIGURE 3: Prototype expert system's layout for the experimental set of particleboard samples.

any of these [26]. It is possible, however, to define a certain threshold for affinity to any of the classes. In that case, samples not passing the threshold may be classified as undefined.

Particleboard panel discrimination at the producer level (model #1) differentiated classes in terms of manufacturing conditions (laboratory or industrial scale), different resources used for panels manufacturing (industrial particles, addition of recycled wood, fast-growing wood species, agricultural wastes, etc.), or different adhesives used for panel production. The discrimination success scores between different panel providers are summarized in Table 2, where selected spectra preprocessing and diverse discrimination algorithms are presented. Four among twenty-four tested combinations resulted in 100% correct classification. The optimal combinations were PLS-DA and SIMCA with 2nd derivative spectra, and SVM with both SNV and EMSC preprocessing. Furthermore, discrimination precision with PLS-DA and SVM was high (>99%), even without any spectra preprocessing.

The second level of the expert system was designed to identify specific batches. Twenty-five lots of particleboards were modelled by four independent chemometric models (#2 to #5), developed for each panel provider separately. The SVM algorithm classifying SNV preprocessed NIR spectra was implemented. The summary of the model's performance is presented in Figure 4, where confusion tables indicate the ability of each model to properly classify (or misclassify) experimented samples from the validation set. All samples of panels manufactured by producer S and G were correctly classified. Industrial panels produced by different manufacturing plants of the producer P were classified with 98.9% success and the panels manufactured by producer D with 63.7% success. It is important to mention that, in the case of the producer D, the same substrate (industrial particles containing 25% of recycled wood) was used for manufacturing all investigated panels. The only difference was the adhesive system, with 0, 10, 15, and 20% share of liquefied wood addition in the UF resin.

3.4. Implementation of the Prototype Expert System. The implementation of the expert system in real conditions is a challenging task. Qian et al. [34] proposed a four-step process that consists of the following:

(i) Knowledge representation (including all variables characterizing production process)

(ii) Database development (including all decision rules, functions, and metadata)

(iii) Machine interface design (influencing the quality of the expert system for fault diagnosis and the real-time response of the system)

(iv) Knowledge maintenance (including update, verification, and correction of the errors in knowledge base)

The experiences gained when developing chemometric models discriminating diverse particleboards stimulated extension of this research to a prototype expert system ready for testing in real-world applications. The hardware used for the prototype (portable NIR spectrometer and computer) was similar to the laboratory tests. The differences were in the software, where the original code was customised. It included integration of tools for the control of the sensor setup, data acquisition, spectra processing, and discriminant analysis. The required chemometric models (#1 to #5) were computed offline with PLS_toolbox and were exported as Matlab scripts to LabView. The algorithm of the prototype expert system for the batch identification is presented in Figure 5. The software allows acquisition of the new NIR

Table 2: The results of the discrimination success rate at the first level of expert system.

Discrimination algorithm	Spectra preprocessing					
	No preprocessing	Normalization	SNV	EMSC	1st derivative	2nd derivative
PLS-DA	99.1	99.1	98.2	98.2	98.2	100
SVM	99.1	63.2	100	100	93.9	63.2
KNN [3]	93.0	96.5	97.4	96.5	97.4	98.2
SIMCA	94.7	99.1	99.1	98.2	99.1	100

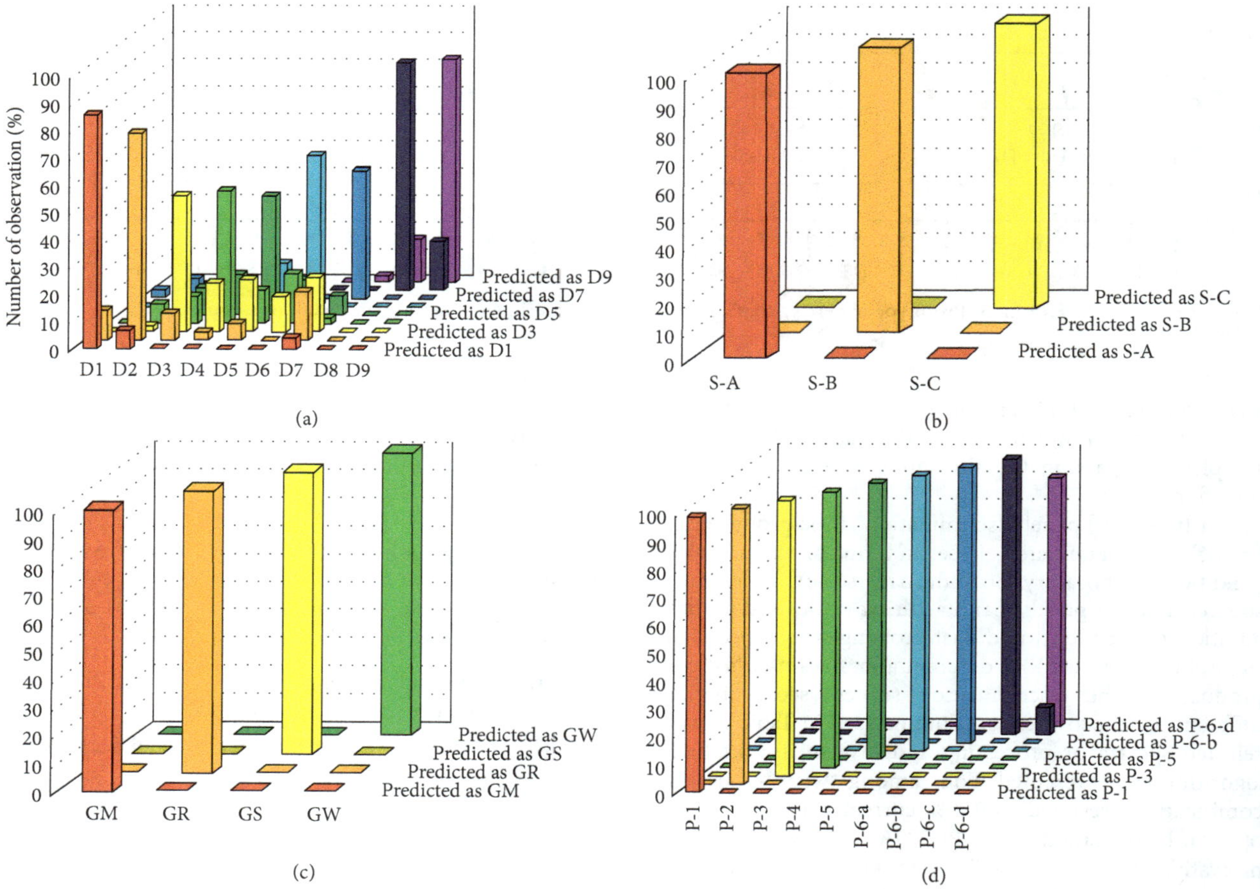

Figure 4: Confusion table of the particleboard batch prediction with the prototype NIR expert system: project LIDER-D (a), OSU-S (b), ITD-G (c), and industry-P (d).

spectra for feeding the database as well as assessment of the unknown spectrum in one of the predefined classes.

4. Conclusions

The traceability of particleboards is an important practical issue affecting economic performance of furniture producers and related industries. It allows control of material quality and ensures high standards of final products. NIR spectroscopy was employed here as a pioneer tool for development of the expert system suitable for assisting such a traceability. Twenty-five batches of particleboards, each containing at least three independent replicas, were used for the original system development and for its performance assessment. The hardware and software developed was working properly, and it was implemented as a ready-to-use prototype system.

A two-stage expert system seems to be optimal for implementation of the particleboard discrimination as it allows a rough (in the first stage) and detailed (in the second stage) classification of particleboard samples to the level of a single batch. In two cases (two producers), 100% correct classification was achieved at the second stage.

It has to be mentioned that a drawback of the system presented is the necessity to condition samples before measurement. It includes both refreshing of the surface by gentle sanding and conditioning of samples at well-defined climatic conditions. It is due to the fact that NIR spectroscopy is extremely sensitive to the water molecules presented in the measured samples. Nevertheless,

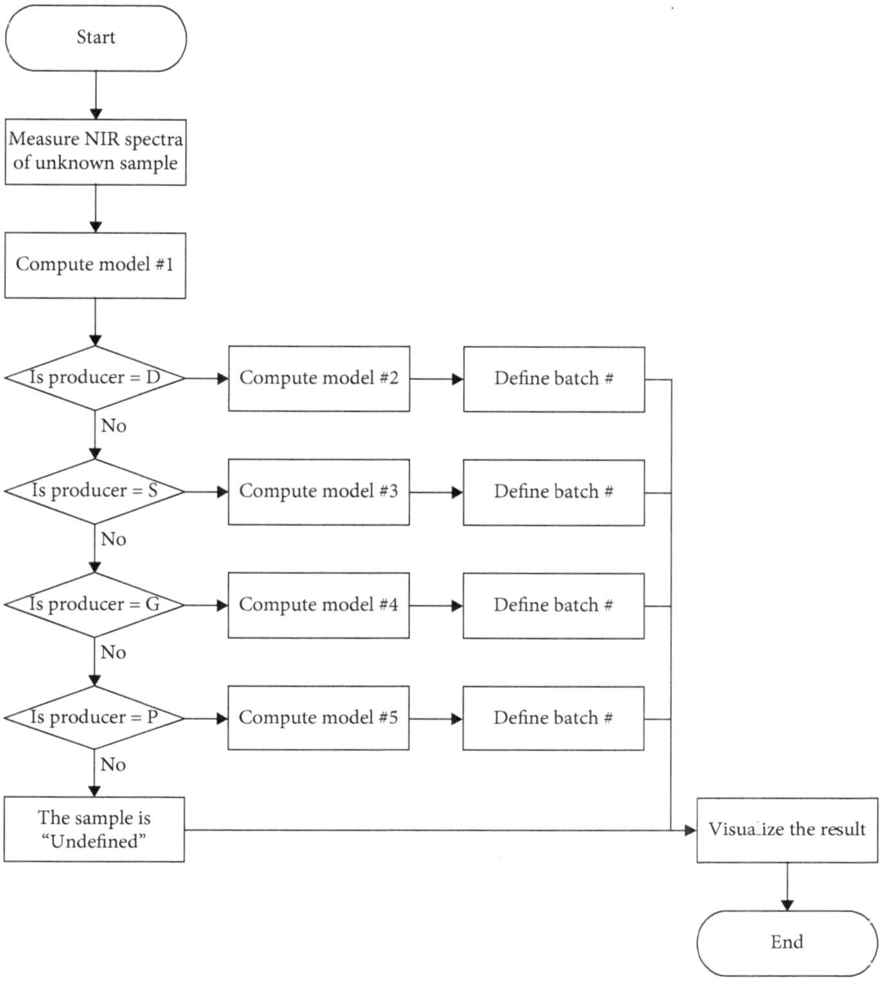

FIGURE 5: The algorithm of the prototype expert system.

implementation of NIR spectroscopy for product traceability and quality control may impact the industry due to the high versatility of the production and wide range of particleboard use.

Conflicts of Interest

The authors declare that there are no conflicts of interest regarding the publication of this paper.

Acknowledgments

The authors are grateful to Dorota Fuczek, Grzegorz Kowaluk, and Ekkehard Brommer for providing part of experimental samples. Special acknowledgments are due to COST Action FP1306 (LIGNOVAL) for funding an STSM that contributed to this work. Part of the work was conducted during the BIO4ever (RBSI14Y7Y4) project funded within SIR call by the Italian Ministry of Education, University and Research (MIUR). The authors gratefully acknowledge the European Commission for funding the InnoRenew CoE project (Grant Agreement #739574) under the Horizon2020 Widespread-Teaming program. The research was carried out within the statutory project "Application of liquefied wood in wood-based materials technology, Part 3. The use of liquefied lignocellulosic wastes as a binder for particleboards" (ST-2-BMD/2015/K) and project LIDER/14/0174/L-7/15/NCBR/2016 "New biopolymer adhesives modified with silanes and ionic liquids for application in wood-based materials technology" funded by the NCBR under the LIDER VII programme.

References

[1] Compilation of Air Pollutant Emissions Factors, Chapter 10: Wood Products Industry Wood Products Industry, 2002, https://www3.epa.gov/ttn/chief/ap42/ch10/index.html, 2018.

[2] The Bioenergy System Planners Handbook—BISY PLAN, http://bisyplan.bioenarea.eu/, 2018.

[3] A. Taramian, K. Dooshoseini, S. A. Mirshokraii, and M. Faezipour, "Particleboard manufacturing: an innovative way to recycle paper sludge," *Waste Management*, vol. 27, no. 12, pp. 1739–1746, 2007.

[4] J. Fiorelli, R. G. Galo, S. L. Castro Jr., U. L. Belini, P. R. O. Lasso, and H. Savastano Jr., "Multilayer particleboard produced with agroindustrial waste and amazonian vegetable fibres," *Waste and Biomass Valorization*, vol. 9, no. 7, pp. 1151–1161, 2017.

[5] L. S. Hua, Z. Ashaari, L. W. Chen et al., "Properties of particleboard with oil palm trunk as core layer in comparison to three-layer rubber wood particleboard," *Journal of Oil Palm Research*, vol. 27, no. 1, pp. 67–74, 2015.

[6] M. C. N. Yemele, P. Blanchet, A. Cloutier, and A. Koubaa, "Effects of bark content and particle geometry on the physical and mechanical properties of particleboard made from black spruce and trembling aspen bark," *Forest Product Journal*, vol. 58, no. 11, pp. 48–56, 2008.

[7] H. Kalaycioglu and G. Nemli, "Producing composite particleboard from kenaf (*Hibiscus cannabinus* L.) stalks," *Industrial Crops and Products*, vol. 24, no. 2, pp. 177–180, 2006.

[8] X. Mo, E. Cheng, D. Wang, and X. S. Sun, "Physical properties of medium-density wheat straw particleboard using different adhesives," *Industrial Crops and Products*, vol. 18, no. 1, pp. 47–53, 2003.

[9] G. Nemli, S. Yildiz, and E. D. Gezer, "The potential for using the needle litter of Scotch pine (*Pinus sylvestris* L.) as a raw material for particleboard manufacturing," *Bioresource Technology*, vol. 99, no. 14, pp. 6054–6058, 2008.

[10] G. A. Ntalos and A. H. Grigoriou, "Characterization and utilisation of vine prunings as a wood substitute for particleboard production," *Industrial Crops and Products*, vol. 16, no. 1, pp. 59–68, 2002.

[11] P. Lertsutthiwong, S. Khunthon, K. Siralertmukul, K. Noomun, and S. Chandrkrachang, "New insulating particleboards prepared from mixture of solid wastes from tissues paper manufacturing and corn peel," *Bioresource Technology*, vol. 99, no. 11, pp. 4841–4845, 2008.

[12] N. Čuk, M. Kunaver, and S. Medved, "Properties of particleboards made by using an adhesive with added liquefied wood," *Materials and technology*, vol. 45, no. 3, pp. 241–245, 2011.

[13] B. Esteves, J. Martins, J. Martins, L. Cruz-Lopes, J. Vicente, and I. Domingos, "Liquefied wood as a partial substitute of melamine-urea-formaldehyde and urea-formaldehyde resins," *Maderas. Ciencia y tecnología*, vol. 17, no. 2, pp. 277–284, 2015.

[14] A. Ugovsek, M. Kariz, and M. Sernek, "Bonding of wood with adhesive mixtures made of liquefied wood combined with tannin or phenolic resin," in *Future with Forests*, R. Ristić, M. Medarević, and Z. Popović, Eds., First Serbian Forestry congress, Beograd, Serbia, 2010.

[15] M. Kunaver, S. Medved, N. Čuk, E. Jasiukaitytė, I. Poljanšek, and T. Strnad, "Application of liquefied wood as a new particle board adhesive system," *Bioresource Technology*, vol. 101, no. 4, pp. 1361–1368, 2010.

[16] D. Janiszewska, I. Frąckowiak, and K. Mytko, "Exploitation of liquefied wood waste for binding recycled wood particleboards," *Holzforschung*, vol. 70, no. 12, pp. 1135–1138, 2016.

[17] D. Janiszewska, I. Frąckowiak, and N. Bielejewska, "Application of selected agents for wood liquefaction and some properties of particleboards produced with the use of liquefied wood," *Drewno*, vol. 59, no. 197, pp. 223–230, 2016.

[18] J. Sandak, A. Sandak, and R Meder, "Assessing trees, wood and derived products with NIR spectroscopy: hints and tips," *Journal of Near Infrared Spectroscopy*, vol. 24, no. 6, pp. 485–505, 2016.

[19] R. Meder, W. Stahl, P. Warburton et al., "At-line validation of a process analytical technology approach for quality control of melamine-urea-formaldehyde resin in composite wood-panel production using near infrared spectroscopy," *Analytical and Bioanalytical Chemistry*, vol. 409, no. 3, pp. 763–771, 2017.

[20] A. Taylor and B. K. Via, "Potential of visible and near infrared spectroscopy to quantify phenol formaldehyde resin content in oriented strandboard," *European Journal of Wood and Wood Products*, vol. 67, no. 1, pp. 3–5, 2009.

[21] A. C. M. Campos, P. R. G. Hein, R. F. Mendes, L. M. Mendes, and G. Chaix, "Near infrared spectroscopy to evaluate composition of agro-based particleboards," *Bioresources*, vol. 4, no. 3, pp. 1058–1069, 2009.

[22] D. Janiszewska, A. Sandak, J. Sandak, I. Frąckowiak, and K. Mytko, "Influence of the raw material properties on the liquefied wood chemical composition," *nnals of WULS, Forestry and Wood Technology*, vol. 94, pp. 298–303, 2016.

[23] M. Bevilacqua, R. Nescatelli, R. Bucci, A. D. Magrì, A. L. Magrì, and F. Marini, "Chemometric classification techniques as a tool for solving problems in analytical chemistry," *Journal of AOAC International*, vol. 97, no. 1, pp. 19–28, 2014.

[24] Y. Tian, Z. Wang, X. Han, H. Hou, and R. Zheng, "Comparative investigation of partial least squares discriminant analysis and support vector machines for geological cuttings identification using laser-induced breakdown spectroscopy," *Spectrochimica Acta Part B: Atomic Spectroscopy*, vol. 102, pp. 52–57, 2014.

[25] D. A. Adeniyi, Z. Wei, and Y. Yongquan, "Automated web usage data mining and recommendation system using K-nearest neighbor (KNN) classification method," *Applied Computing and Informatics*, vol. 12, no. 1, pp. 90–108, 2016.

[26] L. A. Berrueta, R. M. Alonso-Salces, and K. Héberger, "Supervised pattern recognition in food analysis," *Journal of Chromatography A*, vol. 1158, no. 1-2, pp. 196–214, 2007.

[27] J. Liu, T. J. Pan, and Z. Y. Zhang, "Incremental support vector machine combined with ultraviolet-visible spectroscopy for rapid discriminant analysis of red wine," *Journal of Spectroscopy*, vol. 2018, Article ID 4230681, 5 pages, 2018.

[28] J. Sandak, A. Sandak, S. Hiziroglu, and S. Bosak, "Surface characterization of particleboard panels manufactured from eastern redcedar using multi-sensor approach," in *Proceedings of International Panel Products Symposium*, pp. 27–37, Llandudno, North Wales, UK, October 2015.

[29] I. Frąckowiak, D. Fuczek, and G. Kowaluk, "Impact of different lignocellulosic materials used in core of particleboard on modulus of elasticity and bending strength," *Drewno*, vol. 51, no. 180, pp. 5–14, 2008.

[30] J. H. Bradley and R. D. Hauser, "A framework for expert system implementation," *Expert Systems with Applications*, vol. 8, no. 1, pp. 157–167, 1995.

[31] E. P. Paladini, "An expert system approach to quality control," *Expert Systems with Applications*, vol. 18, no. 2, pp. 133–151, 2000.

[32] M. Liukkonen, E. Havia, H. Leinonen, and Y. Hiltunen, "Expert system for analysis of quality in production of electronics," *Expert Systems with Applications*, vol. 38, no. 7, pp. 8724–8729, 2011.

8

Comparison of Methods for Estimating Mechanical Properties of Wood by NIR Spectroscopy

L. R. Schimleck,[1] J. L. M. Matos,[2] R. Trianoski,[2] and J. G. Prata[2]

[1]*Department of Wood Science and Engineering, Oregon State University, Corvallis, OR 97331, USA*
[2]*Department of Wood Science and Technology, Federal University of Paraná, Curitiba, PR, Brazil*

Correspondence should be addressed to L. R. Schimleck; laurence.schimleck@oregonstate.edu

Academic Editor: Davidson Sajan

Near-infrared (NIR) spectroscopy provides a rapid alternative to traditional methods of wood property assessment. For organizations who assess wood properties on a large scale, multisite, multispecies calibrations are of practical interest. We examined NIR spectroscopy for the estimation of density (at 12% moisture content), modulus of elasticity (MOE), and modulus of rupture (MOR) using clear wood samples obtained from several pine species (*Pinus caribaea* var. *bahamensis*, var. *hondurensis*, and var. *caribaea*, *P. chiapensis*, *P. maximinoi*, *P. oocarpa*, *P. taeda*, and *P. tecunumanii*). We compared different methodologies for collecting spectra, that is, benchtop instrument versus benchtop fiber-optic probe and field portable fiber-optic probe, and different wood surfaces (radial and transverse). Calibrations based on the benchtop instrument were superior to those obtained using the fiber-optic probe systems. Difficulty with adequately representing the sample when collecting spectra using a fiber-optic probe and lower quality spectra explain the differences among the data sets. Spectra collected from radial and transverse surfaces provided similar calibration statistics. The calibrations obtained for density ($R^2 = 0.81$, SECV = 38.5 kg/m^3) and MOE ($R^2 = 0.81$, SECV = 1124 GPa) using benchtop instrument spectra demonstrate that it is possible to obtain general calibrations for estimating the wood properties of a number of tropical, subtropical, and temperate pine species.

1. Introduction

For organizations who assess wood properties on a large scale, the cost of traditional methods of analysis limits the number of samples that can be analyzed. The practical implications of such limitations are far reaching in forestry and wood product-related industries. An understanding of wood variability for a given species, that is, within trees, and within and among stands, and among different species is severely limited as is our understanding of how different environments affect wood properties and quality. Similarly, the estimation of genetic parameters for wood properties, critical to the success of tree improvement programs, is limited by the cost and time involved in determining such properties in large numbers [1, 2].

Efforts to lower the cost and time involved in wood property analysis have seen the development of rapid, nondestructive assessment methods. Many studies have explored near-infrared (NIR) spectroscopy as an alternative approach to traditional methods of wood property analysis [3, 4]. Reports can generally be divided into two groups, that is, those related to wood chemistry, for example, cellulose or lignin content and pulp yield, and those related to physical-mechanical properties such as density, modulus of elasticity (MOE), and modulus of rupture (MOR).

In terms of understanding the performance of lumber when used in construction, density (mass divided by volume at a given moisture content), MOE (a measure of the elastic deformation that wood undergoes when subject to an applied load), and MOR (a measure of the breaking strength of wood when subject to an applied load) are critical. Current measurement standards require destructive sampling and careful sample preparation and testing. The samples used for destructive testing, referred to as short, clear samples or static bending samples, have been utilized in a range of NIR-based research over the past twenty years. In these studies, many different NIR instruments have been utilized; however, approaches to spectral acquisition differ in two key aspects.

The first is the spectroscopic method utilized to collect NIR spectra; either a fiber-optic probe (referred to in the following text as "Probe") was used or the surface of the wood sample was placed directly over the NIR beam ("Direct"). The second is the wood surface used for analysis. Owing to the structure of wood, a straight grain, clear wood sample has three distinct surfaces (the radial, tangential, and transverse or cross section) (Figure 1) available for testing. In addition to these differences, test sample size varied as did the size of the area analyzed by NIR. The area analyzed varied both in terms of the number of locations measured for a given piece of wood and the size of the NIR beam, as determined in the Direct approach by the aperture of the NIR instrument or a window placed over the beam and in the Probe method by the diameter of the fiber-optic probe. Table 1 provides a summary of these reports and includes species examined, wood surface utilized, size of the short, clear sample tested, method for collecting NIR spectra, number of spectra collected, and size of the NIR beam.

Comparison of different methodologies (Probe versus Direct or wood surface used for analysis) on the same set of samples is rare as are multiple species calibrations for physical-mechanical properties such as MOE and MOR. Therefore, the objectives of this study were as follows:

(i) To develop density, MOE, and MOR calibrations using a number of tropical, subtropical, and temperate pine species

(ii) To compare wood property calibrations based on NIR spectra collected using a benchtop instrument, a benchtop instrument fitted with a fiber-optic probe, and field portable instrument fitted with a fiber-optic probe

(iii) To compare wood property calibrations based on NIR spectra collected from two different wood surfaces (radial and transverse) using the three different NIR instrument options

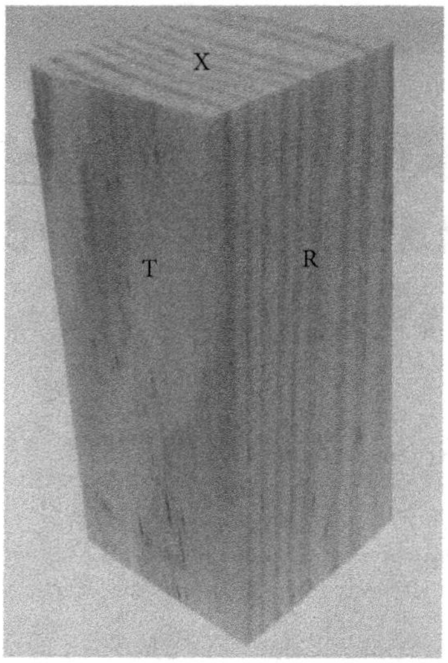

FIGURE 1: The three surfaces (radial, R; tangential, T; and transverse (cross-section), X) of a straight grained short, clear sample available for near-infrared (NIR) analysis. Note the arrangement of the growth rings.

2. Material and Methods

2.1. Sample Origin. The samples utilized in this study were from the plantations managed by Brazilian companies in the Center for Genetic Conservation and Improvement of Tropical Pine and Camcore (http://www.camcore.org). *Pinus caribaea* var. *hondurensis* (Sénéclauze) W.H. Barrett & Golfari, *Pinus chiapensis* (Martínez) Andersen, *Pinus maximinoi* H.E. Moore, *Pinus oocarpa* Schiede ex Schltdl, *Pinus taeda* L., and *Pinus tecunumanii* F. Schwerdtf. ex Eguiluz & J.P. Perry samples were collected from a site located in Ventania, Paraná, Brazil, while a site located in the Municipality of Itararé, São Paulo, Brazil, provided samples from the remaining species (*P. caribaea* var. *bahamensis* (Grisebach) W.H. Barrett & Golfari and *P. caribaea* var. *caribaea* Morelet). Figure 2 shows the locations of the two plantations, while Table 2 provides a summary of site characteristics.

2.2. Sample Collection and Preparation of Samples. Five medium-sized trees of each species were selected, and diameter at breast height and total height were measured (Table 3). After felling, the length from the base of the tree to the merchantable top was measured and the tree cut into four logs of equal length (2.5 m). At a sawmill, a 100 mm thick bark-bark slab 2.5 m in length was cut through the pith from each log.

At the Federal University of Paraná (Curitiba, Brazil) Wood Technology Laboratory, the slabs were cut to provide clear wood samples from pith to bark with dimensions 20 mm (T) × 20 mm (R) × 300 mm (L). The samples were then conditioned to an equilibrium moisture content (MC) of 12%, and MOE and MOR were determined using an EMIC Universal Testing Machine (EMIC Equipamentos e Sistemas de Ensaio LTDA, São José dos Pinhais, PR, Brasil) according to the COPANT standard [24]. Density of the samples was determined using the dimensions and weight of each piece at 12% MC. After completion of wood property determination, a block with dimensions 20 mm (T) × 20 mm (R) × 20 mm (L) was cut from the end of each sample for NIR analysis.

2.3. Near-Infrared Spectroscopy. Diffuse reflectance NIR spectra were collected from the radial and transverse faces of each static bending sample using a Foss NIRSystems (Laurel MD, USA) Model 5000 scanning spectrometer and an ASD AgriSpec backpack portable NIR spectrometer fitted with a contact probe (approximately 25 mm in diameter, but with a spot size of 10 mm). Spectra collected using the Foss instrument were obtained using two different methods: (1) Foss Static—samples were held in a custom-made holder that had been fitted with a 12.5 mm × 10 mm Teflon mask (the mask was used to ensure that an area of constant size was analyzed) and (2) Foss Probe—NIR spectra were collected using an Optiprobe analyzer fiber-optic probe approximately 5 mm in diameter. Owing to the size of the ASD probe, a

TABLE 1: Summary of publications that have utilized near-infrared (NIR) spectroscopy to estimate physical mechanical properties of short, clear samples. Information includes species examined, wood surface utilized, size of the short, clear sample tested, method for collecting NIR spectra, number of spectra collected, and size of the NIR beam. NA denotes that the information was not available in the publication. R = radial surface; T = tangential surface; X = transverse (or cross section) surface; GR = growth rings.

Authors	Species	Wood surface	Test specimen size (R × T × L, mm)	NIR method (number of spectra)	Window size
Hoffmeyer and Pedersen (1995) [5]	*Picea abies*	X	45, 95, 1800	Direct (NA)	NA
Gindl et al. (2001) [6]	*Larix decidua*	R	18, 18, 250	Probe (3)	4 mm diam.
Thumm and Meder (2001) [7]	*Pinus radiata*	R and T	20, 20, 300	Direct, sample moving (1)	20 mm
Schimleck et al. (2001) [8]	*Eucalyptus delegatensis*	R	20, 20, 300	Direct (1)	5 × 20 mm
Schimleck et al. (2002) [9]	*Pinus radiata*	R	20, 20, 300	Direct (1)	5 × 20 mm
Via et al. (2003) [10]	*Pinus palustris*	R	3GR, 12.7, 304.8	Direct (2)	NA
Kelley et al. (2004) [11]	*Pinus taeda*	R	3GR 20–35, 15, 300	Probe (4)	20 mm^2
Kelley et al. (2004) [12]	6 softwood species	R	3GR 20–35, 15, 300 or 20, 15, 200	Probe (4)	20 mm^2
Schimleck et al. (2005) [13]	*Pinus taeda*	R, X	25.4, 25.4, 406	Direct (2)	5 × 12.5 mm
Via et al. (2005) [14]	*Pinus palustris*	R	3GR, 12.7, 304.8	Direct (2)	8 mm
				Probe (2)	5 mm
Andre et al. (2006) [15]	*Populus*	R	50, 50, 1000	Probe (4)	20 mm diam.
Haartveit and Flæte (2006) [16]	*Picea abies*	X	20, 20, 340	Direct (NA)	10 mm diam.
Fujimoto et al. (2007) [17]	Hybrid larch	R	20, 20, 320	Probe (2)	NA
Fujimoto et al. (2008) [18]	Hybrid larch	R and T	20, 20, 320	Probe (2)	NA
Lestander et al. (2008) [19]	*Pinus sylvestris*	X	3GR, 6.3, 120.5	Probe (NA)	4 mm diam.
Zhao et al. (2009) [20]	*Eucalyptus pellita*	R and T	NA	Probe (NA)	NA
Kothiyal and Raturi (2011) [21]	*Eucalyptus tereticornis*	R and T	20, 20, 300	Direct (10)	NA
Schimleck et al. (2011) [22]	*Caesalpinia echinata*	X	20, 20, 300	Direct (4,2,1)	5 × 5, 5 × 10, 15 × 15
				Probe (1)	15 × 15 mm
Kothiyal et al. (2014) [23]	*Eucalyptus tereticornis*	R and T	20, 20, 300	Direct (4)	15 mm
				Probe (4)	10 mm

Instruments from the following companies were utilized in the publications listed: ASD Field Spec [11, 12, 14, 15, 20, 22], Bran and Luebbe [17], Bruker [6, 18, 21, 23], FOSS [7–9, 13, 19, 22], LT industries [5], Perkin Elmer [16], and Thermo Nicolet [10, 14].

single spectrum was collected from each surface, while for the two Foss methods, two spectra were collected and averaged (Figure 3).

Spectra obtained by the Foss NIRSystems instrument were collected at 2 nm intervals over the wavelength range 1100 to 2498 nm, while for the ASD instrument, the sampling interval was 2 nm for 1000 to 2500 nm. The full wavelength range available was used for the Foss Direct method. Owing to noise for the fiber-optic probe systems and also for comparative purposes, data collected by both spectrometers were limited to a wavelength range of 1100–2300 nm.

All measurements were made in a controlled environment of 40% relative humidity and a temperature of 20°C.

2.4. Wood Property Calibrations. Wood property calibrations for three properties: density at 12% MC, MOE, and MOR, were developed using raw spectra, that is, no spectral pretreatments were employed. The Unscrambler software (version 9.2) and partial least squares (PLS) regression (with four cross-validation segments) were used. The software recommended the number of factors reported for each calibration. A total of 334 samples were analyzed; however, two that had very high densities were identified as outliers and were omitted from the final density models.

The standard error of calibration (SEC) (determined from the residuals from the final calibration), standard error of cross-validation (SECV) (determined from the residuals of each cross-validation phase), and the coefficient of determination (R^2) were used to assess calibration performance. The ratio of performance to deviation (RPD_c) [25], calculated as the ratio of the standard deviation of the reference data to the SECV, was also used to assess calibration performance. Determination of the RPD_c allowed comparison of calibrations developed for different wood properties that have different ranges in values. As this study was only concerned in comparing the relative performance of calibrations obtained using different spectroscopic approaches, it was decided to maximize the samples available for calibration and not split the datasets for calibration and prediction.

3. Results

3.1. Measured Wood Properties. A summary of measured wood properties for each species is given in Table 4.

FIGURE 2: Location of the sites sampled: (1) Ventania, Paraná, and (2) Municipality of Itararé, Sao Paolo.

TABLE 2: Location and description of sampling sites.

	Ventania, Paraná	Itararé, São Paulo
Latitude (°)	24°14′45″S	24°06′33″S
Longitude (°)	50°14′34″W	49°19′57″W
Elevation (m a.s.l.)	990	750
Average temperature (°C)	22	25
Average rainfall (mm)	1400	1300
Climate	Humid subtropical	Subtropical

TABLE 3: Average diameter (at breast height) and height for the six species sampled. One species, *Pinus caribaea*, included three varieties.

Species	Age (yr)	Diameter at breast height (mm)	Height (m)
Pinus caribaea var. *bahamensis*	17	370	27.0
Pinus caribaea var. *caribaea*	17	370	26.3
Pinus caribaea var. *hondurensis*	18	420	25.1
Pinus chiapensis	18	460	29.8
Pinus maximinoi	18	470	27.6
Pinus oocarpa	18	410	26.7
Pinus taeda	18	320	18.4
Pinus tecunumanii	18	460	25.9

3.2. Wood Property Calibrations. Density, MOE, and MOR calibrations obtained using transverse and radial face spectra collected using the three instrument options (Foss Direct, Foss Probe, and ASD Probe) are summarized in Tables 5, 6, and 7, respectively. All calibrations reported used the number of factors indicated by the Unscrambler software, hence comparisons are not exact; however, the number of factors used are generally similar for each property and having the number of factors identical throughout did not change the ranking of instrument options or comments related to the two wood faces examined. Some calibrations required up to 10 factors, which may be considered excessive, and is possibly the result of using untreated NIR data.

For all properties, calibrations based on Foss Direct spectra provided the strongest statistics. Density (Figure 4(a)) and MOE (Figure 4(b)) calibrations had similar R^2 (0.78 to 0.81) and RPD_c (2.1 to 2.2) values while those for MOR (0.69 to 0.73 and 1.7 to 1.9, resp.) were lower (Figure 4(c)).

Of the two probe options, NIR spectra obtained using the ASD spectrophotometer consistently provided stronger calibration statistics than those obtained using spectra collected using the Foss Probe. For the ASD probe, R^2 and RPD_c values ranged from 0.64 to 0.75 and 1.6 to 2.0, respectively, while for the Foss Probe they ranged from 0.50 to 0.68 and 1.3 to 1.7. Again, density and MOE calibrations were similar, while those for MOR had weaker statistics.

For a given property, and method of collecting spectra, the transverse face may be marginally better at capturing sample variability than the radial face as RPD_c values are often equivalent even though fewer factors were used for the transverse face calibrations. The only instance where

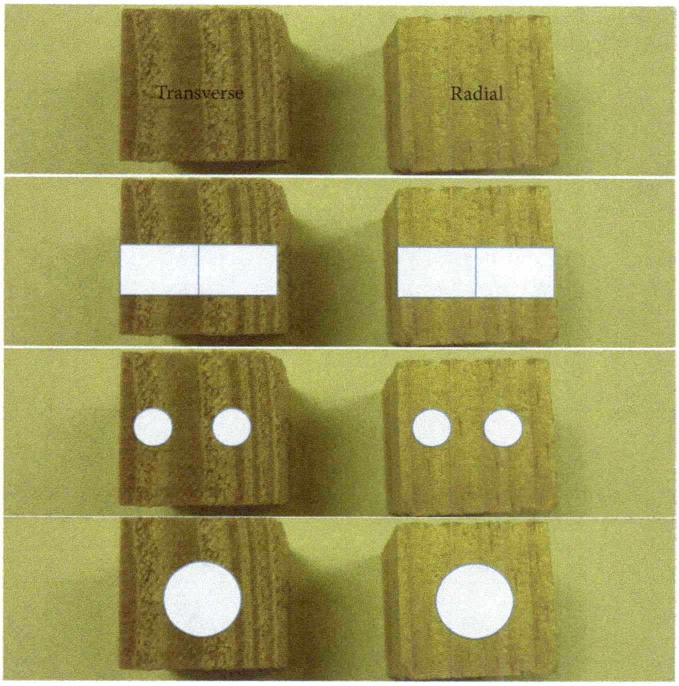

FIGURE 3: A typical pine sample utilized in this study. NIR spectra were collected from the transverse and radial faces, with the approximate areas examined for the different methods (Foss Direct, Foss Probe, and ASD Probe) employed for obtaining spectra highlighted.

TABLE 4: Summary information (average, standard deviation, maximum, and minimum) for density, modulus of elasticity (MOE), and modulus of rupture (MOR) values for each of the six pine species utilized in this study. One species, *Pinus caribaea*, included three varieties.

Species	Samples tested	Density (12%) (kg/m^3)	MOE (MPa)	MOR (MPa)
P. c. bahamensis	47	490, 89 (340, 749)	7048, 2668 (2791, 13,791)	64, 18.6 (33, 104)
P. c. caribaea	37	445, 65 (366, 641)	5930, 2149 (2883, 10,877)	55, 15 (26, 96)
P. c. hondurensis	46	507, 112 (383, 1008)	7027, 2250 (3784, 12,130)	59, 13 (38, 83)
P. chiapensis	50	444, 48 (360, 551)	7377, 1431 (2560, 10,190)	59, 10 (42, 80)
P. maximinoi	43	522, 72 (420, 701)	9011, 2241 (5641, 14,149)	68, 14 (47, 103)
P. oocarpa	38	534, 92 (412, 699)	7699, 2791 (4035, 15,433)	66, 16 (47, 109)
P. taeda	24	515, 73 (371, 655)	8015, 2250 (4219, 11,269)	61, 14 (36, 88)
P. tecunumanii	49	560, 74 (464, 785)	9110, 2448 (5882, 16,032)	70, 13 (48, 119)

Note: *P. c.* = *Pinus caribaea* variety.

a calibration based on transverse face NIR spectra was clearly inferior to the corresponding radial face spectra was for the density calibration based on data provided by the ASD probe.

4. Discussion

Calibrations for density, MOE, and MOR were obtained using multiple pine species grown on two sites in southern

Table 5: Summary of air-dry density calibrations based on radial and transverse face NIR spectra obtained using three different methods (Foss Direct, Foss Probe, and ASD Probe) for collecting spectral data. 332 samples were used to develop the calibrations.

Property/method	Number of factors	R^2	SEC	SECV	RPD_c
	Foss Direct				
Density (kg/m^3)—radial face	8	0.81	35.6	38.5	2.1
Density (kg/m^3)—transverse face	8	0.80	36.8	38.3	2.2
	Foss Probe				
Density (kg/m^3)—radial face	4	0.51	57.7	59.8	1.4
Density (kg/m^3)—transverse face	10	0.68	46.8	49.5	1.7
	ASD Probe				
Density (kg/m^3)—radial face	8	0.73	43.2	45.8	2.0
Density (kg/m^3)—transverse face	10	0.70	45.0	48.5	1.8

Table 6: Summary of modulus of elasticity (MOE) calibrations based on radial and transverse face NIR spectra obtained using three different methods (Foss Direct, Foss Probe, and ASD Probe) for collecting spectral data. 334 samples were used to develop the calibrations.

Property/method	Number of factors	R^2	SEC	SECV	RPD_c
	Foss Direct				
MOE (MPa)—radial face	8	0.78	1154	1215	2.1
MOE (MPa)—transverse face	6	0.81	1095	1124	2.2
	Foss Probe				
MOE (MPa)—radial face	7	0.55	1668	1727	1.4
MOE (MPa)—transverse face	10	0.71	1342	1431	1.7
	ASD Probe				
MOE (MPa)—radial face	9	0.73	1308	1413	1.8
MOE (MPa)—transverse face	6	0.75	1252	1303	1.7

Table 7: Summary of modulus of rupture (MOR) based on radial and transverse face NIR spectra obtained using three different methods (Foss Direct, Foss Probe, and ASD Probe) for collecting spectral data. 334 samples were used to develop the calibrations.

Property/method	Number of factors	R^2	SEC	SECV	RPD_c
	Foss Direct				
MOR (MPa)—radial face	4	0.69	8.4	8.6	1.7
MOR (MPa)—transverse face	6	0.73	7.9	8.0	1.9
	Foss Probe				
MOR (MPa)—radial face	7	0.50	10.6	11.2	1.3
MOR (MPa)—transverse face	10	0.61	9.3	10.3	1.4
	ASD Probe				
MOR (MPa)—radial face	8	0.67	8.6	9.2	1.6
MOR (MPa)—transverse face	5	0.64	9.0	9.2	1.6

Brazil. Calibrations were based on NIR spectra collected using three different approaches (Foss Direct, Foss Probe, and ASD Probe) and from two different wood surfaces (radial and transverse). The strong calibration statistics obtained for models using the Foss Direct and ASD Probe methods demonstrates the potential for general calibrations for estimating wood properties of a number of tropical, subtropical, and temperate pine species. The multiple species calibrations may not be as accurate as those based on a single site or species but are sufficient to rank samples [26].

Comparison of the different spectroscopic approaches used to collect spectra indicated that the Direct approach provided stronger calibration statistics than either of the Probe methods. The advantage of using the Direct approach is that the sample is exposed directly to the NIR beam (whether it is placed directly on the instrument aperture or against a Teflon window) rather than having to transmit the beam through a fiber-optic cable which causes noise in the signal beyond 2300 nm. However, to utilize a Direct approach, subsamples were often removed from the end of

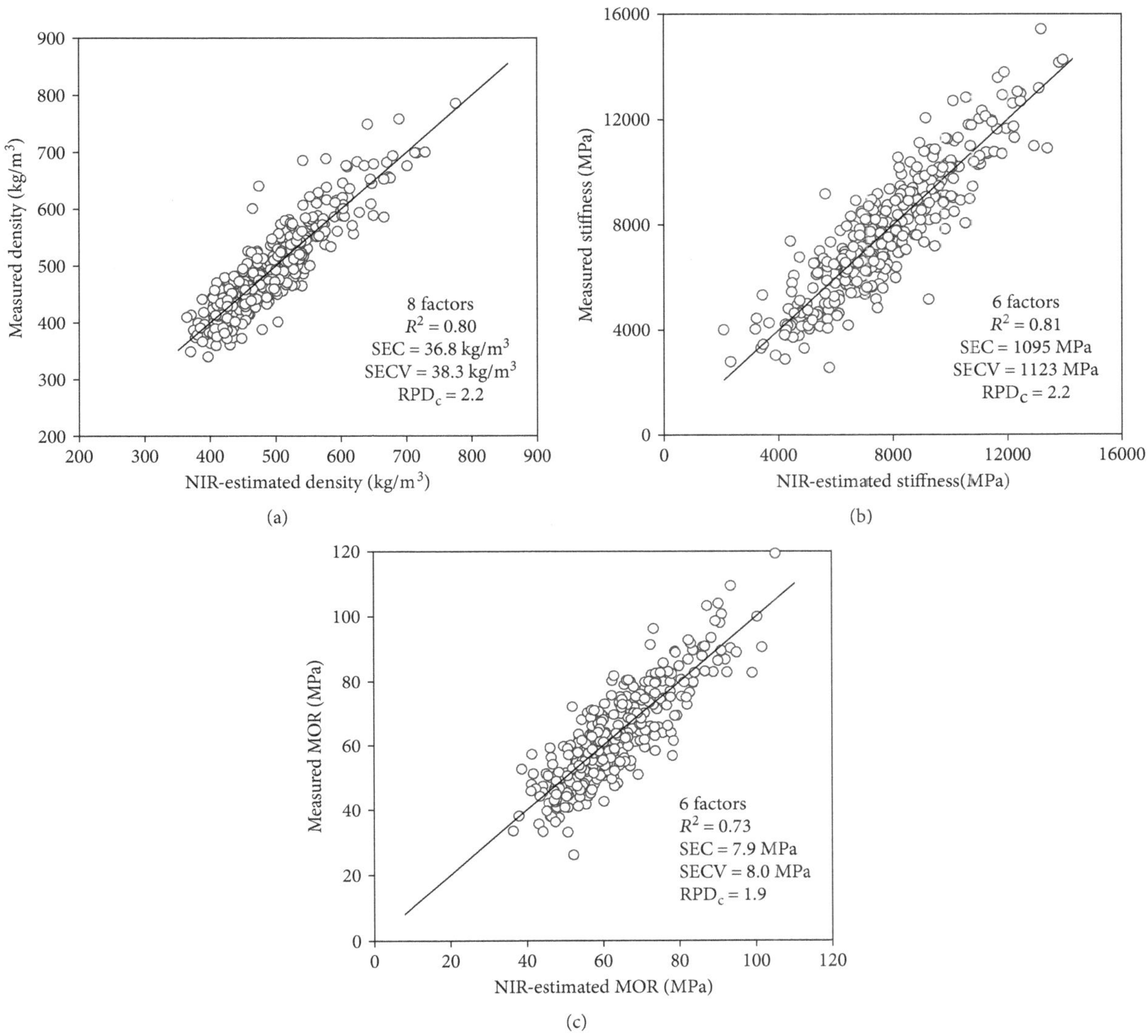

FIGURE 4: Relationship between lab-measured wood properties and NIR-fitted wood properties, (a) density, (b) modulus of elasticity (MOE), and (c) modulus of rupture (MOR). The broken line represents the line of equivalence. 332 samples were used to develop the density calibration, while 334 samples were used to develop the calibrations for MOE and MOR.

the short clear sample to allow the sample to be tested by the NIR instrument [13, 16, 21]. In comparison, utilization of the Probe approach provides much greater flexibility for testing samples [12, 15], and cutting the static bending sample is not required.

When comparing the ASD and Foss Probe spectra, the ASD instrument provided stronger calibrations for all properties. A critical factor in the success of the ASD Probe method was the greater surface area covered by the ASD probe compared to the Foss probe (Figure 3). Owing to its spot size, the ASD probe was able to capture more sample variability, that is, variation in earlywood and latewood components of individual rings, while it was not possible to achieve the same coverage when collecting two spectra using the Foss probe. Using a probe with a large diameter that approaches the width of the static bending sample also offers greater consistency in NIR measurements. For the smaller Foss probe, identification of sampling points was problematic as it always had the potential to over or under selection of latewood, which is critical in determining density and MOE [27].

An examination of related NIR studies shows that many authors either attempted to collect spectra across the complete surface of the sample, for example, Thumm and Meder [7], using a 20 mm wide window for samples having radial and tangential dimensions of 20 mm, or were mindful of the cumulative area they were testing, for example, Gindl et al. [6], noting that three spectra from different locations on the radial face of their samples covered two-thirds of the radial width. In other studies [10–12, 14, 19], radial width of static bending samples varied as the authors wished to include at least three growth rings in their samples introducing the possibility that NIR spectra collected from wider samples underrepresented the sample compared to the same number of spectra collected from narrower samples. While its significance is unclear, a related issue was encountered by Lestander et al. [19] who noted that because of variation in the width of the target rings, spectral information from the nearest adjacent rings were occasionally included in the collected spectra.

The transverse face had marginally better results regardless of the method used to collect spectra. The only other study that attempted to compare mechanical/physical property calibrations based on spectra collected from radial and transverse surfaces was Schimleck et al.'s study [13] who also reported similar results for the two surfaces. Other studies have compared calibrations based on radial and tangential surface NIR spectra. Thumm and Meder [7] reported MOE calibrations for radial and tangential face spectra (moving face) and found much better statistics for radial face data as did Zhao et al. [20] in their study of *E. pellita*. In comparison, Kothiyal and Raturi [21] reported similar calibrations for both radial and tangential face spectra obtained from *E. tereticornis* static bending samples. In a more recent study, Kothiyal et al. [23] reported calibrations based on radial and tangential surface spectra and calibrations based on an average of the two. Radial face spectra provided the strongest calibrations; however, the authors concluded that "the composite model is just as good as the radial and tangential face models and can adequately address the problem where it is difficult to obtain perfect radial or tangential surface, particularly in small girth plantation timber or where surface recognition is difficult." One solution to the issue noted by Kothiyal et al. [23] is to utilize the transverse surface. In this study, tangential surface spectra were not collected owing to doubts about how well the surface would represent samples with straight grain (i.e., clear radial and tangential surfaces, as shown in Figure 1). As noted by Dahlen et al. [28], tangential face NIR spectra could easily under- or overestimate properties based on whether an earlywood or latewood component of a ring was present where the spectrum was collected.

Despite the strength of relationships reported in the various studies over the last 20+ years, it is unlikely that NIR-estimated MOE and MOR will replace the standard method used to determine these properties. However, NIR technology can complement the standard methods, for example, to measure applied loads [15], or in applications where the accuracy of a single value is not critical. Examples include studies of within-tree variation in MOE [29] or regional variation in MOE and MOR [30]. In such studies, the use of NIR-estimated values can greatly improve approximations of how properties vary spatially. The success of research based on static bending samples has also led to studies aimed at estimating properties of lumber [18, 28, 31–34]. The issue of collecting representative spectra again exists, but results are promising and owing to its ability to estimate a range of properties simultaneously [33] NIR could have an important role in sawmills in the future. In related research, the assessment of MOE and load for wood in service has recently been demonstrated [35–37].

5. Conclusions

NIR-based calibrations for the estimation of density, modulus of elasticity (MOE), and modulus of rupture (MOR) were developed using clear wood samples obtained from several tropical, subtropical, and temperate pine species growing on two plantation sites in Brazil. NIR spectra were collected using a benchtop instrument, a benchtop instrument with fiber-optic probe, and a field portable fiber-optic probe instrument and from two different wood surfaces (radial and transverse).

Comparison of calibrations found that the benchtop instrument provided superior results, while for the fiber-optic probe systems, the field portable NIR outperformed the benchtop probe system. Difficulty with adequately representing the sample when collecting spectra using a fiber-optic probe and lower quality spectra explain the differences among the data sets.

NIR spectra collected from the radial and transverse faces provided similar calibration statistics.

Disclosure

The work described in this paper was presented at the 65th Forest Products Society International Convention, Portland, Oregon, USA, 19–21 June 2011.

Conflicts of Interest

The authors declare that they have no competing interests.

Acknowledgments

The authors acknowledge the financial support for this research provided by the Conselho Nacional de Desenvolvimento Científico e Tecnológico (CNPq), Brazil.

References

[1] C. A. Raymond, "Genetics of *Eucalyptus* wood properties," *Annals of Forest Science*, vol. 59, no. 5-6, pp. 525–531, 2002.

[2] L. R. Schimleck, "Near infrared spectroscopy: a rapid, non-destructive method for measuring wood properties and its application to tree breeding," *New Zealand Journal of Forestry Science*, vol. 38, no. 1, pp. 14–35, 2008.

[3] S. Tsuchikawa, "A review of recent near infrared research for wood and paper," *Applied Spectroscopy Reviews*, vol. 42, no. 1, pp. 43–71, 2007.

[4] S. Tsuchikawa and H. Kobori, "A review of recent application of near infrared spectroscopy to wood science and technology," *Journal of Wood Science*, vol. 61, no. 3, pp. 213–220, 2015.

[5] P. Hoffmeyer and J. G. Pedersen, "Evaluation of density and strength of Norway spruce wood by near-infrared reflectance spectroscopy," *Holz Als Roh-Und Werkstoff*, vol. 53, no. 1, pp. 165–170, 1995.

[6] W. Gindl, A. Teischinger, M. Schwanninger, and B. Hinterstoisser, "The relationship between near infrared spectra of radial wood surfaces and wood mechanical properties," *Journal of Near Infrared Spectroscopy*, vol. 9, no. 1, pp. 255–261, 2001.

[7] A. Thumm and R. Meder, "Stiffness prediction of radiata pine clearwood test pieces using near infrared spectroscopy," *Journal of Near Infrared Spectroscopy*, vol. 9, no. 1, pp. 117–122, 2001.

[8] L. R. Schimleck, R. Evans, and J. Ilic, "Estimation of *Eucalyptus delegatensis* wood properties by near infrared spectroscopy," *Canadian Journal of Forest Research*, vol. 31, no. 10, pp. 1671–1675, 2001.

[9] L. R. Schimleck, R. Evans, and A. C. Matheson, "Estimation of *Pinus radiata* D. Don clear wood properties by near-infrared spectroscopy," *Journal of Wood Science*, vol. 48, no. 2, pp. 132–137, 2002.

[10] B. K. Via, T. Shupe, L. Groom, M. Stine, and C.-L. So, "Multivariate modelling of density, strength and stiffness from near infrared spectra for mature, juvenile and pith wood of longleaf pine (*Pinus palustris*)," *Journal of Near Infrared Spectroscopy*, vol. 11, no. 1, pp. 365–378, 2003.

[11] S. S. Kelley, T. G. Rials, R. Snell, L. H. Groom, and A. Sluiter, "Use of near infrared spectroscopy to measure the chemical and mechanical properties of solid wood," *Wood Science and Technology*, vol. 38, no. 4, pp. 257–276, 2004.

[12] S. S. Kelley, T. G. Rials, L. R. Groom, and C.-L. So, "Use of near infrared spectroscopy to predict the mechanical properties of six softwoods," *Holzforschung*, vol. 58, no. 3, pp. 252–260, 2004.

[13] L. R. Schimleck, P. D. Jones, A. Clark, R. F. Daniels, and G. F. Peter, "Near infrared spectroscopy for the nondestructive estimation of clear wood properties of *Pinus taeda* L. from the southern United States," *Forest Products Journal*, vol. 55, no. 12, pp. 21–28, 2005.

[14] B. K. Via, C.-L. So, T. Shupe, L. Eckhardt, M. Stine, and L. Groom, "Prediction of wood mechanical and chemical properties in the presence and absence of blue stain using two near infrared instruments," *Journal of Near Infrared Spectroscopy*, vol. 13, no. 1, pp. 201–212, 2005.

[15] N. Andre, N. Labbé, T. G. Rials, and S. S. Kelley, "Assessment of wood load condition by near infrared (NIR) spectroscopy," *Journal of Materials Science*, vol. 41, no. 7, pp. 1879–1886, 2006.

[16] E. Y. Haartveit and P. O. Flæte, "Rapid prediction of basic wood properties by near infrared spectroscopy," *New Zealand Journal of Forestry Science*, vol. 36, no. 2/3, p. 393, 2006.

[17] T. Fujimoto, H. Yamamoto, and S. Tsuchikawa, "Estimation of wood stiffness and strength properties of hybrid larch by near-infrared spectroscopy," *Applied Spectroscopy*, vol. 61, no. 8, pp. 882–888, 2007.

[18] T. Fujimoto, Y. Kurata, K. Matsumoto, and S. Tsuchikawa, "Application of near infrared spectroscopy for estimating wood mechanical properties of small clear and full length lumber specimens," *Journal of Near Infrared Spectroscopy*, vol. 16, no. 6, pp. 529–537, 2008.

[19] T. A. Lestander, J. Lindeberg, D. Eriksson, and U. Bergsten, "Prediction of *Pinus sylvestris* clear-wood properties using NIR spectroscopy and biorthogonal partial least squares regression," *Canadian Journal of Forest Research*, vol. 38, no. 7, pp. 2052–2062, 2008.

[20] R. J. Zhao, X. M. Huo, and L. Zhang, "Estimation of modulus of elasticity of *Eucalyptus pellita* wood by near infrared spectroscopy," *Spectroscopy and Spectral Analysis*, vol. 29, no. 9, pp. 2392–2395, 2009.

[21] V. Kothiyal and A. Raturi, "Estimating mechanical properties and specific gravity for five-year-old *Eucalyptus tereticornis* having broad moisture content range by NIR spectroscopy," *Holzforschung*, vol. 65, no. 5, pp. 757–762, 2011.

[22] L. R. Schimleck, J. L. M. de Matos, J. T. da Silva Oliveira, and G. I. B. Muniz, "Non-destructive estimation of pernambuco (*Caesalpinia echinata*) clear wood properties using near infrared spectroscopy," *Journal of Near Infrared Spectroscopy*, vol. 19, no. 5, pp. 411–419, 2011.

[23] V. Kothiyal, A. Raturi, Jaideep, and Y. M. Dubey, "Enhancing the applicability of near infrared spectroscopy for estimating specific gravity of green timber from *Eucalyptus tereticornis* by developing composite calibration using both radial and tangential face of wood," *European Journal of Wood and Wood Products*, vol. 72, no. 1, pp. 11–20, 2014.

[24] COPANT, *555 Método de ensayo de flexion estatica*, Comisión Panamericana de Normas Técnicas (Pan American Standards Commission), 1972.

[25] P. C. Williams and D. C. Sobering, "Comparison of commercial near infrared transmittance and reflectance instruments for analysis of whole grains and seed," *Journal of Near Infrared Spectroscopy*, vol. 1, no. 1, p. 8, 1993.

[26] L. R. Schimleck, G. R. Hodge, and W. Woodbridge, "Toward global calibrations for estimating the wood properties of tropical, sub-tropical and temperate pine species," *Journal of Near Infrared Spectroscopy*, vol. 18, no. 6, pp. 355–365, 2010.

[27] L. Jordan, A. Clark, L. R. Schimleck, D. B. Hall, and R. F. Daniels, "Regional variation in wood specific gravity of planted loblolly pine in the United States," *Canadian Journal of Forest Research*, vol. 38, no. 4, pp. 698–710, 2008.

[28] J. Dahlen, I. Diaz, L. Schimleck, and P. D. Jones, "Near-infrared spectroscopy prediction of southern pine no. 2 lumber physical and mechanical properties," *Wood Science and Technology*, vol. 51, no. 2, pp. 309–322, 2017.

[29] C. R. Mora and L. R. Schimleck, "Determination of within-tree variation of *Pinus taeda* wood properties by near infrared spectroscopy. Part 2: whole-tree wood property maps," *Appita Journal*, vol. 62, no. 3, pp. 232–238, 2009.

[30] F. Antony, L. Jordan, L. R. Schimleck, A. Clark III, R. A. Souter, and R. F. Daniels, "Regional variation in wood modulus of elasticity (stiffness) and modulus of rupture (strength) of planted loblolly pine in the United States," *Canadian Journal of Forest Research*, vol. 41, no. 7, pp. 1522–1533, 2011.

[31] R. Meder, A. Thumm, and D. Marston, "Sawmill trial of at-line prediction of recovered lumber stiffness by NIR spectroscopy of *Pinus radiata* cants," *Journal of Near Infrared Spectroscopy*, vol. 11, no. 1, pp. 137–143, 2003.

[32] T. Fujimoto, Y. Kurata, K. Matsumoto, and S. Tsuchikawa, "Feasibility of near-infrared spectroscopy for on-line grading of sawn lumber," *Applied Spectroscopy*, vol. 64, no. 1, pp. 92–99, 2010.

[33] T. Fujimoto, Y. Kurata, K. Matsumoto, and S. Tsuchikawa, "Feasibility of near-infrared spectroscopy for online multiple trait assessment of sawn lumber," *Journal of Wood Science*, vol. 56, no. 6, pp. 452–459, 2010.

[34] H. Kobori, T. Inagaki, T. Fujimoto, T. Okura, and S. Tsuchikawa, "Fast online NIR technique to predict MOE and moisture content of sawn lumber," *Holzforschung*, vol. 69, no. 3, pp. 329–335, 2015.

[35] M. Riggio, J. Sandak, A. Sandak, D. Pauliny, and L. Babiński, "Analysis and prediction of selected mechanical/dynamic properties of wood after short and long-term waterlogging," *Construction and Building Materials*, vol. 68, pp. 444–454, 2014.

[36] A. Sandak, J. Sandak, and M. Riggio, "Estimation of physical and mechanical properties of timber members in service by means of infrared spectroscopy," *Construction and Building Materials*, vol. 101, pp. 1197–1205, 2015.

[37] A. Sandak, J. Sandak, and M. Riggio, "Assessment of wood structural members degradation by means of infrared spectroscopy: an overview," *Structural Control and Health Monitoring*, vol. 23, no. 3, pp. 396–408, 2016.

9

Rapid Identification of Nine Easily Confused Mineral Traditional Chinese Medicines Using Raman Spectroscopy Based on Support Vector Machine

Jing Ming,[1] **Long Chen**,[1,2] **Yan Cao**,[1] **Chi Yu**,[1] **Bi-Sheng Huang**,[1] **and Ke-Li Chen**[1]

[1]*Key Laboratory of Ministry of Education on Traditional Chinese Medicine Resource and Compound Prescription, Hubei University of Chinese Medicine, Wuhan, Hubei 430065, China*
[2]*Xiangyang Central Hospital, Xiangyang, Hubei 441021, China*

Correspondence should be addressed to Bi-Sheng Huang; hbsh1963@163.com and Ke-Li Chen; kelichen@126.com

Academic Editor: Feride Severcan

Mineral traditional Chinese medicines (TCMs) are natural minerals, mineral processing products, and some fossils of animals or animal bones that can be used as medicines. Mineral TCMs are a characteristic part of TCMs and play a unique role in the development of TCMs. Mineral TCMs are usually identified according to their morphological properties such as shape, color, or smell, but it is difficult to separate TCMs that are similar in appearance or smell. In this study, the feasibility of using Raman spectroscopy combined with support vector machine (SVM) for rapid identification of nine easily confused mineral TCMs, i.e., borax, gypsum fibrosum, natrii sulfas exsiccatus, natrii sulfas, alumen, sal ammoniac, quartz, calcite, and yellow croaker otolith, was investigated. Initially, two methods, characteristic intensity data extraction and principal component analysis (PCA), were performed to reduce the dimensionality of spectral data. The identification model was subsequently built by the SVM algorithm. The 3-fold cross validation (3-CV) accuracy of the SVM model established based on extracting characteristic intensity data from spectra pretreated by first derivation was 98.61%, and the prediction accuracies of the training set and validation set were 100%. As for the PCA-SVM model, when the spectra pretreated by vector normalization and the number of principal components (NPC) is 7, the 3-CV accuracy and prediction accuracies all reached 100%. Both models have good performance and strong prediction capacity. These results demonstrate that Raman spectroscopy combined with a powerful SVM algorithm has great potential for providing an effective and accurate identification method for mineral TCMs.

1. Introduction

Mineral traditional Chinese medicines (TCMs) are an important part of TCMs. Nine commonly used mineral TCMs, namely, borax, gypsum fibrosum, natrii sulfas exsiccatus, natrii sulfas, alumen, sal ammoniac, quartz, calcite, and yellow croaker otolith, have great differences in chemical compositions and pharmacological action. For instance, alumen, which is widely used in China for treating hemorrhoids, eczema, and scabies, has therapeutic functions including removing dampness to relieve itching, stopping bleeding, and preventing diarrhea [1]. Natrii sulfas is used to clean away heat, relax the bowels, and treat periappendicular abscesses [2]. Borax is the main ingredient of the famous Chinese patent medicine, Musk hemorrhoids ointment [1]. Sal ammoniac has some degree of toxicity and should be used with caution [3].

However, these nine mineral TCMs are of high similarity in color, lustre, shape, texture, and cut surface character. The characteristics of these nine mineral TCMs were shown in Table 1. It can be seen that most of them are odorless, white or colorless, transparent to translucent, and irregular in shape. Furthermore, mineral TCMs are usually used in powder form, which would make them more easily confused with each other. Traditionally, mineral TCMs are mainly identified relying upon the visual recognition and

Table 1: Characteristics of nine mineral TCMs.

Mineral TCMs	Shape	Color	Diaphaneity	Tenacity	Lustre	Odor	Taste
Borax [4]	Irregular, massive	Colorless to white	Transparent to translucent	Brittle	Vitreous	Odorless	Slightly salty, slightly sweet
Gypsum fibrosum [1]	Fibrous masses, tabular, or irregularly massive	White, gray white, or pale yellow	Translucent	Flexible	Silky	Odorless	Bland
Natrii sulfas exsiccatus [1]	Powder	White	—	—	—	Odorless	Salty
Natrii sulfas [1]	Prismatic, oblong, irregularly shaped, or granular	Colorless or off-white	Transparent or translucent	Brittle	Vitreous	Odorless	Salty
Alumen [1]	Irregular massive or granular	Colorless or yellowish-white	Transparent or translucent	Brittle	Vitreous	Odorless	Sour, slightly sweet, very astringent
Sal ammoniac [5]	Naturally occurring sal ammoniac are irregular platy crystals, and the synthetic one is cuboid	Colorless, white, pale gray, yellowish-white, or brown	Transparent or translucent	Brittle	Vitreous	Sulfury	Salty, bitter
Quartz [4]	6-sided prism or coarse granular aggregation with sharp edges	White or gray white	Translucent to opaque	Hard	Greasy	Odorless	Bland
Calcite [5]	Rhombohedral flat masses, rhombohedral prism	White, pale yellow, or pale red	Transparent to translucent	Brittle	Vitreous	Odorless	Tasteless
Yellow croaker otolith [4]	Oblong, ovoid, trigonous. One end is rounded, and the other is pointed.	White	Opaque	Brittle	Porcellaneous	Odorless	Bland, slightly astringent

physicochemical methods. The former is subjective and may lead to unreliable results, and the latter are time-consuming. Therefore, a faster and more reliable detection technique is required.

Raman spectroscopy is a nondestructive analytical tool that has long been employed for the analysis of minerals since the discovery of the Raman effect. Raman spectroscopy measures inelastic light scattering and is a vibrational spectroscopic technique that can provide arrays of fingerprint assignments of the physical vibration mode. Such information could be used to characterize differences among different materials. In recent years, the technique has been investigated extensively in the field of TCMs because of the advantages such as rapid, convenient, nondestructive, and less sample consumption. Many studies have been reported for identification of TCMs using Raman spectroscopy. In these studies, Raman spectroscopy was successfully applied to authentication of raw materials [6–8], characterization of adulterants [9, 10], detection of counterfeits [11, 12], and geographical origin identification of TCMs [13, 14]. These studies obtained good performance, indicating the excellent discriminative ability of using Raman spectroscopy techniques in TCMs. At present, spectral technologies such as near-infrared spectroscopy and Raman spectroscopy combined with different types of chemometric algorithms such as principal component analysis (PCA) [15], support vector machine (SVM) [16, 17], and artificial neural network (ANN) [18–20] have been successfully used for the qualitative and quantitative analysis of TCMs. Therefore, an identification model was established based on Raman spectroscopy combined with SVM to realize the rapid and accurate identification of nine easily confused mineral TCMs.

2. Materials and Methods

2.1. Sample Collection. A total of 108 batches of samples were collected from Chinese medicine markets of Bozhou in Anhui Province, Yuzhou in Henan, Yulin in Guangxi, Xi'an in Shanxi, and Zhangshu in Jiangxi as well as Jointown Pharmaceutical Group Co., Ltd and Mayinglong pharmaceutical Group Co., Ltd, respectively. Among the samples, there were 14 batches of gypsum fibrosum, 10 batches of alumen, 14 batches of sal ammoniac, 10 batches of quartz, 10 batches of natrii sulfas, 12 batches of natrii sulfas exsiccatus, 14 batches of calcite, 12 batches of yellow croaker Otolith, and 12 batches of borax. All samples were identified to be authentic by X-ray diffraction (XRD) and chemical method according to the Chinese Pharmacopoeia (ChP 2015 I).

2.2. Instruments and Software. The instruments used for this project consisted of an XPertPro X-ray diffraction instrument (PANalytical Company) and a Portable i-Raman 475–785H (B&W Tek) spectrometer. The portable Raman spectrometer was connected to a BAC150B (B&W Tek) Raman sampling accessory by a 150 cm optical fiber probe BAC102-785E (B&W Tek). The instrument was equipped with a CleanLaze (B&W Tek) laser emitting at 785 nm with continuously adjustable power from 0 to 420 mW and a thermoelectrically cooled charged-coupled detector (CCD), covering a spectral range of 65–2700 cm^{-1} at a resolution of approximately 3.5 cm^{-1}.

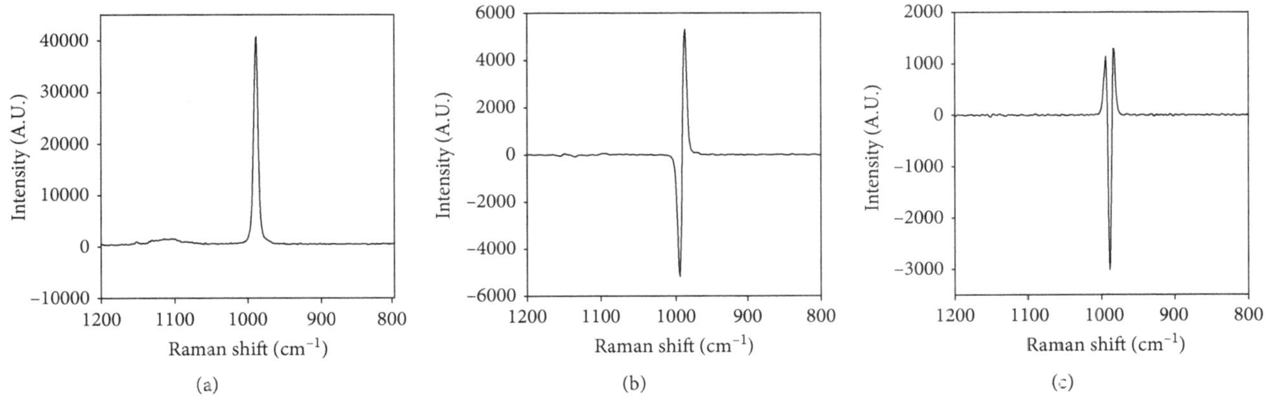

FIGURE 1: Raman spectra of natrii sulfas from 800 to 1200 cm^{-1}: (a) original spectra; (b) spectra pretreated by FD; (c) spectra pretreated by SD.

The software used in this study included BWSpec4™ spectral data-acquisition software, OPUS 7.5 spectrum analysis software (Bruker), IBM SPSS Statistics 19 (SPSS Statistics/IBM Corp), MATLAB R2014b (The MathWorks Inc.), and libsvm toolbox (libsvm-3.1, Faruto Ultimate).

2.3. Raman Spectra Collection. Each sample was crushed into powder and then passed through a 100-mesh sieve; 2 g of each sample powder was put into a specimen cup on the stage of the sampling accessory. The spectra have been recorded in the range of 70–2695 cm^{-1}. The laser output power was 100%. The integration time of each spectrum was adjusted to provide a better Raman signal. The measurements of each sample were replicated three times. The average value of the three spectra was selected as the spectrum information for analyzing.

2.4. Spectra Pretreatment Method. In the process of collecting sample spectrum, the original Raman spectra are often subjected to factors that are unrelated to the test sample properties, such as sample autofluorescence and variation in the measurement condition. These factors will lead to baseline drift and instability. Therefore, it is very necessary to conduct suitable spectrum pretreatment. In this study, the separate pretreatment methods vector normalization (VN), first derivation (FD), second derivation (SD), and combined pretreatment methods VN + FD and VN + SD were employed by OPUS in an effort to optimize model performance.

2.5. Spectral Data Compression Method. The high dimension of Raman spectral space will result in computational complexity and inefficiency in optimization and implementation of the SVM algorithms. It is necessary to extract spectra feature and compress spectral data in order to create effective and robust SVM model.

Principal component analysis (PCA) is a common dimension reduction method of spectrum, which can transform a number of possibly correlated variables (spectrum matrix) into a smaller number of variables called principal components (PCs). The new variables after transformation are not related with each other, which can eliminate the overlapped part of information [21]. Moreover, these new variables include the most informative dimensions of the original variables and as can as possible without losing useful information. In this research, PCA was conducted in MATLAB 2014b to reduce dimension of original spectra and pretreated spectra, respectively.

The other method to compress spectral data in this research is extracting intensity data of characteristic peaks from complete spectra. The detailed method is as follows. The Raman shifts of strong Raman peaks of each mineral TCM were selected, and then the intensity data at these Raman shifts in every sample spectra were extracted as input variables for the SVM model instead of the complete spectra. As shown in Figure 1, one Raman peak was transformed into two peaks after being pretreated with FD, which is equal in intensity and opposite in direction. One Raman peak was transformed into three peaks after being pretreated with SD: two of which were toward y-axis positive direction, and one towards y-axis negative direction. In consideration of reducing the amount of data, when selecting characteristic peaks, we selected only the peaks toward the y-axis positive direction for the spectra pretreated with FD and selected only the peaks toward the y-axis negative direction for the spectra pretreated with SD.

2.6. Support Vector Machine. Support vector machine is a powerful supervised learning algorithm that was first proposed by Vapnik [22] and successfully extended by a number of other researchers in recent years, which is based on the principle of minimization of structural risk in constructing an optimally separating hyperplane that separates different classes of data. In the process, input vectors are mapped to a newly constructed high dimensional space, and then parallel hyperplanes are constructed to maximize the interplane distance which separates the data. Details about the SVM classifier can be found in [23, 24]. In this research, SVM was performed in MATLAB 2014b to build qualitative models.

Radial basis function (RBF) was adopted as the kernel function of SVM due to its strong ability in processing

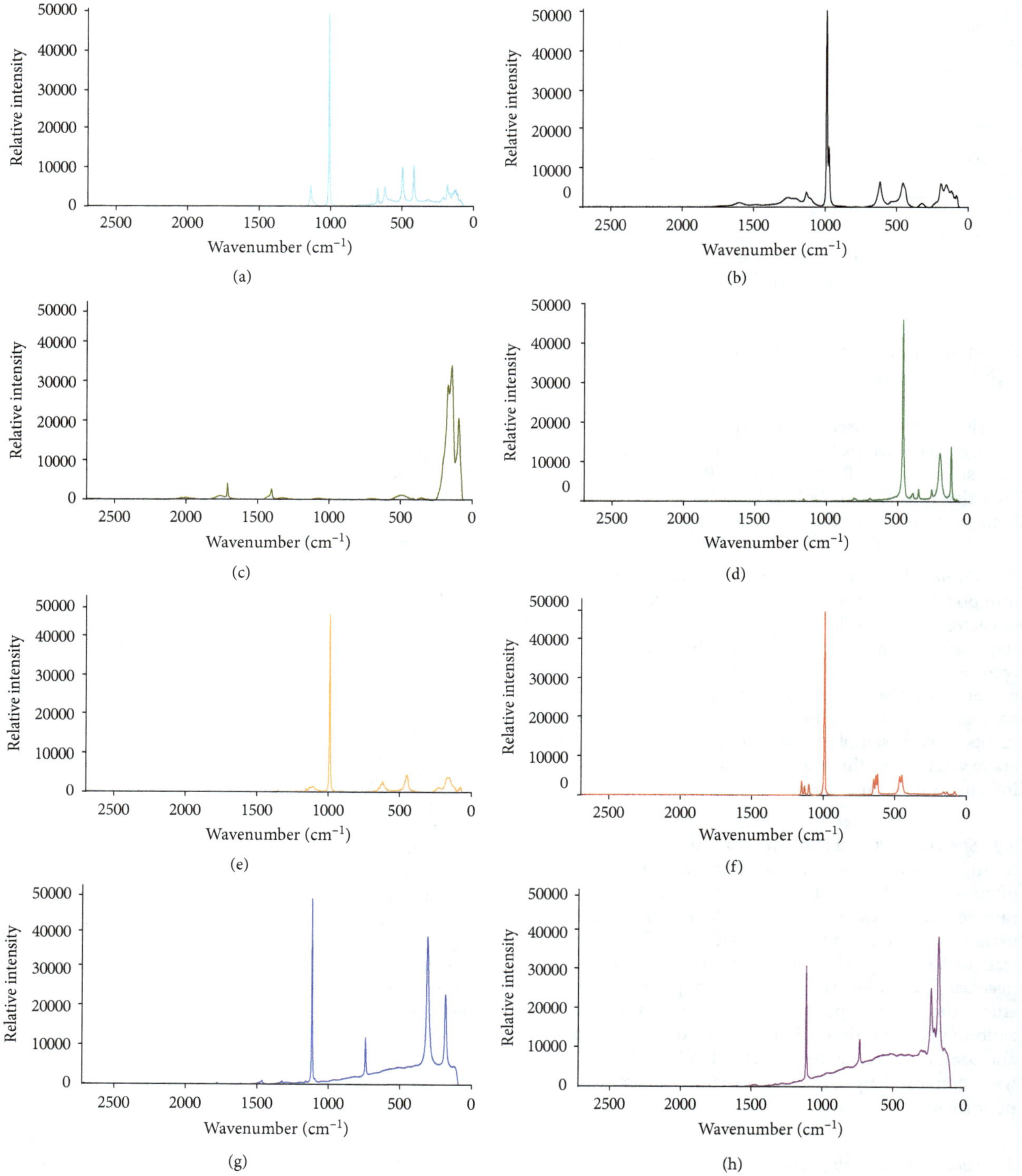

FIGURE 2: Continued.

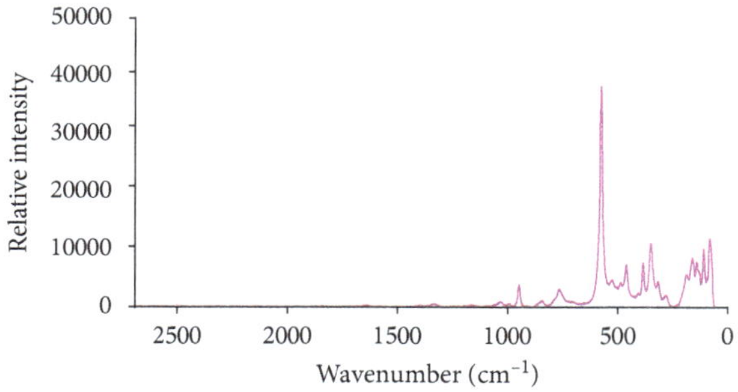

(i)

FIGURE 2: The normalized, baseline corrected average spectra of each medicine: (a) gypsum fibrosum; (b) alumen; (c) sal ammoniac; (d) quartz; (e) natrii sulfas; (f) natrii sulfas exsiccatus; (g) calcite; (h) yellow croaker otolith; (i) borax.

TABLE 2: Peak positions and assignments of Raman bands.

Mineral TCMs	Chemical composition	Raman shifts (cm^{-1})	Assignment	References
Gypsum fibrosum	$CaSO_4 \cdot 2H_2O$	92~210	Lattice modes	[29]
		415, 494	ν_2 SO_4 bending	
		619, 672	ν_4 SO_4 bending	
		1009	ν_1 SO_4 stretching	
		1137	ν_3 SO_4 asymmetric stretching	
Alumen	$KAl(SO_4)_2 \cdot 12H_2O$	86~191	Lattice modes	[30]
		443, 461	ν_2 SO_4^{2-} symmetric bending	
		617	ν_4 SO_4^{2-} asymmetric bending	
		969, 990	ν_1 SO_4^{2-} symmetric stretching	
		1097, 1133	ν_3 SO_4^{2-} asymmetric stretching	
Sal ammoniac	NH_4Cl	93~167	Lattice modes	[31]
		1402	ν_4 NH_4 bending	
		1708	ν_2 NH_4 bending	
Quartz	SiO_2	128~264	Lattice modes	[32]
		355, 395, 410, 464	Si-O bending	
		696, 795, 808	Si-O-Si symmetric stretching	
		1161	Si-O asymmetric stretching	
Natrii sulfas	$Na_2SO_4 \cdot 10H_2O$	169, 225	Lattice modes	[33]
		450	ν_2 SO_4^{2-} symmetric bending	
		620, 633, 648	ν_4 SO_4^{2-} asymmetric bending	
		989	ν_1 SO_4^{2-} symmetric stretching	
		1100~1160	ν_3 SO_4^{2-} asymmetric stretching	
Natrii sulfas exsiccatus	Na_2SO_4	80~160	Lattice modes	[34]
		452, 466	ν_2 SO_4^{2-} symmetric bending	
		620, 633, 648	ν_4 SO_4^{2-} asymmetric bending	
		993	ν_1 SO_4^{2-} symmetric stretching	
		1101, 1131, 1152	ν_3 SO_4^{2-} asymmetric stretching	
Calcite	$CaCO_3$	154, 280	Lattice modes	[35, 36]
		712	CO_3 ν_4 symmetric deformation	
		1086	CO_3 ν_1 symmetric stretching	
		1437	CO_3 ν_3 asymmetric stretching	
Yellow croaker otolith	$CaCO_3$	150, 178, 202	Lattice modes	[37]
		704	CO_3^{2-} ν_4 symmetric deformation	
		1083	CO_3^{2-} ν_1 symmetric stretching	
Borax	$Na_2B_4O_7 \cdot 10H_2O$	82~317	Lattice modes	[38]
		352, 386	ν_2 BO_4 symmetric bending	
		462, 487	ν_4 BO_4 asymmetric bending	
		576	ν_1 BO_4 symmetric stretching	
		761	ν_2 BO_3 symmetric bending	
		950	ν_1 BO_3 symmetric stretching	
		1035	ν_3 BO_4 asymmetric stretching	
		1338	ν_3 BO_3 asymmetric stretching	

Table 3: Sample sets.

Name	Label	Total	Training set	Validation set
Gypsum fibrosum	1	14	10	4
Alumen	2	10	6	4
Sal ammoniac	3	14	10	4
Quartz	4	10	6	4
Natrii sulfas	5	10	6	4
Natrii sulfas exsiccatus	6	12	8	4
Calcite	7	14	10	4
Yellow croaker otolith	8	12	8	4
Borax	9	12	8	4
Total	—	108	72	36

Table 4: The correction rates of cluster analysis based on complete spectra and intensity data extracted from original spectra and pretreated spectra (%).

Label	Complete spectra	Intensity data extracted from original spectra and pretreated spectra					
		Original spectra	VN	FD	SD	VN + FD	VN + SD
1	90 (9/10)	90 (9/10)	100	100	100	100	100
2	0 (0/6)	50 (3/6)	100	66.67 (4/6)	66.67 (4/6)	100	100
3	100	100	100	100	100	100	100
4	83.3 (5/6)	100	100	100	100	100	100
5	0 (0/6)	83.33 (5/6)	50 (3/6)	50 (3/6)	33.33 (2/6)	50 (3/6)	33.33 (2/6)
6	100	87.5 (7/8)	100	100	100	100	100
7	40 (4/10)	60 (6/10)	30 (3/10)	100	100	70 (7/10)	70 (7/10)
8	37.5 (3/8)	75 (6/8)	75 (6/8)	75 (6/8)	75 (6/8)	25 (2/8)	25 (2/8)
9	62.5 (5/8)	100	100	100	100	100	0 (0/8)
Total	61.11	83.33	83.3	90.28	88.89	86.11	70.83

nonlinear problems but also is the standard kernel applied in the majority of SVM applications [25]. RBF has two important parameters: the soft margin parameter C and kernel function parameter g [26]. The values of both parameters should be determined during the model optimization process. Common optimization methods include grid-search method (GS), particle swarm optimization algorithm (PSO), and genetic algorithm (GA). PSO was derived from the simulation of the birds in finding foods [27]. In the PSO system, each alternative solution is regarded as a "particle," multiple particles coexist and collaborate with each other (approximately like the foraging of birds), and each particle "flies" to a better position in the space of problems according to its own "experience" or the optimal "experience" of adjacent particle, so that the optimal solution can be searched. GA is an operation based on biological natural selection and genetic mechanism, in which selection, exchange, and catastrophe are considered as the operation method. With continuous genetic iterations, variables with good target values are retained, and thus, the optimal result can be eventually realized [28].

3. Results and Discussion

The normalized, baseline corrected average spectra of each medicine are shown in Figure 2, and all observed Raman bands were summarized in Table 2.

3.1. Sample Sets Partitioning. Each spectrum was assigned a label from 1 to 9 according to its name, and each spectrum accompanied by its label becomes a sample for modeling. Four samples were selected randomly from each label as validation set samples (a total of 36 samples), and the rest of 72 samples were used as training set samples (shown in Table 3). The training set is used to build the identification model, while the validation set is used for validation.

Table 5: The sensitivity and specificity of cluster analysis based on intensity data extracted from spectra pretreated by FD.

Label	Sensitivity (%)	Specificity (%)
1	100	100
2	66.67	100
3	100	96.77
4	100	100
5	50	100
6	100	100
7	100	100
8	75	100
9	100	100

3.2. Spectral Data Compression. Because Raman peaks are distinct and sharp, the intensity data of characteristic peaks could be regarded as characteristic input variables for the SVM model instead of the complete spectra. In order to investigate the feasibility of this method, cluster analysis of the extracted intensity data of training set samples was performed based on the between-groups linkage method and square Euclidean distances by SPSS. The correction rate

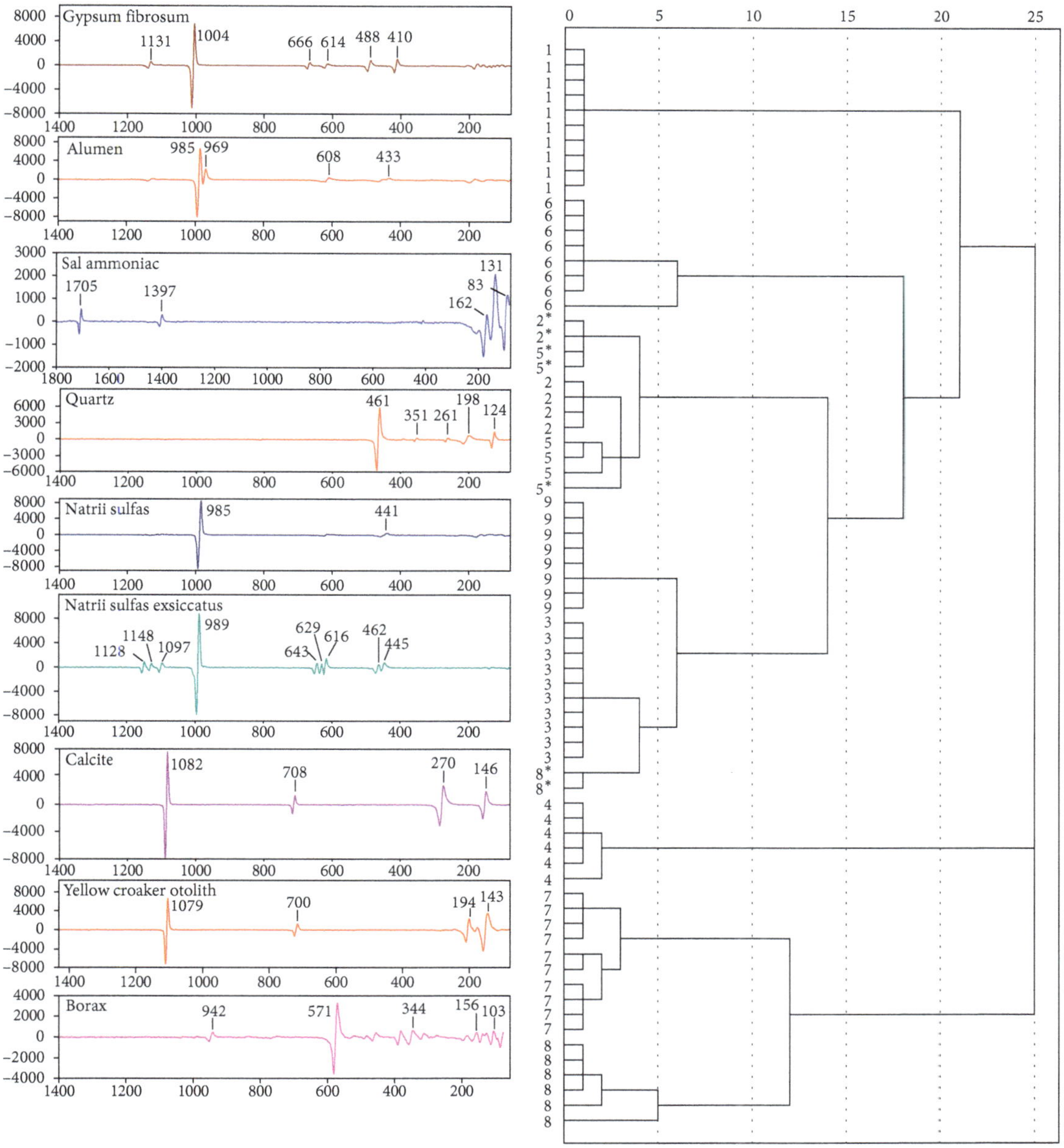

FIGURE 3: (a) Peaks selected from spectra pretreated by FD. (b) The dendrogram (*samples which were clustered falsely).

of cluster analysis based on the extracted intensity data of original spectra and pretreated spectra was compared in order to determine the optimal pretreatment method.

The correction rate of cluster analysis based on complete spectra, the intensity data of original spectra, and intensity data of pretreated spectra is shown in Table 4. The correction rate based on complete spectra was just 61.11%. The value improved significantly based on the intensity data. Above all, the correction rate based on intensity data extracted from spectra pretreated by FD was highest (90.28%), and the corresponding sensitivity and specificity values of each mineral TCM are shown in Table 5. It can be seen that the results of cluster analysis based on intensity data extracted from spectra pretreated by FD were good. Figure 3 shows the selected strong peaks of spectra pretreated by FD and the corresponding dendrogram. There are seven samples clustered falsely, which are mainly alumen and natrii sulfas. Both of them are sulfate mineral TCMs, so their Raman spectra are of high similarity.

The result of cluster analysis shows the feasibility of the spectral compression method based on extracting intensity data. However, some samples whose Raman spectra are of high similarity, such as alumen and natrii sulfas, cannot be differentiated by cluster analysis. Therefore, the SVM algorithm was introduced to establish an identification model on the basis of extracted intensity data.

3.3. Establishment of SVM Identification Model. Because the clustering correction rate based on intensity data extracted from spectra pretreated by FD was highest, SVM models were established by using the intensity data extracted from spectra pretreated by FD as input variables and the labels as output variables. All the input variables should be normalized firstly. The kernel parameters were optimized by GS, GA, and PSO simultaneously. The model performance was evaluated by 3-fold cross validation (3-CV) of the training set. In the process of optimization, the optimal values of C and g were determined when the 3-CV accuracy reached the maximum. After SVM models were established, the training set and validation set were predicted. The prediction ability was evaluated by the prediction accuracies of the training set and validation set.

The modeling results show that the best combination of C and g ($C = 1$, $g = 0.5744$) was determined by GS, the 3-CV accuracy is 98.61%, and both the prediction accuracy of the training set and validation set are 100%. The optimization process is shown in Figure 4.

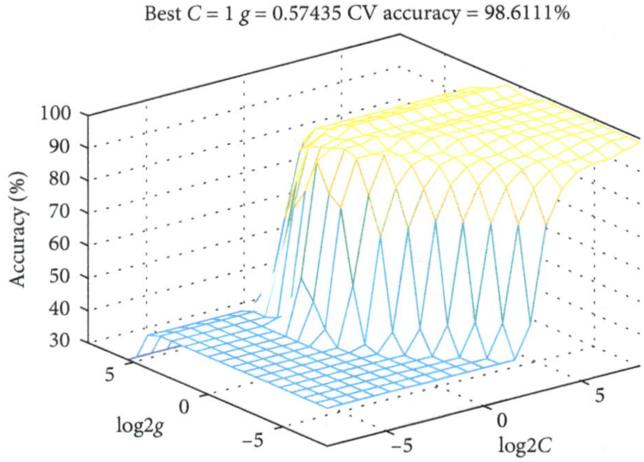

FIGURE 4: The GS optimization process of SVM models.

3.4. Spectral Compression Based on PCA. Spectra data points were compressed from more than 2000 into less than 50 by using the spectral compression method based on extracting intensity data. The other spectral data compression method investigated in this research is PCA. PCA was conducted in MATLAB 2014b to reduce dimension of original spectra and pretreated spectra. The accumulative contribution rates (ACR) of PCs are shown in Figure 5.

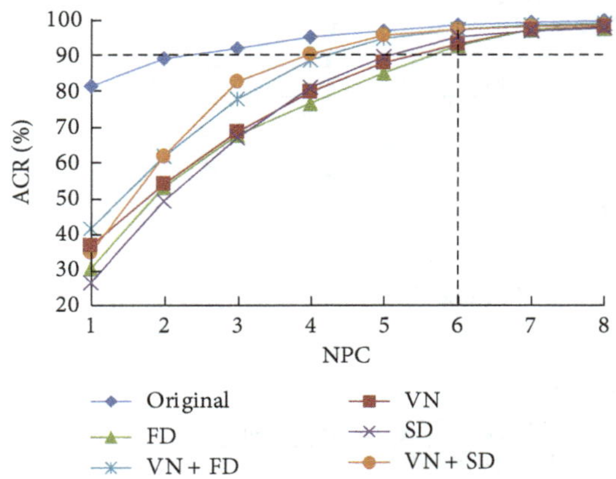

FIGURE 5: ACR of PCs of original spectra and pretreated spectra.

As shown in Figure 5, for original spectra and pretreated spectra, the ACR of the first 6 PCs reached 90%, which means that the first 6 PCs can represent most of the information provided by the complete spectrum basically. Therefore, PCA-SVM models were established using the first 6 PCs of the original spectra and pretreated spectra to investigate the effect of different spectra pretreatment methods to the model performances.

3.5. Establishment of PCA-SVM Identification Model. PCA-SVM models were established using the scores of the first 6 PCs as input variables and the labels as output variables. Table 6 shows the results of PCA-SVM models based on original spectra and pretreated spectra. It can be found that the 3-CV accuracy and the prediction accuracies of the PCA-SVM model based on VN were the highest, so VN is considered as the most suitable pretreatment method for the PCA-SVM model.

However, under the pretreatment of VN, the first 6 PCs may not be optimal for modeling. If NPC is too less, the established model would be unable to reflect the relation between sample characteristics and the spectra information, and the phenomenon of "under-fitting" will occur. However, if NPC is too much, the prediction accuracy and generalization ability of the model will be affected, and "over-fitting" will occur. Therefore, model performance should be investigated under a different NPC so as to determine the optimal NPC. In this study, PCA-SVM models were established using the scores of the first 1, 2, 3, . . ., 8 PCs, respectively, and the results are shown in Table 7.

As shown in Table 7, the best performance and prediction ability of the PCA-SVM model were obtained when the NPC is 7 and optimized by GA. The 3-CV accuracy and the prediction accuracies of the training set and validation set were just reached 100%. Prediction accuracy declined when less PCs were loaded, and the prediction ability of models kept unchanged when more PCs were loaded. The GA parameters-optimized process is shown in Figure 6.

The models established on the two spectra compression method have good accuracies. Compared with PCA, the spectra compression method based on extracting intensity data has some advantages; for instance, it is easy to learn and to popularize because it avoids complicated calculation. What is more, the extracted Raman shifts and intensity data of Raman peaks could be used to establish Raman spectra database. However, the disadvantage is that the errors of the peak position and intensity may cause the reduction of prediction accuracy.

The overall results indicated that Raman spectroscopy with the SVM method could efficiently identify the mineral TCMs. Compared with other characterization techniques,

TABLE 6: Results of PCA-SVM models based on different pretreatment methods.

Pretreatment method	NPC	Parameter optimization	C	g	3-CV	Accuracy (%)																			
						Training set										Validation set									
						1	2	3	4	5	6	7	8	9	Total	1	2	3	4	5	6	7	8	9	Total
Original	6	GS	9.1896	5.2780	93.06	100	100	100	50	83.33	100	100	100	100	94.44	100	100	100	25	100	100	100	100	100	91.67
VN	6	GS	3.0314	9.1896	98.61	100	100	100	100	83.33	100	100	100	100	98.61	100	100	100	100	100	100	100	100	100	100
FD	6	GA	0.3272	16.5711	93.06	100	100	100	66.67	66.67	100	100	100	87.5	93.06	100	100	100	50	100	100	100	100	100	94.44
SD	6	PSO	1.0047	7.1087	88.89	100	100	100	83.33	100	100	100	100	100	98.61	100	100	100	25	75	100	100	100	100	88.89
VN + FD	6	GA	1.4940	9.4271	95.83	100	100	100	83.33	83.33	100	100	100	100	97.22	100	100	100	100	100	100	100	100	100	100
VN + SD	6	PSO	6.4480	4.2768	94.44	100	100	100	66.67	66.67	100	100	100	100	94.44	100	100	100	100	100	100	100	100	100	100

Table 7: Results of PCA-SVM models with a different NPC.

NPC	Parameter optimization	C	g	3-CV	Accuracy (%)																			
					Training set										Validation set									
					1	2	3	4	5	6	7	8	9	Total	1	2	3	4	5	6	7	8	9	Total
8	GS	1	5.2780	100	100	100	100	100	100	100	100	100	100	100	100	100	100	100	100	100	100	100	100	100
7	GA	10.6841	6.3515	100	100	100	100	100	100	100	100	100	100	100	100	100	100	100	100	100	100	100	100	100
6	GS	3.0314	9.1896	98.61	100	100	100	100	83.33	100	100	100	100	98.61	100	100	100	100	100	100	100	100	100	100
5	PSO	16.2224	11.7130	98.61	100	100	100	100	83.33	100	100	100	100	98.61	100	100	100	100	100	100	100	100	100	100
4	GS	256	0.1088	98.61	90	100	100	100	66.67	87.5	100	100	100	94.44	100	100	100	100	75	100	100	100	100	97.22
3	GA	37.7015	136.7847	87.50	100	100	100	100	100	100	100	100	100	100	100	100	100	100	100	100	100	100	50	94.44

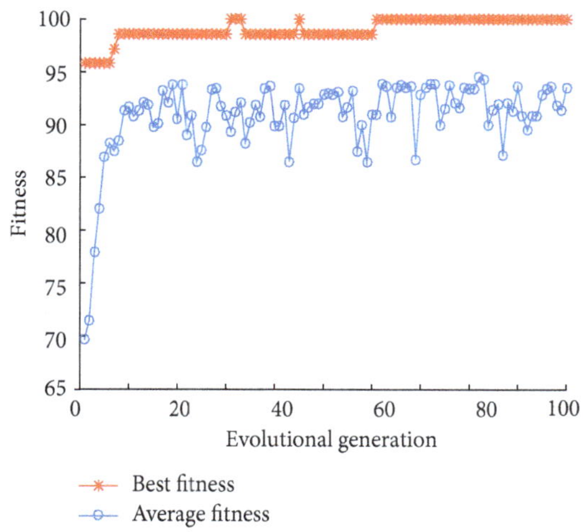

FIGURE 6: The GA optimization process of the optimal PCA-SVM models.

Raman spectroscopy shows significant advantages. It is more objective and effective than the morphological description method in identifying TCMs that are similar in appearance. It is of high chemical specificity and high accuracy than the chemical method. For an unknown mineral TCM sample, the chemical method requires at least two experiments, and Raman spectroscopy only needs one detection to obtain the identification results. Moreover, since Raman spectra can reflect the structural characteristics of crystals, it can distinguish mineral TCMs with the same chemical composition but a different crystal structure, such as calcite and aragonite whose main components are both $CaCO_3$. Furthermore, compared with infrared spectroscopy (IR) or near-infrared spectroscopy (NIR), Raman spectroscopy has the advantages of sharp peaks, high resolution, and high specificity and are not susceptible to water or other interference. X-ray diffraction (XRD) is one of the most effective modern technology for identification of mineral TCMs. It is of high accuracy and good specificity, but the requirement of sample preparation is high, and the spectra analysis is complicated. The advantages of nondestructive and no sample pretreatment of Raman spectroscopy make it more convenient and efficient than XRD.

4. Conclusion

The models established using Raman spectroscopy based on the SVM method have excellent performance and strong prediction capacity for identifying these nine easily confused mineral TCMs. It is rapid and convenient to collect spectra of mineral TCMs without sample pretreatment using a portable Raman spectrometer, and the result of identification could be obtained immediately when the spectra of samples were loaded into the SVM model. It means that the SVM model can not only meet the need of rapid inspection in the Chinese medicine markets but also be used in the production process to control the quality of raw materials to ensure the safe use of the medication.

The result shows the feasibility of the SVM algorithm in the rapid identification and quality analysis by Raman spectroscopy. Raman spectroscopy in combination with chemometric algorithms such as PCA, SVM, PSO, and GA has the potential to contribute significantly to the analysis of other mineral TCMs. In the next work, more samples should be used for improving the reliability and applicability of the model. Raman spectra database of mineral TCMs will be created with the increase of samples. Other algorithms will also be investigated in the future to improve the model performance and expand the application of Raman spectroscopy in the field of TCMs.

Conflicts of Interest

The authors declare that there are no conflicts of interest regarding the publication of this paper.

Acknowledgments

This study was supported by the Wuhan Special Fund of Biotechnology and New Medicine from Development Action Plan of High-Tech Industry in 2012 (201260523193).

References

[1] National Pharmacopoeia Committee, *China Pharmacopoeia of 2015 Edition*, Vol. 1, China Medical Science and Technology Press, Beijing, China, 2015.

[2] Y. Sohn, H. C. Kang, K. S. Kim et al., "Protective effects of natrii sulfas on cerebral focal ischemia induced by MCAO in rats," *American Journal of Chinese Medicine*, vol. 37, no. 2, pp. 273–293, 2009.

[3] J. X. Yu, T. L. Lu, C. Q. Mao, and Q. Chen, "Research on anti-inflammation and acute toxicity test in sal ammoniac and halite violaceous," *Journal of Nanjing University of Chinese Medicine*, vol. 28, pp. 77–79, 2012.

[4] *Editorial Committee of Zhong Yao Da Ci Dian, Zhong Yao Da Ci Dian*, Shanghai Scientific and Technical Publishers, Shanghai, China, 2nd edition, 2006.

[5] Editorial Committee of Zhong Hua Ben Cao, *State Administration of Traditional Chinese Medicine of China*, Zhong Hua Ben Cao, Shanghai Scientific and Technical Publishers, Shanghai, China, 1999.

[6] Y.-H. Liao, C.-H. Wang, C.-Y. Tseng, H.-L. Chen, L.-L. Lin, and W. Chen, "Compositional and conformational analysis of yam proteins by near infrared fourier transform Raman spectroscopy," *Journal of Agricultural and Food Chemistry*, vol. 52, no. 26, pp. 8190–8196, 2004.

[7] X. Qi, C. C. Zeng, H. P. Liu, and S. H. Liu, "Rapid quantitative analysis of puerarin using Raman spectroscopy," *Spectroscopy*, vol. 28, pp. 34–39, 2013.

[8] Z. J. Zhang, Q. Zhou, J. Z. Wei et al., "Validation of the crystal structure of medicinal realgar in China," *Guang Pu Xue Yu Guang Pu Fen Xi*, vol. 31, pp. 291–296, 2011.

[9] S. Li, Y. Shan, X. Zhu, X. Zhang, and G. Ling, "Detection of honey adulteration by high fructose corn syrup and maltose

syrup using Raman spectroscopy," *Journal of Food Composition and Analysis*, vol. 28, no. 1, pp. 69–74, 2012.

[10] Y. Zhang, X. Huang, W. Liu, Z. Cheng, C. Chen, and L. Yin, "Analysis of drugs illegally added into Chinese traditional patent medicine using surface-enhanced Raman scattering," *Analytical Sciences*, vol. 29, no. 10, pp. 985–990, 2013.

[11] Q. E. Wan, H. P. Liu, H. M. Zhang, and S. H. Liu, "Identification of ginseng and its counterfeit by laser Raman spectroscopy," *Guang Pu Xue Yu Guang Pu Fen Xi*, vol. 32, pp. 989–992, 2012.

[12] W. N. Wang, D. L. Chen, M. F. Zhu, and H. M. Zhang, "The analysis and identification of Fritillaria cirrhosa by Raman spectra," *Guang Pu Xue Yu Guang Pu Fen Xi*, vol. 33, pp. 2109–2111, 2013.

[13] H. G. M. Edwards, T. Munshi, and K. Page, "Analytical discrimination between sources of ginseng using Raman spectroscopy," *Analytical and Bioanalytical Chemistry*, vol. 389, no. 7-8, pp. 2203–2215, 2007.

[14] H. Huang, J. Li, R. Chen et al., "Discrimination of Huangqi (Radix Astragali Seu Hedysari) from different producing areas using Raman spectroscopy and statistical analysis," *Journal Fuzhou University*, vol. 42, pp. 646–652, 2014.

[15] J. Ming, L. Chen, K. L. Chen, and B. S. Huang, "Qualitative and quantitative analysis of amber adulteration using near infrared spectroscopy based on support vector machine," *Chinese Medicinal Materials*, vol. 40, pp. 32–37, 2017.

[16] S. Kokot, Y. Lai, and Y. Ni, "Classification of raw and roasted Semen Cassiae samples with the use of Fourier transform infrared fingerprints and least squares support vector machines," *Applied Spectroscopy*, vol. 64, no. 6, pp. 649–656, 2010.

[17] M. Lei, L. Chen, B. Huang, and K. Chen, "Determination of magnesium oxide content in mineral medicine talcum using near-infrared spectroscopy integrated with support vector machine," *Applied Spectroscopy*, vol. 71, no. 11, pp. 2427–2436, 2017.

[18] L. Chen, J. Ming, M. Y. Yuan, B. S. Huang, and K. L. Chen, "Quantitative model of Raman spectra for CaF2 in fluoritum based on BP-ANN algorithm," *Chinese Journal of Experimental Traditional Medical Formulae*, vol. 22, pp. 77–82, 2016.

[19] Y. Sun, L. Chen, B. Huang, and K. Chen, "A rapid identification method for calamine using near-infrared spectroscopy based on multi-reference correlation coefficient method and back propagation artificial neural network," *Applied Spectroscopy*, vol. 71, no. 7, pp. 1447–1456, 2017.

[20] X. D. Zhang, L. Chen, Y. B. Sun, Y. Bai, B. S. Huang, and K. L. Chen, "Determination of zinc oxide content of mineral medicine calamine using near-infrared spectroscopy based on MIV and BP-ANN algorithm," *Spectrochimica Acta Part A: Molecular and Biomolecular Spectroscopy*, vol. 193, pp. 133–140, 2017.

[21] K. R. Beebe, R. J. Pell, and M. B. Seasholtz, *Chemometrics: A Practical Guide*, John Wiley & Sons, London, UK, 1988.

[22] V. Vapnik, *The Nature of Statistical Learning Theory*, Springer Berlin Heidelberg, New York, NY, USA, 1995.

[23] S. R. Amendolia, G. Cossu, M. L. Ganadu, B. Golosio, G. L. Masala, and G. M. Mura, "A comparative study of K-nearest neighbour, support vector machine and multi-layer perceptron for thalassemia screening," *Chemometrics and Intelligent Laboratory Systems*, vol. 69, no. 1-2, pp. 13–20, 2003.

[24] N. Cristianini and J. Shawe-Taylor, *An Introduction to Support Vector Machines: And Other Kernel-Based Learning Methods*, Cambridge University Press, Cambridge, UK, 2000.

[25] B. Tudu, A. Jana, A. Metla, D. Ghosh, N. Bhattacharyya, and R. Bandyopadhyay, "Electronic nose for black tea quality evaluation by an incremental RBF network," *Sensors and Actuators B: Chemical*, vol. 138, no. 1, pp. 90–95, 2009.

[26] L.-Q. Hu, C.-L. Yin, and Z.-P. Zeng, "Detection of adulteration in acetonitrile using near infrared spectroscopy coupled with pattern recognition techniques," *Spectrochimica Acta Part A: Molecular and Biomolecular Spectroscopy*, vol. 151, pp. 34–39, 2015.

[27] J. Kennedy and R. Eberhart, "Particle swarm optimization," in *Proceedings of the IEEE International Conference on Neural Networks*, vol. 4, pp. 1942–1948, Honolulu, Hawaii, May 2002.

[28] M. Jalali-Heravi and A. Kyani, "Application of genetic algorithm-kernel partial least square as a novel nonlinear feature selection method: activity of carbonic anhydrase II inhibitors," *European Journal of Medicinal Chemistry*, vol. 42, no. 5, pp. 649–659, 2007.

[29] B. J. Berenblut, P. Dawson, and G. R. Wilkinson, "The Raman spectrum of gypsum," *Spectrochimica Acta Part A: Molecular Spectroscopy*, vol. 27, no. 9, pp. 1849–1863, 1971.

[30] H. Kishimura, Y. Imasu, and H. Matsumoto, "Thermal dehydration of potash alum studied by Raman spectroscopy and X-ray diffraction analysis," *Materials Chemistry and Physics*, vol. 149-150, pp. 99–104, 2015.

[31] J. Ming, L. Chen, K. L. Chen, Y. Cao, and B. S. Huang, "Identification of six white crystal mineral medicines," *Chinese Journal of Experimental Traditional Medical Formulae*, vol. 22, pp. 33–38, 2016.

[32] E. Zu, M. Li, and P. Zhang, "Study on jades of SiO_2 by Raman spectroscopy," *Journal of Kunming University of Science and Technology*, vol. 25, pp. 77-78, 2000.

[33] A. Hamilton and R. I. Menzies, "Raman spectra of mirabilite, $Na_2SO_4{\cdot}10H_2O$ and the rediscovered metastable heptahydrate, $Na_2SO_4{\cdot}7H_2O$," *Journal of Raman Spectroscopy*, vol. 41, no. 9, pp. 1014–1020, 2010.

[34] S. Gunasekaran, G. Anbalagan, and S. Pandi, "Raman and infrared spectra of carbonates of calcite structure," *Journal of Raman Spectroscopy*, vol. 37, no. 9, pp. 892–899, 2010.

[35] B.-K. Choi and D. J. Lockwood, "Raman spectrum of Na2SO4 (phase V)," *Solid State Communications*, vol. 72, no. 1, pp. 133–137, 1989.

[36] W. P. Griffith, "Raman spectroscopy of minerals," *Nature*, vol. 224, no. 5216, pp. 264–266, 1969.

[37] R. Frech, E. C. Wang, and J. B. Bates, "The i.r. and Raman spectra of $CaCO_3$ (aragonite)," *Spectrochimica Acta Part A: Molecular Spectroscopy*, vol. 36, no. 10, pp. 915–919, 1980.

[38] S. A. Devi, D. Philip, and G. Aruldhas, "Infrared, polarized Raman, and SERS spectra of borax," *Journal of Solid State Chemistry*, vol. 113, no. 1, pp. 157–162, 1994.

10

The Chemical States of Color-Induced Cations in Tourmaline Characterized by X-Ray Photoelectron Spectroscopy

Ming Li,[1,2] Hanlie Hong,[1] Ke Yin,[1] Chaowen Wang,[3] Feng Cheng,[1] and Qian Fang[1]

[1]*School of Earth Sciences, China University of Geosciences, Wuhan, China*
[2]*Department of Gemmology, Jinling Institute of Technology, Nanjing, China*
[3]*Gemmological Institute, China University of Geosciences, Wuhan, China*

Correspondence should be addressed to Hanlie Hong; honghl8311@aliyun.com

Academic Editor: Vincenza Crupi

In order to better understand the effect of transition metal cations on color of tourmaline, X-ray photoelectron spectroscopy was used to investigate the species, chemical state, site occupancy, and chemical environment of color-induced metal cations in colorful tourmaline samples from Minas Gerais State, Brazil. Our results showed that the colorful tourmalines usually contained a small amount of transition metal elements, and a colorful tourmaline sample had several transition metal cations; however, the color of tourmaline resulted from the transition metal cations in the Y site of the crystal structure. The pink color of tourmaline was associated with Mn^{2+} in the Y site coordinating with F; the yellow color was derived from Ni^{2-} in the Y site binding to O; the green color was associated with Fe^{3+} in the Y site coordinating with O, OH, and F; the rose red color originated from Mn^{2+} and Ni^{2+} in the Y site in which Mn^{2+} coordinated with O and F, and Ni^{2+} coordinated with O; and the blue color was derived from Fe^{3+} and Mn^{2+} in the Y site in which Fe^{3+} binded to O, OH, and F and Mn^{2+} binded to F. Additionally, other transition metal cations were also observed in colorful tourmalines, but all these species occupied the Z site of the structure. In the pink and yellow samples, Fe and Cr were observed in Fe^{3+} and Cr^{3+}; in the rose red sample, Fe was also found in Fe^{3+}; in the blue sample, Cr was present in Cr^{3+}; in the green sample, Mn, Ni, and Cu were found in Mn^{2+}, Ni^{2+}, and Cu^{2+}, respectively. The color of tourmaline was induced from the absorption of the d-d transition of transition metals in the crystal structure, as charge transfer tended to occur between cations occupying different coordination positions.

1. Introduction

Tourmaline, commonly known as "Bixi" in China, is a precious medium-high-grade gem material. It is termed the "fallen rainbow" and valued by consumers for its extremely rich and unique colors. Tourmaline is a polar silicate mineral crystal with no center of symmetry and has complex chemical compositions. Fluorine, hydroxyl, and other anions in addition to the silicate root are often found with boric acid and cations in the structure, such as Cr, Mn, and Fe, which have a wide range of isomorphic replacement. Tourmaline colors are extremely rich and variable [1–8]. Tourmaline can be colorless, pink, rose red, red, yellow, brown, green, dark green, light blue, blue, purple, and black, with color changes associated with variable chemical compositions. The mechanism that determines the color of tourmaline has been a perennial problem in mineralogy. For example, some researchers believe that pink originated from the absorption of d-d transition of Mn in the octahedron, while others hypothesize it is the result of a color heart, similar to smoky quartz [9]. However, Chaudhry and Howie [10] did not find Mn when analyzing pink tourmaline from Devonshire, which rebuts the hypothesis that pink is caused by the absorption of the d-d transition of Mn^{3+}. Babińska et al. [11] used electron paramagnetic resonance (EPR) to identify Mn^{2+} in elbaite, Fe^{3+} in dravite and schorl. They also found that the color and spectral characteristic of tourmaline that contains Fe did not change regularly with varied Fe contents. This phenomenon suggests that the color of those tourmaline results from the charge transfer between Fe^{2+} and Fe^{3+} [12, 13]. Through research on pink and green tourmaline using the XPS (X-ray photoelectron spectroscopy) method, Hong et al. [14] hypothesized that transition metal cations in different colored tourmalines have the same chemical state, for example, Fe^{3+}, Mn^{4+}, Ti^{4+}, and Cr^{3+}, but with different coordination ions. That may be the primary reason for the color difference. Petrov [15] showed that the violet color in tourmaline is caused by the absorption of the d-d transition for Cr^{3+}. A large number of studies have shown that the varied color in tourmaline is related to the various transition metal cation species, especially the state of these cations. However, researchers still have differing opinions on the attribution of the absorption belt, chemical state of the ions, and color mechanism [16, 17]. This study researched the chemical states of metal cations using X-ray photoelectron spectroscopy and explored the species, valence state, and especially coordination ion and site occupancy of color-induced metal cations, in pink, yellow, green, rose red, and blue tourmaline.

2. Materials and Methods

Six pieces of tourmaline, representative of different colors, were selected as samples from Minas Gerais, Brazil (Figure 1). Five samples were colored pink, yellow, green, rose red, and blue, and one colorless sample was chosen as the baseline sample.

Photoelectron spectroscopy can provide both qualitative and quantitative results, including the chemical state of the surface elements of the sample by testing the kinetic energy of the photoelectron and Auger electrons emitted from the solid surface from the photoelectric and Auger effects. This series of experiments was conducted on an AXIS-ULTRA DLD-600W X-ray photoelectron spectrometer at the Huazhong University of Science and Technology Analysis and Testing Center. In order to avoid contamination, the crystal sample was enwrapped with tinfoil and was then broken by a pair of pincers, and the fresh surface was subsequently used for XPS measurement. Test conditions were as follows: Al Kα ($E = 1486.6$ eV) X-ray radiation source, instrument vacuum better than 5×10^{-9} Torr, scanning step length of 0.05 eV, and counting time of 500 ms. The charge displacement is fixed, and the C1s electron binding energy value (285 eV) was used for equipment calibration. The range of energy spectrum was 0–1200 eV. Because the subject of analysis was primarily trace elements, we carefully scanned the photoelectron spectra of these trace elements and repeated 15 times to improve the resolution.

For X-ray diffraction (XRD) analysis, the tourmaline sample was broken and further ground manually to powder using an agate mortar. XRD measurement was performed in a Philips PW 3710 diffractometer operated at CuKα radiation (45 kV, 35 mA) and scan rate of 4° 2θ/min at the State Key Laboratory of Geological Processes and Mineral Resources, China University of Geosciences (Wuhan).

3. Results and Discussion

3.1. Crystal Chemistry of Tourmaline. The X-ray photoelectron spectroscopy results for the tourmaline samples are shown in Figure 2. Based on the standard electron binding energy data of the elements, although the tourmaline samples

FIGURE 1: Tourmaline samples.

differ in color, their main chemical components are similar; they all contain Al, Si, Na, O, F, B, and other elements. Results from the further slow scan confirm that these tourmaline samples also contain a small amount of Cr, Mn, Fe, Ni, and Cu, and other elements, but no Li or Be.

Tourmaline is a ring borosilicate mineral; its crystal chemical formula is generally recognized as [18] $XY_3Z_6[T_6O_{18}][BO_3]_3V_3W$, where the X site is primarily occupied by large radius metal cations, such as Na^+, K^+, and Ca^{2+}. Sometimes, Mg^{2+} or a vacancy can be found in the X site. Its coordination number is 9. The Y site is occupied by Al^{3+}, Fe^{3+}, V^{3+}, Cr^{3+}, Mg^{2+}, Mn^{2+}, Fe^{2+}, Cu^{2+}, Zn^{2+}, Li^+, and Ti^{4+}; it can also contain some vacancies and has a coordination number of 6 [19]. The Z site is occupied by Al^{3+}, Fe^{3+}, V^{3+}, Cr^{3+}, Mg^{2+}, and Fe^{2+} and has a coordination number of 6. The T site is occupied by Si, can be partly replaced by Al, B, and Be, and has a coordination number of 4. The coordination number of B is 3, and there is no clear substitution. The W (O1) site is occupied by OH, F, and O and has a coordination number of 3. The V (O3) site is occupied by OH and O and has a coordination number of 3 [20].

In the tourmaline structure, the $[TO_4]$ tetrahedron forms $[T_6O_{18}]^{12-}$ a hexagonal ring. In comparison, the cation occupying the Y position of the crystal structure forms a brucite structure with O^{2-}, OH^-, and F^-. Three $[Y\text{-}O_4VW]$ coordination octahedrons are connected to the hexagonal ring by sharing an O^{2-} on the top of $[TO_4]$. The intersection of the three octahedrons lies on the central axis of the hexagonal ring, which is the W position. $[BO_3]$ is found between the six square rings and octahedral layer, which shares one O^{2-} with the octahedral layer. These complex anions are connected to each other by Z ions. The $[Z\text{-}O_5V]$ octahedron and $[Y\text{-}O_4VW]$ octahedron share one edge [21]. The X ions are located in the upper space of the hexagonal. Bosi [7] has shown that the W (O1) anion is mainly bonded to the metal cation in the Y site of the tourmaline crystal structure. Therefore, the metal cation in the Y site is mainly coordinated with the W (O1) anion, while the metal cation in the Z site does not bind to the W (O1) anion.

3.2. The Chemical State of Trace Elements. The tourmaline color is often associated with a very small amount of metal cations [22]. The fine analysis results showing the characteristic energy spectra for the metal cations are shown in Figures 3–7.

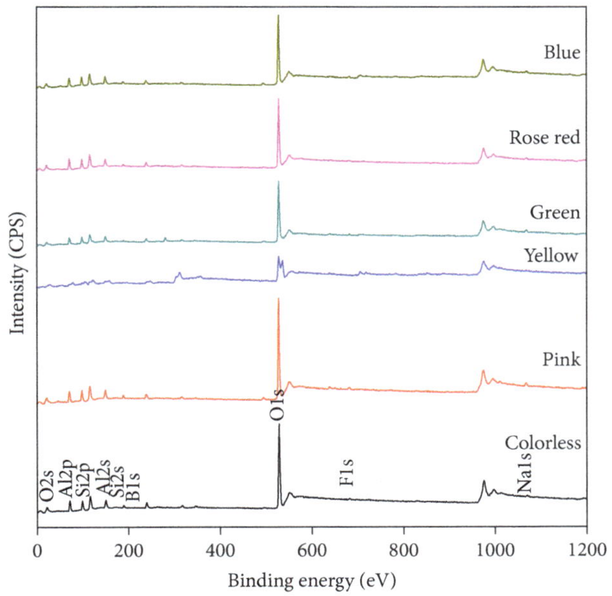

FIGURE 2: X-ray photoelectron spectra of tourmaline samples.

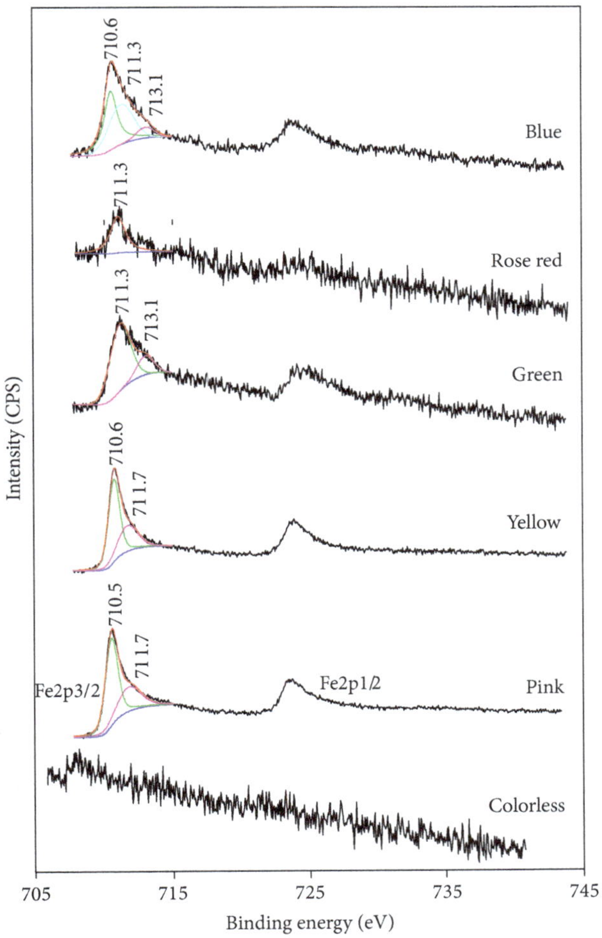

FIGURE 3: XPS spectra for Fe2p.

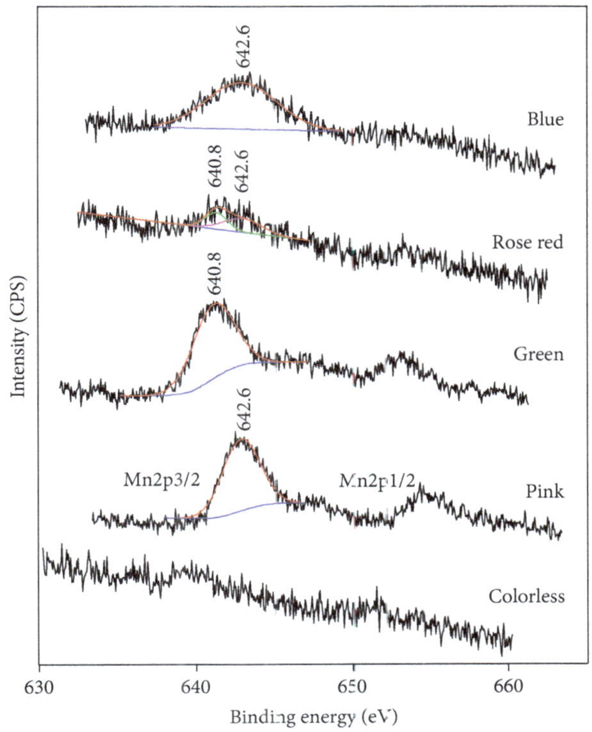

FIGURE 4: XPS spectra for Mn.

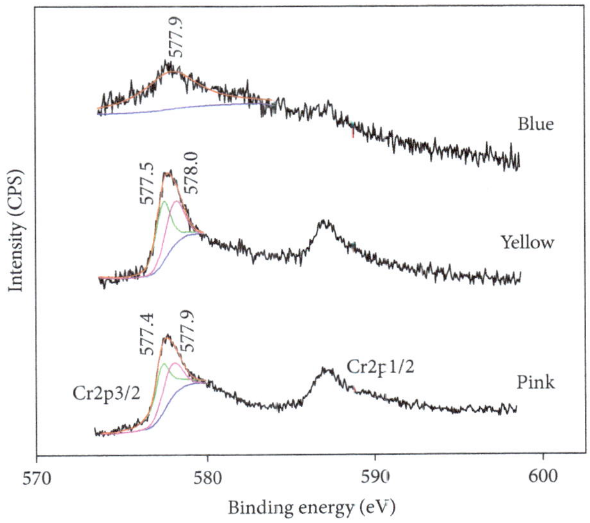

FIGURE 5: XPS spectra for Cr2p.

Figure 3 shows the XPS spectrum for Fe2p. The peaks of the energy spectrum line for the pink, yellow, and blue samples are clearly higher than those of colorless, green, and rose red samples. The peak heights reflect the abundance of Fe in tourmaline crystal. Therefore, the abundance of Fe in the pink, yellow, and blue samples is clearly higher than in the colorless, green, and rose red samples. For the colorless contrast sample, there is no significant spectral peak, indicating the lack of Fe. Furthermore, different samples have different

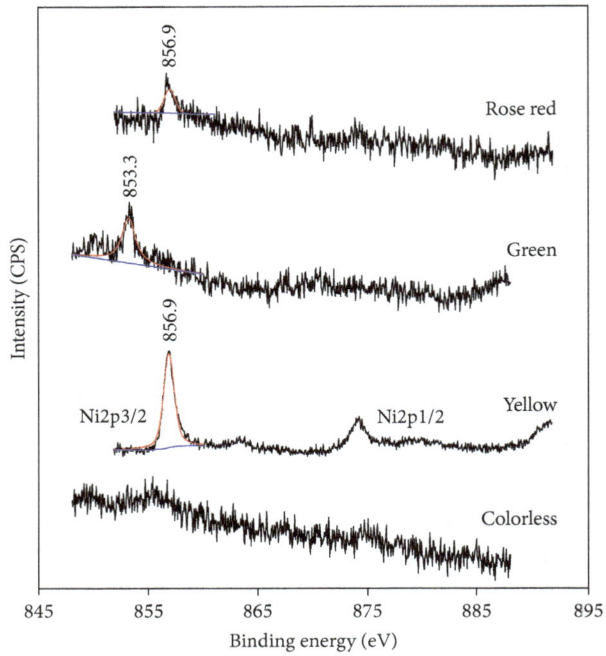

FIGURE 6: XPS spectra for Ni2p.

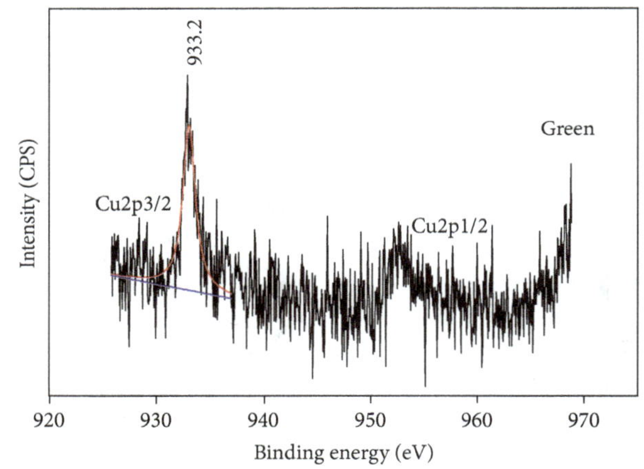

FIGURE 7: XPS spectrum for Cu2p.

peak positions. The Fe2p3/2 peaks of these samples show clearly an asymmetric shape, with one shoulder peak in the higher binding energy side, suggesting that the Fe2p3/2 line is an overlapping peak. By fitting calculation [23], the peak was decomposed into two peaks at 710.5 and 711.7 eV for the pink sample, at 710.6 and 711.7 eV for the yellow sample, at 711.3 and 713.1 eV for the green sample, and at 710.6, 711.3, and 713.1 eV for the blue sample, respectively. For the rose red sample, the Fe2p3/2 signal is significantly weak and is present as a single peak at 711.3 eV.

The different binding configurations and estimated percentages of Fe are reported in Table 1. The electron binding energies of Fe2p3/2 in Fe_2O_3, FeOOH, and $NaFeO_2$ are 710.7, 711.3, and 711.8 eV, respectively. In the pink, yellow, and blue samples, the electron binding energy of the strongest Fe2p3/2 component are 710.5, 710.6, and 710.6 eV, respectively, coinciding with the electron binding energy of Fe2p3/2 in Fe_2O_3, indicating that Fe is mainly in the Fe^{3+} chemical state and coordinates with O. The Fe2p3/2 peak at 711.7 eV for pink and yellow samples is approximately consistent with the electron binding energy of Fe2p3/2 in $NaFeO_2$, indicating that part of Fe coordinates with O and O also binds with Na in the crystal structure of these samples. The binding energy of Fe2p3/2 in the rose red sample is 711.3 eV, in agreement with the electron binding energy of Fe2p3/2 in FeOOH, indicating that Fe mainly bonds with O and OH. However, the electron binding energy of Fe2p3/2 at 713.1 eV in the green and blue samples is slightly lower than the value 713.9 eV for Fe2p3/2 in FeF_3 [24], indicating that Fe is in the Fe^{3+} chemical state and mainly bonds with F, O, and OH. The relative proportion of Fe in different coordinate states is listed in Table 1.

TABLE 1: Different binding configurations, estimated atomic percentages, and occupancy site of Fe, Mn, Cr, and Ni.

(a)

	Fe^{3+}—O	Fe^{3+}—O +OH	Fe^{3+}—O(—Na)	Fe^{3+}—F	Site
Pink	66%	/	34%	/	Z
Yellow	65%	/	35%	/	Z
Green	/	82%	/	18%	Y
Rose red	/	100%	/	/	Z
Blue	48%	43%	/	9%	Y

(b)

	Mn^{2+}—O	Mn^{2+}—F	Site
Pink	/	100%	Y
Green	100%	/	Z
Rose red	41%	59%	Y
Blue	/	100%	Y

(c)

	Cr^{3+}—O	Cr^{3+}—O(—B)	Site
Pink	64%	36%	Z
Yellow	60%	40%	Z
Blue	/	100%	Z

(d)

	Ni^{2+}—O	Ni^{2+}—O(—Al)	Site
Yellow	/	100%	Y
Green	100%	/	Z
Rose red	/	100%	Y

The colorless, rose red, and blue samples show extremely weak Mn2p peaks, while the pink and green samples have an obvious Mn2p peak, indicating that the colorless, rose red, and blue samples contain significantly a small amount of Mn in the crystals compared to the pink and green samples

(Figure 4). The Mn2p peaks exhibit generally symmetric shapes for pink, green, and blue samples, and the Mn2p electron binding energy is 640.8 eV for the green sample and 642.6 eV for pink and blue samples. On the contrary, the Mn2p3/2 peak of the rose red sample is asymmetric, which is decomposed into two peaks at 640.8 and 642.6 eV, respectively, by peak fitting.

The different binding configurations and estimated percentages of Mn in various chemical states are reported in Table 1. According to the energy spectrum for Mn standard compounds, in bivalent manganese oxides and bivalent manganese fluorides, the electron binding energies of Mn2p3/2 are 640.8 and 642.6 eV, respectively. In the pink and blue samples, the electron binding energy of Mn2p3/2 approximately agrees with that of bivalent manganese fluorides, indicating that Mn is in the Mn^{2+} state and mainly coordinated with F. In the green sample, the electron binding energy of Mn2p3/2 agrees with that of bivalent manganese oxides, indicating that Mn is in the Mn^{2+} chemical state and mainly coordinates with O. In the rose red sample, the Mn2p3/2 peak is asymmetric and could be decomposed into two peaks at 640.8 and 642.6 eV, consistent with those of bivalent manganese oxides and bivalent manganese fluorides, respectively, suggesting that Mn is in the Mn^{2+} chemical state and coordinates with both O and F. Also, for pink, rose red, and blue samples, Mn^{2+} occupies the Y site, whereas for the green sample it occupies the Z site in the crystal structure.

Figure 5 shows the XPS spectrum for Cr2p. Colorless, green, and rose red samples do not show the energy spectrum peak for Cr. In the blue sample, the energy spectrum peak of Cr is significantly weak, while it is relatively strong in the spectra of the pink and yellow samples. In the blue sample, the Cr2p displays a symmetric shape and situates at the position of 577.9 eV. In the pink and yellow samples, the Cr2p3/2 peak show a notably asymmetric shape, which was decomposed into two peaks at 577.4 and 577.9 eV for the pink sample and at 577.5 and 578.0 eV for the yellow sample, respectively. The electron-binding energies of Cr2p3/2 in Cr_2O_3 and $CrBO_3$ are about 577.4 and 577.95 eV. Obviously, Cr in tourmaline is present in the Cr^{3+} state and coordinates with O; however, in pink and yellow samples, part of Cr also binds with O and B (Table 1). From the Cr2p3/2 electron-binding energies of the color tourmaline samples, it can be inferred that Cr^{3+} occupies the Z site of the crystal structure.

Pink and blue samples do not show energy peaks of Ni in the energy spectrum (Figure 6). In the colorless sample, the electron binding energy peak of Ni is extremely weak, while it is quite strong for the yellow, green, and rose red samples, which all show a good symmetric shape. The yellow and rose red samples have the same Ni2p3/2 electron-binding energy peak at 856.9 eV, while the green sample has the Ni2p3/2 peak at 853.3 eV. The Ni2p3/2 peak position at 856.9 eV for the yellow and rose red samples is consistent with that of the Ni standard compound Al_2NiO_4 (857.0 eV), suggesting that Ni is present in the form of Ni^{2+} and coordinates with O, and O also binds to Al in the structure [14]. Therefore, Ni clearly occupies the Y site and mainly coordinates with O in yellow and rose red samples [18]. In the green sample, the Ni2p3/2 electron-binding energy is 853.3 eV, in good agreement with that in NiO, indicating that Ni in green tourmaline is present in Ni^{2+} state and is predominantly bonded to O and only occupies the Z site of the crystal structure (Table 1).

Figure 7 shows the XPS spectrum for Cu2p. Only the green sample shows a weak energy peak for Cu at 933.2 eV. The peak is generally symmetric, in good agreement with the Cu2p3/2 electron-bonding energy (933.2 eV) in CuO. This result indicates that Cu is present in the form of Cu^{2+} and mainly coordinates with O in the green sample.

3.3. Influence of Chemical State on the Tourmaline Color. The mineral crystal color is primarily caused by the transition metal cation in the crystal composition, impurity defects, intrinsic defects, structural distortions, color centers, and charge transfer. The effects of the reflection, diffraction, and diffuse and interference of light can also result in coloration [22, 25]. In addition, a pigment ion with different coordination numbers in the crystal structure may generate different colors. For example, Co^{2+} makes minerals appear red when the coordination number is 6, whereas it makes minerals show a special blue when the coordination number is 4. It is widely believed that Fe-rich tourmaline appears dark green, dark blue, dark brown, or black. Mg-rich tourmaline appears yellow or brown. Li-rich and Mn-rich tourmaline appears rose red or light blue, and Cr-rich tourmaline is dark green [26].

The XPS results for colorless, pink, yellow, green, rose red, and blue tourmaline samples show that a color tourmaline sample usually contains various kinds of transition metal ions; however, the transition metal cations in the Y site are obviously different between the color tourmaline samples. For example, the pink sample contains Fe^{3+}, Cr^{3+}, and Mn^{2+}, while the yellow sample contains Fe^{3+}, Cr^{3+}, and Ni^{2+}. However, in the crystal structure, both Fe^{3+} and Cr^{3+} occupy the Z site, and Mn^{2+} and Ni^{2+} occupy the Y site. The previous investigation showed that the color of tourmaline resulted from the transition metal cations in the Y site of the structure [9]. Thus, the difference in color between the pink and yellow samples is probably dominated by the presence of Mn^{2+} or Ni^{2+} in the Y site. Although Fe^{3+}, Mn^{2+}, Ni^{2+}, and Cu^{2+} were present in the green tourmaline sample, only Fe^{3+} occupies the Y site and coordinates with O, OH, and F, while other components Mn^{2+}, Ni^{2+}, and Cu^{2+} occupy the Z site and coordinate with O. Therefore, the green color of tourmaline is dominantly related to Fe^{3+} in the Y site coordinating with O, OH, and F. The rose red tourmaline sample contains Mn^{2+}, Ni^{2+}, and Fe^{3+}, with the occupation of Fe^{3+} in the Z site and both Mn^{2+} and Ni^{2+} in the Y site in the crystal structure. Fe^{3+} coordinates primarily with O and OH, and Mn^{2+} coordinates with O and F, while Ni^{2+} binds to O. Obviously, the rose red color of tourmaline is derived from both Mn^{2+} and Ni^{2+}. The blue tourmaline sample contains Fe^{3+}, Mn^{2+}, and Cr^{3+}, but Cr^{3+} occupies the Z site and both Fe^{3+} and Mn^{2+} occupy the Y site of the crystal structure. Cr^{3+} coordinates primarily with O, Fe^{3+} coordinates with O, OH, and F, while Mn^{2+} coordinates with F. The blue color of tourmaline is derived from both Fe^{3+} and Mn^{2+} in the crystal.

According to crystal field theory, the characteristic color of many gem minerals is related to the crystal field transition

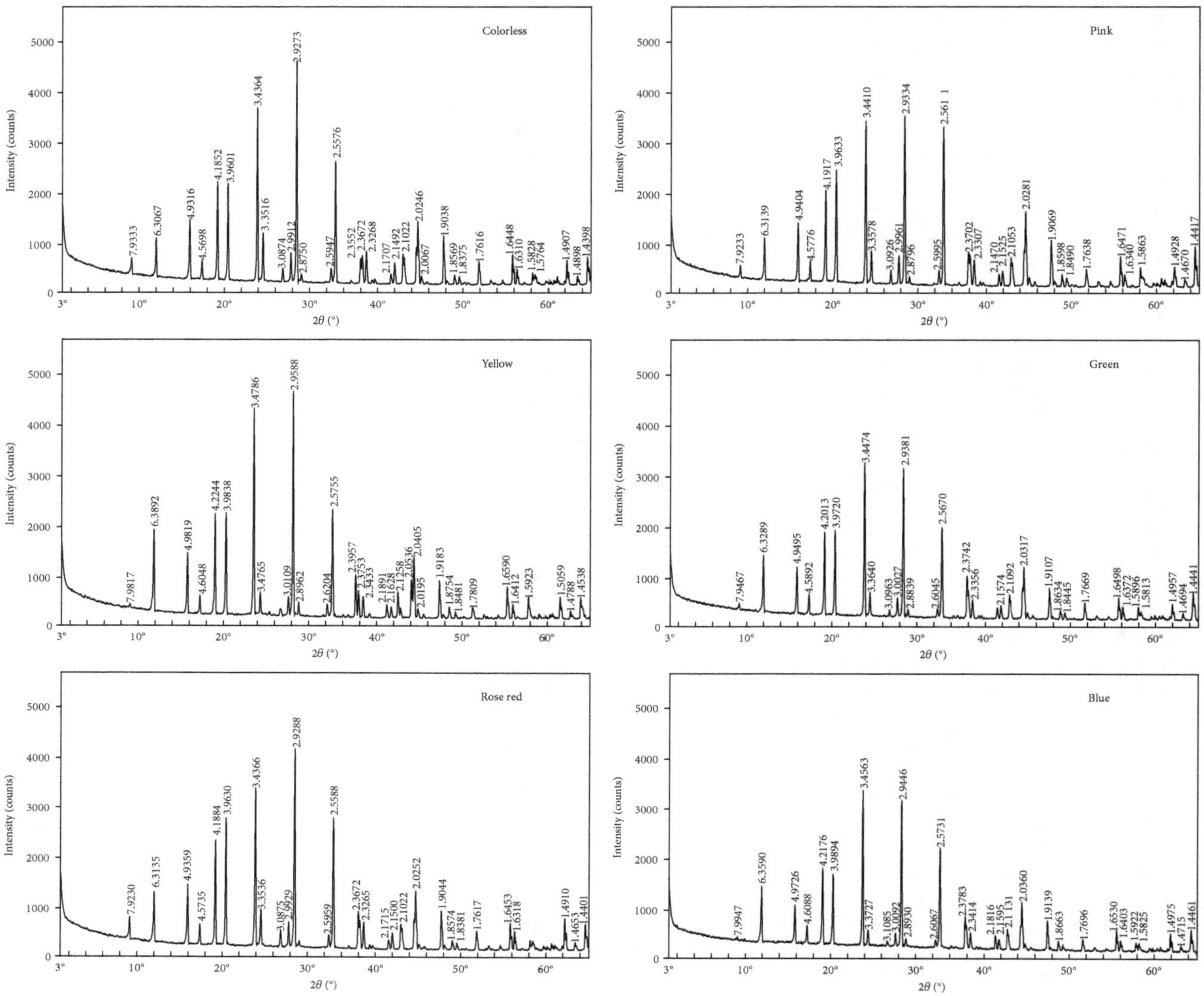

FIGURE 8: XRD spectra of tourmaline samples.

of Cr [27]. The electron configuration of Cr^{3+} is $3d^3$. Splitting the $3d^3$ electron energy level in the octahedron can lead to an energy-level transition, which is directly associated with mineral colors [25]. Furthermore, the distance between atoms directly exerts on the colors of minerals. With a long distance between Cr^{3+} and O, the splitting parameter of the octahedron crystal field is small, and thus the mineral is green; otherwise, the mineral is red. However, in the crystal structure of tourmaline, the transition metal cations related to color are located mainly in the Y site [9]. The charge transfer is the electron exchange between ions in the crystal activated by photons, which may occur between metal cations (M-M), or between O^{2-} and a metal cation. The charge transfers between O and a metal cation often involves high energy; for example, the long-wave tail in the charge transfer between O^{2-} and Fe^{3+} can occur under visible light, thus affecting color of the mineral. The peak position of the charge transfers between O and a metal cation depends on the species, coordination number, and coordination symmetry of the cation. The color of many green silicate minerals is often associated with the peak position of the charge transfer between O^{2-} and Fe^{3+}. The charge transfer between transition metal cations often occurs between coedge or coplanar coordination octahedrons; however, the color-induced transition metal cations occupy the same Y site in the structure and charge transfer between metal cations (M-M) is not expected [28].

The XRD results are shown in Figure 8. The sharp peaks of all the tourmaline samples show that all samples are generally well crystallized; however, the colorless tourmaline sample with no detected trace elements in the structure displays the most intensive reflections compared to other samples especially for the weak peaks in the patterns, suggesting that the colorless sample has the most well-crystallized crystal and ordered structure [29]. On the contrary, the colorful tourmaline samples with certain amounts of trace elements show relatively lower peak intensity and/or different peak ratios of characteristic reflections, indicating the presence of impurity defects and structural distortion in the crystals due to the substitutions of trace elements. Therefore, differences in the chemical environment, including the types of coordination anions, impurity defects, and structural distortion, can lead to different electronic transitions, which may also be the reason for color differences in tourmaline.

4. Conclusions

A color tourmaline sample usually contains several species of transition metal elements. However, there is distinctive difference in element of the transition metals in the Y site between different color samples. The pink color of tourmaline is caused by Mn^{2+} occupying the Y site of the crystal structure, while green is attributed to Fe^{3+} occupying the Y site. The yellow color of tourmaline is related to Ni^{2+} bonding primarily with O, while rose red is derived from both Mn^{2+} and Ni^{2+} in the Y site. The blue color is attributed to Fe^{3+} and Mn^{2+} in the Y site. The charge transfer between cation pairs is unlikely to take place in the color-induced cations of the sole Y site, and the color of tourmaline is probably induced from the absorption of the d-d transition of transition metals in the crystal structure instead.

Conflicts of Interest

The authors declare that there is no conflict of interest with any institution or funding body.

Acknowledgments

This work was supported by the Natural Science Foundation of China (Grant nos. 41472041 and 41772032), National Natural Science Youth Foundation of China (Grant no. 41602037), Natural Science Youth Foundation of Hubei (Grant no. 2016CFB183), the Postdoctoral Science Foundation of China (Grant no. 2015M582301), and Fundamental Research Funds for the Central Universities, China University of Geosciences (Wuhan) (CUG160848).

References

[1] F. F. Foit Jr., "Crystal chemistry of alkali-deficient schorl and tourmaline structural relationships," *American Mineralogist*, vol. 74, no. 3-4, pp. 422–431, 1989.

[2] F. C. Hawthorne, D. J. MacDonald, and P. C. Burns, "Reassignment of cation occupancies in tourmaline: Al-Mg disorder in the crystal structure of dravite," *American Mineralogist*, vol. 78, no. 3-4, pp. 265–270, 1993.

[3] F. Cámara, L. Ottolini, and F. C. Hawthorne, "Crystal chemistry of three tourmalines by SREF, EMPA, and SIMS," *American Mineralogist*, vol. 87, no. 10, pp. 1437–1442, 2002.

[4] J. J. Wang, X. F. Tao, and W. J. Wang, "Color characteristics of green tourmaline from Xinjiang," *Acta Petrologica et Mineralogica*, vol. 24, no. 4, pp. 319–323, 2005.

[5] Y. J. Hou, H. L. Hong, H. J. Xue, and X. L. Zhang, "Gemmological and mineralogical characteristics of colored tourmaline from Minas Gerais State, Brazil," *Journal of Gems & Gemmology*, vol. 10, no. 3, pp. 8–13, 2008.

[6] F. Bosi, T. Balić-Žunić, and A. A. Surour, "Crystal structure analyses of four tourmaline specimens from the Cleopatra's Mines (Egypt) and Jabal Zalm (Saudi Arabia), and the role of Al in the tourmaline group," *American Mineralogist*, vol. 95, no. 4, pp. 510–518, 2010.

[7] F. Bosi, "Stereochemical constraints in tourmaline: from a short-range to a long-range structure," *The Canadian Mineralogist*, vol. 49, no. 1, pp. 17–27, 2011.

[8] W. Han, J. Ke, H. Chen, and K. Yin, "Diffuse reflectance spectroscopy of red colored 'Laowo Stone'," *Spectroscopy and Spectral Analysis*, vol. 36, no. 8, pp. 26–34, 2016.

[9] P. G. Manning, "An optical absorption study of the origin of colour and pleochroism in pink and brown tourmalines," *The Canadian Mineralogist*, vol. 9, no. 5, pp. 678–690, 1969.

[10] M. N. Chaudhry and R. A Howie, "Lithium tourmalines from the Meldon aplite, Devonshire, England," *Mineralogical Magazine*, vol. 40, no. 315, pp. 747–751, 1976.

[11] J. Babińska, K. Dyrek, A. Pieczka, and Z. Sojka, "X and Q band EPR studies of paramagnetic centres in natural and heated tourmaline," *European Journal of Mineralogy*, vol. 20, no. 2, pp. 233–240, 2008.

[12] S. M. Mattson and G. R. Rossman, "Fe^{2+}-Fe^{3+} interactions in tourmaline," *Physics and Chemistry of Minerals*, vol. 14, no. 2, pp. 163–171, 1987.

[13] J. Fan, X. Feng, S. Guo, and X. Liu, "Optical absorption spectra of tourmaline crystals," *Journal of the Chinese Ceramic Society*, vol. 37, no. 4, pp. 523–531, 2009.

[14] H. L. Hong, J. Li, D. W. Du, Z. Q. Zhong, K. Yin, and C. W. Wang, "Chemical states of colour-induced cations in colourful tourmaline," *Journal of Gems & Gemmology*, vol. 13, no. 2, pp. 6–12, 2011.

[15] I. Petrov, "Role of natural radiation in tourmaline coloration: discussion," *American Mineralogist*, vol. 75, no. 1-2, pp. 237–239, 1990.

[16] J. E. S. Bradley and O. Bradley, "Observations on the colouring of pink and green zoned tourmaline," *Mineralogical Magazine*, vol. 30, no. 220, pp. 26–38, 1953.

[17] W. A. d. Fonseca-Zang, J. W. Zang, and W. Hofmeister, "The Ti-influence on the tourmaline color," *Journal of the Brazilian Chemical Society*, vol. 19, no. 6, pp. 1186–1192, 2008.

[18] F. C. Hawthorne and D. J. Henrys, "Classification of the minerals of the tourmaline group," *European Journal of Mineralogy*, vol. 11, no. 2, pp. 201–216, 1999.

[19] F. F. Foit Jr. and P. E. Rosenberg, "Coupled substitutions in the tourmaline group," *Contributions to Mineralogy and Petrology*, vol. 62, no. 2, pp. 109–127, 1977.

[20] M. Benvenuti, P. Costagliola, P. Lattanzi, and G. Tanelli, "Mineral chemistry of tourmalines from the Bottino mining district, Apuane Alps (Italy)," *European Journal of Mineralogy*, vol. 3, no. 3, pp. 537–548, 1991.

[21] G. B. Andreozzi, F. Bosi, and M. Longo, "Linking Mössbauer and structural parameters in elbaite-schorl-dravite tourmalines," *American Mineralogist*, vol. 93, no. 4, pp. 658–666, 2008.

[22] R. I. Mashkovtsev, S. Z. Smirnov, and J. E. Shigley, "The features of the Cu^{2+}-entry into the structure of tourmaline," *Journal of Structural Chemistry*, vol. 47, no. 2, pp. 252–257, 2006.

[23] M. C. Biesinger, B. P. Payne, A. P. Grosvenor, L. W. M. Lau, A. R. Gerson, and R. S. C. Smart, "Resolving surface chemical states in XPS analysis of first row transition metals, oxides and hydroxides: Cr, Mn, Fe, Co and Ni," *Applied Surface Science*, vol. 257, no. 7, pp. 2717–2730, 2011.

[24] C. D. Wagner, W. M. Riggs, L. E. Davis, J. F. Moulder, and G. E. Muilenberg, *Handbook of X-Ray Photoelectron Spectroscopy*, Perkin-Elmer Corporation, Eden Prairie, MN, USA, 1979.

[25] K. Nassau, "The origins of color in minerals," *American Mineralogist*, vol. 63, no. 3-4, pp. 219–229, 1978.

[26] B. L. Zhang, *Systematic Gemmology*, Geological Publishing House, Beijing, China, 2006.

[27] D. L. Wood and K. Nassau, "The characterization of beryl and emerald by visible and infrared absorption spectroscopy," *American Mineralogist*, vol. 53, no. 5, pp. 777–800, 1968.

[28] F. Chen, C. Y. Lin, H. F. Zhang, and H. S. Xie, *Introduction to Mineral Physics*, Science Press, Beijing, China, 1995.

[29] F. D. Bloss, *Crystallography and Crystal Chemistry: An Introduction*, Mineralogical Society of America, Washington, DC, USA, 1994.

11

Development and Integration of a CO Detection System Based on Wavelength Modulation Spectroscopy Using Near-Infrared DFB Laser

Bin Li, Shuo-Cheng Zhang, and Yao-Dan Chi

Institute of Electrical & Computer Engineering, Jilin Jianzhu University, Changchun 130012, China

Correspondence should be addressed to Yao-Dan Chi; 2484670561@qq.com

Academic Editor: Stephane Schilt

A wavelength modulation spectroscopy- (WMS-) based gas sensing system was established to measure concentration of carbon monoxide (CO) in the range 0–100%. The CO absorption line at 1563.06 nm was scanned with a tunable distributed feedback (DFB) laser, and two InGaAs photodiodes were applied to perform optic-electric conversion. Without using commercial instruments, essential electrical circuits were self-developed and integrated, including laser temperature controller, laser current driver, signal generator, and digital lock-in amplifier. The gas cell deployed in the system was fiber coupled with a total effective optical path length of 50 cm. The second-order harmonic signal was extracted, and experiments of gas detection were carried out to investigate the performance of the sensor, including detection repeatability, detection accuracy, response time, and limit of detection (LoD). Experiment results show that the sensor is reliable and has acceptable probing performance. The maximum relative detection error is less than 3.8%, suggesting good detection stability. Benefiting from the self-developed sensor, the whole CO detection system has small size, affordable expense, and application potential.

1. Introduction

Carbon monoxide, which is odorless and colorless, is an important indicator gas of preventing spontaneous combustion of coal seam [1–4]. In order to avoid disasters and guarantee safety production in coal mines, CO concentration monitoring must be performed effectively in real time. Among the numerous gas sensing techniques, optical methods have been studied worldwide in recent years due to its advantages including high sensitivity, fast response time, nonintrusive nature, wide sensing range, and long lifespan [5–7]. Tunable diode laser absorption spectroscopy (TDLAS) using lasers as light sources is a powerful and widely used technique for gas sensing applications including environmental monitoring, industrial process control, chemical analysis, and combustion diagnostics [8–12]. In order to achieve highly gas sensing ability, wavelength modulation spectroscopy technique is widely adopted to achieve high signal-to-noise ratio (SNR). This technique using harmonic detection is one of the most effective methods to suppress the background noises. Therefore, TDLAS-WMS technology has been widely studied and reported in the trace gas detection field [13–16].

Distributed feedback (DFB) laser has numerous advantages including compact size, high spectral resolution, and narrow line width [17, 18]. By modulating DFB laser's driving current and working temperature, the center wavelength can be tuned across the absorption line of a target gas. Thus, DFB lasers coupled with optical fiber can be used as light sources in gas sensing systems. Generally, a typical WMS gas detection system using a DFB laser as light source can be divided into three parts: (1) DFB laser and its driving devices—in order to meet the sensing requirements, the laser's output must be precisely controlled. Thus, commercial devices including laser current driver, laser temperature controller, and signal generator are widely applied in such systems; (2) gas cell and optical structure—gas cell, as the container of target gas, is also coupled with optical parts to provide the path for laser beam; and (3) the signal detection part is the backend of a WMS system. Analog or digital

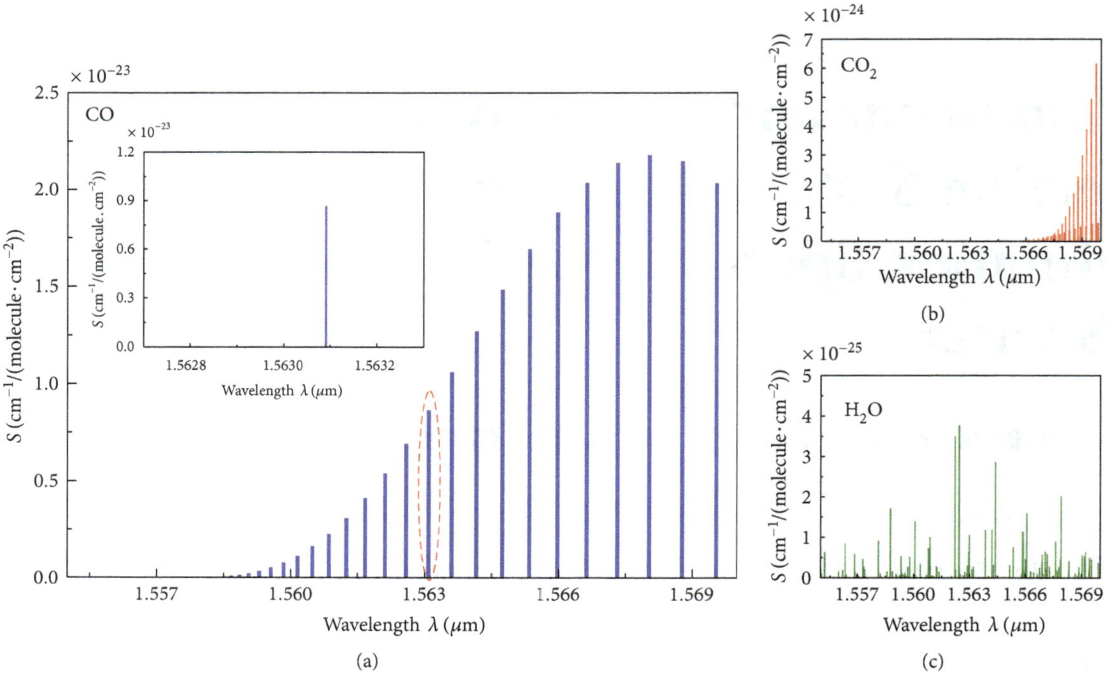

FIGURE 1: Absorption line of carbon monoxide (a), carbon dioxide (b), and H_2O and target absorption line (c) in this gas detection system.

lock-in amplifiers are adopted to extract 1f and 2f harmonic signals which is related to gas concentration [19–22]. However, systems involving commercial instruments usually have some inescapable drawbacks including large size, high expense, and inconvenience to operate. These drawbacks are barriers for practical applications.

In view of the above considerations, a near-infrared CO sensing system based on the TDLAS-WMS technique was experimentally demonstrated. In this system, commercial instruments are replaced by self-developed circuits. A 14-pin DFB laser with 1563.06 nm center wavelength is applied as the light source in this system. The laser's driving circuits and signal detection circuits were self-developed and integrated as a stand-alone instrument. Firstly, the gas sensing principle and system structure are introduced in this paper to guide the reader. Then, design details are shown with photos and descriptions. Finally, experiments were carried out and system performance was demonstrated.

2. Structure and Theory

2.1. CO Absorption Characteristics around 1.563 μm. The center wavelength of the DFB laser applied in this system is around 1.563 μm. The target CO absorption line in the near-infrared spectral region is shown in Figure 1(a). The absorption line strengths are obtained from HITRAN database [23]. CO detection around this band is affected very slightly by other major gases such as CO_2 and H_2O. The absorption intensity of CO_2 and H_2O is shown in Figures 1(b) and 2(c). As shown in Figure 1(d), the target CO absorption line is clearly illustrated. In addition, another advantage of using a DFB laser in the near-infrared spectral region is that reliable and accessible telecommunication optical fiber can be applied for signal transmission in this system. Detailed optical signal transmission and reception will be described later in this paper.

2.2. System Structure. The schematic of this WMS-based CO detection system is shown in Figure 2. The whole system can be divided into four parts of hardware and their connections. The first part is the WMS-based sensor, which is the self-developed instrument contents electrical circuit boards, DFB laser, fiber optic beam splitter (FOBS), optical attenuator (OA), InGaAs detectors, and input/output interfaces. The DFB laser, which is 14-pin butterfly-packaged, has a central wavelength of around 1.563 μm. A digital signal processor (DSP) chip (TMS320F28335) is applied as the main controller of this instrument. It controls the laser's temperature, driving current, and modulating signals by sending signals to corresponding circuits such as the TEC controlling circuit and DAC circuit. Meanwhile, it also processes signals which are related to gas concentration. Two InGaAs detectors are applied to convert the optical signal to electrical signal. The yellow lines in Figure 2 represent optical fibers. The second part of the system is a gas cell. Compared to our previous work [24], the effective optical absorption length has increased from 40 cm to 50 cm by utilizing a recently developed gas cell. The gas cell is fiber coupled and has a better gas tightness feature than our previous gas cells. Thirdly, a mass flow controller is applied in the system to perform dynamic gas mixing. It has a maximum 10 L/min capability of distribution and three ports for connecting gas cylinders. These gas cylinders are the final part of the system and contain standard gases such as pure nitrogen and standard CO gas. The blue and red lines represent gas tubes and optical beam in Figure 2.

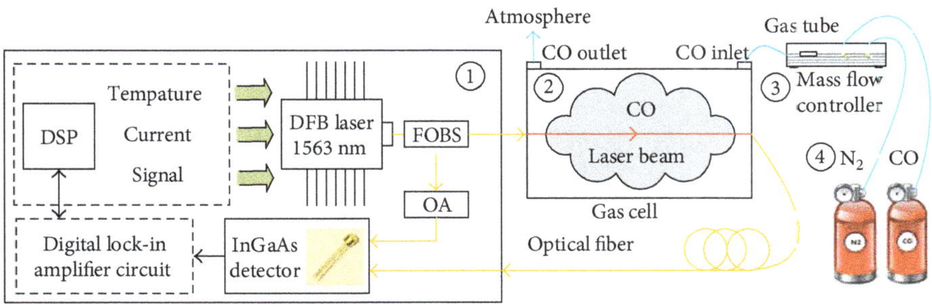

FIGURE 2: Schematic of the WMS-based CO detection system using a DFB laser at 1563 nm.

2.3. Detection Theory. The fundamentals of the WMS technique have been reported widely by many research groups. In this section, only a brief background is described to familiarize the reader with various terms and explain certain experimental choices in this detection system. When the emitting wavelength of the DFB driver has been stabilized to the absorption peak of gas, a saw tooth wave signal $u_{\text{saw}}(t) = A_{\text{saw}} + (A_{\text{saw}}/T_{\text{saw}})(t - T_{\text{saw}})$ is generated to tune the central wavelength to scan the absorption line periodically. In this system, the frequency of the saw tooth signal is 10 Hz. Meanwhile, in order to ensure the sensitivity and accuracy of detection, a sine signal $u_{\sin}(t) = (A_{\sin}/2)\sin(\omega_{\sin}t)$ is generated to superimpose on the saw tooth signal. The driving signal of DFB laser can be expressed as $u(t) = u_{\text{saw}}(t) + u_{\sin}(t)$. The below equation expresses the reference signal of light intensity:

$$I_r(t) = nI_0[1 + mu(t)], \tag{1}$$

where m, n, and I_0 are the light modulation coefficient, the attenuation coefficient of OA, and central emitting light intensity, respectively. According to the Beer-Lambert law, the absorption signal can be expressed below:

$$\begin{aligned} I_t(t) &= I_0[1 + mu(t)]\exp[-\alpha(t)LC] \\ &\equiv I_0[1 + mu(t)][1 - \alpha(t)LC], \end{aligned} \tag{2}$$

where $\alpha(v, t)$ is the absorption coefficient dependent on the emitting wavelength of the DFB laser in the above equation. L is the effective optical path length and C is gas concentration. After optical-to-electrical conversion and amplification, signals from the gas cell channel and reference channel can be expressed below:

$$\begin{aligned} u_t(t) &= K_t D_{\text{oe}}^t I_0[1 + mu(t)][1 - \alpha(t)LC], \\ u_r(t) &= K_r D_{\text{oe}}^r n I_0[1 + mu(t)], \end{aligned} \tag{3}$$

where K_t, D_{oe}^t, K_r, and D_{oe}^r are the amplifying factor for the detection channel, optical-to-electrical conversion coefficient for the detection channel, amplifying factor for the reference channel, and optical-to-electrical conversion coefficient for the reference channel, respectively. Then, the differential signal of the absorption signal and the reference signal can be obtained below:

$$\begin{aligned} u(t) &= u_r(t) - u_t(t) \\ &= I_0[1 + mu(t)]\left[K_r D_{\text{oe}}^r n - K_t D_{\text{oe}}^t + K_t D_{\text{oe}}^t \alpha(t)LC\right]. \end{aligned} \tag{4}$$

By tuning n of OA, let $K_r D_{\text{oe}}^r n = K_t D_{\text{oe}}^t$. Thus,

$$u(t) = I_0[1 + mu(t)]\left[K_t D_{\text{oe}}^t \alpha(t)LC\right]. \tag{5}$$

Considering that $mu(t) << 1$, the function below is obtained:

$$u(t) = I_0 K_t D_{\text{oe}}^t \alpha(t)LC. \tag{6}$$

Extracting the second harmonic signal of $u(t)$,

$$\begin{aligned} S_2(\omega) &= \int_{t-T_{\sin}}^{t} u(\tau)e^{-j2\omega_{\sin}\tau}d\tau \\ &= \int_{t-T_{\sin}}^{t} I_0 K_t D_{\text{oe}}^t \alpha(\tau)LCe^{-j2\omega_{\sin}\tau}d\tau \\ &= K_{\text{eq}}C\int_{t-T_{\sin}}^{t} \alpha(\tau)e^{-j2\omega_{\sin}\tau}d\tau, \end{aligned} \tag{7}$$

where $K_{\text{eq}} = I_0 K_t D_{\text{oe}}^t L$ and $T_{\sin} = 2\pi/\omega_{\sin}$.

$S_2(\omega)$ is the intensity of the second harmonic signal on the frequency of $2\omega_{\sin}$. The frequency of the saw signal and sine signal is set to be $T_{\text{saw}} = 10$ Hz and $T_{\sin} = 5$ KHz, respectively. Then, an inverse Fourier transform is performed for the transformation from frequency domain to time domain. The 2f signal in each saw wave period can be obtained by the summation of a continuous second-order harmonic wave component in the time domain as shown in (8). The amplitude of the 2f signal, which is expressed as $\text{Amp}[S_2(t)]$, can be obtained from the peak-to-peak value as shown in (9).

$$S_2(t) = \sum_{t=0}^{T_{\text{saw}}} S_2(\omega), \tag{8}$$

$$\text{Amp}[S_2(t)] = \max[S_2(\omega)] - \min[S_2(\omega)]. \tag{9}$$

In order to simplify the expression of the amplitude of the 2f signal, $A_2(\omega)$ and $A_2(t)$ are introduced in (10) and (11), respectively. By combining the equations from (7) to (12), the amplitude of the 2f signal can be obtained and simplified in the time domain as shown in (13).

$$A_2(\omega) = \int_{t-T_{\sin}}^{t} \alpha(\tau)e^{-j2\omega_{\sin}\tau}d\tau, \quad (10)$$

$$A_2(t) = \sum_{t=0}^{T_{\text{saw}}} A_2(\omega), \quad (11)$$

$$\text{Amp}[A_2(t)] = \max[A_2(\omega)] - \min[A_2(\omega)], \quad (12)$$

$$\text{Amp}[S_2(t)] = K_{\text{eq}}C\text{Amp}[A_2(t)]. \quad (13)$$

Therefore, the relation of the gas concentration and the amplitude of the 2f signal is obtained. Also, it can be seen that the 2f signal is determined by the gas concentration.

3. Design Details of Key Modules

3.1. DFB Laser and NIR Detector. The DFB laser applied in this system, which is a butterfly-packaged fiber-coupled laser, is provided by the Institute of Semiconductors, Chinese Academy of Sciences (ISCAS). The center wavelength of this laser is around 1.563 μm which can be tuned by modulating the laser's temperature and injection current. The maximum optical output power is 10 mW when the operating temperature is set as 45.18°C and the operating current is set as 73.4 mA. The laser's threshold current and slope efficiency are 21.8 mA and 0.19 W/A, respectively. The laser's temperature is determined by a thermoelectric cooler (TEC) module which is packaged inside the laser. The TEC module can be controlled to adjust the laser's working temperature in order to locate the center wavelength to the absorption line. In addition, there are two InGaAs photodiodes (J22-18I-R500U) applied in this system for optical-electrical conversion. The peak responsivity is 0.9 A/W and the diameter of the photosensitive area is 0.5 mm. The two photodiodes are applied for the absorption channel and reference channel in order to obtain the differential signal. The differential channel structure was applied to obtain the differential signal between the absorption signal and the reference signal. The obtained differential signal is for 2f signal extraction. By using the differential channel structure, the fluctuation of the light source and intrinsic noise can be eliminate in order to achieve better sensing accuracy.

3.2. Power Supply Module. Normal switching power supplies cannot meet the requirement of the laser temperature control circuit because the power ripple is too high. In this way, a low-ripple linear power supply is developed. This linear power supply is indispensable for the temperature control module. Moreover, electrostatic discharge (ESD) protection and surge protection are also put into use to protect the DFBL and to guarantee system stability. The photo of the self-developed power supply is shown in Figure 3. The custom-made transformer is packaged in a black case. The power supply provides four channels of DC voltage which are 5 V, −5 V, 12 V, and −12 V.

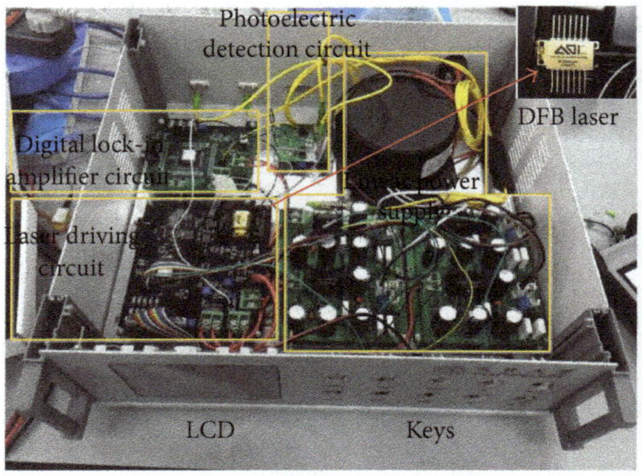

FIGURE 3: Inside view of the self-developed CO sensor based on the WMS technique.

3.3. Laser Driving Module. Temperature control for the DFB laser is necessary because the central wavelength depends on the laser's operation temperature while injection current is fixed. Thus, a temperature controller is developed to meet the requirements of control features including high stability, high accuracy, and quick response. Based on Peltier effect, the laser's temperature can be controlled by the TEC module packaged inside. The real-time temperature can be obtained by monitoring the inside-packaged thermistor with negative temperature coefficient (NTC). A high-efficiency and reliable monolithic TEC controller, ADN8831, is used to drive the TEC. It sets the temperature of the laser in a programmable way and maintains the temperature stable by using an analog PID control loop.

In this system, the temperature scope is set 35–55°C according to the laser test report. The center wavelength will be tuned around the absorption line when temperature is set to about 41.7°C while the injection current is 60 mA. Experiments were carried out to investigate the performance of the self-developed temperature control circuit. As shown in Figure 4, the center wavelength is shifted linearly while the temperature is changed linearly from 39 to 43°C and the injection current is fixed as 80 mA. This proves the good linearity of the temperature control circuit.

The V-I curve of the DFB laser applied in this system is provided in its test report. In this way, a DFB laser driving circuit is developed which is shown in Figure 3. The laser is mounted on a special socket which is soldered on the driving board. Under the condition of fixed temperature, the center wavelength is varied while changing the driving current. As shown in Figure 5, the laser temperature is fixed at 45°C and the driving current is increasing from 40 to 80 mA. It can be observed that the center wavelength is shifted linearly. Meanwhile, the optical power is also increased observably.

A high-accuracy digital-to-analog converter (DAC, AD5060) chip is applied to set constant voltage or periodical changing voltage. In CO detection experiments, the laser temperature is set at first to locate the center wavelength at the absorption line. Then, a saw tooth wave signal at 10 Hz generated by the DAC will act on the laser's output in order to make the center wavelength scan across the absorption line periodically. The DAC's output is controlled by DSP's

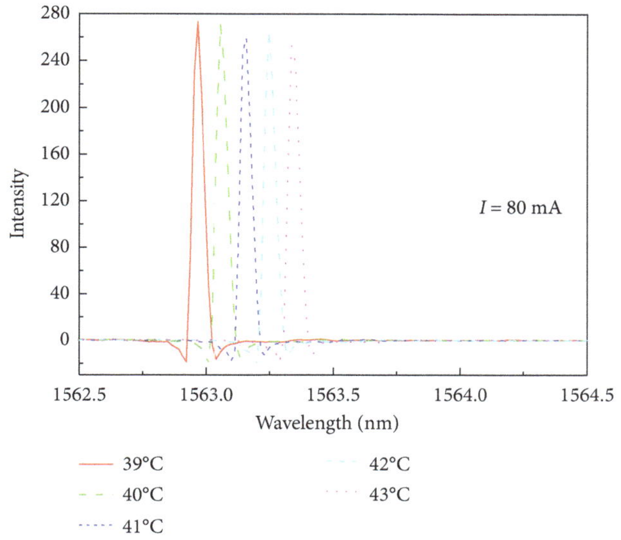

FIGURE 4: Center wavelength shifts while laser temperature is tuned linearly under constant injection current.

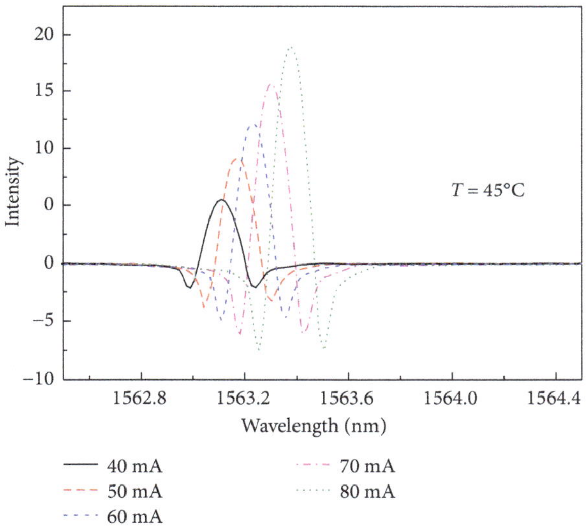

FIGURE 5: Under constant laser temperature, the center wavelength and optical power are controlled by injection current.

CPU timer. In WMS, there is also a sine wave signal required to modulate the laser's output. In this system, this is performed by a direct digital synthesizer (DDS, AD9851). Then, the saw tooth signal and sine wave signal are combined by AD8513, which is a precision and low-noise JFET amplifier.

3.4. Digital Lock-In Amplifier. In WMS, harmonic signals such as 1f and 2f are usually extracted by using commercial lock-in amplifiers. In this system, a digital lock-in amplifier based on DSP is developed and the photo is shown in Figure 3. The 2f harmonic signal is isolated by the digital lock-in amplifier to indicated CO concentration. Experiments were carried out to evaluate its performance. As shown

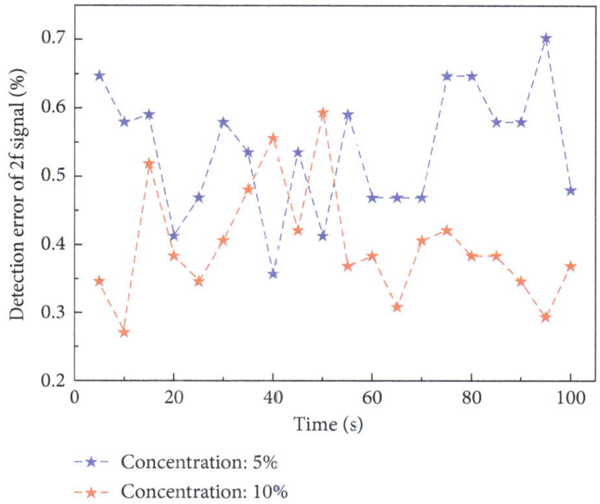

FIGURE 6: Performance test of the digital lock-in amplifier.

in Figure 6, under the concentration of 5% and 10% of carbon monoxide, the detection errors of the 2f signal are shown in blue and red data, respectively. It can be seen that the detection error is measured 20 times in a period of 100 seconds and the maximum error is under 0.7% which suggests good stability.

3.5. Gas Cell. According to the Beer-Lambert law, effective absorption optical path length L is a very important parameter in a TDLAS system. This is due to the fact that the incensement of L can lead to clearer absorption phenomenon and lower detection limit. Thus, the gas cell was redesigned and developed in this system. Compared to our previous work, the path length is increased from 40 cm to 50 cm. This gas cell is integrated with a pair of optical fiber collimator. The gas cell is a single-pass cell. It is stable and suitable for gas sensing experiments.

4. Experiments and Discussion

4.1. Experimental Setup. The actual experimental setup is shown in Figure 7. The DFB laser and essential self-developed circuits are integrated in a stand-alone instrument as the CO sensor. The details of this sensor is described in the last section, and the inside of the instrument is shown in Figure 3. The gas cell, which has a total effective optical path length of 50 cm, is connected with the CO sensor and the mass flow controller by the optical fiber and air tube, respectively. The sample gas for concentration measurements was supplied from a combination of two channels of the mass flow controller. One is connected with pure N_2 and the other is connected with the standard CO gas. For the measurements, a total flow of 1 L/min was used. There are manual knobs on the front panel of the mass flow controller for flow control. Each time the mass flow controller is powered on, there is an initializing process of the mass flow controller to achieve better accuracy and this period of time lasts about 15 minutes.

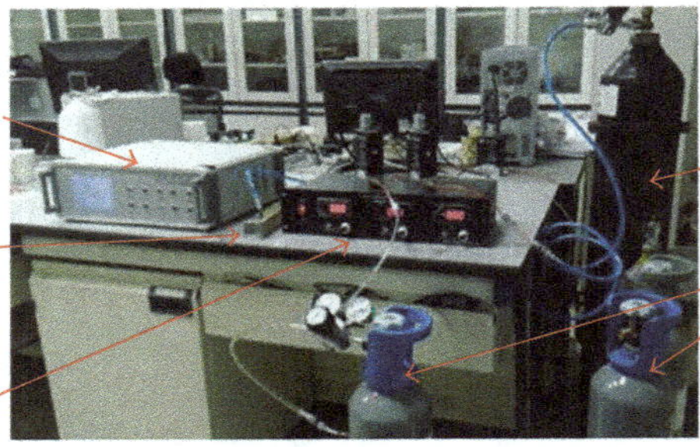

FIGURE 7: Laboratory experimental setup of the WMS-based CO sensor.

TABLE 1: Results of the detection repeatability test under 70% concentration.

N	1	2	3	4	5	6
Ai	70.56%	70.41%	71.12%	70.06%	69.88%	70.25%

During the CO detection experiments, the DFB laser temperature was set to 42°C to locate the center wavelength at the absorption line. The laser driving current is shifting in the range of 40–80 mA, and the frequency is 10 Hz. The 5 kHz sine wave signal is generated to be the modulation signal. These operations can be done by pressing the keys embedded on the front panel of the instrument. These experimental conditions were kept constant for CO concentration measurement carried out for performance evaluation which are discussed below.

4.2. Detection Repeatability. Experiments were carried out to investigate the repeatability of the performance firstly. Pure N_2 was injected into the gas cell, and the concentration was set to zero of the CO sensor (instrument) while the gas is filled full of N_2. Then, a concentration of 70% CO gas was filled into the gas cell and the results were recorded. The whole process was repeated six times (as shown in Table 1) and the results were recorded for analysis. The repeatability S_r is calculated by using the equation below:

$$S_r = \frac{1}{\bar{A}}\sqrt{\frac{\sum_{i=1}^{n}\left(A_i - \bar{A}\right)^2}{n-1}} \times 100\%, \tag{14}$$

where A_i is the concentration result displayed on the LCD, n is the number of measured times, $n = 6$, $\bar{A}$ is the arithmetic average of results, and S_r is the detection repeatability. Thus, $S_r = 1.38\%$ was calculated and it satisfies the project's requirement of $S_r < 2\%$.

4.3. Response Time. Distinguished with chemical gas sensors, infrared optical sensors do not require time for chemical reaction and the sensing process is in real time. However, there is still a short period of time consumed by signal transmitting, data calculation, delays in program code, and LCD refreshing rate. Besides, it is not convenient to read the display on LCD if data is varying too fast. In this way, the display cycle is set as 2 seconds in this system.

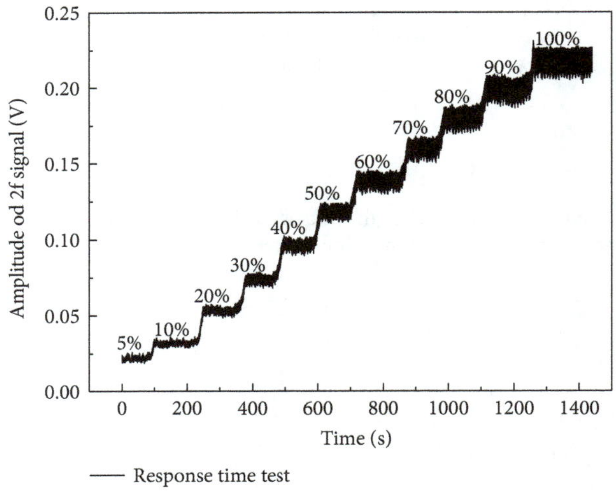

FIGURE 8: Response time test of the system under 1 L/min flow speed.

Due to the advantage of quick response, the sensor can be used in fast-changing applications. For example, experiments were performed to acquire the variation of concentration in the gas cell and the result is shown in Figure 8. By changing the CO concentration from 5% to full scale, the 2f signal was recorded. As shown in Figure 8, there are slopes between adjacent concentrations. These slopes were corresponding to time which was consumed by manual operations and gas injecting to the gas cell. Therefore, compared to traditional gas sensors, the developed sensor can be utilized to applications demanding quick response.

4.4. Detection Accuracy. Experiments were carried out to investigate the detection accuracy. Measured concentrations as well as the standard concentrations of 10 CO samples are comparatively shown in Figure 9(a). The relative error of each gas sample was calculated, as shown in Figure 9(b). Under the concentration of 10×10^4 ppm, the relative error becomes the maximum, which is about 3.8%. As shown in Figure 10, the absolute error increases while the concentration raises. However, with the increasing of concentration, the relative error tends to be decreasing and within the whole detection range, the relative detection error is within the range of −2.9%~3.8%. This confirms that the developed system reveals acceptable performances.

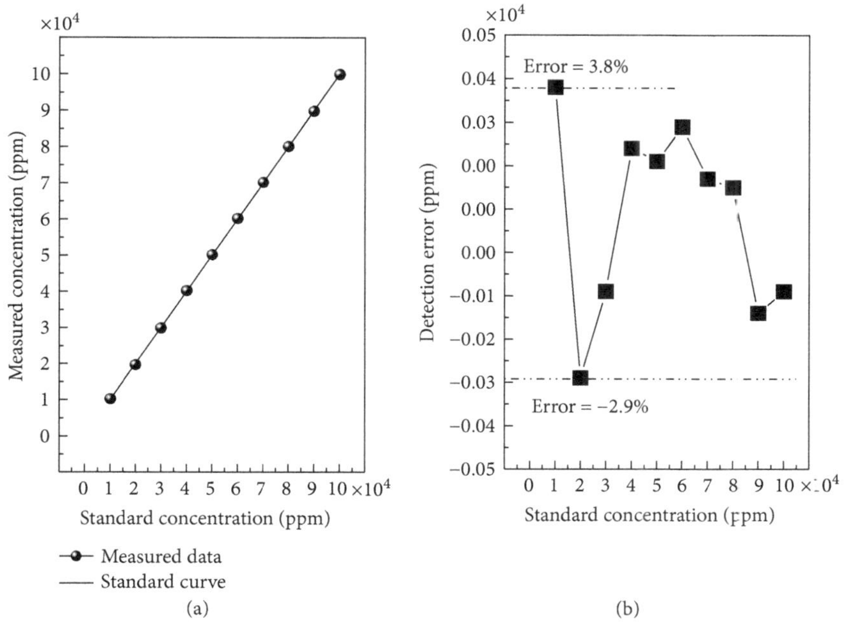

FIGURE 9: (a) Measured concentrations as well as standard concentrations and (b) relative detection error test.

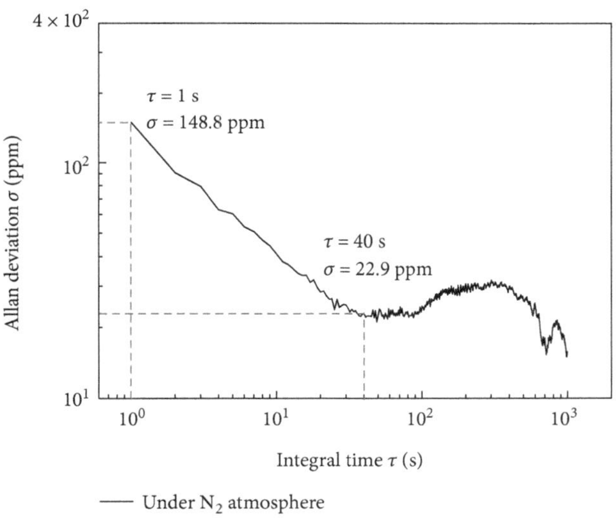

FIGURE 10: The Allan deviation curve of the detection system.

4.5. Limit of Detection. The Allan deviation curve of the detection system is shown in Figure 10, where the integral time is within the range of 1–1000 s. With 1 s integral time, the Allan deviation value of the system is about 148.8 ppm. The system shows the best stability with an integral time of around 40 s, where the corresponding Allan deviation value is about 22.9 ppm.

5. Conclusion

Based on the TDLAS technique, a CO detection system was experimentally demonstrated. Without using commercial instruments, the self-developed electrical part includes a linear power supply module, a laser temperature control and modulation module, and a reliable digital lock-in amplifier. Under the operation of the developed DFB laser driver, the laser temperature can be controlled effectively. The laser's emitting wavelength varies linearly with the temperature and injection current. The developed digital lock-in amplifier can effectively extract the harmonic signals, and the maximum extraction errors on the harmonic amplitude for two gas samples are less than 1% for 2f. As the CO concentration increases from 10% to 100%, the amplitudes of the 2f harmonic signals were obtained. The limit of detection is estimated to be about 22.9 ppm, and the relative detection error is within the range of −2.9%~3.8%. The whole CO detection system has high sensitivity, small size, affordable expense, and application potential.

Conflicts of Interest

The authors declare that they have no conflicts of interest.

Acknowledgments

The authors wish to express their gratitude to Dr. Xiaotian Yang, National Natural Science Foundation of China (no. 51672103), Jilin Jianzhu University, the National Key Technology R&D Program of the Ministry of Science and Technology of China, the Education Department of Jilin Province of China, and the State Key Laboratory of Integrated Optoelectronics, Jilin University for their generous support of this work.

References

[1] A. A. I. Khalil, M. A. Gondal, and N. Al-Suliman, "Resonant photo-acoustic detection of carbon monoxide with UV Laser at 213 nm," *Applied Physics B: Lasers and Optics*, vol. 103, no. 2, pp. 441–450, 2011.

[2] X. H. Li, X. M. Liu, Y. K. Gong, H. B. Sun, L. R. Wang, and K. Q. Lu, "A novel erbium/ytterbium co-doped distributed feedback fiber laser with single-polarization and unidirectional output," *Laser Physics Letters*, vol. 7, no. 1, pp. 55–59, 2010.

[3] Y. X. Meng, T. G. Liu, K. Liu et al., "A modified empirical mode decomposition algorithm in TDLAS for gas detection," *IEEE Photonics Journal*, vol. 6, no. 6, pp. 1–7, 2014.

[4] J. Vingerhoets, J. Snoeys, and S. Minten, "Detection of smouldering fires by carbon monoxide gas concentration measurement," *Chemical Engineering Transactions*, vol. 48, p. 457, 2016.

[5] H. Moser, A. Genner, J. Ofner, C. Schwarzer, G. Strasser, and B. Lendl, "Application of a ring cavity surface emitting quantum cascade laser (RCSE-QCL) on the measurement of H_2S in a CH_4 matrix for process analytics," *Optics Express*, vol. 24, no. 6, pp. 6572–6585, 2016.

[6] R. Ghorbani and F. M. Schmidt, "Real-time breath gas analysis of CO and CO_2 using an EC-QCL," *Applied Physics B: Lasers and Optics*, vol. 123, no. 5, p. 144, 2017.

[7] G. D. Banik, S. Som, A. Maity et al., "An EC-QCL based N_2O sensor at 5.2 μm using cavity ring-down spectroscopy for environmental applications," *Analytical Methods*, vol. 9, no. 15, pp. 2315–2320, 2017.

[8] G. Zhang, J. Liu, Z. Xu, Y. He, and R. Kan, "Characterization of temperature non-uniformity over a premixed CH_4–air flame based on line-of-sight TDLAS," *Applied Physics B: Lasers and Optics*, vol. 122, no. 1, p. 3, 2016.

[9] B. Buchholz, S. Kallweit, and V. Ebert, "SEALDH-II—an autonomous, holistically controlled, first principles TDLAS hygrometer for field and airborne applications: design–setup–accuracy/stability stress test," *Sensors*, vol. 17, no. 1, p. 68, 2017.

[10] D. Choi, M. Jeon, G. Cho, T. Kamimoto, Y. Deguchi, and D. Doh, "Performance improvements in temperature reconstructions of 2-D tunable diode laser absorption spectroscopy (TDLAS)," *Journal of Thermal Science*, vol. 25, no. 1, pp. 84–89, 2016.

[11] Y. Krishna and S. Obyrne, "Tunable diode laser absorption spectroscopy as a flow diagnostic tool: a review," *Journal of the Indian Institute of Science*, vol. 96, p. 17, 2016.

[12] X. Zhou, J. Yu, L. Wang, Q. Gao, and Z. Zhang, "Sensitive detection of oxygen using a diffused integrating cavity as a gas absorption cell," *Sensors and Actuators B: Chemical*, vol. 241, pp. 1076–1081, 2017.

[13] A. Upadhyay, D. Wilson, M. Lengden, A. L. Chakraborty, G. Stewart, and W. Johnstone, "Calibration-free WMS using a cw-DFB-QCL, a VCSEL, and an edge-emitting DFB laser with in-situ real-time laser parameter characterization," *IEEE Photonics Journal*, vol. 9, no. 2, pp. 1–17, 2017.

[14] W. Wei, J. Chang, Y. Y. Liu et al., "Eliminating the effect of phase shift between injection current and amplitude modulation in DFB-LD WMS for high-precision measurement," *Applied Optics*, vol. 55, no. 13, pp. 3526–3530, 2016.

[15] S. S. Choi, A. Mandelis, X. Guo, B. Lashkari, S. Kellnberger, and V. Ntziachristos, "Wavelength-modulated differential photoacoustic spectroscopy (WM-DPAS) for noninvasive early cancer detection and tissue hypoxia monitoring," *Journal of Biophotonics*, vol. 9, no. 4, pp. 388–395, 2016.

[16] Z. Du, J. Li, X. Cao, H. Gao, and Y. Ma, "High-sensitive carbon disulfide sensor using wavelength modulation spectroscopy in the mid-infrared fingerprint region," *Sensors and Actuators B: Chemical*, vol. 247, pp. 384–391, 2017.

[17] H. Virtanen, T. Uusitalo, and M. Dumitrescu, "Simulation studies of DFB laser longitudinal structures for narrow linewidth emission," *Optical and Quantum Electronics*, vol. 49, no. 4, p. 160, 2017.

[18] T. Yang, Y. Song, W. Zhang, and F. Li, "Acoustic emission detection using intensity-modulated DFB fiber laser sensor," *Chinese Optics Letters*, vol. 14, no. 12, article 120602, 2016.

[19] L. M. Song, L. W. Liu, Y. G. Yang, Q. H. Guo, and J. T. Xi, "An optical sensor for hydrogen sulfide detection in open path using WMS-$2f/1f$ technique," *Optoelectronics Letters*, vol. 12, no. 6, pp. 465–468, 2016.

[20] S. Neethu, R. Verma, S. Kamble, J. K. Radhakrishnan, P. P. Krishnapur, and V. C. Padaki, "Validation of wavelength modulation spectroscopy techniques for oxygen concentration measurement," *Sensors and Actuators B: Chemical*, vol. 192, pp. 70–76, 2014.

[21] Q. Wang, J. Chang, F. P. Wang, and W. Wei, "Recovery of pure wavelength modulation second harmonic signal waveforms in distributed feedback diode laser-based photoacoustic spectroscopy," *Sensors and Actuators A: Physical*, vol. 245, pp. 54–62, 2016.

[22] T. Benoy, M. Lengden, G. Stewart, and W. Johnstone, "Recovery of absorption line shapes with correction for the wavelength modulation characteristics of DFB lasers," *IEEE Photonics Journal*, vol. 8, pp. 1–17, 2016.

[23] L. S. Rothman, D. Jacquemart, A. Barbe et al., "The HITRAN 2004 molecular spectroscopic database," *Journal of Quantitative Spectroscopy and Radiative Transfer*, vol. 96, pp. 139–204, 2003.

[24] B. Li, C. T. Zheng, H. F. Liu et al., "Development and measurement of a near-infrared CH4 detection system using 1.654μm wavelength-modulated diode laser and open reflective gas sensing probe," *Sensors and Actuators B: Chemical*, vol. 225, pp. 188–198, 2016.

12

Spectroelectrochemical Investigation of the Interaction of Adenine with Pyridoxine at Physiological pH

Yasmin Roye,[1] Uche Udeochu,[1] Maraizu Ukaegbu,[2] and Jonathan Onuegbu[2]

[1]*Department of Natural Sciences, University of Maryland Eastern Shore, 1 Backbone Rd., Princess Anne, Maryland 21853, USA*
[2]*Department of Chemistry, Howard University, 525 College St. NW, Washington, DC 20059, USA*

Correspondence should be addressed to Uche Udeochu; charlesudeochu@yahoo.com

Academic Editor: Rizwan Hasan Khan

Spectroelectrochemical techniques were used to probe the interaction of adenine with pyridoxine at pH 7.0. Analysis of UV-visible absorption of the adenine-pyridoxine complex at 260 nm using the Lineweaver–Burk double reciprocal plot produced a linear graph indicating a 1 : 1 mode of interaction between the compounds and a binding constant of 29.1. Change in the background current and broadening of adenine and pyridoxine cyclic voltammetry (CV) oxidation peaks at 1.0 V and 0.8 V, respectively, compared to the CV of the individual compounds is indicative of an interaction. The Raman shift of the coupled $-C(11)H_2$-OH bending and in-plane N(7)-H mode at 1235 cm^{-1} to 1215 cm^{-1} of pyridoxine and the shift to the lower wavenumber of the adenine $-N(10)H_2$ rocking band from 942 to 906 cm^{-1} confirm that the adenine exocyclic amino group and its purine nitrogen atom N(7) interacts with pyridoxine O(12) via the formation of hydrogen bonds. Strong enhancement of surface-enhanced Raman spectroscopy (SERS) bands pertaining to adenine and the bathochromic shift of the normal Raman band due to the adenine ring breathing mode observed at 722 cm^{-1} in the spectrum of adenine, to 732 cm^{-1} in the SERS spectrum of aqueous adenine-pyridoxine indicates that the complex adsorbs onto the Ag nanoparticle surface with the adenine portion possessing a perpendicular orientation.

1. Introduction

The interaction of deoxyribonucleic acid (DNA) with small molecules has remained a subject of interest in many biochemical studies owing to its role in many cellular activities [1–3]. The obvious starting point for understanding the activity of DNA is the study of the components of its structure. Adenine (Figure 1(a)), one of the components of DNA, participates significantly in many life processes of the human body. It is involved in cellular respiration and protein synthesis, and its interaction with various molecules (as in its pairing with thymine within the DNA double helix and uracil in RNA) is essential for the transfer and expression of genetic materials and stabilization of the nucleic acid structure [4–7]. Furthermore, the level of adenine in the human body can serve as a marker for the body's susceptibility to metabolic disorders and diseases. Therefore, understanding the kinetic characteristics, structure, and conformational behavior of adenine under physiological conditions especially in its interaction with small molecules is essential for understanding biochemical processes and drug discovery involving DNA nucleobases.

In recent years, there has been growing interest in characterizing how DNA interacts with small molecules using spectroelectrochemical techniques and this is with respect to establishing interaction models. Models relating to electrostatic interaction and hydrophobic binding within the minor groove and intercalation between stacked base pairs have been proposed for the interaction of small molecules with DNA [3, 8, 9]. Base groups on DNA exhibit supramolecular interaction with pyridoxine (Figure 1(b)) via the formation of hydrogen bonding. The choice of pyridoxine, a 4-methanol derivative of vitamin B6, as a small molecule for this interactional study is due to its presence as active ingredients in various food, dietary supplements, and drugs. The deficiency of pyridoxine in the human body has been

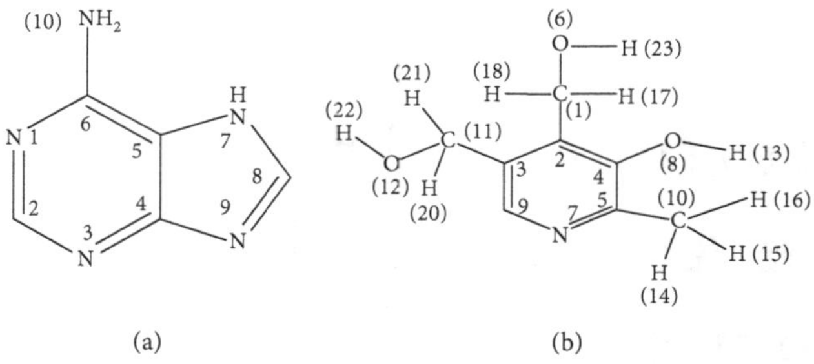

Figure 1: Chemical structure: (a) adenine; (b) pyridoxine.

linked to anemia, nerve damage, skin problems, and various neurological disorders [10, 11]. The ease of pyridoxine absorption within the small intestine makes it readily available for important roles in red blood cell regeneration, nervous and immune systems, and food regulatory functions. Pyridoxine is transformed to pyridoxal phosphate, a coenzyme for synthesis of amino acids, neurotransmitters, and sphingolipids in biological systems [12, 13].

Spectroelectrochemical methods have been shown to be sensitive for characterizing DNA interaction with small molecules [8, 14, 15] and for confirming DNA-pyridoxine interaction [8]. However, in this study, an in-depth investigation of the binding characteristics of adenine with pyridoxine using cyclic voltammetry, UV-visible spectra, and Raman spectroscopy is explored to gain better understanding on the binding characteristics between the two compounds at the molecular level. Adenine and pyridoxine are electrochemically active [8, 16–20]. The relevance of the Raman effect of adenine in DNA studies cannot be over emphasized because of the sensitivity of its vibrational modes in SERS studies [21, 22]. The SERS characteristic of pyridoxine adsorbed on Ag colloidal particles has been examined in aqueous solution, and the results have shown that pyridoxine exhibits a pH-dependent orientation on the Ag surface with a parallel or tilted orientation suggested as a more probable conformation [23]. In this work, changes in the UV-visible absorbance property of the adenine and pyridoxine system are used as the basis for determining the binding constant between the compounds at the physiological hydrogen ion condition. The interaction is investigated in aqueous solution using CV and Raman spectroscopy with the aim of identifying potential changes and vibrational modes associated with the interaction. Furthermore, SERS is used to probe the adenine-pyridoxine interaction on Ag nanoparticles to determine possible orientation of the complex on the surface. Experimental Raman and SERS data are analyzed with reference to computed fundamental and assigned bands of adenine and pyridoxine in literature.

2. Experimental

2.1. Chemicals. Adenine and pyridoxine were obtained from Sigma-Aldrich and used with no further purification. Sodium phosphate buffer solution prepared in deionized water was used to keep the pH value at 7.0, while buffer-electrolyte solution containing 0.1 M phosphate buffer and 5 mM NaCl supporting electrolyte at pH of 7.0 were prepared from analytical grade reagents and distilled deionized water for electrochemical measurements. A small amount of aqueous solution of adenine-pyridoxine was dropped on the Ag colloid that was prepared according to the procedure reported in an earlier study [24].

2.2. Spectroscopic Measurements. Solutions of adenine and pyridoxine of different concentrations meant for UV-visible measurements were prepared from stock solutions of adenine (0.01 mol/L) and pyridoxine (0.01 mol/L) prepared in the sodium phosphate buffer solution. Samples were scanned between 200 and 800 nm using a UV-visible light DU 800 Spectrophotometer equipped with a quartz cuvette of 1 cm path length. The maximum absorbance of adenine was plotted against various concentrations (1.00×10^{-3} M–6.00×10^{-3} M) at pH 7.0 at fixed wavelength. For the determination of the binding constant of adenine-pyridoxine, the Lineweaver–Burk double reciprocal plot, which is a linear transformation of the Michaelis–Menten equation, is especially useful. By using a method similar to what was used by Charak et al. [25], the concentration of adenine was kept constant (1×10^{-3} M) with the concentration of added pyridoxine varied between 4.00×10^{-5} and 4.00×10^{-4} M. The data obtained were treated according to the following equations:

$$\text{Adenine} + \text{pyridoxine} \leftrightharpoons \text{adenine pyridoxine}$$

$$K = \frac{\text{adenine} - \text{pyrdoxine}}{[\text{adenine}]\text{uncomplexed}\ [\text{pyridoxine}]\text{uncomplexed}},$$

$$\frac{A_0}{A - A_0} = \frac{\varepsilon_A}{(\varepsilon_{AP})(K)[\text{pyridoxine}]} + \frac{\varepsilon_A}{\varepsilon_{AP}}, \tag{1}$$

where A_0 is the absorbance of adenine at its absorption maxima, in the absence of pyridoxine, A is the absorbance of adenine observed after adding different concentrations of pyridoxine to it, and A and AP are the molar extinction coefficients of adenine and adenine-pyridoxine, respectively. By plotting $1/(A - A_0)$ versus 1/[pyridoxine] and obtaining a linear graph, a 1 : 1 stoichiometry between the two compounds is justified. The binding constant (K) for the interaction is deduced from the ratio of the intercept to the slope.

2.3. Electrochemical Measurements. For the CV, stock solutions of adenine (1.0×10^{-3} mol/L) and pyridoxine (1×10^{-2} mol/L) were prepared using the pH 7 electrolyte-buffer solution. Electrochemical measurements were performed on a Princeton potentiostat/galvanostat Model 263 A. A three-electrode system incorporating a 1.5 mm-diameter glassy carbon working electrode, platinum wire counterelectrode, and a standard calomel electrode (SCE) reference was employed. The glassy carbon electrode (GCE) was polished with alumina oxide (particle size 0.3 μm) before every electrochemical assay, and for every GCE cleaning sequence, the electrode was immersed in solutions of distilled deionized water and sonicated for 5 minutes.

2.4. Raman and SERS Measurements. Raman spectra of saturated solutions of adenine, pyridoxine hydrochloride, and adenine-pyridoxine and the SERS of 1.0×10^{-3} M adenine-pyridoxine in the sodium phosphate buffer were recorded using a DeltaNU micro-Raman microscope having 633 nm laser line and a laser power of 200 mW. SERS measurements of adenine-pyridoxine were accomplished by first dropping 1.0×10^{-3} M solution of adenine-pyridoxine on Ag nanoparticles and then recording with the micro-Raman instrument. The 633 nm laser line of an Ar ion laser was employed for the acquisition of data. The resolution of the instrument is 2 cm^{-1}, and measurements were performed at room temperature. Ag nanoparticles were prepared according to the procedure reported in literature [24, 26].

3. Results and Discussion

3.1. UV-Visible Spectral Studies on the Adenine-Pyridoxine Complex. The UV-visible absorption spectra of adenine, pyridoxine, and the adenine-pyridoxine complex were measured to preliminarily investigate the extent of the interaction of adenine with pyridoxine. In the UV-visible spectrum of adenine shown in Figure 2, there is a presence of an absorption peak at 260 nm. In literature, the presence of this peak in the UV-visible spectrum of DNA has been attributed to presence of adenine and guanine [27]. Beer's plot of absorbance versus concentration of adenine at λ_{max} yielded a linear graph ($R^2 = 0.9818$) with a molar absorptivity of 7.710×10^{-4} $M^{-1}\cdot cm^{-1}$ (Figure 3). The UV-visible absorption spectrum of pyridoxine between 240 and 280 nm has been reported since this is the wavelength range of interest for the double reciprocal plot. However, in this study, it was found that pyridoxine absorbs at 252 and 323 nm as reported in literature [28]. The spectrum of the pyridoxine absorption band at 252 nm is shown in Figure 4). The absorption of UV-visible light by pyridoxine increases with concentration, and Beer's plot of pyridoxine at 252 nm shows that the molar absorptivity of pyridoxine is 6.800×10^{-4} $Lmol^{-1}\cdot cm^{-1}$.

UV-visible spectra recorded for adenine-pyridoxine (Figure 5) show that adenine absorbance at 260 nm increases with pyridoxine concentration. Performing a double reciprocal plot using absorbance maxima values of adenine, that is, at wavelength of 260 nm as shown in Figure 6, generates a linear graph with $R^2 = 0.9138$. The linearity of the graph indicates that the binding of adenine to pyridoxine is characterized by a 1 : 1 stoichiometric relationship. Computing the ratio of intercept to slope as determined from Figure 6, the binding constant is estimated as 29.1. This value

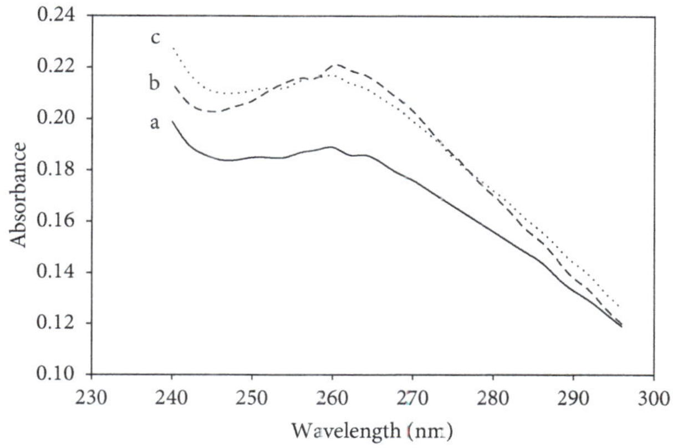

FIGURE 2: UV-visible absorption spectrum of aqueous solution of adenine at pH 7.0: (a) 1×10^{-3} M; (b) 2×10^{-3} M; (c) 3×10^{-3} M.

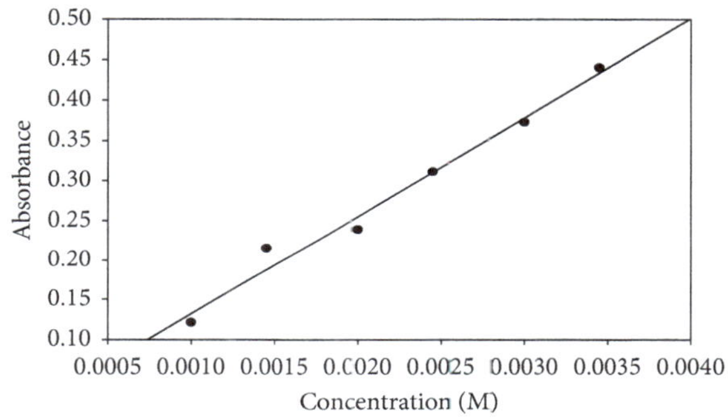

FIGURE 3: Plot of absorbance versus concentration of aqueous solution of adenine at 260 nm, pH 7.0, and $R^2 = 0.9818$.

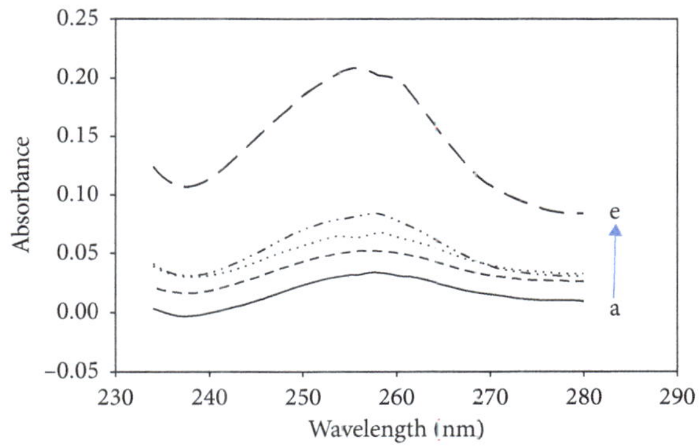

FIGURE 4: UV-visible absorption spectrum of aqueous solution of pyridoxine at pH 7.0. The lowest to the highest spectrum corresponds with (a) 4.0×10^{-5} M, (b) 5.0×10^{-5} M, (c) 2.0×10^{-4} M, (d) 3.0×10^{-4} M, and (e) 4.0×10^{-4} M pyridoxine.

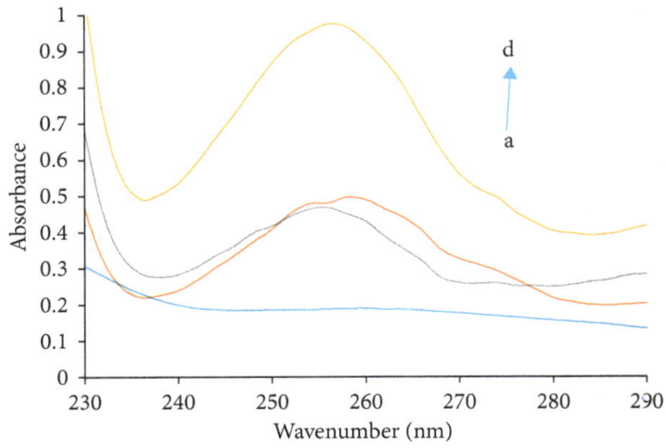

FIGURE 5: UV-visible absorption spectrum of aqueous solution of pyridoxine added to adenine at pH 7.0 with constant concentration of adenine (1.0×10^{-3} M). The lowest to highest spectrum corresponds with (a) 0.0 M, (b) 3.0×10^{-5} M, (c) 4.0×10^{-5} M, and (d) 5.0×10^{-5} M pyridoxine.

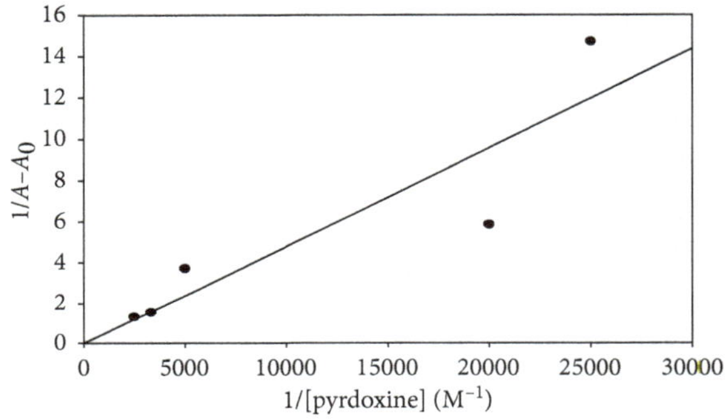

FIGURE 6: Lineweaver–Burk plot at 260 nm.

suggests a moderate interaction between adenine and pyridoxine.

3.2. Electrochemical Study of Adenine-Pyridoxine Interaction. Figure 7 presents the CV electrochemical activity of adenine (1.0×10^{-3} M) in the sodium chloride electrolyte/sodium phosphate buffer (pH 7.0) on the glassy carbon electrode (GCE). In the forward scan at a scan rate of 0.05 V/s, there is a one-step anodic peak at 1.0 V (versus SCE) corresponding with the oxidation of the hydroxyl group on adenine. The absence of CV peaks in the reverse sweep is an indication that the oxidation of adenine is irreversible. The CV recorded at different potential scan rates (0.05–0.7 V/s) shows an increase in the peak current with scan rate, and by plotting the peak current against the square root of the scan rate, a liner graph was generated. (inset in Figure 7, $R^2 = 0.9979$). This is typical of a diffusion-controlled electrochemical oxidation.

A broad irreversible oxidation peak at 0.8 V (versus SCE) is a characteristic of the CV of pyridoxine (2.0×10^{-2} M) in sodium chloride electrolyte-sodium phosphate buffer (pH 7.0) on GCE (Figure 8). A linear plot (inset in Figure 8, $R^2 = 0.9979$) of peak current versus square root of scan rate

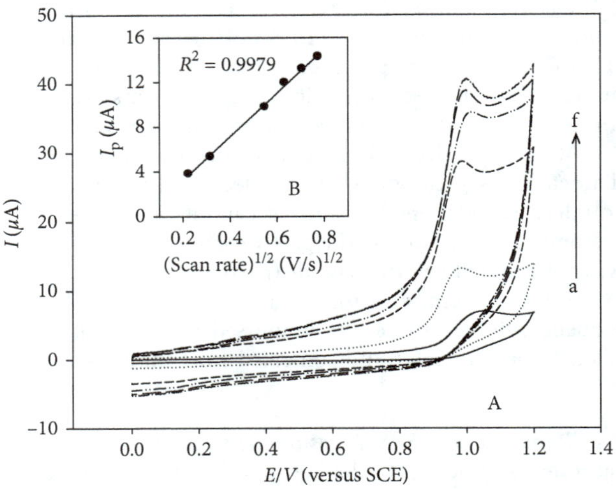

FIGURE 7: CV of 1.0×10^{-3} M adenine on the glassy carbon electrode in 0.1 M phosphate buffer/5 mM NaCl supporting electrolyte (pH 7.0) at different scan rates. Scan rate: (a) 0.05 V/s; (b) 0.1 V/s; (c) 0.2 V/s; (d) 0.3 V/s; (e) 0.4 V/s 0.5; (f) 0.6 V/s.

showed that the oxidation of pyridoxine is diffusion controlled. Other authors observed similar electrochemical behavior for pyridoxine on carbon electrodes stating that it is oxidized via a one-electron diffusion-controlled process [29–32].

Further investigation of the adenine-pyridoxine interaction at pH 7.0 performed by preparing a solution containing adenine and pyridoxine (1.0×10^{-3} M and 1.0×10^{-2} M, respectively) on GCE at 0.1 V/s (Figure 9) showed peaks associated with the individual compounds were present in the CV of the complex. Very noticeable is the imprecise nature and broadness of the adenine and pyridoxine peaks. Also obvious is the slight shift to the higher potential of the adenine peak at 1.2 V from 1.0 V.

3.3. Raman Study of the Interaction of Adenine with Pyridoxine. Figure 10(a) shows the Raman spectrum of adenine at 633 nm excitation. The Raman vibrational assignment of adenine vibrational modes has been reported in the literature [21, 33]. In this study, the observed bands in the Raman spectrum of adenine are assigned according to literature report and are presented in Table 1. The four most prominent bands in the Raman spectrum of adenine are 722, 1250, 1335, and 1485 cm^{-1}. The Raman band at 1485 cm^{-1} is assigned to the N7-C8 stretching coupled to C8-H bending modes. The vibrational motions that contribute the most to the different normal modes of the purine ring in the structure of adenine are observed at 1128, 1250, and 1335 cm^{-1}. The Raman band at 1335 cm^{-1} is assigned to the C8-N9 and C4-N3 stretching coupled to the N9-H and C2-H bending modes, respectively. Also, the band at 1250 cm^{-1} is assigned to the in-plane N7-C8 stretching coupled to the C8-H and N9-H bending mode. The Raman band at 1165 cm^{-1} has been assigned to the in-plane $N10H_2$ rocking mode coupled to the C6-N10, N3-C4, and C4-N9 stretching modes [21, 33]. A strong Raman band observed at 1128 cm^{-1} which

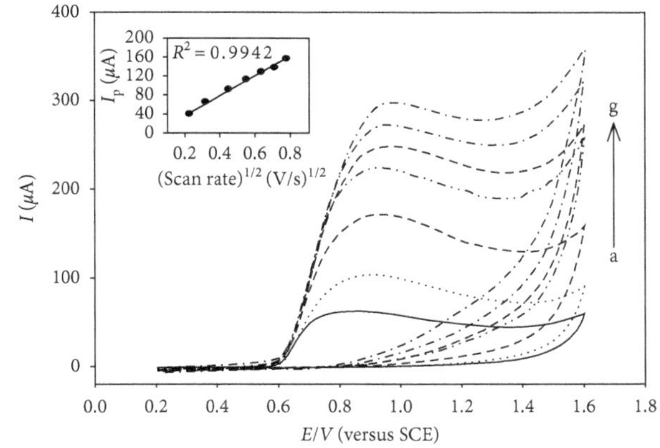

FIGURE 8: CV of 2.0×10^{-2} M pyridoxine on the glassy carbon electrode in the 0.1 M phosphate buffer/5 mM NaCl supporting electrolyte at pH 7.0. Scan rate: (a) 0.05 V/s; (b) 0.1 V/s; (c) 0.2 V/s; (d) 0.3 V/s; (e) 0.4 V/s; (f) 0.5 mV/s; (g) 0.6 V/s.

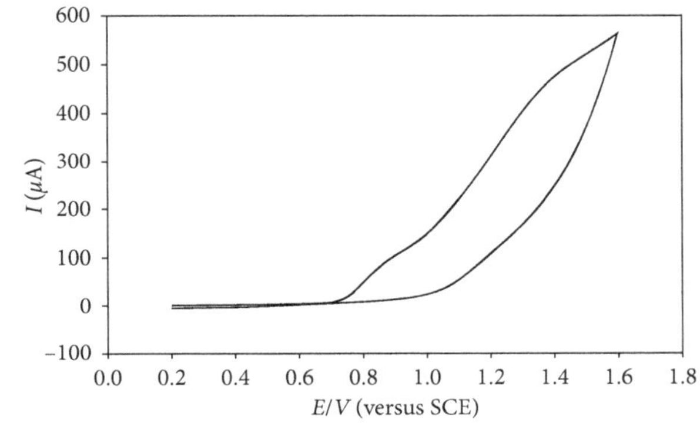

FIGURE 9: CV of a solution of 1.0×10^{-3} M adenine + 1.0×10^{-2} M pyridoxine on the glassy carbon electrode in the 0.1 M phosphate buffer/5 mM NaCl supporting electrolyte at pH 7.0. Scan rate is 100 mV/s.

contributed to the normal modes of the purine ring is assigned to C8-N9 stretching coupled to the C8-N9 and N9-H bending mode. A relative medium intense Raman band is observed at 1025 cm^{-1}, and it is specifically assigned to the NH_2 rocking mode. The Raman band at 722 cm^{-1} is assigned to the in-plane ring breathing mode of the whole molecule [33].

The Raman spectrum of aqueous pyridoxine in the spectral region of 350–1800 cm^{-1} is presented in Figure 10(b), and the vibrational assignment is shown in Table 2. In this study, the vibrational assignment is consistent with the vibrational assignment of pyridoxine reported by Srivastava et al. [36] and Atak-Bulbul and Akyuz [35]. The following Raman bands at 518, 865, 929, and 1091 cm^{-1} were not observed by Srivastava et al., but the Raman bands at 929 and 518 cm^{-1} were observed by Atak-Bulbul et al. The Raman bands at 518, 865, 929, and 1091 cm^{-1} are assigned based on the DFT and PED calculations on pyridoxine reported by Atak-Bulbul and Akyuz [35]. The most intense Raman bands which characterize the Raman spectrum of pyridoxine are observed at 1630, 1327, 1235, 1052, 967, and 690 cm^{-1}. The Raman band at 1235 cm^{-1} is an asymmetrical band due to the contributions of several vibration modes, and it is assigned to the C-O-H/C-C-H bending mode coupled to the N-H in-plane bending mode. The band at 690 cm^{-1} is assigned to the C-C-O-H/C-C-C-H/C-C-C-C torsion mode [23, 35].

The Raman spectrum of aqueous solution of adenine-pyridoxine reveals noticeable changes, and this is shown in Figure 10(c). Examining the Raman spectrum of adenine-pyridoxine solution, there is essentially a sharp decrease in the relative intensity of the Raman bands at 690, 751, 1235, 1361, 1452, and 1630 cm^{-1} when compared to their counterparts in the Raman spectrum of pyridoxine. These bands are very strong in the Raman spectrum of pyridoxine but very weak in the spectrum of adenine-pyridoxine solution. In a study that focused on understanding the binding characteristics of adenine, Otto et al. reported that vibrations originating from molecular groups located close to interacting surfaces are the strongly enhanced Raman modes [37]. Therefore, the decrease in the relative intensity of the mentioned vibration modes (690, 751, 1235, 1297, and 1327 cm^{-1}) suggests that groups associated with these modes in pyridoxine are less involved in the interaction with adenine. Also obvious is the enhancement of the band at 590 cm^{-1} in the Raman spectrum of adenine-pyridoxine, when compared to its counterpart band in the Raman spectrum of pyridoxine. This band is assigned to the C-C-C-N torsion bending modes of the pyridoxine structure.

Shifts in the position of bands at 900 and 1084 cm^{-1} were observed in the Raman spectrum of adenine-pyridoxine solution. The band at 1084 cm^{-1} in the Raman spectrum of adenine-pyridoxine solution is assigned to υ (C11–O12) stretching coupled to deformation of C11–H and C-O-H. The Raman band at 1084 cm^{-1} is observed in the Raman spectrum of pyridoxine at 1091 cm^{-1}, with the red shift suggesting interaction between the two compounds.

It is expected that the interaction between the two molecules will result in shifts or enhancement of Raman bands in the spectrum of the interacting molecules. The electromagnetic enhancement model suggests that vibrational modes that involve a large change in polarizability of bonds that are perpendicular to surfaces are enhanced the most [38]. The spectral region of 350–1700 cm^{-1} in the Raman spectrum of adenine-pyridoxine solution shows that there is an increase in the relative intensity of the Raman bands at 624, 770, 1250, 1334, and 1484 cm^{-1}. The increase in intensity suggests that the adenine bonds associated with these vibrations are more involved in the intermolecular interaction with pyridoxine [37].

Adenine interacts with molecules preferentially using the nitrogen atom on position 7 (N7) in the adenine structure [39] and the exocyclic amino group (NH_2) [22], and the most intense lines in its spectrum are associated with atoms in closest proximity to the SERS surface. In Watanabe's report, the presence of the strong Raman band at 1334 cm^{-1} in the SERS spectrum of adenine was used to characterize the adsorption of adenine to a surface via the nitrogen atom on the position 7 of the adenine structure. Since this is a

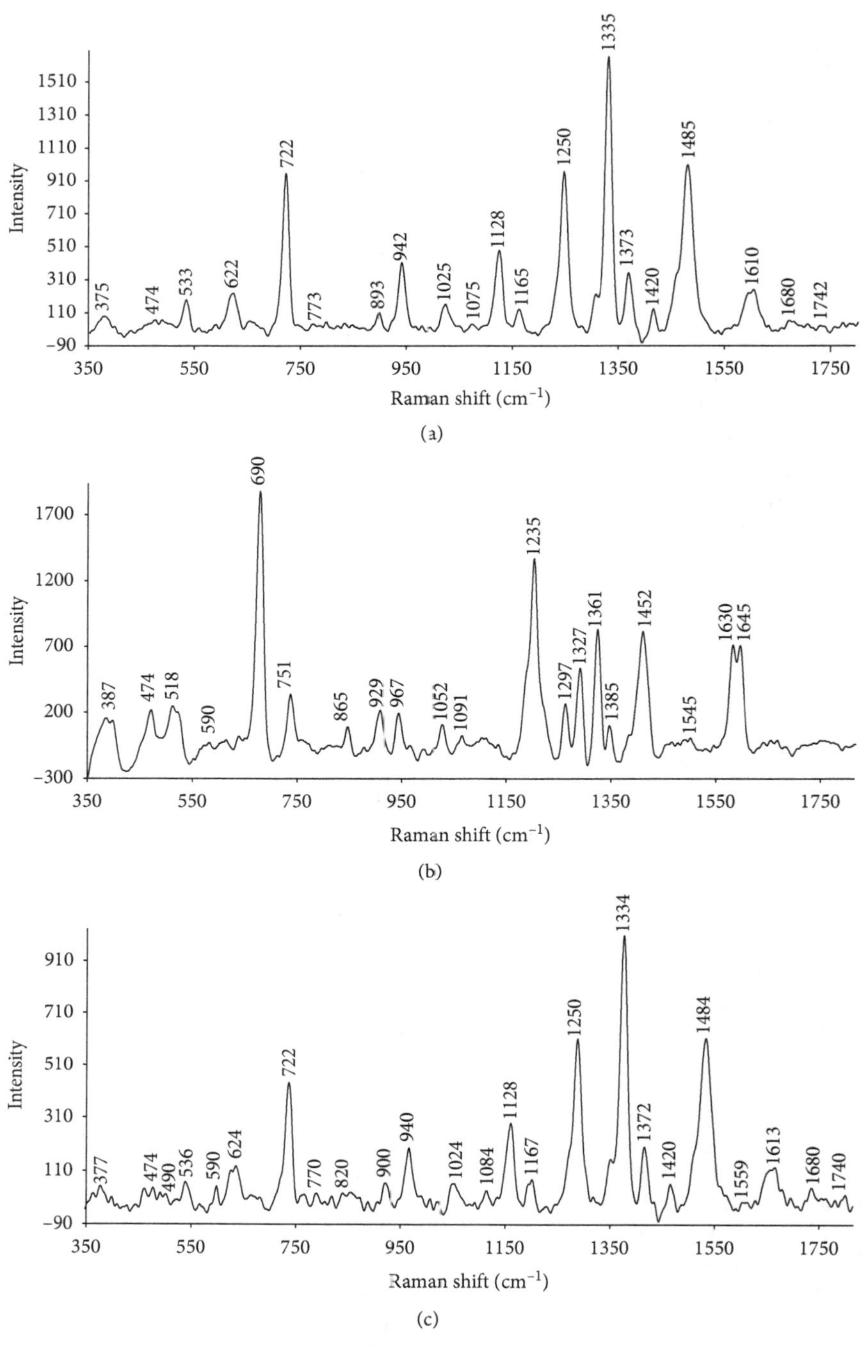

FIGURE 10: (a) Raman spectrum of aqueous solution of adenine at pH 7.0 (laser output: 633 nm, $P = 200$ mW). (b) Raman spectrum of aqueous solution of pyridoxine at pH 7.0 (laser output: 633 nm, $P = 200$ mW). (c) Raman spectrum of aqueous solution of adenine-pyridoxine at pH 7.0 (laser output: 633 nm, $P = 200$ mW).

solution-based study, pyridoxine is thought to produce a surface-like effect that enhances the vibrational mode associated with the Raman band at 1334 cm^{-1}, which is the most intense band in the Raman spectrum of adenine-pyridoxine. The enhancement of this band highlights the involvement of the endocyclic nitrogen atom (N7) of adenine in its interaction with pyridoxine.

Furthermore, Cinta et al. [23] reported the Raman and SERS of aqueous pyridoxine in the pH range of 1 to 10. In their study, it was observed that there were changes in the relative intensities of the SERS bands at 978, 1240, and 1330 cm^{-1} in the basic pH region and vibrational modes arising from ring deprotonation at 1240 and 1330 cm^{-1}. The vibrational modes at 1240 and 1330 cm^{-1} have been assigned

Table 1: Experimental Raman shift values (cm^{-1}) for adenine in solution.

Experimental Raman (cm^{-1}) [34]	B3LYP/6-31G (d, p) [34]	Experimental Raman values (cm^{-1}) [21]	B3LYP/6-31^{2+}G (d, p) [21]	This work, Raman values (cm^{-1})	Vibrational assignment [21]
1639	1641	1674	1665	1680	sciss NH2, υ (C6-N10), υ (C5-C6)
1612	1617	1613	1643	1610	υ (N3-C4), υ (N1-C6), υ (C5-N7), υ (N7-C8), δ (N9-H)
1599	1584	1597	1613	1597	sciss NH2
1482	1502	1483	1524	1485	υ (N7-C8), δ (C8-H), sciss NH2
1474	1487	1463	1510		υ (C2-N3), υ (N1-C6), δ (C2-H), sciss NH2
1419	1416	1419	1441	1420	υ (C4-N9), (C4-C5), (C6-N10), (N7-C8), δ (C2-H)
1389	1400	1372	1423	1373	δ (C2-H, N9-H), υ (C8-N9, C4-N9)
1345	1350	1333	1372	1335	δ (C2-H, C8-H, N9-H), υ (C6-N1, C8-N9, N3-C4)
1328	1342	1333	1365	1335	υ (C5-N7, N1-C2), δ (C2-H, C8-H),
1290	1317	1308	1341	1308	υ (C2-N3, N1-C2, C5-C6, C5-N7)
1240	1250	1248	1272	1250	δ (C8-H, N9-H), υ (N7-C8)
1229	1228	1234	1246	1234	rock NH2, υ (C5-N7, N1-C2, C2-N3)
1127	1129	1162	1148	1165	δ (C8-H, N10-H11), υ (C4-N9, N3-C4, C6-N10)
1032	1065	1126	1085	1128	υ (C8-N9), δ (N9-H, C8-H),
1005	1000	1025	1012	1025	rock NH2
958	953		974		wag (C2-H) o.p.
927	925	942	942	942	def R5
887	882	899	899	893	def R6, R5 (υ C5-N7)
848	831	839	849		wag (C8-H) o.p.
802	793	797	805	773	def R6 (wag C4-C5-C6), wag (C8-H) o.p.
717	713	723	726	722	Ring breath whole molecule (distorted)
	672		685		def R5, R6 (τ C4-C5-C6, wag N3-C4-N9) o.p.
655	658		666		def R5 (wag C5-N7-C8), wag C8-H, N9-H,) o.p.
610	607	623	618	622	def R6 and R5
566	568	560	576	560	wag (C2-H, N9-H) o.p.
	528		540		ι (NH2) o.p.
	521	536	530	533	def R6—i.p./o.p.
513	514	536	520	533	wag (N9-H), def R6, R5—i.p./o.p.
	504		518	474	wag (N9-H), τ (NH2) o.p.
	298		300	375	def R6, R5 (wag N1-C2-N3-C4, C5-N7-C8-N9), wag C6-N10) o.p.

υ = stretching; δ = bending; τ = torsion; rock = rocking; def = deformation; sciss = scissoring; wag = wagging; R5 = five-membered ring; R6 = six-membered ring; breath = breathing; i.p. = in plane; o.p. = out plane.

Table 2: Raman shift values (cm^{-1}) for aqueous solution of pyridoxine.

Experimental Raman values (cm^{-1}) [23]	IR (ab initio) [23]	Experimental Raman values (cm^{-1}) [35]	DFT/6-311++G (d, p) [35]	This work, Raman values (cm^{-1})	Vibrational assignment [21, 35]
1646		1618	1611	1645	υ (C-C)/ring stretch
1630	1600	1565	1576	1630	υ (C-C), δ (H15–C10–H16), δ (C9–N7–H20)
1520	1542		1501	1545	δ (C-H ring) + υ (C-C)
1451	1465	1460	1459	1452	δ (C4–O8–H13) + δ (C-C-O) + υ_{as} (CH3), δ (C3–C9–H19)
1389		1378	1376	1385	υ (C1–C2, C2–C4) + δ (CH3)
1360	1351		1362	1361	δ (C11–O12–H23) + δ (C-O-H) + (CH2) twist
1326	1325	1316	1341	1327	δ (C9–N7–H20) + δ (C3–C9–H19) + δ (C–H) i.p. + ring τ + δ (H17-C1-O6), υ (C-N)
1295	1288	1280	1287	1297	υ (C5–N7)/(C-C) + τ (C5–CH3) + δ (C10–H)
1234	1256	1236	1242	1235	δ (C-O-H)/(C-C-H), δ (N–H) i.p.
1219	1178		1178	1091	υ (C11–O12) + δ (C11–H), δ (C-O-H)
1054	1061	1054	1067	1052	Ring stretch + τ (H17–C1–O6–H24) + υ (C-C-O)
966	972	986	988	967	δ (N–H) o.p. + δ (C–C) + υ (C-O)
		928	904	929	τ (C-C-N-H)
			895	865	τ (C-C-N-H)
753	756	751	751	751	τ (H19–C9–N7–H20) + δ (C6–C1–O6) + υ (C1–C2–C3)
692	655	695	706	690	τ (C2–C3–C9–H19) + stretch quadrant (C2–C4–C5–N7) + τ (C3-C2-C4-C5)
			585	590	τ (C-C-C-N)
533	539	519	520	518	δ (OH–C4–C5–CH_3) + τ (H19–C9–N7–H20), υ (C-C) + υ (C-O)
477	453	482	494	474	τ (H19–C9–N7–H20) + δ (C-C-O)/(C-C-H)
403	404	391	411	387	δ (H17–C1–O6–H24) + τ (H21–C11–O12-H23), υ (C-O)

υ = stretch; δ = bending; τ = torsion; twist = twisting; i.p. = in plane; o.p. = out plane.

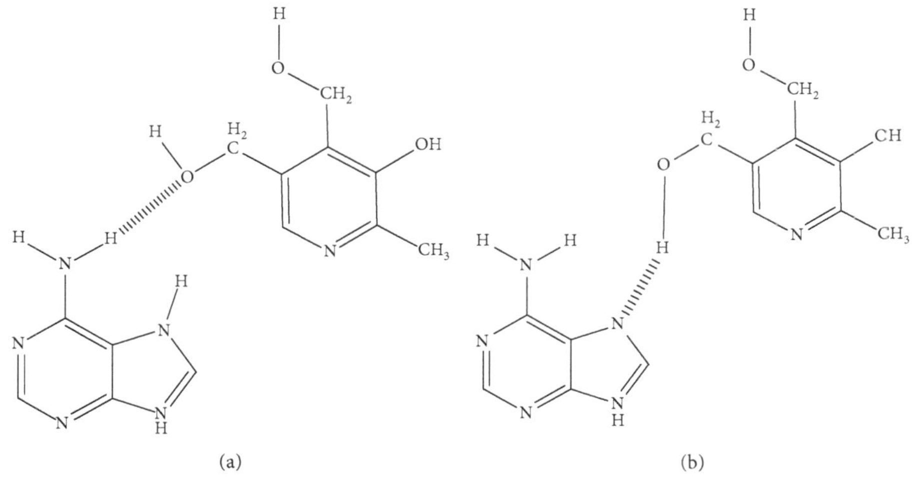

FIGURE 11: Proposed chemical structures of adenine-pyridoxine interaction.

to ring stretching modes of pyridoxine. Cinta et al. [23] also observed the deprotonation of the N-ring atom leading to a SERS response typical of pyridine derivatives. In this study, a difference is observed in the relative intensity of bands at 1235 and 1327 cm^{-1} in the Raman spectrum of pyridoxine when compared to corresponding bands (1250 and 1334 cm^{-1}) in the spectrum of adenine-pyridoxine solution. A very strong band at 1334 cm^{-1} in the spectrum of adenine-pyridoxine solution suggests a strong interaction between the adenine and pyridoxine molecule via the N7 atom, as shown in Figure 11(a).

Otto et al. observed that the vibrational modes at 626, 960, 1028, and 1194 cm^{-1}, linked to the motion of C6-NH_2, gave rise to strong contributions in the SERS spectrum of adenine, leading them to conclude that adenine was bound to the surface via the exocyclic amino group [22]. In the Raman spectrum of adenine-pyridoxine solution, it is evident that the bands at 624, 940, 1024, 1128, and 1167 cm^{-1} are enhanced when examined with respect to their equivalents (622, 942, 1025, 1128, and 1165 cm^{-1}) in the Raman spectrum of adenine. This suggests that adenine is coordinated to pyridoxine via the exocyclic amino group (Figure 11(a)). This is not surprising because adenine can undergo protonation at the endocyclic nitrogen atoms N1, N3, and N7 [40]. Considering that the adenine exocyclic amino group has the least sterically hindered nitrogen atom, this portion of the ring is likely the preferred site for intermolecular interactions. Furthermore, the delocalization of electrons on the exocyclic amino group N10 into the heterocyclic ring due to the resonance effect confers a quaternary structure on the nitrogen, thereby increasing the possibility of having an intermolecular interaction between it and pyridoxine O12.

3.4. SERS Spectrum of Adenine-Pyridoxine on Ag Nanoparticles. The SERS spectrum adenine-pyridoxine is characterized by the presence of strong SERS bands at 732 and 1332 cm^{-1}. The in-plane breathing mode at 733 cm^{-1} as reported in literature is the most prominent band in the SERS of adenine adsorbed onto the Ag surface [21, 33]. This band that occurs at 732 cm^{-1} in this study is also observed as the most prominent band in the SERS spectrum of adenine-pyridoxine.

In this study, more vibrational information on the interaction of adenine with pyridoxine is gained by examining the SERS spectrum of adenine-pyridoxine adsorbed onto the surface of Ag nanoparticles. The SERS spectrum of adenine-pyridoxine is shown in Figure 12, and vibrational assignments are presented in Table 3. The SERS spectrum of adenine-pyridoxine is well resolved, and it shows a noticeable change in the intensity of the bands when compared with the normal Raman bands of the complex. There are shifts in the SERS bands of adenine-pyridoxine at 1463, 1033, 962, 732, and 455 cm^{-1}. The SERS band at 732 cm^{-1} is observed in the Raman spectrum of adenine-pyridoxine solution at 722 cm^{-1}. There is also the enhancement of bands at 455, 732, 962, 1033, 1332, and 1463 cm^{-1} in the SERS spectrum of adenine-pyridoxine when compared to the corresponding normal Raman bands of adenine-pyridoxine (1484, 1334, 1024, 940, 722, and 474 cm^{-1}). This suggests that these vibration modes more related to the adsorption of adenine-pyridoxine on the Ag nanoparticles. The SERS band at 1332 cm^{-1} is assigned to the C-N stretching mode of adenine [21]. This SERS band (1332 cm^{-1}) is relatively enhanced when compared to its counterpart Raman band at 1335 cm^{-1}. The stronger enhancement of SERS bands at 1332 and 732 cm^{-1} strongly suggests a perpendicular orientation of the adenine-pyrimidine-purine ring portion of the complex close to the Ag nanoparticle surface. Furthermore, the SERS spectrum of adenine-pyridoxine shows a pattern of enhancement of other vibrational modes pertaining to adenine, at 1171, 1120, 1033, 962, and 623 cm^{-1}, as was observed by Otto et al. [22].

In conclusion, spectroelectrochemical analyses show evidence of adenine interaction with pyridoxine. Analyses of UV-visible data using the Lineweaver–Burk double reciprocal plot reveal that adenine interacts with pyridoxine in

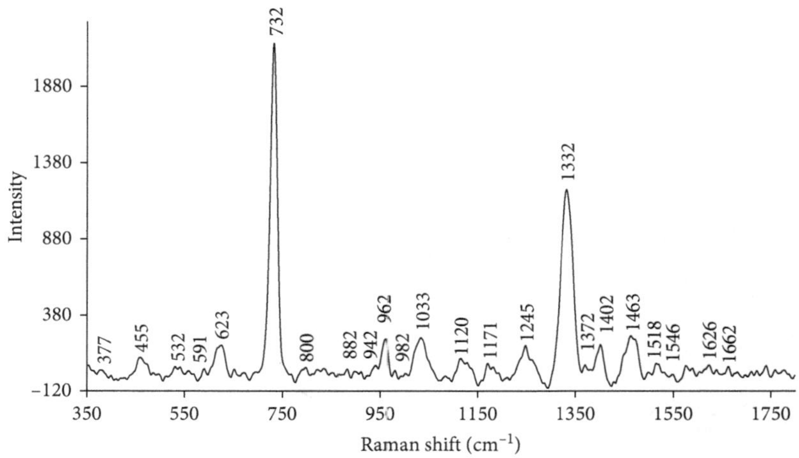

FIGURE 12: SERS spectrum of aqueous solution of adenine-pyridoxine at pH 7.0 (laser output: 633 nm, P = 200 mW).

TABLE 3: Experimental Raman and SERS shift values (cm^{-1}) for aqueous solution of adenine-pyridoxine.

Experimental Raman values (cm^{-1})	SERS values (cm^{-1})	Vibrational assignment [34, 35]
1680	1662	sciss NH2, υ (C-N), υ (C-C)/ring stretch
1610	1626	υ (N-C), υ (C-C), δ (H-C-H), δ (C–N–H)
1597	1589	sciss NH2, υ R6 (C=C)
	1518	δ (C-H ring) + υ (C-C)
1485	1463	υ (N-C), δ (C-H), sciss NH2
1420	1402	υ (C-N), δ (C-H), υ (C–C) + δ (CH3)
1373	1372	δ (C-H, N-H), υ (C-N) and δ (C–O-H) + δ (C-O-H) + (CH2) twist
1335	1332	δ (C2-H, C8-H, N9-H), υ (C6-N1, C8-N9, N3-C4)
1335		δ (C–N–H) + δ (C–C–H) + δ (C–H) i.p., ring τ + δ (H-C-O), υ (C-N)
1308		υ (C-N), υ (C-C)
1250	1245	δ (C-H, N-H) i.p., υ (N7-C8), δ (C-O-H)/(C-C-H)
1234		rock NH2, υ (C-N, N-C)
1165	1171	δ (C-H, N-H), υ (C-N, N-C)
1128	1120	υ (C8-N9), δ (N9-H, C8-H),
1025	1033	rock NH2, ring stretch, τ (H–C–O–H) + υ (C-C-O)
	982	wag (C-H) o.p., δ (N–H) o.p. + δ (C–C), υ (C-O)
942	962, 942 sh.	def R5
893	882	def R6, R5 υ (C-N)
773	800	def R6 (wag C-C-C), wag (C-H) o.p.
722	732	Ring breath whole molecule (distorted)
622	623	def R6 and R5, τ (C–C–C–H) + stretch quadrant (C-C-C-N), τ (C-C-C-C)
560	591	wag (C2-H, N9-H) o.p.
533	532	def R6—i.p./o.p.
533	532	wag (N-H), def R6, R5—i.p./o.p., δ (OH–C–C–CH3) + τ (H–C–N–H), υ (C-C) + υ (C-O)
474	455	wag (N-H), τ (NH2) o.p.
375	377	def R6, R5 wag (N-C-N-C, C-N-C-N), wag (C-N) o.p.

υ = stretching; δ = bending; τ = torsion; rock = rocking; def = deformation; sciss = scissoring; wag = wagging; R5 = five-membered ring; R6 = six-membered ring; breath = breathing; i.p. = in plane; o.p. = out plane; sh. = shoulder.

a 1 : 1 mode with a binding constant of 29.1. The broadening of the CV oxidation peak of pyridoxine and the shift to a higher potential of the adenine anodic peak provides evidence supporting the interaction between the two compounds. Additional evidences obtained from experimental and theoretical analysis of Raman data showed that adenine interacts with pyridoxine O(12) using its exocyclic amino and purine nitrogen N(7) groups. Shifts of spectral peaks to the lower wavenumber especially with the pyridoxine hydrochloride -CH_2-OH bending and in-plane -N-H mode at

1235 cm^{-1} and adenine $-NH_2$ rocking band at 942 cm^{-1} confirm that there is the formation of hydrogen bonds as the dominant intermolecular force between the two compounds. Shifts in the SERS bands of adenine-pyridoxine, particularly the band at 732 cm^{-1}, observed in the Raman spectrum of adenine-pyridoxine solution at 722 cm^{-1} and strong enhancement of SERS bands pertaining to adenine suggest that the complex interacts with the Ag nanoparticle surface possessing a perpendicularly orientation, with the adenine moiety closest to the surface. The results obtained in this study validate the utility of spectroelectrochemical techniques in investigating the interaction of small molecules like pyridoxine with nucleobases and for understanding how various parts of DNA interact with vitamins on substrates in biological systems.

Conflicts of Interest

The authors declare that they have no conflicts of interest.

Acknowledgments

The authors thank the University of Maryland Eastern Shore, Department of Natural Sciences, and Louis Stokes Alliances for Minority Participation program for, respectively, providing funds for the purchase of reagents and support of the undergraduate student who participated in this project. The authors also wish to appreciate the research group of Professor Charles Hosten of Howard University, Department of Chemistry, for granting access for the use of the CV and Raman instrument for analysis.

References

[1] M. Aleksić and V. Kapetanović, "An overview of the optical and electrochemical methods for detection of DNA-drug interactions," *Acta Chimica Slovenica*, vol. 61, no. 3, pp. 555–573, 2014.

[2] B. H. Geierstanger and D. E. Wemmer, "Complexes of the minor groove of DNA," *Annual Review of Biophysics and Biomolecular Structure*, vol. 24, no. 1, pp. 463–493, 1995.

[3] H. Li, W.-J. Mei, Z. Xu, D.-W. Pang, L.-N. Ji, and Z.-H. Lin, "Electrochemistry of a novel monoruthenated porphyrin and its interaction with DNA," *Journal of Electroanalytical Chemistry*, vol. 600, no. 2, pp. 243–250, 2007.

[4] D. E. Metzler, "Vitamins and coenzymes," in *Encyclopedia of Physical Science and Technology*, pp. 509–528, Academic Press, Cambridge, MA, USA, 2003.

[5] P.-O. Löwdin, "Proton tunneling in DNA and its biological implications," *Reviews of Modern Physics*, vol. 35, no. 3, pp. 724–732, 1963.

[6] P. O. Lowdin, "Quantum genetics and the aperiodic," *Advances in Quantum Chemistry*, vol. 2, pp. 213–360, 1966.

[7] S. W. Ebbinghaus, "Site-selective DNA binding drugs," *Chemistry & Biology*, vol. 10, no. 10, pp. 895–897, 2003.

[8] S.-Q. Liu, M.-L. Cao, and S.-L. Dong, "Electrochemical and ultraviolet-visible spectroscopic studies on the interaction of deoxyribonucleic acid with vitamin B6," *Bioelectrochemistry*, vol. 74, no. 1, pp. 164–169, 2008.

[9] T. R. Krugh, "Drug-DNA interactions," *Current Opinion in Structural Biology*, vol. 4, no. 3, pp. 351–364, 1994.

[10] G. Arena, R. Purrello, E. Rizzarelli, A. Gianguzza, and L. Pellerito, "Thermodynamics of hydroxo complex formation of dialkyltin(IV) ions in aqueous solution," *Journal of the Chemical Society, Dalton Transactions*, no. 5, pp. 773–777, 1989.

[11] D. A. Bender, "Non-nutritional uses of vitamin B6," *British Journal of Nutrition*, vol. 81, no. 1, pp. 7–20, 1999.

[12] R. Casasnovas, A. Salvà, J. Frau, J. Donoso, and F. Muñoz, "Theoretical study on the distribution of atomic charges in the schiff bases of 3-hydroxypyridine-4-aldehyde and alanine: the effect of the protonation state of the pyridine and imine nitrogen atoms," *Chemical Physics*, vol. 355, no. 2-3, pp. 149–156, 2009.

[13] S. Dakshinamurti and K. Dakshinamurti, "Antihypertensive and neuroprotective actions of pyridoxine and its derivatives," *Canadian Journal of Physiology and Pharmacology*, vol. 93, no. 12, pp. 1083–1090, 2015.

[14] E. Paleček, M. Fojita, F. Jelen, and V. Vetterl, "Electrochemical analysis of nucleic acids," in *The Encyclopedia of Electrochemistry, Bioelectrochemistry, 9*, A. J. Bard and M. Stratmann, Eds., pp. 365–429, Wiley-VCH Verlag, Weinheim, Germany, 2002.

[15] A. M. Oliveira-Brett, M. Vivian, I. R. Fernandes, and J. A. P. Piedade, "Electrochemical detection of in situ adriamycin oxidative damage to DNA," *Talanta*, vol. 56, no. 5, pp. 959–970, 2002.

[16] A. M. Oliveira-Brett, J. A. P. Piedade, L. A. Silva, and V. C. Diculescu, "Voltammetric Determination of all DNA nucleotides," *Analytical Biochemistry*, vol. 332, no. 2, pp. 321–329, 2004.

[17] M. L. Wang, Y. Y. Zhang, Q. J. Xie, and S. Z. Yao, "In situ FT-IR spectroelectrochemical study of electrooxidation of pyridoxol on a gold electrode," *Electrochimica Acta*, vol. 51, no. 6, pp. 1059–1068, 2005.

[18] T. Pineda, J. M. Sevilla, A. J. Román, and M. Blázquez, "Electrooxidation of pyridoxal (PL) on a polycrystalline gold electrode in alkaline solutions," *Journal of Electroanalytical Chemistry*, vol. 492, no. 1, pp. 38–45, 2000.

[19] W. Qu, K. Wu, and S. Hu, "Voltammetric determination of pyridoxine (vitamin B6) by use of a chemically-modified glassy carbon electrode," *Journal of Pharmaceutical and Biomedical Analysis*, vol. 36, no. 3, pp. 631–635, 2004.

[20] R. C. Barthus, L. H. Mazo, and R. J. Poppi, "Simultaneous determination of vitamins C, B6 and PP in pharmaceutics using differential pulse voltammetry with a glassy carbon electrode and multivariate calibration tools," *Journal of Pharmaceutical and Biomedical Analysis*, vol. 38, no. 1, pp. 94–99, 2005.

[21] B. Giese and D. McNaughton, "Surface-enhanced Raman spectroscopic and density functional theory study of adenine adsorption to silver surfaces," *Journal of Physical Chemistry B*, vol. 106, no. 1, pp. 101–112, 2002.

[22] C. Otto, F. F. M. De Mul, A. Huizinga, and J. Greve, "Surface enhanced Raman scattering of derivatives of adenine: the importance of the external amino group in adenine for surface binding," *Journal of Physical Chemistry*, vol. 92, no. 5, pp. 1239–1244, 1988.

[23] S. Cinta, C. Morari, E. Vogel et al., "Vibrational studies of B6 vitamin," *Vibrational Spectroscopy*, vol. 19, no. 2, pp. 329–334, 1999.

[24] V. Ramakrishnan, N. Krishnamurthy, M. Gurunathan, and V. J. P. Srivatsavoy, "SERS studies of some α-amino-anthraquinones in silver sol," *Spectrochimica Acta Part A: Molecular Spectroscopy*, vol. 46, no. 11, pp. 1615–1619, 1990.

[25] S. Charak, M. Shandilya, G. Tyagi, and R. Mehrotra, "Spectroscopic and molecular docking studies on chlorambucil interaction with DNA," *International Journal of Biological Macromolecules*, vol. 51, no. 4, pp. 406–411, 2012.

[26] S. Shrivastava, T. Bera, A. Roy, G. Singh, P. Ramachandrarao, and D. Dash, "Characterization of enhanced antibacterial effects of novel silver nanoparticles," *Nanotechnology*, vol. 18, no. 22, article 225103, 2007.

[27] N. Li, Y. Ma, C. Yang, L. Guo, and X. Yang, "Interaction of anticancer drug mitoxantrone with DNA analyzed by electrochemical and spectroscopic methods," *Biophysical Chemistry*, vol. 116, no. 3, pp. 199–205, 2005.

[28] F. Manouchehri, Y. Izadmanesh, E. Aghaee, and J. B. Ghasemi, "Experimental, computational and chemometrics studies of BSA-vitamin B6 interaction by UV-vis, FT-IR, fluorescence spectroscopy, molecular dynamics simulation and hard-soft modeling methods," *Bioorganic Chemistry*, vol. 68, pp. 124–136, 2016.

[29] M. F. S. Tixeira, G. Marino, E. R. Dockal, and E. T. G. Cavalheiro, "Voltammetric determination of pyridoxine (vitamin B6) at a carbon paste electrode modified with vanadyl(IV)-salen complex," *Analytica Chimica Acta*, vol. 508, no. 1, pp. 79–85, 2004.

[30] H.-Y. Gu, A.-M. Yu, and H.-Y. Chen, "Electrochemical behavior and simultaneous determination of vitamin B2, B6, and C at electrochemically pretreated glassy carbon electrode," *Analytical Letters*, vol. 34, no. 13, pp. 2361–2374, 2001.

[31] Y. Wu and F. Song, "Voltammetric investigation of vitamin B6 at a glassy carbon electrode and its application in determination," *Bulletin of the Korean Chemical Society*, vol. 29, no. 1, pp. 38–42, 2008.

[32] G. Chen, X. Ding, Z. Cao, and J. Ye, "Determination of melatonin and pyridoxine in pharmaceutical preparations for health-caring purposes by capillary electrophoresis with electrochemical detection," *Analytica Chimica Acta*, vol. 408, no. 1-2, pp. 249–256, 2000.

[33] F. Madzharova, Z. Heiner, M. Gühlke, and J. Kneipp, "Surface-enhanced hyper-Raman spectra of adenine, guanine, cytosine, thymine, and uracil," *Journal of Physical Chemistry C*, vol. 120, no. 28, pp. 15415–15423, 2016.

[34] M. J. Nowak, L. Lapinski, J. S. Kwiatkowski, and J. Leszczyński, "Molecular structure and infrared spectra of adenine. Experimental matrix isolation and density functional theory study of adenine ^{15}N isotopomers," *The Journal of Physical Chemistry*, vol. 100, no. 9, pp. 3527–3534, 1996.

[35] B. Atak-Bulbul and S. Akyuz, "Ab initio density functional theory calculations on pyridoxine and its water clusters," *Asian Journal of Chemistry*, vol. 26, pp. S299–S304, 2014.

[36] M. Srivastava, P. Rani, N. P. Singh, and R. A. Yadav, "Experimental and theoretical studies of vibrational spectrum and molecular structure and related properties of pyridoxine (vitamin B6)," *Spectrochimica Acta Part A: Molecular and Biomolecular Spectroscopy*, vol. 120, pp. 274–286, 2014.

[37] C. Otto, T. J. J. van den Tweel, F.F.M. de Mul, and J. Greve, "Surface-enhanced Raman spectroscopy of DNA bases," *Journal of Raman Spectroscopy*, vol. 17, no. 3, pp. 289–298, 1986.

[38] J. A. Creighton, "Surface Raman electromagnetic enhancement factors for molecules at the surface of small isolated metal spheres: the determination of adsorbate orientation from SERS relative intensities," *Surface Science*, vol. 124, no. 1, pp. 209–219, 1983.

[39] T. Watanabe, O. Kawasaki, H. Katoh, K. Honda, Y. Nishimura, and M. Tsuboi, "SERS study of molecular adsorption: some nucleic acid bases on Ag electrodes," *Surface Science*, vol. 158, no. 1–3, pp. 341–351, 1985.

[40] J. E. Šponer, J. Leszczynski, F. Glahé, B. Lippet, and J. Šponer, "Protonation of platinated adenine nucleobases: gas phase vs condensed phase picture," *Inorganic Chemistry*, vol. 40, no. 14, pp. 3269–3278, 2001.

13

Determination of Heating Value of Estonian Oil Shale by Laser-Induced Breakdown Spectroscopy

M. Aints,[1] P. Paris,[1] M. Laan,[1] K. Piip,[1] H. Riisalu,[2] and I. Tufail[1]

[1]Institute of Physics, University of Tartu, Tartu, Estonia
[2]Virumaa College, Tallinn University of Technology, Kohtla-Järve, Estonia

Correspondence should be addressed to P. Paris; peeter.paris@ut.ee

Academic Editor: Alessandro De Giacomo

The laser-induced breakdown spectroscopy (LIBS) combined with multivariate regression analysis of measured data were utilised for determination of the heating value and the chemical composition of pellets made from Estonian oil shale samples with different heating values. The study is the first where the oil shale heating value is determined on the basis of LIBS spectra. The method for selecting the optimal number of spectral lines for ordinary multivariate least squares regression model is presented. The correlation coefficient between the heating value predicted by the regression model, and that measured by calorimetric bomb, was $R^2 = 0.98$. The standard deviation of prediction was 0.24 MJ/kg. Concentrations of oil shale components predicted by the regression model were compared with those measured by ordinary methods.

1. Introduction

Oil shale is used for production of fuel oil and various chemical products and is also used as a fuel in Estonian power plants. New boilers of oil shale power plants as well as oil retorts require a relatively constant quality of raw materials and fuel. The basic parameter of oil shale quality is the heating value. Heating value can vary considerably within the location in a deposit and depends substantially on concretions and limestone content [1]. Optimal operation of boilers requires continuous monitoring of heating value of incoming fuel. Up to now, the heating value of oil shale in laboratories of Estonian enterprises is measured using oxygen bomb calorimeters, which are the standard instruments for measuring heating values of solid and liquid combustible samples. However, this precise method is time consuming and thus not useable for online monitoring of the continuous flow of raw material.

Laser-induced breakdown spectroscopy (LIBS) as a fast diagnostic method has been widely used for characterization of coals and lignite. Overview of activities in this field can be found for example in [2] and references therein. LIBS adapted with advanced data processing methods provides real-time information required by boiler operators nowadays. In addition to elemental composition, other important properties of coal, such as ash content and ash fusion temperature or even heating value that have complicated relation to elemental composition, are determined on the basis of LIBS spectra using multivariate analysis methods [3–5]. Contrary to numerous studies dealing with LIBS diagnostic of coals and lignite, the number of papers related to LIBS diagnostics of oil shale is rather limited.

In [6, 7], laser-induced pyrolysis was used for determination of oil shale characteristics. Directing a number of laser shots with high repetition rate to the same spot of the sample under an argon purge, the changes in the spectral line intensities in LIBS spectra over the course of measurements were related to H/C ratios determined on isolated kerogens from the rock samples. Predicted kerogen H/C ratios from the LIBS measurements of whole rock samples were well correlated ($R^2 = 0.99$) to values determined for kerogen isolates measured by elemental analysis [6] An important advantage of the method is that the decay of intensities of H and C lines with the growth of laser shot number allowed a clear

Table 1: Composition of dry samples.

Sample number	1	2	3	4	5	6	7	8	9	10
Heating value by bomb, Q^b (MJ/kg)	6.36	6.60	7.28	7.98	9.10	9.08	9.86	10.36	10.78	11.62
Analytical moisture (%)	0.6	0.6	0.7	0.7	0.8	0.8	0.8	0.8	0.8	0.8
C (%)	22.4	22.3	23.4	24.3	26.4	26.4	27.2	27.8	29.5	30.9
C inorganic (%)	7.2	7.2	6.9	6.6	6.1	6.3	5.6	5.1	5.8	5.6
C organic (%)	15.2	15.1	16.5	17.7	20.3	20.1	21.6	22.7	23.7	25.3
S (%)	1.0	1.2	1.2	1.3	1.4	1.4	1.3	1.5	1.2	1.3
N (%)	0.07	0.07	0.07	0.08	0.08	0.08	0.09	0.1	0.09	0.09
H (%)	1.9	1.9	2.1	2.2	2.6	2.6	2.7	3.0	2.9	3.0
Fe_2O_3 (%)	2.0	1.9	2.1	2.2	2.3	2.5	2.2	2.5	2.1	1.9
SiO_2 (%)	11.2	10.1	10.8	11.1	13.0	13.3	15.9	16.6	12.7	12.2
MgO (%)	3.2	3.1	2.1	3.0	3.2	3.5	2.8	2.7	2.8	2.7
CaO (%)	32.9	32.1	31.2	26.4	27.7	32.1	26.6	22.7	31.6	30.4
K_2O (%)	1.5	2.2	2.2	3.8	3.6	1.7	1.7	1.4	2.9	1.4
Al_2O_3 (%)	1.9	2.4	3.6	3.7	3.5	3.0	2.6	2.9	3.4	2.4
Na_2O (%)	4.4	5.4	8.6	8.2	4.3	5.6	1.7	2.5	7.9	5.7

separation of H and C in the organic part of oil shale from that in the inorganic part. However, it is not realistic to apply the above-described laser-induced pyrolysis method for online monitoring of raw material on a running belt.

In [8], elemental composition of Marcellus Shale was determined with a high precision using LIBS spectra and multivariate analysis. However, for calculation of heating value on the basis of known elemental composition, additional information about the content of elements in the organic fraction is needed.

In [9], we applied LIBS diagnostics on Estonian oil shale samples for testing the moisture-caused matrix effects. The mineral and elemental composition of tested oil shale and waste rock (kerogenic limestone) samples were very similar, except the low percentage of organics in waste rock. Changes of both, content of kerogen and content of moisture, had strong effect on intensities of analytical spectral lines. A proper choice of spectral lines allowed reliable distinction of oil shale and limestone independently of moisture content.

The oil shale chemical composition and physical properties differ significantly from those of coal and lignite. The oil shale heating value is correlated with the concentrations of elements C, H, O, S, N, and Cl in the organic fraction of oil shale [10]. However, these elements are also present in large quantities in inorganic fraction of oil shale. Besides, as it is shown in [10], calculation of heating value using only elemental composition did not work well for Estonian oil shale. Mendeleev's and Dulong's formulae are widely used and accepted for calculation of heating value on the basis of elemental composition of organic part of many solid fuels, but situation with Estonian oil shale is more complicated than the applications for which these formulae were intended. "There is also the possibility that describing the heat of formation of a solid fuel using only elemental composition (as is inherent in all these methods) simply does not work quite as well for oil shale organics as for other organics" [10]. These reasons could complicate also the oil shale diagnostics by LIBS. Furthermore, elements belonging to surrounding air constituents cause an extra interference to LIBS spectra. According to the best of our knowledge, no investigation has been done to clarify the feasibility of LIBS for online determination of oil shale heating value.

The main aim of the present study was the evaluation of the LIBS applicability for determination of the heating value of Estonian oil shale using statistical methods of data processing. As an extra task, the LIBS-based estimation of the oil shale chemical composition was done. Experiments were done in conditions close to real industrial conditions: measurements were done in ambient air and spectra were recorded from different spots at the moving/rotating pellet surface.

2. Experimental

2.1. Samples. 10 samples of ground oil shale with different heating values were prepared by the laboratory of Mines of Estonian Energy. The samples represent commercial oil shale from concentration plant with heating value ranging from 6.3 to 11.6 MJ/kg. Samples, each about 1 kg, had grain size below 0.1 mm. The heating value Q^b and the mass percentage of the main composites of samples (Table 1) were determined in the laboratory of Virumaa College of Tallinn University of Technology using standardised methods. Concentration of C, H, and N was determined according to Standard Test Methods for Determination of Carbon, Hydrogen, and Nitrogen in Analysis Samples of Coal ASTM D5373; MgO, CaO, and Al_2O_3 according to EVN-EN ISO 7980; and K_2O and Na_2O according to ISO 9964.

About 200 g of each sample was additionally grinded using mortar and pestle to particle size smaller than 60 μm (i.e., 10 times smaller than the laser spot diameter) assuring sample homogeneity for different laser shots [11, 12]. Samples were pelletised without any binder, whereas before pressing 5 wt% of distilled water was added to the room dry

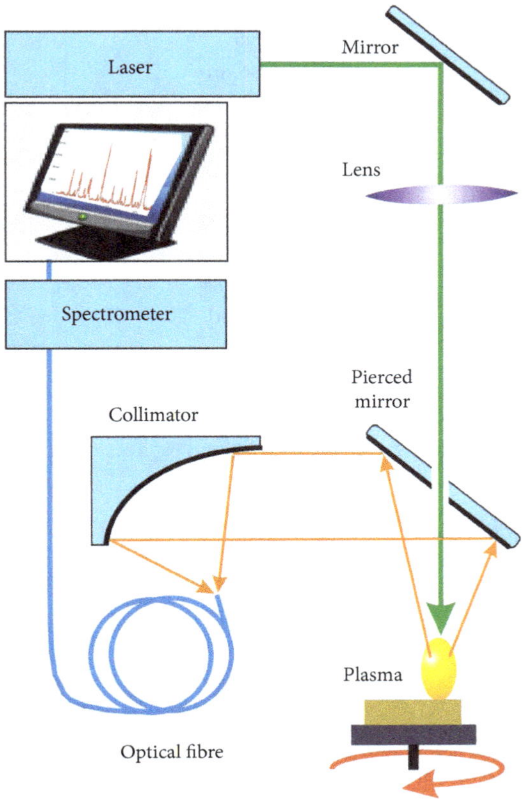

FIGURE 1: Sketch of the setup of measurements. Q-switched Nd:YAG laser YG980 and spectrometer Mechelle5000 were used.

powders. Powders were pressed into 25 mm in diameter and approximately 6 mm thick pellets under 18 tons held for 5 min. Four pellets from each sample were pressed. Thus, altogether 40 pellets were prepared. These pellets were stored in lab conditions at room temperature 21°C and relative humidity about 30% during a week to balance the moisture content of samples with the humidity of ambient air resulting approximately in about 2 wt% of moisture in pellets.

2.2. LIBS Measurements. A sketch of the laboratory LIBS setup is presented in Figure 1. The frequency-doubled Nd:YAG (YG980 Quantel) laser pulse of 532 nm wavelength and of 9 ns FWHM was focused onto the pellet surface with a plano-convex lens of 40 cm focal length in air. The focal plane of the lens was set about 3 cm under the samples' surface to avoid air breakdown in the focal point. The laser spot diameter at the sample surface was about 0.6 mm. The laser pulse energy at the plane of sample was measured with the pyroelectric laser energy meter Ophir Nova II. The pulse energy was set to 75 mJ making fluence at the pellet surface about 27 J/cm^2. The plasma emission was collected collinearly along the laser beam axis with the off-axis parabolic mirror collimator CC52 (Andor Technology) with focal length of 52 mm and delivered via the optical fibre of 50 μm diameter to the Mechelle5000 (Andor Technology) spectrometer equipped with an intensified charge-coupled device (ICCD) camera (Andor iStar DH334T-18F-03). The spectrum in 220–850 nm range with $\lambda/\Delta\lambda = 5000$ resolution was recorded. The delay time $t_d = 800$ ns between the laser pulse and the beginning of data acquisition was used. The recording time-gate Δt was 1 μs. These timing parameters were optimal for the signal-to-background ratio for carbon (C I) line at the wavelength 247.8 nm. Samples were fastened on a rotating stage driven with stepper motor supplying a fresh sample surface to each successive laser pulse avoiding possible pyrolysis-caused effects to the recorded spectra. The repetition rate of laser pulses was 0.55 Hz and the distance between the consecutive laser spots was more than 3 mm. Measurements were carried out under ambient atmospheric conditions. A continuous airflow of about 3 m/s was provided above the sample to remove laser-produced aerosols and dust between laser pulses [13]. To suppress the effect of possible sample heterogeneity, each pellet was characterised by a sum of 100 spectra. The relative standard deviation (RSD) of the shot-to-shot fluctuations of laser energy was 0.03.

3. Data Processing and Results

3.1. Selection of Spectral Lines. More than a hundred emission lines belonging to 12 major elements in oil shale (C, N, H, Fe, Mg, Si, Ca, K, Al, Na, Li, and Ti) were identified in recorded spectra using NIST atomic spectra database [14]. Spectral lines strongly overlapping with neighbouring lines and those with known strong self-absorption such as Ca II lines at 393.4 and 396.8 nm were removed from further analysis after preliminary investigations.

The heating value of oil shale is determined mainly by its kerogen concentration. It is known that carbon and hydrogen are the main constituent elements of kerogen [15]. In 220–850 nm spectral range, the only reliably detectable carbon line was the one at 247.8 nm. In [16, 17], molecular compounds C_2 and CN formed in the plasma plume were additionally used for carbon detection. Due to strong overlapping by other spectral lines, in our case, the usage of these bands was infeasible. Segments of oil shale LIBS spectra containing carbon line at 247.8 nm and hydrogen line at 656.3 nm are presented in Figure 2.

Integral intensity of a spectral line was characterised by a sum of counts belonging to pixels which cover approximately the half-width of the line (Figure 2(a)). For most of the spectral lines, the number of pixels, p, covering approximately the full width at half maximum (FWHM) comprises 9 pixels of the ICCD camera, whereas for H_α line, which was strongly broadened by Stark effect, 60 pixels were used, and for whole triplet of O at 777 nm, 20 pixels were used. Usually, especially when recording with ungated detectors, for getting the actual spectral line intensities, the background should be subtracted. In the present study with gated camera and long enough delay time between the laser pulse and the recording gate, the background intensity was comparatively low. Besides, because of the presence of a number of overlapped low-intensity lines nearby some analytical lines, it was difficult to reliably measure the intensity of the background. As we checked (Section 3.4), the final result was not remarkably

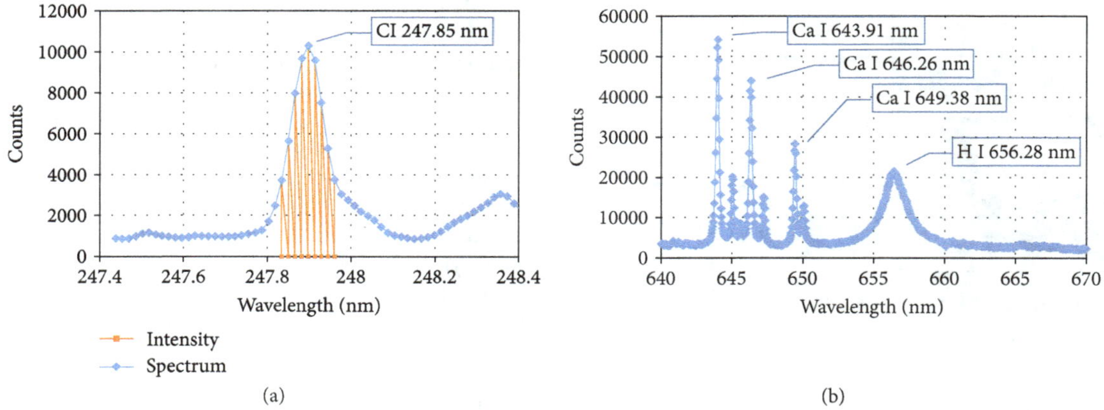

FIGURE 2: (a) A segment of the spectrum with carbon line (sample 1). (b) A segment with Ca and H_α lines.

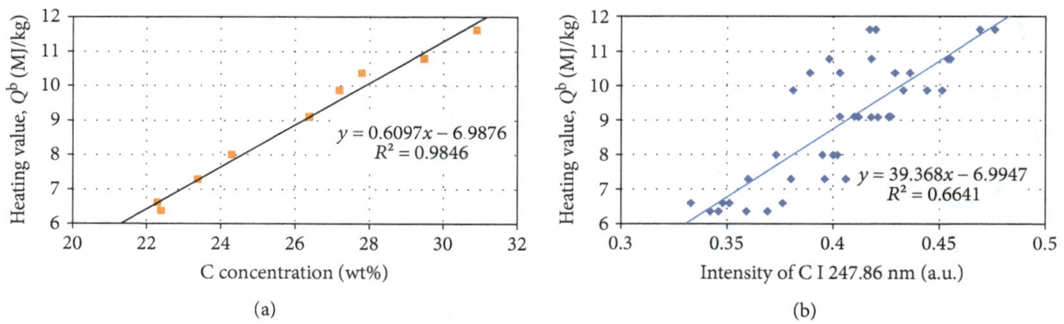

FIGURE 3: (a) Heating value of oil shale versus concentration of total C, data from Table 1. (b) Heating value as a function of the normalised intensity of C I line at 247.86 nm.

influenced by background subtraction and therefore we dropped this procedure.

Poor repeatability of LIBS signal is reported in many works [12, 18, 19]. Guidelines to reduce fluctuations of measured signals are reviewed in [12]. In case of oil shale, the chemical element with a constant concentration is missing and this is why the normalisation by internal standard [18] was not applicable.

In the present study, similarly to the papers [20–24], the role of laser pulse energy and plasma fluctuations was diminished by normalising the line intensity with the total intensity recorded between 220 and 850 nm.

Figure 3(a) based on Table 1 shows a high correlation coefficient $R^2 = 0.9846$ between the mass fraction of total carbon and the heating value. At the same time, the LIBS measurements gave a correlation coefficient between the carbon line intensity and heating value only $R^2 = 0.664$ (Figure 3(b)).

Similarly, the correlation coefficients between the line intensities and the heating values were found for all identified spectral lines. Calculated correlation coefficients between the heating value and normalised intensities of 43 preselected lines are presented in Figure 4. Lines of low absolute value of the correlation coefficient and lines having low signal-to-background ratio were excluded from further usage.

As one can see in Figure 4, the maximum value of correlation coefficient $R = 0.861$ belongs to H line. Absolute values for most of spectral lines of these correlation coefficients are less than 0.6. Since Ca and Mg lines represent mostly carbonates of waste rock, the values of correlation coefficients for Ca I, Ca II, and Mg I lines are negative. Indeed, the higher percentage of waste rock leads to a lower heating value.

It is known that one reason for the low values of correlation coefficients is matrix effect [25]. Low-value correlation coefficients indicate that the prediction of heating value by only a single spectral line intensity will have low accuracy. The use of multivariate analysis helps to minimise the matrix effect of complex geochemical materials and increase the accuracy of prediction [20, 26].

3.2. Multivariate Linear Regression Model. We assumed that the heating value of an oil shale sample could be expressed as a linear combination of intensities of measured spectral lines like it was done for lignite and coal [5, 27]:

$$Q = \mathbf{X} \cdot \mathbf{I}. \tag{1}$$

Here, Q denotes the heating value of a sample, vector $\mathbf{I}$ elements I_n are intensities of spectral lines, and vector $\mathbf{X}$ elements x_n are coefficients which values must be estimated in the course of calibration. The index n has values $0, \ldots, N$ where N is the number of spectral lines chosen for calibration.

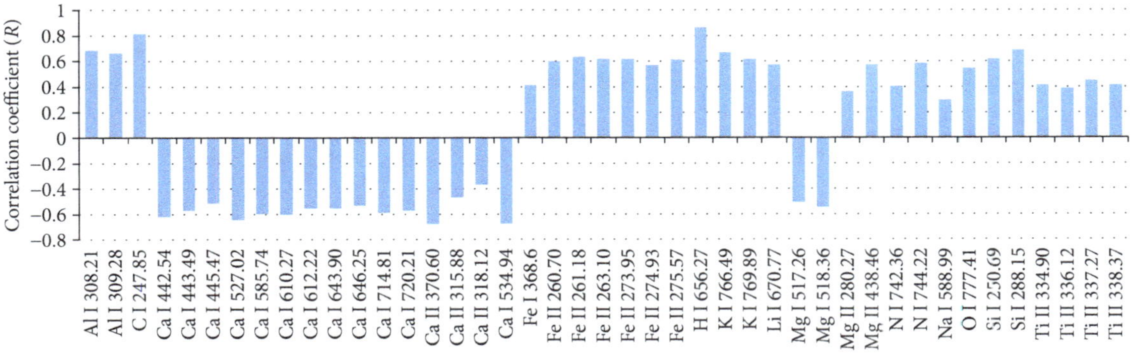

FIGURE 4: Correlation coefficients between normalised spectral line intensity and heating value.

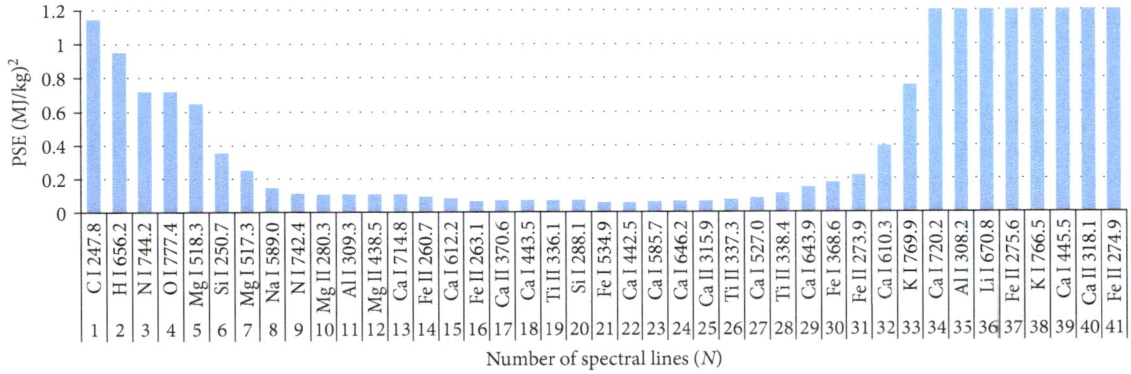

FIGURE 5: Predictive square error as a function of number of used spectral lines (N).

Coefficients of **X** were calculated according to the ordinary least square regression [12, 28] as

$$\mathbf{X} = \left(\mathbf{A}^{\mathbf{T}} \cdot \mathbf{A}\right)^{-1} \cdot \mathbf{A}^{\mathbf{T}} \cdot \mathbf{B}, \tag{2}$$

where **A** is a $M \times N$ matrix which elements $I_{m,n}$ are normalised intensities of spectral lines obtained in the course of calibration and M is the number of pellets used for calibration. **B** is a M element vector which elements are heating values determined by calorimetric bomb method (Table 1) for samples used for calibration. Superscript **T** denotes the transposed matrix.

3.3. Cross-Validation. The leave-one-out (LOO) cross-validation method [29–32] was used to find out the optimal set of spectral lines and to estimate the regression model predictivity. We used spectra of 36 pellets with nine different heating values for calibration, and for testing, we left out a subset of spectra of four pellets of the same sample having the known heating value from calorimetric measurements. This procedure was repeated leaving out one by one each of 4-pellet subsets of 10 samples.

The predictive square error (PSE) was used to evaluate the predictivity of the model:

$$\mathrm{PSE} = \frac{1}{K}\sum_{i=1}^{K}\left(Q_i^{\mathrm{p}} - Q_i^{\mathrm{b}}\right)^2. \tag{3}$$

Here Q_i^{b} denotes the heating value of the ith pellet measured with bomb calorimeter (Table 1) and Q_i^{p} is calculated using relationships (1) and (2), whereby the ith spectrum is left out from the calibration set. K is the number of pellets tested ($K = 40$).

As the first step, PSE was calculated using only the C line for calculation of the heating value. The C line was selected because the carbon concentration is directly related to the heating value (Figure 3(a)). H_α line intensity correlation with the heating value is even higher than that of C line (see Figure 4), but H_α line intensity has also a strong correlation with the moisture content [33]. The next step was the calculation of the heating value and PSE using additionally to the C I line one of the remaining 42 preselected lines (Figure 4). This step was repeated 42 times using one by one all remaining lines. A two-line combination with the minimal value of PSE was selected and the procedure of seeking the minimal value of PSE was repeated for combinations of 3 lines, two of them being selected from the previous step. Increasing step by step the number of lines, the minimal PSE at a certain number of lines was obtained. The dependence of minimal PSE on the number of involved lines is presented in Figure 5.

As one can see in Figure 5, the initial fast decrease of the PSE value continues until N reaches the value of about 10. Intensities of the first two lines, C and H_α, in the succession of spectral lines in Figure 5 have the highest value of correlation with the heating value (Figure 4). The presence of O, Mg,

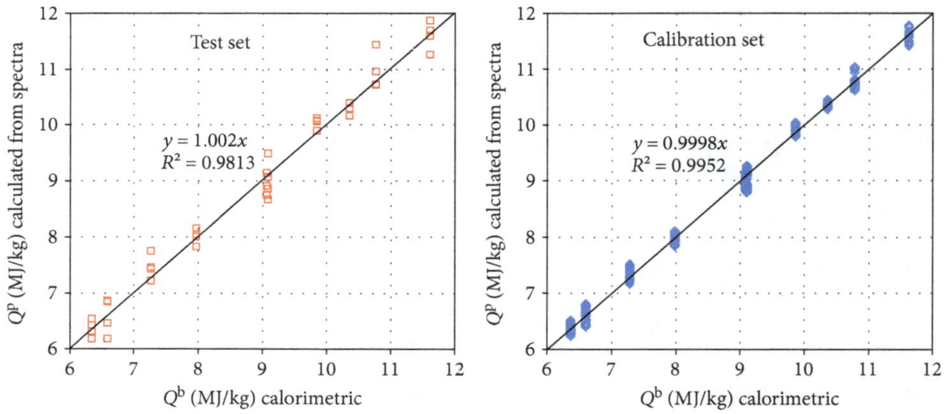

FIGURE 6: Heating value of oil shale Q^p determined by LIBS versus calorimetric heating value Q^b. Solid lines are linear fits for data points, whereby the intercept is set to zero.

Si, Na, Ca, and Al in the initial part of progression of lines with fast decrease of PSE is also expectable as they belong to compounds presented in oil shale in relatively large quantities (Table 1, composition of samples). However, the presence of nitrogen lines there is somewhat surprising as the nitrogen content in the oil shale is very low. At the same time, nitrogen is the main component of the ambient air and its presence in the initial part of sequence could be related to the air breakdown on the surface of the samples as proposed by [34]. PSE in Figure 5 has its minimum value, $PSE_{min} = 0.056$, at $N_{min} = 21$. Fast growth of PSE up to very high values takes place when N becomes close to the number of pellets used for calibration which is a common behaviour for ordinary least square regression.

The heating value predicted by the set of first 21 lines in Figure 5 is presented in Figure 6.

The value of the correlation coefficient, $R^2 = 0.981$, between the heating value Q^p predicted by LIBS and the calorimetric heating value Q^b, is close to the correlation coefficient between Q^b and the total carbon content of oil shale (Figure 3(a)).

Figure 7 presents the distribution of the difference between the "true" heating value and the heating value predicted by LIBS: $\Delta Q = Q^b - Q^p$ for the test set.

Achieved standard deviation value SD = 0.24 MJ/kg differs from the uncertainty 0.13 MJ/kg of the calorimetric measurements only by a factor of two. According to Figure 5, near the vicinity of the PSE_{min}, the value of PSE depends weakly on N. Indeed, using instead of 21 lines only the first 10 lines, gives for SD = 0.33 MJ/kg and $R^2 = 0.963$.

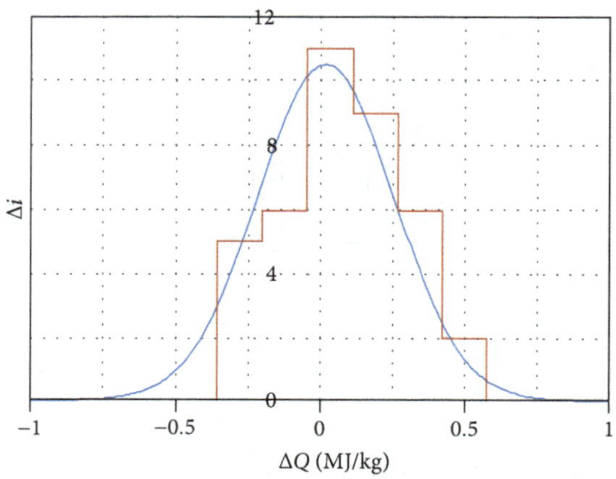

FIGURE 7: Histogram of $\Delta Q = Q^b - Q^p$. Solid line—Gaussian fit for data points. The width of Gaussian curve (standard deviation (SD)) is 0.24 MJ/kg.

3.4. Effect of Initial Data Variation. We tested the effect of the preselection of spectral lines as well as the procedure of the determination of integral intensity of spectral lines on the final results. Both the succession of lines and the number of spectral lines N_{min} resulting to the minimum of PSE value (Figure 5) PSE_{min} depended on those factors. The succession changed when for determination of line intensities, the background was taken into account, or the number of pixels, p, which was considered for integral intensity determination was varied. Similarly, the succession of lines altered with the variation in the list of preselected lines. Different successions gave a little bit different values of PSE_{min}. Figure 8 shows how the value of PSE_{min} and N_{min} depend on the used numbers of pixels, p. For most of lines, the value of p was varied from 3 to 15. For wide lines O at 777 nm and H at 656 nm, the number of pixels remained unchanged.

PSE_{min} and N_{min} as a function of p was calculated for two preselections, one of them containing 54 lines, and another, being the subset of the first one, containing 43 lines (Figure 8). For both cases, intensity of lines was calculated with and without subtraction of the background. The background for each line was determined as the lowest intensity in spectral region 50 pixels left and right from the line.

Figure 8 demonstrates that in the case of 54 preselected lines, PSE_{min} is always below 0.1 $(MJ/kg)^2$, and in the case of 43 lines, PSE_{min} is below 0.15 $(MJ/kg)^2$. Variations of PSE_{min} and N_{min} with p seem to have a random character.

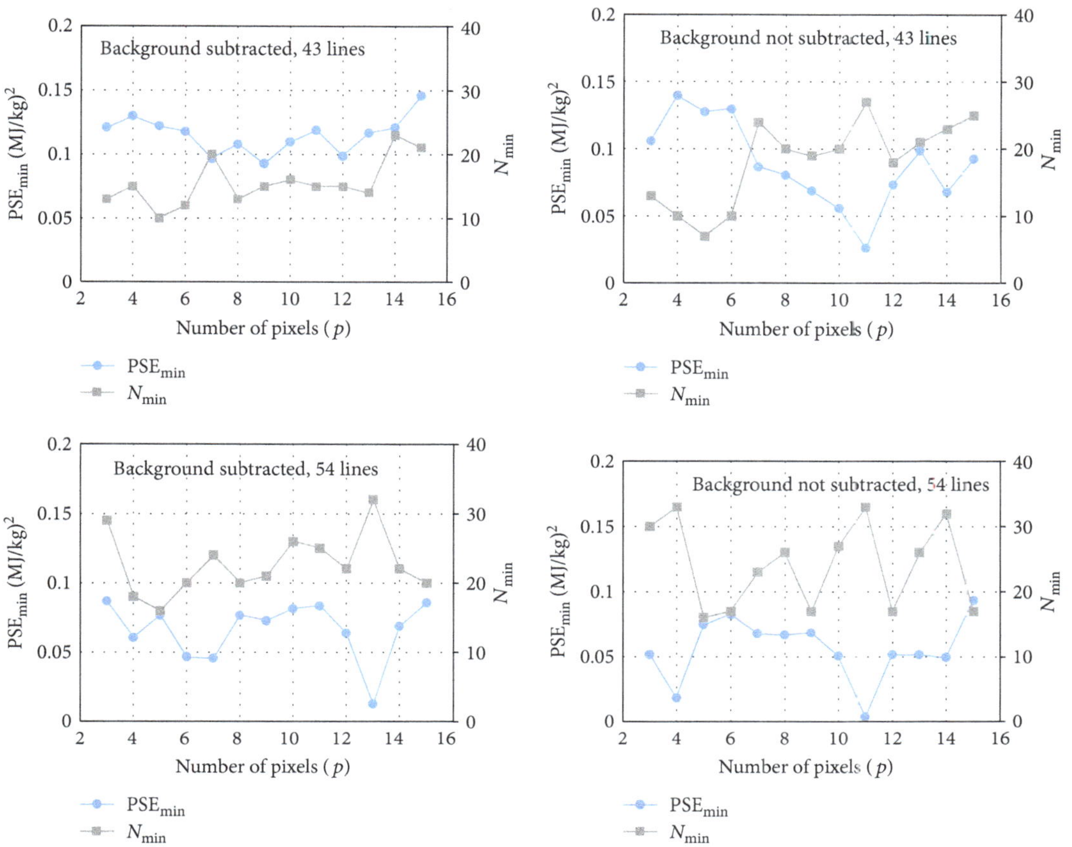

FIGURE 8: PSE_{min}: minimum value of PSE as a function of number of pixels accounted for determination of integral intensities; N_{min}: value of N at minimum PSE.

The shape of PSE $=f(N)$ curves was independent of choice of p and stayed similar to that in Figure 5. The initial fast decrease of the PSE took place until $N \approx 10$ followed by much slower decrease until N_{min}. When N became close to the number of pellets used for calibration, a fast increase of PSE started. N_{min} was below 35 in the case of 54 lines, and below 25 for 43 lines.

Like in Figure 5, when the background was not subtracted, lines of C, O, Si, Mg, H, and N were always present within the first ten lines of the succession. When the background was subtracted, the H line was not always in the list of the first ten lines.

Results presented in Sections 3.3 and 3.5 are obtained under the following conditions:

(i) $p = 9$ because this number of pixels corresponded best to the full width at half maximum of most of spectral lines.

(ii) The background was not subtracted, because the determination of real value of background was complicated.

(iii) We use the preselection of 43 lines because the calculations are less time consuming.

TABLE 2: RSD and R^2 for ingredients of oil shale.

	RSD	R^2
C	0.015	0.982
N	0.016	0.984
H	0.027	0.973
CaO	0.03	0.933
SiO_2	0.031	0.964
Fe_2O_3	0.049	0.729
MgO	0.136	0.281
Al_2O_3	0.137	0.629
Na_2O	0.251	0.691
K_2O	0.296	0.533
S	0.041	0.869

However, as Figure 8 shows, most of tested values for p and other tested factors give a satisfactory ($PSE_{min} < 0.1$) results as well.

3.5. Prediction of Ingredients Concentrations. The procedure of the ordinary least square regression applied for prediction of the heating value was applied also to predict the

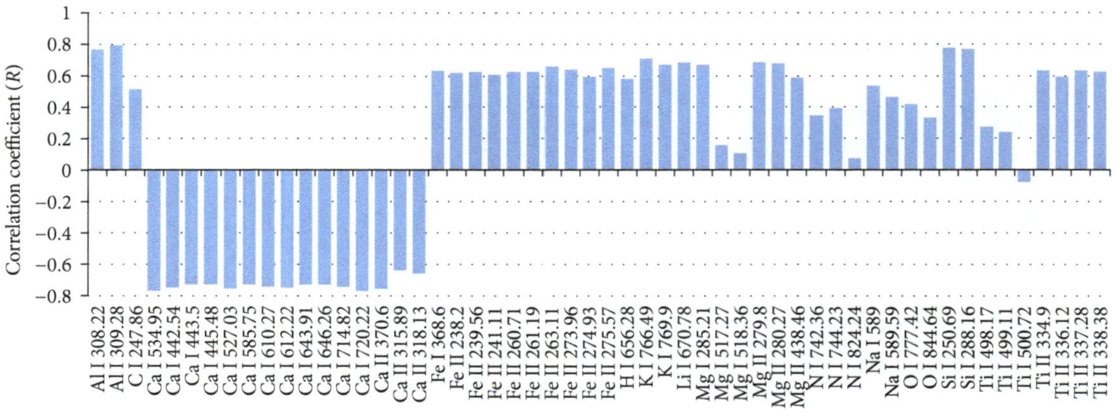

FIGURE 9: Correlation of spectral line intensity with concentration of sulphur.

concentration of the ingredients determined at the laboratory of Virumaa College of Tallinn University of Technology (Table 1). The only change introduced to the method described above was the replacing in relationships (1) and (3) the heating value, Q, by the concentration of a component and finding by (2) the set of coefficients which corresponds to this component. For each of the oil shale constituents determined in the laboratory, the set of spectral lines giving minimum of PSE was determined and the relative standard deviation (RSD) of prediction and correlation coefficient (R^2) between prediction by LIBS and laboratory measured data were calculated. Values of RSD and R^2 are presented in Table 2.

The minimum RSD = 1.5% was obtained for the total carbon. The highest values of RSD were found for constituents containing Na, K, and Al. The likely reason is that these elements were characterised by resonant lines which have a strong self-absorption [35] and as a result, the intensity of these lines is not a linear function of element concentration.

Strong sulphur lines at 921.3, 922.8, and 923.8 nm were out the wavelength range of our spectrometer. However, intensities of spectral lines of other elements are correlated with the sulphur concentration determined at the laboratory. According to Figure 9, a number of lines have values of the correlation coefficient $R = 0.6 - 0.8$. This indicates that the presence of sulphur influences spectral line intensities of other elements. By our opinion, this influence can be attributed to the matrix effect. It was possible to predict the sulphur concentration on the basis of other lines (Figure 10) with $RSD_S = 4.1\%$.

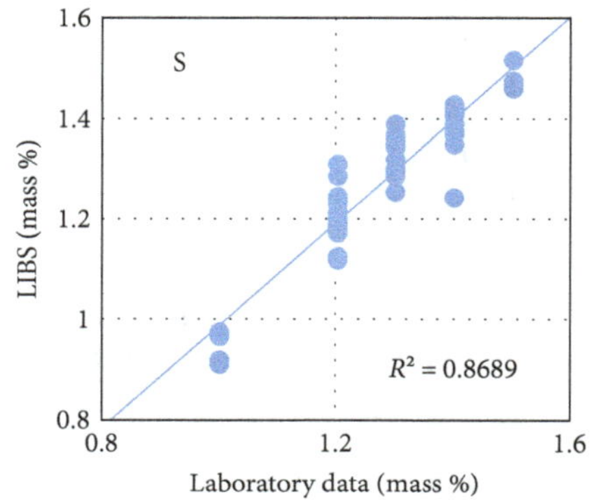

FIGURE 10: Sulphur content determined by LIBS versus laboratory data. Solid line is a linear fit for data points.

4. Conclusions

The study showed that LIBS diagnostics gives reliable results for oil shale heating value in laboratory conditions using homogenised pellets of samples.

The most important results are as follows:

(i) The multivariate linear regression is applicable for determination of the oil shale heating value. The used method of selection of spectral lines enables to find out minimal number of spectral lines needed for determination of the heating value and concentrations of constituents with an acceptable accuracy.

(ii) The root mean square error of predicted heating value of commercial oil shale was 0.24 MJ/kg. This value is comparable with uncertainty of determination of heating value with calorimetric bomb (0.13 MJ/kg).

(iii) The method allows to determine the dominating constituents of oil shale (C, H, CaO, SiO_2, and Fe_2O_3) with the value of relative standard deviation < 5%.

(iv) Determination of sulphur concentration is possible using a proper statistical model for description of relation between LIBS spectrum and sample composition.

Conflicts of Interest

The authors declare that they have no conflicts of interest.

Acknowledgments

This work was financially supported by Environmental Investment Centre of Estonia (Project no. 7594).

References

[1] I. Valgma, E. Reinsalu, S. Sabanov, and V. Karu, "Quality control of oil shale production in Estonian mines," *Oil Shale*, vol. 27, no. 3, pp. 239–249, 2010.

[2] C. E. Romero and R. De Saro, "LIBS analysis for coal," in *Laser-Induced Breakdown Spectroscopy*, S. Musazzi and U. Perini, Eds., vol. 182 of Springer Series in Optical Sciences, Springer, Berlin, Heidelberg, 2014.

[3] S. Yao, J. Lu, M. Dong, K. Chen, J. Li, and J. Li, "Extracting coal ash content from laser-induced breakdown spectroscopy (LIBS) spectra by multivariate analysis," *Applied Spectroscopy*, vol. 65, no. 10, pp. 1197–1201, 2011.

[4] J. Craparo, R. De Saro, C. Romero, Z. Yao, A. Whitehouse, and A. Weisberg, "Measuring thermal properties of coal with a commercial bench top LIBS system," in *Applied Industrial Optics: Spectroscopy, Imaging and Metrology 2012*, pp. 4–6, Monterey, California United States, 2012.

[5] T. Yuan, Z. Wang, S.-L. Lui et al., "Coal property analysis using laser-induced breakdown spectroscopy," *Journal of Analytical Atomic Spectrometry*, vol. 28, no. 7, p. 1045, 2013.

[6] J. E. Birdwell and K. E. Washburn, "Rapid analysis of kerogen hydrogen-to-carbon ratios in shale and mudrocks by laser-induced breakdown spectroscopy," *Energy and Fuels*, vol. 29, no. 11, pp. 6999–7004, 2015.

[7] K. E. Washburn, "Rapid geochemical and mineralogical characterization of shale by laser-induced breakdown spectroscopy," *Organic Geochemistry*, vol. 83-84, pp. 114–117, 2015.

[8] H. K. Sanghapi, J. Jain, A. Bol'Shakov, C. Lopano, D. McIntyre, and R. Russo, "Determination of elemental composition of shale rocks by laser induced breakdown spectroscopy," *Spectrochimica Acta Part B: Atomic Spectroscopy*, vol. 122, pp. 9–14, 2016.

[9] P. Paris, K. Piip, A. Lepp, A. Lissovski, M. Aints, and M. Laan, "Discrimination of moist oil shale and limestone using laser induced breakdown spectroscopy," *Spectrochimica Acta Part B: Atomic Spectroscopy*, vol. 107, pp. 61–66, 2015.

[10] H. Arro, A. Prikk, and T. Pihu, "Calculation of qualitative and quantitative composition of Estonian oil shale and its combustion products," *Fuel*, vol. 82, no. 18, pp. 2179–2195, 2003.

[11] R. B. Anderson, R. V. Morris, S. M. Clegg et al., "The influence of multivariate analysis methods and target grain size on the accuracy of remote quantitative chemical analysis of rocks using laser induced breakdown spectroscopy," *Icarus*, vol. 215, no. 2, pp. 608–627, 2011.

[12] E. Tognoni and G. Cristoforetti, "[Invited] Signal and noise in laser induced breakdown spectroscopy: an introductory review," *Optics & Laser Technology*, vol. 79, pp. 164–172, 2016.

[13] R. Wisbrun, I. Schechter, R. Niessner, H. Schroder, and K. L. Kompa, "Detector for trace elemental analysis of solid environmental-samples by laser-plasma spectroscopy," *Analytical Chemistry*, vol. 66, no. 18, pp. 2964–2975, 1994.

[14] A. Kramida, Y. Ralchenko, J. Reader, and NIST ASD Team, *NIST Atomic Spectra Database (Ver. 5.2)*, National Institute of Standards and Technology (NIST), Gaithersburg, MD, 2013.

[15] U. Lille, I. Heinmaa, and T. Pehk, "Molecular model of Estonian kukersite kerogen evaluated by C-13 MAS NMR spectra," *Fuel*, vol. 82, no. 7, pp. 799–804, 2003.

[16] M. Dong, X. Mao, J. J. Gonzalez, J. Lu, and R. E. Russo, "Time-resolved LIBS of atomic and molecular carbon from coal in air, argon and helium," *Journal of Analytical Atomic Spectrometry*, vol. 27, pp. 2066–2075, 2012.

[17] X. Li, H. Yin, Z. Wang, Y. Fu, Z. Li, and W. Ni, "Quantitative carbon analysis in coal by combining data processing and spatial confinement in laser-induced breakdown spectroscopy," *Spectrochimica Acta Part B: Atomic Spectroscopy*, vol. 111, pp. 102–107, 2015.

[18] D. Díaz, D. W. Hahn, and A. Molina, "Evaluation of laser-induced breakdown spectroscopy (LIBS) as a measurement technique for evaluation of total elemental concentration in soils," *Applied Spectroscopy*, vol. 66, no. 1, pp. 99–106, 2012.

[19] J. M. Mermet, P. Mauchien, and J. L. Lacour, "Processing of shot-to-shot raw data to improve precision in laser-induced breakdown spectrometry microprobe," *Spectrochimica Acta Part B: Atomic Spectroscopy*, vol. 63, no. 10, pp. 999–1005, 2008.

[20] J. B. Sirven, B. Bousquet, L. Canioni, and L. Sarger, "Laser-induced breakdown spectroscopy of composite samples: comparison of advanced chemometrics methods," *Analytical Chemistry*, vol. 78, no. 5, pp. 1462–1469, 2006.

[21] B. Bousquet, J. B. Sirven, and L. Canioni, "Towards quantitative laser-induced breakdown spectroscopy analysis of soil samples," *Spectrochimica Acta Part B: Atomic Spectroscopy*, vol. 62, no. 12, pp. 1582–1589, 2007.

[22] J. Feng, Z. Wang, Z. Li, and W. Ni, "Study to reduce laser-induced breakdown spectroscopy measurement uncertainty using plasma characteristic parameters," *Spectrochimica Acta Part B: Atomic Spectroscopy*, vol. 65, no. 7, pp. 549–556, 2010.

[23] A. Ciucci, M. Corsi, V. Palleschi, S. Rastelli, A. Salvetti, and E. Tognoni, "New procedure for quantitative elemental analysis by laser-induced plasma spectroscopy," *Applied Spectroscopy*, vol. 53, no. 8, pp. 960–964, 1999.

[24] D. Body and B. L. U. Chadwick, "Optimization of the spectral data processing in a LIBS simultaneous elemental analysis system," *Spectrochimica Acta Part B: Atomic Spectroscopy*, vol. 56, no. 6, pp. 725–736, 2001.

[25] A. S. Eppler, D. A. Cremers, D. D. Hickmott, M. J. Ferris, and A. C. Koskelo, "Matrix effects in the detection of Pb and Ba in soils using laser-induced breakdown spectroscopy," *Applied Spectroscopy*, vol. 50, no. 9, pp. 1175–1181, 1996.

[26] J. M. Tucker, M. D. Dyar, M. W. Schaefer, S. M. Clegg, and R. C. Wiens, "Optimization of laser-induced breakdown spectroscopy for rapid geochemical analysis," *Chemical Geology*, vol. 277, no. 1-2, pp. 137–148, 2010.

[27] J. Feng, Z. Wang, L. West, Z. Li, and W. Ni, "A PLS model based on dominant factor for coal analysis using laser-induced breakdown spectroscopy," *Analytical and Bioanalytical Chemistry*, vol. 400, no. 10, pp. 3261–3271, 2011.

[28] H. Martens and T. Naes, *Multivariate calibration*, John Wiley & Sons, Chichester, 2014.

[29] M. Shahlaei, "Descriptor selection methods in quantitative structure – activity relationship studies: a review study," *Chemical Reviews*, vol. 113, no. 10, pp. 8093–8103, 2013.

[30] P. Filzmoser, B. Liebmann, and K. Varmuza, "Repeated double cross validation," *Journal of Chemometrics*, vol. 23, no. 4, pp. 160–171, 2009.

[31] M. Forina, S. Lanteri, and M. Casale, "Multivariate calibration," *Journal of Chromatography A*, vol. 1158, no. 1-2, pp. 61–93, 2007.

[32] J. El Haddad, L. Canioni, and B. Bousquet, "Good practices in LIBS analysis: review and advices," *Spectrochimica Acta Part B: Atomic Spectroscopy*, vol. 101, pp. 171–182, 2014.

[33] Y. Zhao, L. Zhang, S.-X. Zhao et al., "Review of methodological and experimental LIBS techniques for coal analysis and their application in power plants in China," *Frontiers of Physics*, vol. 11, no. 6, article 114211, 2016.

[34] L. Zhang, Z. Y. Hu, W. B. Yin et al., "Recent progress on laser-induced breakdown spectroscopy for the monitoring of coal quality and unburned carbon in fly ash," *Frontiers of Physics*, vol. 7, no. 6, pp. 690–700, 2012.

[35] R. Yi, L. Guo, C. Li et al., "Investigation of the self-absorption effect using spatially resolved laser-induced breakdown spectroscopy," *Journal of Analytical Atomic Spectrometry*, vol. 31, no. 4, pp. 961–967, 2016.

14

Stability-Indicating Photochemical Method for the Assay of Thiamine by Spectrophotometry

Iqbal Ahmad,[1] Syed Haider Abbas,[1] Zubair Anwar,[1] Muhammad Ali Sheraz,[1] Sofia Ahmed,[1] Muhammad Furqan Mobeen,[2] Nafeesa Mustaan,[1] and Wajiha Gul[3]

[1]*Baqai Institute of Pharmaceutical Sciences, Baqai Medical University, Toll Plaza, Super Highway, Gadap Road, Karachi 74600, Pakistan*
[2]*Pharmatec Pakistan (Pvt.) Ltd., D-86/A S.I.T.E., Karachi, Pakistan*
[3]*Department of Pharmaceutical Chemistry, Dow College of Pharmacy, Dow University of Health Sciences, Ojha Campus, Karachi 74200, Pakistan*

Correspondence should be addressed to Sofia Ahmed; sofia.ahmed@baqai.edu.pk

Academic Editor: Rizwan Hasan Khan

A stability-indicating photochemical method has been developed for the assay of thiamine (TH) salts in aqueous solution and in fresh and aged vitamin preparations. It is based on the photooxidation of TH by UV irradiation to form thiochrome (TC) in alkaline solution. The TC : TH ratio under controlled conditions of light intensity, temperature, pH, exposure time, and irradiation distance is constant and can be used to determine the concentration of UV irradiated TH solutions. TC, on extraction with isobutanol from the photodegraded solution of TH, has been determined by the UV spectrophotometric method at 370 nm. It exhibits a high intensity of absorption in the UV region that can be used for the assay of even low concentrations of TH. Under optimum conditions, Beer's law is obeyed in the concentration range of 0.20–2.00 mg/100 ml ($R^2 = 09998$). The limit of detection (LOD) and limit of quantification (LOQ) are 0.0076 and 0.0231 mg/100 ml, respectively. The method has been validated and applied to aqueous solutions and vitamin preparations. The results have statistically been compared with the United States Pharmacopeia liquid chromatography method. It has been found that there is no significant difference between the two methods at 95% confidence level.

1. Introduction

Thiamine (vitamin B_1) (TH) is a water-soluble vitamin and is used in the prevention and cure of the disease beriberi [1]. It is extensively used as a component of single vitamin B complex and multivitamin preparations [2–4]. Several analytical methods, including UV and visible spectrophotometric and multivariate analysis [5–7], fluorimetric [8–21], high performance liquid chromatography (HPLC) [22–35], UHPLC/MS-MS [36], turbidimetric and nephelometric [37], amperometric [38], and voltammetric methods [39–45], have been used for the assay of thiamine and its salts in pharmaceutical preparations, food materials and biological fluids. Some of these methods have previously been reviewed [46–49]. A flow injection analysis- (FIA-) spectrophotometric method for the determination of TH after UV irradiation, based on difference spectrophotometry, has been reported [50]. The official methods for the assay of thiamine salts are based on non-aqueous titration [2] and liquid chromatography [3]. Thiamine undergoes chemical [1, 8, 9, 11, 14, 16, 17, 19, 51–54] and photochemical oxidation [55, 56] to form thiochrome (TC) which exhibits a blue fluorescence (440 nm). This has been made as the basis of fluorimetric determination of TH [8, 9, 11, 14, 16, 17, 19, 54–60]. The photochemical dehydrogenation of TH (1) to TC (3) probably proceeds through an intermediate (2) [52] as shown in Figure 1. The formation of TC is pH-dependent, and its yield is maximum at pH 10 [52]. TC exhibits strong absorption in aqueous solution in the UV region [61].

The objective of this study is to develop and validate a simple, economical, and stability-indicating photochemical method for the assay of TH on the basis of the formation

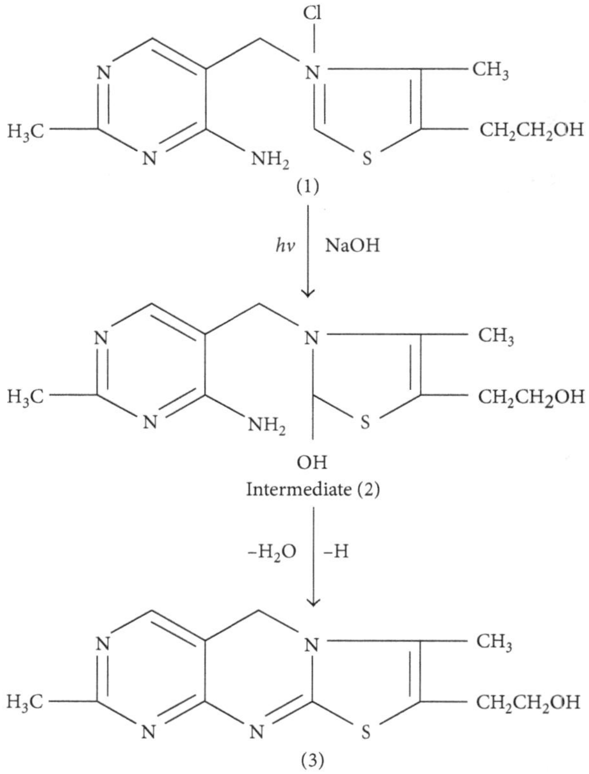

FIGURE 1: Photochemical dehydrogenation of thiamine (1) to thiochrome (3).

of TC, its spectrophotometric determination, and to establish conditions for its quantitative relationship with TH. The method would be applied to the assay of TH in pure solutions and in vitamin preparations.

2. Experimental

2.1. Materials. Thiamine hydrochloride (TH) and thiocrome (TC) were purchased from Sigma-Aldrich (USA). All reagents and solvents were of the purest form available from Merck (Germany). The buffer systems used were $Na_2B_4O_7$-NaOH (pH 9–10) and Na_2HPO_4-NaoH (pH 11); the ionic strength was maintained at 0.002 M.

2.2. Precautions. The experimental work was carried out in a dark chamber under subdued light. The TH solutions were protected from light to avoid the effect of any photochemical change before UV irradiation.

2.3. Photooxidation of TH. Aqueous solutions containing 4 mg/100 ml of TH were prepared at pH 9–11 using appropriate buffers with the addition of 2 ml of a 30% H_2O_2 solution. A 50 ml quantity of the solution spread in a Petri dish was placed in a thermostat bath at 25 ± 1°C and irradiated using a Philips 36 W TUV tube (100% emission at 254 nm) for 30 min. The tube was fixed horizontally at a distance of 25 cm from the centre of the Petri dish in a radiation chamber. Samples were withdrawn at appropriate intervals and subjected to chromatographic examination and spectrophotometric assay.

2.4. Thin-Layer Chromatography (TLC). The formation of TC in UV irradiated solutions of TH was confirmed by TLC using 250 μm silica gel GF_{254} plates and solvent system: (a) pyridine-isobutanol-water (66 : 17 : 17, v/v) and (b) 1-butanol-acetic acid-water (40 : 10 : 50, v/v, organic layer) [62].

The compounds were detected on comparison of their R_f values and fluorescence colour under 365 nm (TC, blue) and fluorescence quenching under 254 nm excitation (TH, dark), with those of the reference standards.

2.5. pH Measurements. The pH of TH solutions was measured on an Elmetron LCD display pH meter (Model CP 501, sensitivity ± 0.01 pH units, Poland) using a combination pH electrode. The calibration of the electrode was carried out using commercially available buffer tablets of pH 7 and 9 (Merck).

2.6. Spectral Measurements. The spectral measurements on TH and TC solutions were performed on a Thermo Fisher Scientific UV-Vis spectrophotometer (Evolution 201, USA) using quartz cells of 10 mm path length.

2.7. Light Intensity Measurement. The intensity of the Philips 36 W TUV tube was determined by potassium ferrioxalate actinometry [63], and a value of $5.50 \pm 0.11 \times 10^{18}$ quanta s^{-1} was obtained.

2.8. Spectrophotometric Assay of TH

2.8.1. Extraction of TC. A 5 ml aliquot of the photooxidized solution of TH was placed in a 100 ml separating funnel, 2 ml of dehydrated ethanol was added, and the solution extracted with 10 ml of isobutanol to isolate TC. The isobutanol layer was dried over Na_2SO_4 and used for the determination of the concentration of TC in the solution at 370 nm using 20,400 $M^{-1}\cdot cm^{-1}$ as the value of molar absorptivity [61] or 654 as the value of A (1% 1 cm) at that wavelength. The same procedure was used for the determination of the concentration of TC in diluted solutions of photooxidized vitamin preparations.

2.8.2. Determination of TC : TH Ratios. The TC : TH ratios are determined by the formation of a certain amount of TC on the photooxidation of a fixed amount of TH under controlled experimental conditions (i.e., concentration of TH, concentration of oxidizing agent, pH of the solution, irradiation source, and irradiation time). These values are determined from the concentrations of TC and TH and have been used for the assay of TH in different samples.

2.8.3. Assay of TH. The spectrophotometric assay of TH in pure solutions and in vitamin preparations has been carried

TABLE 1: R_f values of thiamine HCl (TH) and thiochrome (TC).

Compound	Solvent system	
	a	*b*
TH	0.03	0.27
TC	0.97	0.52

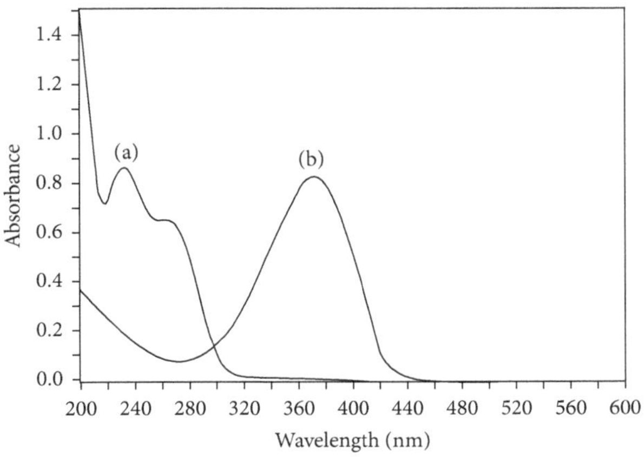

FIGURE 2: Absorption spectra of (a) thiamine HCl and (b) thiochrome, at pH 7.0.

out on the basis of the TC : TH ratios determined under controlled conditions as described in the above section. For example, if the TC : TH ratio obtained on the irradiation of a vitamin preparation containing 30 mg TH per 100 ml is 0.405 : 1, the corresponding TH concentration can be calculated by the following formula:

$$TH = \frac{\text{amount of TC formed}}{\text{TC : TH ratio}} \times \text{dilution factor of vitamin preparation}, \quad (1)$$

$$= \frac{1.600}{0.405} \times 7.5,$$

$$= \frac{29.7\ \text{mg}}{100\ \text{ml}}.$$

3. Results and Discussion

3.1. Identification of Photoproduct of TH. The photooxidation of TH in alkaline solution leads to the formation of TC. The identification of TC in UV irradiated solutions of TH at pH 9–11 has been carried out by TLC using the solvent systems (a) and (b) (Section 2.4). TC was detected by comparison of its characteristic fluorescence (blue) and R_f values with those of the authentic compounds (Table 1). TC was only detected in TH solutions irradiated at pH 9–11 since it is known to be formed by photooxidation in alkaline solution [52]. Judging from the intensity of the fluorescence of TC in the pH range studied, the maximum yield was found for the reactions carried out at pH 10.

TABLE 2: Analytical parameters for the assay of thiochrome (TC).

Parameter	Value
λ_{max}	370 nm
Linearity[a]	
Concentration range	0.20–2.00 mg%
Correlation coefficient	0.9999
Slope	654
SE of slope	0.0048
Intercept	0.0031
SE of intercept	0.00325
SD of intercept	0.0103
Recovery range (%)	99.3–100.7
Accuracy (%)[b] ± SD	99.71 ± 0.50
Precision (% RSD)[c]	0.50
LOD[d] (mg)	0.052
LOQ[e] (mg)	0.157

[a]Linearity, $n = 5$. [b]Accuracy, $n = 6$ at 100% concentration. [c]Precision, $n = 6$ at 100% concentration. [d]Limit of detection (LOD) = $3.3 \times \sigma/S$, where σ is the standard deviation and S is the slope of the standard curve. [e]Limit of quantitation (LOQ) = $10 \times \sigma/S$, where σ is the standard deviation and S is the slope of the standard curve.

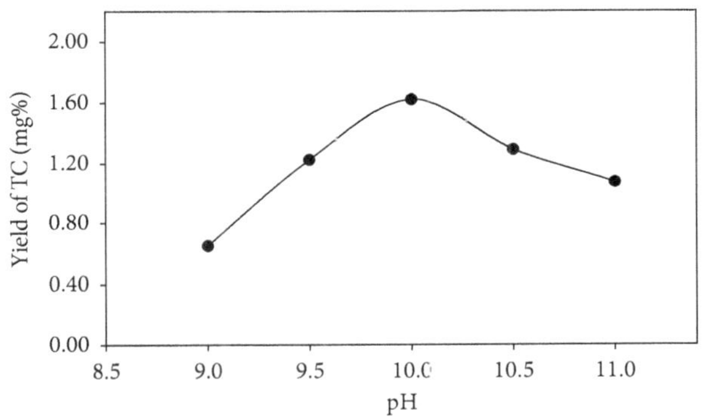

FIGURE 3: Yield of TC as a function of reaction pH.

3.2. Spectral Characteristics of TH and TC. TH exhibits absorption maximum at pH 7 at 246 nm with a molar absorptivity value of 15,200 $M^{-1}\cdot cm^{-1}$ and an A (1% 1 cm) value of 450 [64]. TC shows an absorption maximum at 370 nm with a molar absorptivity of 20,400 $M^{-1}\cdot cm^{-1}$ [61] and an A (1% 1 cm) value of 654 (Figure 2). In the present study, a molar absorptivity value of 19,800 $M^{-1}\cdot cm^{-1}$ and an A (1%, 1 cm) value of 635 for TC at 370 nm have been obtained (Table 2). In view of the very high value of the molar absorptivity of TC at 370 nm, it can be conveniently used for the assay of TH in vitamin preparations on its photooxidation to TC under controlled conditions as described in Section 3.5. The absorption maximum of TC at 370 nm provides maximum specificity and sensitivity for absorbance measurements in the method. The validation data for the test method is reported in Table 2.

3.3. Yield of TC. The photooxidation reaction of TH has been carried out at pH 9–11. It has been observed that the yield of TC on UV irradiation for a fixed period of time is maximum at pH 10 compared to that of pH 9 and 11 (Figure 3). Therefore, pH 10 was considered suitable for the photooxidation reaction of TH under the conditions used in this

Table 3: TC : TH ratios for UV irradiated TH solutions (4 mg %) at pH 9.0–11.0.

pH	TH concentration (mg %)	Time (min)	$A_{370\,nm}$	TC concentration (mg %)[a]	TC : TH ratio
		0	0	0	—
9.0	4.0	15	0.221	0.334	0.084 : 1
		30	0.425	0.650	0.162 : 1
		0	0	0	—
10.0	4.0	15	0.512	0.782	0.196 : 1
		30	1.06	1.621	0.405 : 1 (optimum value)
		0	0	0	—
11.0	4.0	15	0.411	0.628	0.157 : 1
		30	0.702	1.073	0.268 : 1

[a]The concentration of TC in photooxidized solution of TH has been calculated from the increase in absorbance at 370 nm of the extracted solution at 30 min irradiation using 654 as the value of A (1% 1 cm) [61].

Table 4: Assay of TH in vitamin preparations (25 ± 1°C).

Method	Labeled[a] (mg/100 ml)	Vitamin preparations stored in original package (pH 3-4): Fresh			Aged (6 months)		
		Found[b] (mg/100 ml)	Recovery[c] (%)	RSD (%)	Found (mg/100 ml)	Recovery (%)	RSD (%)
USP LC method							
Sample 1	30	30.2	100.7	1.7	28.8	96.0	1.6
Sample 2	50	49.8	99.6	1.9	46.5	93.0	1.8
Sample 3	83	83.6	100.7	1.6	78.5	94.6	1.7
Proposed method							
Sample 1	30	30.0	100.0	1.2	27.5	91.7	1.6
Sample 2	50	48.2	98.6	1.4	45.9	91.8	1.5
Sample 3	83	82.5	99.4	1.7	73.6	88.6	1.8

[a]The labeled amount of thiamin is the amount (mg/100 ml) stated on the label of the vitamin preparations. [b]$n = 3$. [c]Percent recovery of TH in vitamin preparation = ((amount found in mg/100 ml)/(labeled amount in mg/100 ml)) × 100.

study (Table 3). It has earlier been reported that the formation of TC depends on the pH of the medium, and its yield is maximum at pH 10 [52]. Above pH 10, TH is ionized to the negatively charged thiol form (pK_a 11.6) which undergoes hydrolytic degradation to thiazole and pyrimidine moieties [53]. Therefore, there is a decrease in the formation of TC with an increase in pH above 10.

3.4. Application of the Assay Method. The photochemical method has been applied to the assay of TH in fresh and aged pharmaceutical vitamin preparations, and the results are compared with the United States Pharmacopeia liquid chromatography method (Table 4). At 95% confidence level, it has been found that the calculated t value for each concentration is less than the tabulated t value, indicating no significant difference between the two methods. The accuracy and precision of the method has been determined, and its sensitivity has been calculated (Table 2). The method is stability-indicating for the assay of TH in vitamin preparations on the basis of the amount of TC photochemically formed under controlled conditions and its quantitative relationship with TH, that is, TC : TH ratio. A similar photochemical method for the assay of riboflavin (RF) on the basis of the formation of its photoproduct, lumichrome (LC) and LC : RF ratio, has been reported [65].

3.5. Experimental Conditions

3.5.1. Radiation Source. A 36 W TUV tube (100% emission at 254 nm) has been found to be suitable for the photooxidation of TH. The absorption maximum of TH (246 nm) [64] overlaps the emission band of the TUV tube that provides maximum radiation energy for the photochemical reaction and the formation of TC.

3.5.2. Radiation Source-Vessel Distance. The distance between the radiation source and the reaction vessel must be constant for a fixed number of quanta to be absorbed to perform a uniform photochemical reaction. A distance of 25 cm between the radiation source and the reaction vessel has been found to be appropriate for this work.

3.5.3. Temperature. The control of temperature is essential for photochemical studies to avoid any thermal effect on the rate of reaction. This has been achieved by irradiation of the solutions in a thermostat bath maintained at a constant temperature, that is, 25 ± 1°C during the reaction.

3.5.4. Irradiation Time. The irradiation of the solutions for a fixed period of time is necessary to achieve uniform results in the formation of TC. In the present study, a time period of 30 min has been found to be sufficient to avoid any photodegradation of TC [52].

3.5.5. Oxidizing Agent. The oxidation of TH in alkaline solutions occurs in the presence of an oxidizing agent, that is, potassium ferricyanide [66] or hydrogen peroxide [67]. In the present study, the photooxidation of TH has been carried out in the presence of hydrogen peroxide to avoid any interference of the oxidizing agent in the absorbance of TC.

4. Conclusion

A simple photochemical method has been developed and validated for the assay of TH in aqueous solution and pharmaceutical vitamin preparations. It is based on the photooxidation of TH to TC, determination of their quantitative relationship (TC : TH ratio) under controlled conditions and on this basis the assay of TH in different pharmaceutical samples. TC has a high value of A (1%, 1 cm) and can be conveniently determined by spectrophotometry at 370 nm. The method is comparable to the United States Pharmacopeia method.

Conflicts of Interest

The authors declare that they have no conflicts of interest.

Acknowledgments

The authors gratefully acknowledge the Higher Education Commission of Pakistan for financial support to conduct this study through an NRPU Grant (Research Project no. 20–3968) to Iqbal Ahmad.

References

[1] C. J. Bates, "Thiamine," in *Handbook of Vitamins*, Chapter 8 J. Zempleni, R. B. Rucker, D. B. McCormick, and J. W. Suttie, Eds., CRC Press, Boca Raton, FL, USA, 2007.

[2] British Pharmacopoeia, *Monograph on Thiamine*, Her Majesty's Stationary Office, London, UK, 2016.

[3] United States Pharmacopeia 29, *Monograph on Thiamine*, United States Pharmacopeial Convention, Rockville, MD, USA, 2016.

[4] S. C. Sweetman, *Martindale: The Complete Drug Reference*, Pharmaceutical Press, London, UK, 36th edition, 2009.

[5] S. T. Abdel Rehman, A. A. Elbashir, M. El-Mukhtar, and M. M. Ibrahim, "Application of spectrophotometric methods for the determination of thiamine (vb1) in pharmaceutical formulations using 7-chloro-4-nitrobenzoxadiazole (NBD–Cl)," *Journal of Analytical & Pharmaceutical Research*, vol. 2, no. 3, p. 0018, 2016.

[6] P. L. Lopez-de-Alba, L. Lopez-Martinez, V. Cerda, and J. Amador-Hernandez, "Simultaneous determination and classification of riboflavin, thiamine, nicotinamide and pyridoxine in pharmaceutical formulations, by UV-visible spectrophotometry and multivariate analysis," *Journal of Brazilian Chemical Society*, vol. 17, no. 4, pp. 715–722, 2006.

[7] M. Abdel-Maaboud, M. Horria, M. A. Niveen, and E.-Z. Marwa, "Chemometric methods for the simultaneous determination of some water-soluble vitamins," *Journal of AOAC International*, vol. 94, no. 2, pp. 467–481, 2011.

[8] Q.-Y. Chen, D.-H. Li, H.-H. Yang, Q.-Z. Zhu, H. Zheng, and J.-G. Xu, "Novel spectrofluorimetric method for the determination of thiamine with iron(iii) tetrasulfonatophthalocyanine as a catalyst," *Analyst*, vol. 124, no. 5, pp. 771–775, 1999.

[9] H. Zhu, H. Chen, and Y. Zhou, "Determination of thiamine in pharmaceutical preparations by sequential injection renewable surface solid-phase spectrofluorometry," *Analytical Sciences*, vol. 19, no. 2, pp. 289–294, 2003.

[10] A. Alonso, M. J. Almendral, M. J. Porras, and Y. Curto, "Flow-injection solvent extraction without phase separation. fluorimetric determination of thiamine by the thiochrome method," *Journal of Pharmaceutical and Biomedical Analysis*, vol. 42, no. 2, pp. 171–177, 2006.

[11] A. B. Tabrizi, "A cloud point extraction-spectrofluorimetric method for determination of thiamine in urine," *Bulletin of Korean Chemical Society*, vol. 27, no. 10, pp. 1604–1608, 2006.

[12] M. Amjadi, J. L. Manzoori, and M. Orooji, "The use of crude extract of kohlrabi (brassica oleracea gongylodes) as a source of peroxidase in the spectrofluorimetric determination of thiamine," *Bulletin of Korean Chemical Society*, vol. 28, no. 2, pp. 246–250, 2007.

[13] J. Sun, L. Liu, C. Ren, X. Chen, and Z. Hu, "A feasible method for the sensitive and selective determination of vitamin B1 with CdSe quantum dots," *Microchimica Acta*, vol. 163, no. 3-4, pp. 271–276, 2008.

[14] M. A. Khan, S. O. Jin, S. H. Lee, and H. Y. Chung, "Spectrofluorimetric determination of vitamin B_1 using horseradish peroxidase as catalyst in the presence of hydrogen peroxide," *Luminescence*, vol. 24, no. 2, pp. 73–78, 2009.

[15] Y. Li, P. Wang, M. Cao et al., "An immediate luminescence enhancement method for determination of vitamin B_1 using long-wavelength emitting water-soluble CdTe nanorods," *Microchimica Acta*, vol. 169, no. 1-2, pp. 65–71, 2010.

[16] M. Zeeb, M. R. Ganjali, and P. Norouzi, "Dispersive liquid-liquid microextraction followed by spectrofluorimetry as a simple and accurate technique for determination of thiamine (vitamin B_1)," *Microchimica Acta*, vol. 168, no. 3-4, pp. 317–324, 2010.

[17] A. M. Mohamed, H. A. Mohamed, N. M. Abdel-Latif, and M. R. Mohamed, "Spectrofluorimetric determination of some water-soluble vitamins," *Journal of AOAC International*, vol. 94, no. 6, pp. 1758–1769, 2011.

[18] G. D. Tarigh and F. Shemirani, "Simultaneous in situ derivatization and ultrasound-assisted dispersive magnetic solid phase extraction for thiamine determination by spectrofluorimetry," *Talanta*, vol. 123, pp. 71–77, 2014.

[19] J. Zhu, S. Liu, Z. Liu, Y. Liu, M. Qiao, and X. Hu, "Enhanced spectrofluorimetric determination of hypochlorite based on the catalytic oxidation of thiamine to thiochrome in the presence of trace ferrocyanide," *RSC Advances*, vol. 4, no. 12, pp. 5990–5994, 2014.

[20] H. Tan, Q. Li, Z. Zhou et al., "A sensitive fluorescent assay for thiamine based on metal-organic frameworks with intrinsic peroxidase-like activity," *Analytica Chimica Acta*, vol. 856, pp. 90–95, 2015.

[21] Y. Luo, H. Miao, and X. Yang, "Glutathione-stabilized Cu nanoclusters as fluorescent probes for sensing pH and vitamin B_1," *Talanta*, vol. 144, pp. 488–495, 2015.

[22] X. Tang, D. A. Cronin, and N. P. Brunton, "A simplified approach to the determination of thiamine and riboflavin in meats using reverse phase HPLC," *Journal of Food Composition and Analysis*, vol. 19, no. 8, pp. 831–837, 2006.

[23] M. Ake, Y. N. Soko, and K. A. Malan, "Liquid chromatographic determination of free thiamine and its esters in whole blood," *Dakar Medical*, vol. 51, no. 3, pp. 172–177, 2006.

[24] A. Zafra-Gomez, A. Garballo, J. C. Morales, and L. E. Garcia-Ayuso, "Simultaneous determination of eight water-soluble vitamins in supplemented foods by liquid chromatography," *Journal of Agriculture and Food Chemistry*, vol. 54, no. 13, pp. 4531–4536, 2006.

[25] A. Lebiedzinska, M. L. Marszall, J. Kuta, and P. Szefer, "Reversed-phase high-performance liquid chromatography method with coulometric electrochemical and ultraviolet detection for the quantification of vitamins B(1) (thiamine), B(6) (pyridoxamine, pyridoxal and pyridoxine) and B(12) in animal and plant foods," *Journal of Chromatography A*, vol. 1173, no. 1-2, pp. 71–80, 2007.

[26] N. Yanith, D. Widowati, Wartini, and T. Aryani, "Validation of HPLC method for determination of thiamine hydrochloride, riboflavin, nicotinamide, and pyridoxine hydrochloride in syrup preparation," *Canadian Journal on Scientific and Industrial Research*, vol. 2, no. 7, pp. 269–278, 2011.

[27] X. X. Wang, Y. Hu, Y. Z. Sun, M. J. Bie, and C. J. Sun, "Simultaneous determination of five water-soluble vitamins in human serum by high performance liquid chromatography," *Sichuan Da Xue Xue Bao Yi Xue Ban, Journal of Sichuan University, Medical Science Edition*, vol. 41, no. 1, pp. 158–161, 2010.

[28] J. Tan, R. Li, and Z.-T. Jiang, "Determination of thiamine (vitamin B_1) in pharmaceutical tablets and human urine by titania-based ligand-exchange hydrophilic interaction chromatography," *Analytical Methods*, vol. 3, pp. 1568–1573, 2011.

[29] E. Dinc, G. Kokdil, and F. Onur, "A comparison of matrix resolution method, ratio spectra derivative spectrophotometry and HPLC method for the determination of thiamine HCl and pyridoxine HCl in pharmaceutical preparation," *Journal of Pharmaceutical and Biomedical Analysis*, vol. 22, no. 6, pp. 915–923, 2000.

[30] P. Jin, L. Xia, Z. Li, N. Che, D. Zou, and X. Hu, "Rapid determination of thiamine, riboflavin, niacinamide, pantothenic acid, pyridoxine, folic acid and ascorbic acid in vitamins with minerals tablets by high-performance liquid chromatography with diode array detector," *Journal of Pharmaceutical and Biomedical Analysis*, vol. 70, pp. 151–157, 2012.

[31] R. S. J. Rodriguez, V. Fernandez-Ruiz, M. Camara, and M. C. Sanchez-Mata, "Simultaneous determination of vitamin B_1 and B_2 in complex cereal foods, by reverse phase isocratic HPLC-UV," *Journal of Cereal Science*, vol. 55, no. 3, pp. 293–299, 2012.

[32] I. Marquez-Sillero, S. Cardenas, and M. Valcarcel, "Determination of water-soluble vitamins in infant milk and dietary supplement using a liquid chromatography on-line coupled to a corona-charged aerosol detector," *Journal of Chromatography A*, vol. 1313, pp. 253–258, 2013.

[33] T. Miyazaki, T. Hori, Y. Aso, and H. Okuda, "Ion-pair HPLC analysis of B vitamins in syrup products in Indonesia," *Kokuritsu Iyakuhin Shokuhin Eisei Kenkyusho Hokoku, Bulletin of National Institute of Health Sciences*, vol. 131, pp. 58–65, 2013.

[34] N. Michlig, F. V. de Velde, M. R. Freyre, M. Cecilia, and H. Bernardi, "Determination of thiamine in wheat flours using a validated isocratic HPLC-fluorescence method previously optimized by box-behnken design," *Food Analytical Methods*, vol. 7, no. 4, pp. 828–835, 2014.

[35] B. Basiri, J. M. Sutton, B. S. Hanberry, J. A. Zastre, and M. G. Bartlett, "Ion pair liquid chromatography method for the determination of thiamine (vitamin B1) homeostasis," *Biomedical Chromatography*, vol. 30, no. 1, pp. 35–41, 2016.

[36] D. Hampel, E. R. York, and L. H. Allen, "Ultra-performance liquid chromatography tandem mass-spectrometry (UPLC–MS/MS) for the rapid, simultaneous analysis of thiamin, riboflavin, flavin adenine dinucleotide, nicotinamide and pyridoxal in human milk," *Journal of Chromatography B Analytical Technologies in the Biomedical and Life Sciences*, vol. 903, pp. 7–13, 2012.

[37] V. B. Santos, T. B. Guerreiro, W. T. Suarez, R. C. Faria, and O. Fatibello-Filho, "Evaluation of turbidimetric and nephelometric techniques for analytical determination of *n*-acetylcysteine and thiamine in pharmaceutical formulations employing a lab-made portable microcontrolled turbidimeter and nephelometer," *Journal of the Brazilian Chemical Society*, vol. 22, no. 10, pp. 1968–1978, 2011.

[38] E. Akyilmaz, I. Yasa, and E. Dinckaya, "Whole cell immobilized amperometric biosensor based on *Saccharomyces cerevisiae* for selective determination of vitamin B_1 (thiamine)," *Analytical Biochemistry*, vol. 354, no. 1, pp. 78–84, 2006.

[39] X. Jiang and T. Sun, "Indication ion square wave voltammetric determination of thiamine and ascorbic acid," *Electrochemistry*, vol. 40, no. 13, pp. 2589–2596, 2007.

[40] P. Norouzi, T. M. Garakani, H. Rashedi, H. A. Zamani, and M. R. Ganjali, "Ultrasensitive flow-injection electrochemical method using fast fourier transform square-wave voltammetry for detection of vitamin B_1," *International Journal of Electrochemical Science*, vol. 5, pp. 639–652, 2010.

[41] B. Bas, M. Jakubowska, and L. Gorski, "Application of renewable silver amalgam annular band electrode to voltammetric determination of vitamins C, B_1 and B_2," *Talanta*, vol. 84, no. 4, pp. 1032–1037, 2011.

[42] K. Tyszczuk-Rotko, "New voltammetric procedure for determination of thiamine in commercially available juices and pharmaceutical formulation using a lead film electrode," *Food Chemistry*, vol. 134, no. 2, pp. 1239–1243, 2012.

[43] P. K. Brahman, R. A. Dar, and K. S. Pitre, "DNA-functionalized electrochemical biosensor for detection of vitamin B1 using electrochemically treated multiwalled carbon nanotube paste electrode by voltammetric methods," *Sensors and Actuators B: Chemical*, vol. 177, pp. 807–812, 2013.

[44] I. P. Antal, Y. R. Bazel, and Z. A. Kormosh, "Electrochemical methods for determining group B vitamins," *Journal of Analytical Chemistry*, vol. 68, no. 7, pp. 565–576, 2013.

[45] I. G. David, M.-A. Florea, O. G. Cracea et al., "Voltammetric determination of B_1 and B_6 vitamins using a pencil graphite electrode," *Chemical Papers*, vol. 69, no. 7, pp. 901–910, 2015.

[46] R. R. Eitenmiller and W. O. Landen Jr., "Thiamin," in *Modern Analytical Methodologies in Fat-and Water-oluble Vitamins*, Chapter 6, W. O. Song, G. R. Becker, and R. R. Eitenmiller, Eds., Wiley, New York, NY, USA, 2000.

[47] G. F. M. Ball, "Thiamin," in *Vitamins in Food Analysis, Bioavailability and Stability*, Chapter 7, CRC Press, Boca Raton, FL, USA, 2006.
[48] R. R. Eitenmiller, L. Ye, and W. O. Landen Jr., "Thiamin," in *Vitamin Analysis for the Health and Food Science*, Chapter 6, CRC Press, Boca Raton, FL, USA, 2nd edition, 2008.
[49] C. J. Bates, Vitamins: Fat and Water Soluble: Analysis, Encyclopedia of Analytical Chemistry, R. A. Meyers, Ed., John Wiley, Chichester, UK, 2000.
[50] A. F. Danet and J. M. Calatayud, "FIA-spectrophotometric determination of thiamine after UV-irradiation," *Talanta*, vol. 41, no. 12, pp. 2147–2151, 1994.
[51] B. C. P. Jansen, "A chemical determination of aneurin (vitamin B_1) by thiochrome reaction," *Recueil des Travaux Chimiques des Pays-Bas*, vol. 55, no. 12, pp. 1046–1052, 1936.
[52] H. R. Rosenberg, *Chemistry and Physiology of the Vitamins*, Interscience, New York, NY, USA, 1945.
[53] K. A. Connors, G. L. Amidon, and V. J. Stella, *Monograph on Thiamine Hydrochloride, Chemical Stability of Pharmaceuticals: A Handbook for Pharmacist*, John Wiley, New York, NY, USA, 1986.
[54] United States Pharmacopoeia Convention, *United States Pharmacopeia 29–National Formulary 24*, United States Pharmacopeial Convention, Inc., Rockville, MD, USA, 2006.
[55] Q. Guo, J.-G. Xu, Y.-Z. Wu, Y.-B. Zhao, X.-Z. Huang, and G.-Z. Chen, "Determination of thiamine (vitamin B_1) by in situ sensitized photochemical spectrofluorimetry," *Analytica Chimica Acta*, vol. 276, no. 1, pp. 151–160, 1993.
[56] A. Wasielczuk, M. Catala Icardo, J. V. Carcia Mateo, and J. Martinez Calatayud, "Flow-injection chemiluminescent determination of thiamine in pharmaceutical samples by online photodegradation," *Analytical Letters*, vol. 37, no. 15, pp. 3205–3218, 2004.
[57] D. F. Clausen and R. E. Brown, "Determination of thiamine by the thiochrome method. Effects of temperature and dissolved oxygen on fluorescence of quinine standard and of thiochrome," *Industrial and Engineering Chemistry*, vol. 16, no. 9, pp. 572–574, 1944.
[58] G. A. Leveille, "Modified thiochrome procedure for the determination of urinary thiamin," *American Journal of Clinical Nutrition*, vol. 25, no. 3, pp. 273-274, 1972.
[59] M. H. Dong, M. D. Green, and H. E. Sauberlich, "Determination of urinary thiamin by the thiochrome method," *Clinical Biochemistry*, vol. 14, no. 1, pp. 16–18, 1981.
[60] G. R. Skurray, "A rapid method for selectively determining small amount of niacin, riboflavin and thiamin in foods," *Food Chemistry*, vol. 7, no. 2, pp. 77–80, 1981.
[61] N. N. Bubeshko, V. I. Stsiapura, and I. I. Stepuro, "Fluorescent properties of thiochrome in solvents of different polarity," *Journal of Applied Spectroscopy*, vol. 78, p. 337, 2011.
[62] H. R. Bolliger and A. Konig, "Vitamin, including carotenoids, chlorophyll and biologically active quinones," in *Thin-Layer Chromatography*, E. Stahl, Ed., p.295, Springer, New York, NY, USA, 2nd edition, 1969.
[63] C. G. Hatchard and C. A. Parker, "A new sensitive chemical actinometer," *Proceedings of Royal Society of London Series A, Mathematical and Physical Sciences*, vol. 235, no. 1203, pp. 518–536, 1956.
[64] A. C. Moffat, M. D. Osselton, and D. Widdop, *Clarke's Analysis of Drugs and Poisons in Pharmaceuticals, Body Fluids and Postmortem Material*, Pharmaceutical Press, London, UK, 3rd edition, 2004.
[65] I. Ahmad, S. H. Abbas, Z. Anwar et al., "Stability-indicating photochemical method for the assay of riboflavin: lumichrome method," *Journal of Chemistry*, vol. 2015, Article ID 256087, 8 pages, 2015.
[66] G. Barger, F. Bergel, and A. R. Todd, "A crystalline fluorescent dehydrogenation product from vitamin B_1," *Nature*, vol. 136, no. 3433, p. 259, 1935.
[67] H. N. Haugen, "The determination of thiamine by the thiochrome method," *Scandinavian Journal of Clinical and Laboratory Investigation*, vol. 12, no. 3, pp. 253–260, 1960.

Evolution Mechanism of Macromolecular Structure in Coal during Heat Treatment: Based on FTIR and XRD In Situ Analysis Techniques

Dun Wu[1,2,3] and Wenyong Zhang[3]

[1]*School of Earth and Space Sciences, University of Science and Technology of China, CAS Key Laboratory of Crust-Mantle Materials and Environment, Hefei, Anhui 230026, China*
[2]*Research Center of Coal Resources Safe Mining and Clean Utilization, LiaoNing Technical University, Fuxin 125105, Liaoning, China*
[3]*Exploration Research Institute, Anhui Provincial Bureau of Coal Geology, Hefei, Anhui 23008, China*

Correspondence should be addressed to Dun Wu; wudun@ustc.edu.cn

Academic Editor: Christoph Krafft

Owing to the complexity and heterogeneity of coal during pyrolysis, the ex situ analytical techniques cannot accurately reflect the real coal pyrolysis process. In this study, according to the joint investigation of Fourier-transform infrared spectroscopy (FTIR) and X-ray diffraction (XRD), the structural evolution characteristics of lignite-subbituminous coal-bituminous coal-anthracite series under heat treatment were discussed in depth. The results of the infrared spectrum of coal show that the different functional groups of coal show different changes with the increase of coal rank before pyrolysis experiment. Based on in situ infrared spectroscopy experiments, it was found that the infrared spectrum curves of the same coal sample have obvious changes at different pyrolysis temperatures. As a whole, when the pyrolysis temperature is between 400 and 500°C, the coal structure can be greatly changed. By fitting the infrared spectrum curve, the infrared spectrum parameters of coal were obtained. With the change of temperature, these parameters show regular changes in coal with different ranks. In the XRD study of coal, the absorption intensity of the diffraction peak (002) of coal increases with increasing coal rank. The XRD patterns of coal have different characteristics at different pyrolysis temperatures. Overall, the area of (002) diffraction peak of the same coal sample increases obviously with the increase of temperature. The XRD structural parameter of coal was obtained by using the curve fitting method. The changing process of two parameters (interlayer spacing (d_{002}) and stacking height (L_c)) can be divided into two main stages, but the average lateral size (L_a) does not change significantly and remains at the 2.98 ± 0.09 nm. In summary, the above two technologies complement each other in the study of coal structure. The temperature range of both experiments is different, but the XRD parameters of coal with different ranks are reduced within the temperature range of less than 500°C, which reflects that the size of coal-heated aromatic ring lamellae is reduced and the distance between lamellae is also reduced, indicating that the degree of condensation of coal aromatic nuclei may be increased. Correspondingly, the FTIR parameters of coal also reflect that, with increasing temperature, the side chains of coal are constantly cracked, the oxygen-containing functional groups are reduced, and the degree of aromatization of coal may be increased.

1. Introduction

The integrally structural arrangement of coal increases with proceeding coalification up to the formation of organic macromolecules of flat shape with carbon atoms in the central part [1–3]. During coalification, the size of coal molecules increases as well as the degree of their ordering and macromolecules arrange themselves laterally or vertically to form a lamellar system [3]. It has long been supposed that the aggregate structure of coal plays a vital role during coalification [4]. Therefore, it is very necessary to deeply understand the structural evolution mechanism of coal during pyrolysis [5].

In general, the pyrolysis process of coal can be divided into two main stages. One is the depolymerization or decomposition stage where gaseous (e.g., gas and water vapor) and liquid products (e.g., tar) were generated. The other is the condensation or repolymerization stage where the turbostratic lamellar system took place. Amounts of heterogeneous pyrolysis reactions can take place in the aforementioned two types of competitive processes. A better knowledge of coal structure alteration during pyrolysis could promote to comprehend the reaction process of coalification [6, 7]. There have been a number of methods to provide new insights into the relationship between coal structure and its pyrolysis process.

In the past couple of decades, the ex situ XRD and FTIR analytical methods were used widely to supply important information on the changes of coal structure under heat treatment [5, 8, 9]. Mae et al. [10] reported the relationship of the yield of pyrolysis products with coal structure. Wang et al. [11] studied the structure and pyrolysis characteristics of Chinese western coals, demonstrating the relationship between evolution of coal composition and its physicochemical structure. On the basis of X-ray, studies of coal reported by many researchers [12–14] show that the X-ray diffraction pattern of coal resembles a very blurred graphite pattern similar to that obtained with activated carbons and carbon black. Of note, Hirsch [15], Watanabe et al. [16], and Sonibare et al. [17] proposed that the XRD structural parameters (interlayer spacing of crystalline structure (d_{002}) and crystallite sizes (L_a and L_c)) can be used to evaluate the structure of coal and other carbonaceous materials with lower carbon crystallinity. These parameters can be obtained using the Scherrer equation and fitting parameters. For the XRD study of coal structure under pyrolysis conditions, Zhang et al. [18] presented the correlation between coal element composition and coal structure and the evolution of XRD structural parameters of coal with the change of temperature. However, they did not discuss the changes of XRD structural parameters of coal with temperature but also did not study the changes of macromolecular functional groups of coal with temperature, especially oxygen-containing functional groups. Li et al. [19] characterized the structural parameters of coal at different pyrolysis temperatures using in situ XRD technology and obtained the relationship between the corresponding structural parameters and temperature. However, they did not use FTIR technology to analyze the change of functional groups of coal with temperature.

The FTIR technique focuses on determining composition on the physicochemical structures of coal. Kister et al. [20], Christy et al. [21], and Wang et al. [22] obtained that the C=O bonds were almost totally lacking in coal, which had isolated C=C bonds and C≡C bonds, and aliphatic groups CH, CH_2, CH_3, and aromatic ring groups were abundant. Wu et al. [23] used FTIR and TG analysis methods to study the chemical structure evolution and small-molecule gas yield of coal during pyrolysis. However, the FTIR they used is not in situ analysis technology, which is different from this study.

In conclusion, although ex situ analysis technology can also obtain information on coal structure and its reactivity, but because of the complexity of coal structure transformation during pyrolysis, this analysis method cannot characterize the instantaneous information of coal structure at a certain temperature [24]. For that reason, in the present study, the in situ XRD and FTIR analytical techniques were employed at real temperature without cooling coal samples and to assist in quantitative analysis for expounding the structural evolution characteristics of coal during pyrolysis.

The purpose of this study was (1) probing deeply the structural evolution signatures of different ranked coals during pyrolysis using in situ XRD and FTIR analytical techniques; (2) a better understanding of the relationship between the evolution of coal structural parameters and pyrolysis temperature of coal; (3) the reference standard of structure parameters of coal with different ranks under real-time pyrolysis temperature was proposed.

2. Experimental

2.1. Samples and Preparation. A set of five different ranked coals having different geological ages was selected from five coal basins located on North China and South China coal-bearing region (Figure 1). These coal samples were denoted according to their location and geological age as: GZ (Guizhou province, Longtan formation in Late Permian epoch), HB (Hubei province, Chuyanken formation in Late Triassic epoch), HN (Anhui province, Upper Shihezi formation in Late Permian epoch), KL (Hebei province, Lower Shihezi formation in Early Permian epoch), and SX (Shanxi province, Taiyuan formation in Late Carboniferous epoch). Their coal ranks are lignite for SX coal, subbituminous coal for KL coal, bituminous gas coal for HN coal, bituminous coking coal for HB coal, and anthracite for GZ coal in turn.

Each coal sample weighed approximately 200 g. These samples were pulverized and sieved to obtain particles of <56 μm in diameter and dried in a desiccator at 60°C for 12 h. After that, they were demineralized to avoid the interference of mineral matter in coal before subjecting them to in situ XRD and FTIR experiments. Detailed removal process of minerals from coal and preparation of coal KBr pellets were described in previous studies [25].

2.2. Proximate and Ultimate Analyses. The proximate and ultimate analyses of five coal samples before pyrolysis were executed according to ISO 625: 1975 (E) and ISO 333: 1983 (E), respectively. The results are presented in Table 1.

2.3. In Situ FTIR Spectroscopy Measurement. For each set of experiments, after the reactor grasped approximately 20 g coal sample, the reactor was sealed and heated in a temperature-programmed furnace. The heat treatment experiment of coal was carried out under oxygen-free conditions. Five rounds of experiments were conducted at difference temperatures, including 100, 200, 300, 400, and

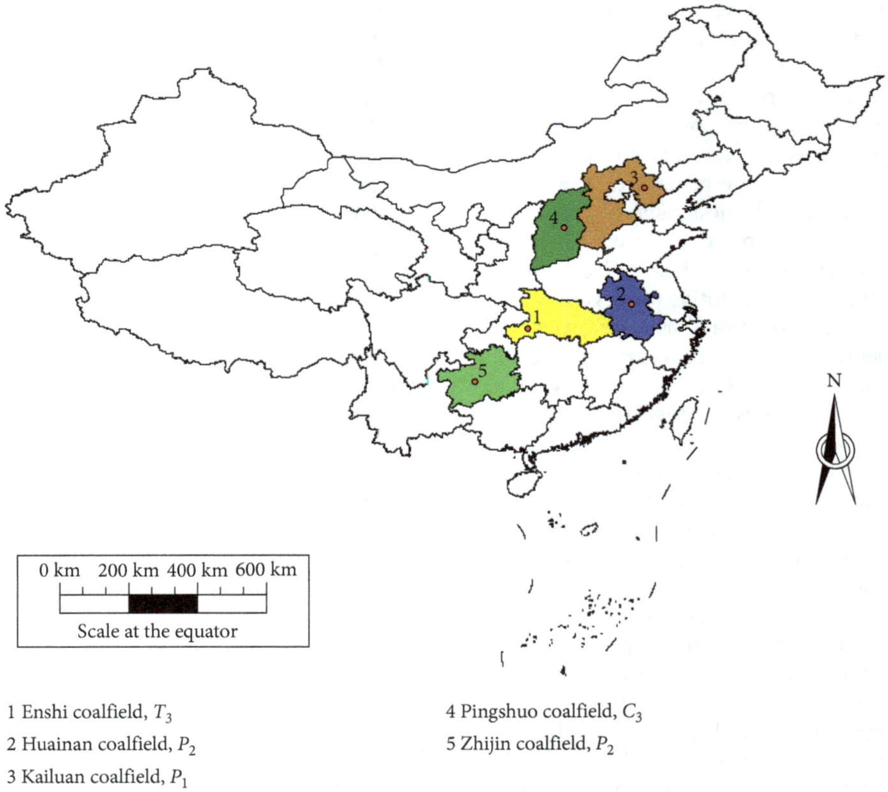

Figure 1: Location and geological ages of Huainan coalfield, Kailuan coalfield, and Pingshuo coalfield in the North China coal-bearing region and Enshi coalfield and Zhijin coalfield in the South China coal-bearing region.

Table 1: Results of proximate and ultimate analyses for five coal samples.

Sample number	Proximate analysis (wt.%)			Ultimate analysis (wt.% daf)					Atomic ratio		O + N + S	Coal rank
	M_{ad}	A_{ad}	V_{daf}	C	H	O	N	S	H/C	O/C		
GZ	1.8	21.3	17.5	91.4	3.1	4.2	0.2	1.0	0.41	0.03	5.4	Anthracite
HB	1.2	12.4	35.0	85.8	5.0	8.1	0.3	0.7	0.70	0.07	9.1	Bituminous coal
HN	1.9	21.6	40.7	80.1	4.5	10.5	1.4	0.3	0.67	0.10	12.2	Bituminous coal
KL	1.2	13.2	33.5	78.5	5.6	13.7	1.2	0.3	0.85	0.13	15.2	Subbituminous coal
SX	2.3	19.9	36.5	71.1	6.1	19.8	1.6	0.4	1.03	0.21	21.8	Lignite

ad: air-dried basis; daf: dried ash-free basis.

500°C. The furnace temperature increased at 5°C/min and was held for 24 hours isothermally after reaching a chosen temperature. Finally, the in situ FTIR spectrograms of coal were measured at each characteristic temperature.

In this work, the infrared spectral bands selected were 2700 to 3000 cm^{-1} and 1300 to 1800 cm^{-1}. In the range of 2700 to 3000 cm^{-1}, two characteristic absorption peaks (2920 and 2860 cm^{-1}) were obtained using fitting parameters. Overlapping peaks in 1300 to 1800 cm^{-1} [25–27] also were separated using the same method. Taking HB as an example, the specific fitting results are exhibited in Figure 2.

The changing features of the infrared spectral structural parameters with respect to different functional groups were previously summarized by José et al. [26]. Nevertheless, the main concern is the influence of heat treatment on the structural parameters of coal in this study. Many spectral parameters based on FTIR have been used to illuminate the structural characteristics of coal [26, 27]. According to the selection of parameters reported in the previous studies, in this study, we focus on the intensity and area ratio of characteristic peaks. The specific parameters and their corresponding functional groups are described as follows:

(i) P_1 and P_2 represent the ratio of the integrated areas of aromatic hydrocarbon CH_x stretching to aliphatic CH_x stretching

(ii) P_3 reveals the variation of oxygen-containing functional groups or the atomic ratio of oxygen to carbon (O/C)

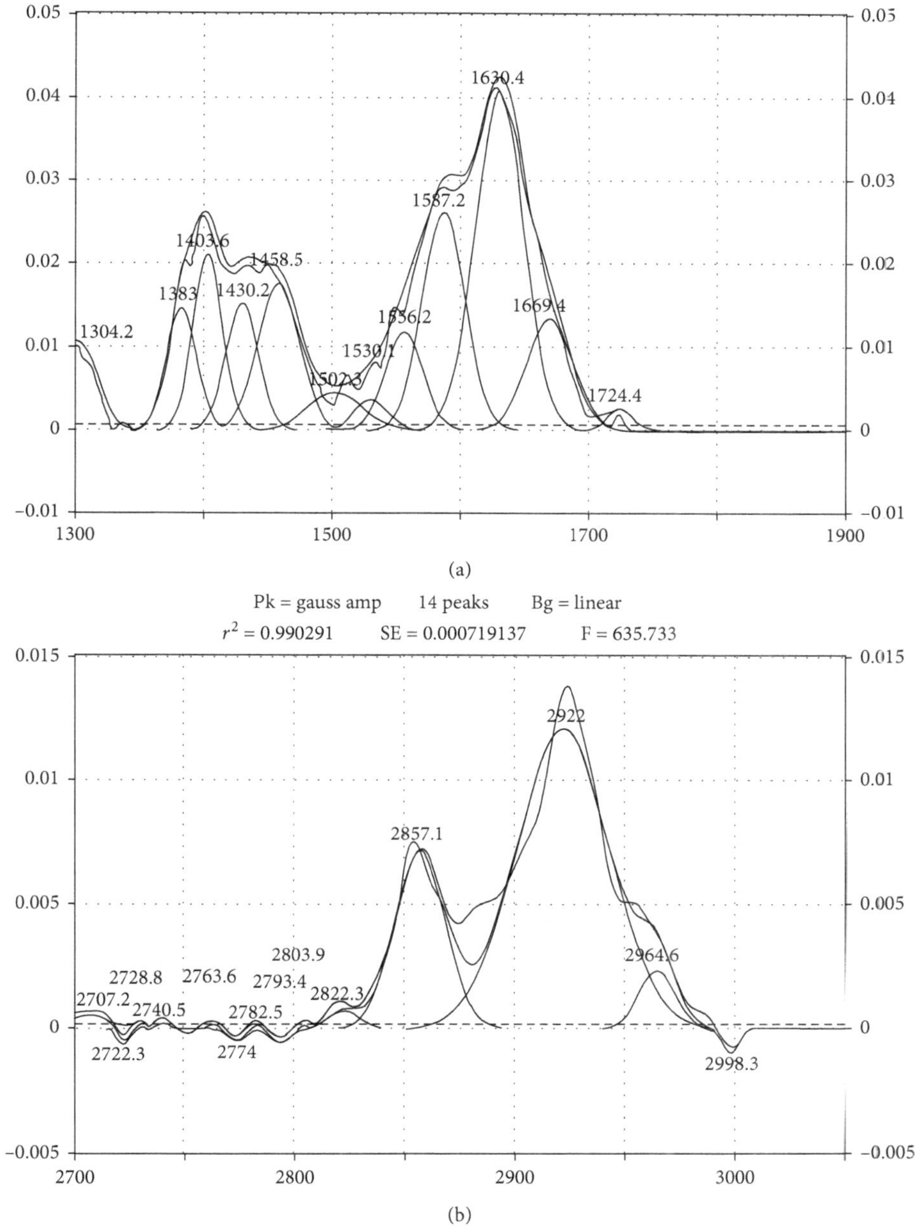

FIGURE 2: Curve-fitted results of 3000–2800 cm^{-1} and 1300–1800 cm^{-1} band for HB coal sample before pyrolysis.

(iii) P_4 and P_5 estimate the degree of enrichment of CH_3 from residual aliphatic structures

2.4. In Situ X-Ray Diffraction Measurement. The in situ XRD data collection was performed by Philips X'Pert PRO X-ray powder diffraction, using Ni-filtered Cu Kα radiation and a scintillation detector. The XRD pattern was recorded over a 2θ interval of 10–60°, with a step size of 0.034 and 2 s/step counter time. The heat treatment experiment of coal was carried out under oxygen-free conditions. Five powered coal samples were heated from room temperature to 900°C with the heating rate of 10°C/min and were kept at constant temperature (25, 100, 200, 300, 400, 500, 600, 700, 800, and 900°C) for 20 min. Meanwhile, the in situ XRD diffractograms were monitored at each characteristic temperature. Structural parameters (L_a, L_c, and d_{002}) were calculated according to Scherrer equations (1)–(3), respectively:

$$\text{crystalline lateral size: } L_a = \frac{1.84\lambda}{B_{100}\cos(\varphi_{100})}, \quad (1)$$

$$\text{stacking height: } L_c = \frac{0.89\lambda}{B_{002}\cos(\varphi_{002})}, \quad (2)$$

$$\text{interlayer spacing: } d_{002} = \frac{\lambda}{2\sin(\theta_{002})}, \quad (3)$$

where λ refers to the wavelength of the radiation in the experiment (λ is 1.5406 Å in this study). B_{100} and B_{002} represent the width of (100) and (002) peaks at half-maximum height, respectively. φ_{100} and φ_{002} represent the peak position of (100) and (002) peaks, respectively. The θ_{002} refers to the Bragg angle of (002) peak.

3. Results and Discussion

3.1. Relationship between Coal Element Composition and Its Structure. Some information about the molecular structures of coal can be obtained from an analysis for the elements carbon (C), hydrogen (H), sulfur (S), and nitrogen (N) [28]. The atomic ratio of hydrogen to carbon (H/C) varies in different ranked coals. From Table 1, it is seen that lignite (SX), subbituminous coal (KL), and bituminous coals (HB and HN) have higher H/C ratios (from 1.03 to 0.70), whereas anthracite (GZ) has the lowest H/C ratio (0.41). The results show the H/C ratio decreases with the degree of coalification. Furthermore, the total content of O + N + S in coal differs appreciably. With the increase of coal rank, the O + N + S content in coal decreases significantly (Table 1), because the O, N, and S would be taken off during coalification, finally forming a graphite-like structure which would have a lower atomic H/C ratio. According to Hirsch [29], in this study, SX and KL, and HN coals (<85% C_{daf}) have an "open" structure with lamellae rarely oriented and connected by cross-links, and HB coal (85% < C_{daf} ≤ 91%) has a "liquid" structure with lamellae some oriented and many cross-links interrupted, and GZ coal (>91% C_{daf}) has an "anthracitic" structure with both lamellae and pores orientated.

3.2. FTIR Investigation

3.2.1. FTIR Characteristics of Coal before Pyrolysis. The infrared spectra of coal with the increase of coal rank are shown in Figure 3. Assignments of characteristic peak positions corresponding to different functional groups are also depicted in Figure 3 based on [26, 30–33]. As shown in Figure 3, the regular change of infrared characteristic peak of coal was closely related to the increase of C_{daf} content in coal. According to Van Krevelen [34], the absorbance of functional groups decreases considerably depending on the degree of coalification.

The following conclusions can be drawn in the coal series lignite-subbituminous coal-bituminous coal-anthracite based on Figure 3:

(1) The dashed line 2 between 2850 cm^{-1} and 3000 cm^{-1} reveals that, with the increase of coal rank, the absorbance of aliphatic CH_2 or CH_3 stretching vibration dwindles considerably, suggesting the decrease of hydrocarbon content in coal. The change in dashed 1 (aromatic hydrocarbon) was similar to that in dashed line 2, which indicates that hydrogen was gradually removed in higher-rank coal, leading to the decrease of intensity of aromatic ring stretching vibration in coal.

(2) In lignite (SX) and subbituminous coal (KL), the absorbance of aromatic carboxyl and carbonyl groups (dashed line 3) was obvious. However, it was absent in bituminous coal (KL and HN) and anthracite (GZ). According to Van Krevelen [34], it can be deduced that as coal rank increases, most of the carboxyl and carbonyl groups in coal can be gradually incorporated into the orthoquinone system structure.

(3) Because of the presence of polar substituents in aromatic rings [30, 31], the absorbance of aromatic nucleus C=C groups (dashed line 4) starts to decrease from lignite (SX) to bituminous coal (HN) and then increases from bituminous coal (HB) to anthracite (GZ).

(4) In the range of lignite (SX) to bituminous coal (HN), the absorbance of phenolic deformation C–O–C groups (dashed line 5) decreases obviously. Up to anthracite (GZ), it disappears progressively.

(5) The absorbance of aromatic out-of-plane bending (dashed line 6) increases slightly between lignite (SX) and subbituminous coal (KL). Reaching the stage of bituminous coal (HB), it decreases moderately. Up to anthracite (GZ), it decreases to a less content.

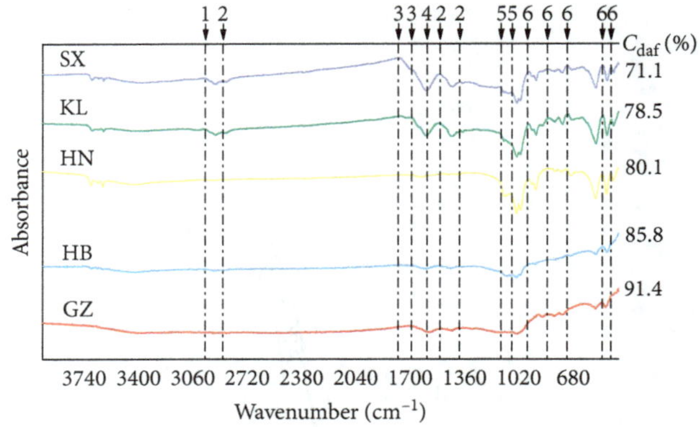

FIGURE 3: Infrared spectrogram of five coal samples. Attribution of characteristic peak positions is descried as follows: (1) aromatic hydrocarbon CH; (2) aliphatic CH_2 or CH_3 stretching vibration; (3) aromatic carbonyl/carboxyl groups (C=O); (4) aromatic nucleus (C=C); (5) phenolic deformation C–O–C (stretching); (6) aromatic CH (out-plane bending).

3.2.2. In Situ FTIR Evolution Characteristics of Different Rank Coals during Pyrolysis. As the pyrolysis temperature increases, in situ infrared spectra of five coal samples are shown in Figure 4. In the range of 3800 to 3200 cm^{-1}, this absorption band is assigned to the –OH groups, which stands for the presence of phenols, ethers, and alcohols.

According to Sonibare et al. [17], this absorption band can also reflect the vibration intensity of water molecules in coal. In comparison to low-ranked coal (Figures 4(a)–4(c)),

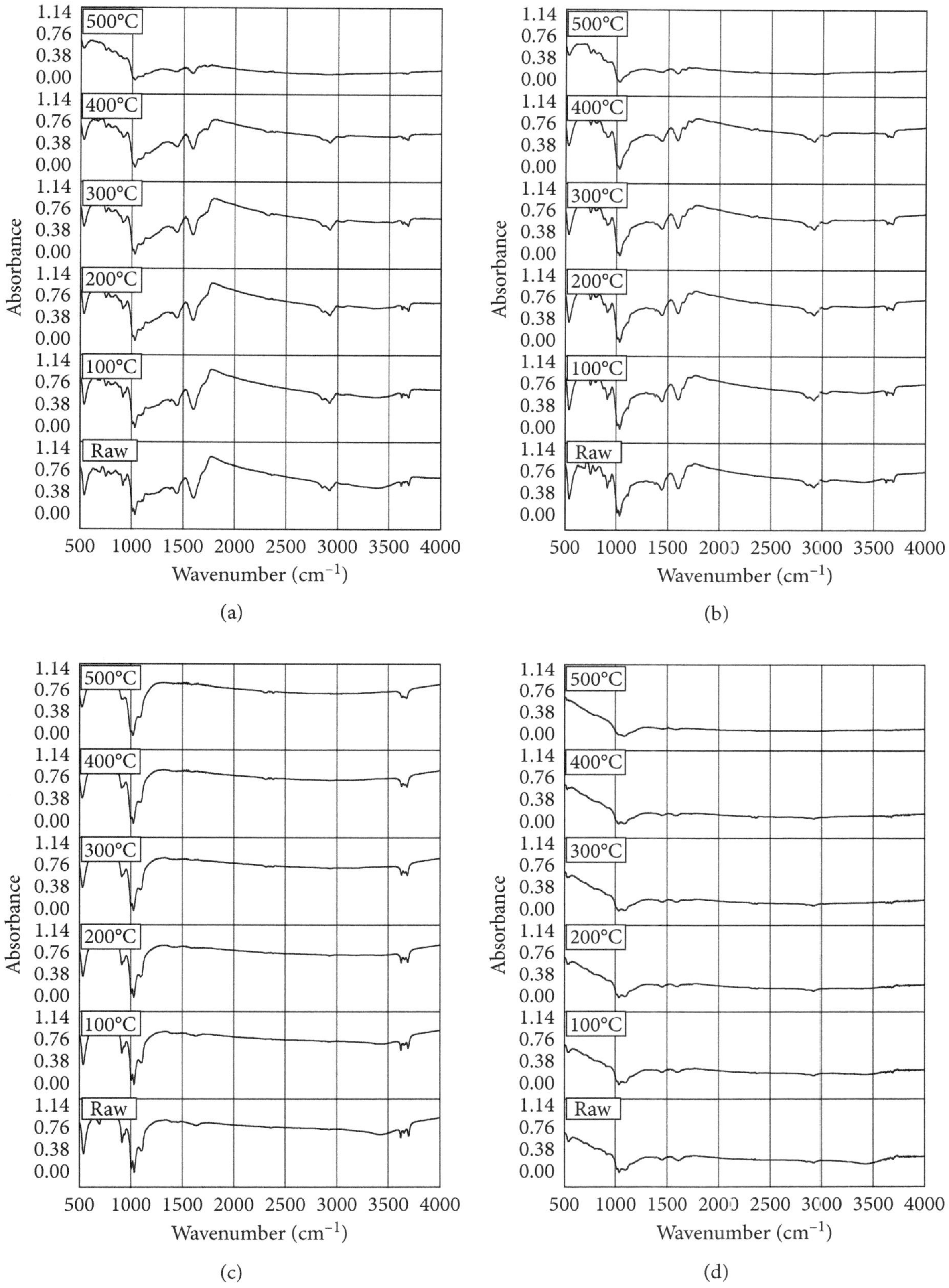

Figure 4: Continued.

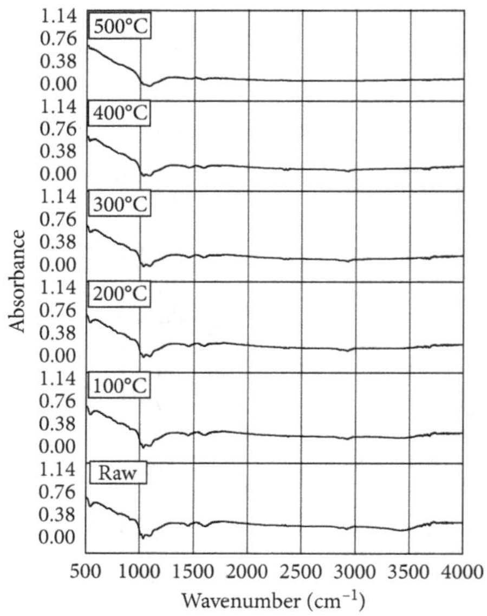

(e)

Figure 4: In situ infrared spectra of five coal samples at different temperatures: (a) SX; (b) KL; (c) HN; (d) HB; (e) GZ.

it can be clearly observed that the absorbance of this characteristic band disappeared almost in anthracite (Figures 4(d) and 4(e)), suggesting that the high-ranked coal contains no intrinsic water molecules. Only lignite (SX) and subbituminous coal (KL) between 3000 cm^{-1} and 2800 cm^{-1} reflects strong absorption peaks at 2920 and 2850 cm^{-1}, representing aliphatic and aromatic stretching vibration including $-CH_2$ and $-CH_3$. However, in bituminous coal (HN and HB) and anthracite (GZ), this characteristic band has no response. Lignite (SX) also presents a characteristic band at approximately 1440 cm^{-1}, corresponding to aliphatic C=C bending vibration. Noteworthy, a characteristic band at approximately 610 cm^{-1} appears in all coal samples. It was attributed to the C–S stretching vibration based on the study by Angoni [13] and Sonibare et al. [17]. Likewise, some peaks between 900 cm^{-1} and 700 cm^{-1} ascribed to the existence of low intensity aromatic $(C-H)_{ar}$ and its out-of-plane bending modes are also present in all coal samples. This result means that the existence of this characteristics band was independent on the increase of coal rank.

Especially, significant changes in the oxygen-containing functional groups are observed in the range of 1600 to 1800 cm^{-1} band (Figure 4). The curve-fitting results (Figure 2) reveal the presence of three characteristics peaks at approximately 1724, 1669, and 1630 cm^{-1}, which were attributed to phenolic esters, carboxylic acids, and conjugate ketonic structures, respectively [35]. Given that the importance of this band, we examine the changes in oxygen-containing functional groups from two perspectives.

(1) Effect of Temperature on Oxygen-Containing Functional Groups. For lignite-subbituminous coal-bituminous coal-anthracite series, with the increase of temperature from 100 to 500°C, the absorption at ~1724 cm^{-1} shifts to ~1694 cm^{-1} because the vibration intensity of aliphatic esters and phenol esters was reduced [34, 35]; the band of carboxylic acids shifts from ~1669 cm^{-1} to ~1644 cm^{-1}, which reflects a progressive conversion of aliphatic carboxylic acids into aromatic carboxyls [35]; the slight excursion of conjugated structures was detected from ~1630 to ~1620 cm^{-1}. On the other hand, the intensity of the above characteristic band decreases slightly from 100 to 400°C and drops abruptly at 500°C. More generally, coal pyrolysis can be divided into a three-step process associated with different chemical reactions that take place in three temperature range intervals, <300, 300–600, and 600–800°C. According to Arenillas et al. [30, 31], the pyrolysis process of coal in the corresponding temperature ranges can be described as follows: (1) degassing; (2) depolymerization; (3) polycondensation. The full pyrolysis of coal usually takes place in "depolymerization." Based on this, it can be deemed that when the temperature was higher than 400°C, the studied coal began to "depolymerize," leading to the decomposition of the relatively high molecular weight substances [30, 31].

(2) Effect of Coal Rank at the Same Temperature on Oxygen-Containing Functional Groups. Figure 5 shows in situ infrared spectra of five coal samples at the same temperature. As can be seen from Figure 5, under the same temperature conditions, in the range of lignite (SX) to bituminous coal (HN), the absorbance of carboxyl groups at ~1700 cm^{-1} and phenolic ester groups at ~1500 cm^{-1} decreases slightly. These two characteristic bands between bituminous coal (HB) and anthracite (GZ) almost disappear. The result shows that as the coal rank increases, the intensity of decarboxylation and decarbonylation becomes deeper correspondingly [30–32]. Interestingly, once the temperature reaches 500°C, their absorbance can drop immediately, which indicates that the

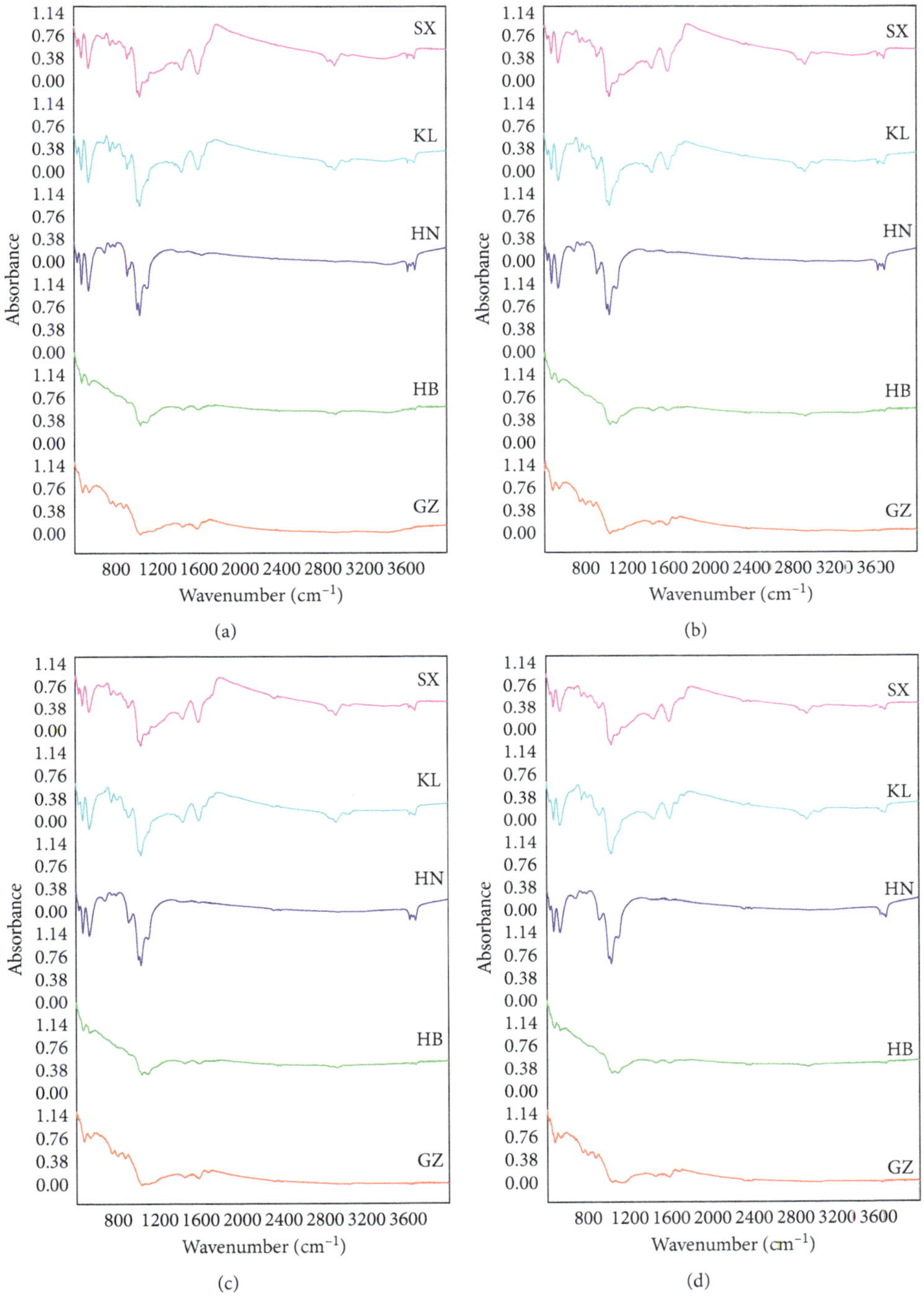

FIGURE 5: Continued.

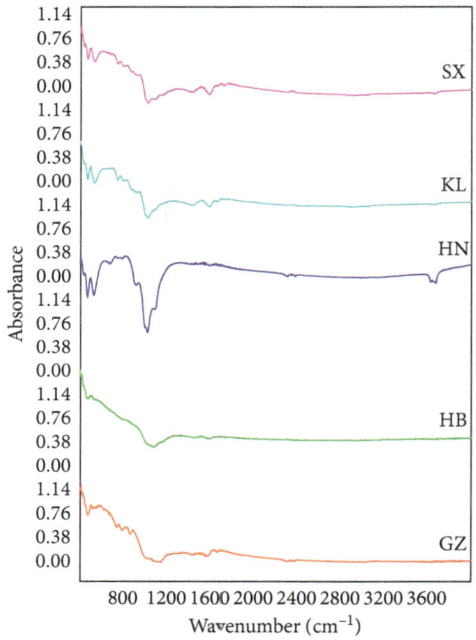

(e)

FIGURE 5: In situ infrared spectra of five coal samples at the same temperature: (a) 100°C; (b) 200°C; (c) 300°C; (d) 400°C; (e) 500°C.

degree of coal pyrolysis is not related to the degree of coalification and may be related to the bond energy of covalent bond of oxygen-containing groups in coal [33].

3.2.3. Effect of Temperature on Structural Parameters. The calculated infrared spectral structural parameters are presented in Table 2. For raw coal, with the increase of coal rank, the structural parameters (P_1, P_2, P_3, P_4, and P_5) of coal increase accordingly, which means that high-ranked coal (GZ and HB) containing less H, less O, and more C becomes more aromatic than lower-ranked coal (HN, KL and SX). However, these parameters change in a different way after coal pyrolysis. The detailed conclusions can be drawn as follows:

(1) For these samples (GZ, HB, and HN), the contraction vibration intensity of the $-CH_3$ band (2920 cm^{-1} and 2860 cm^{-1}) is very weak at different temperatures and the peak area of this band cannot be obtained so that the P_1 value cannot be calculated. For KL and SX samples, the P_1 value shows a decreasing trend with the increase of temperature at less than 500°C. Above 500°C, the P_1 value cannot be calculated. This indicates that the absorption peaks of hydrogen on aliphatic hydrocarbon and naphthenic hydrocarbon groups appear at 2920 cm^{-1} and 2860 cm^{-1}, and the absorption intensity of these peaks decreases significantly with the deepening of the coalification degree (after KL coal). For low-rank coals (KL and SX), the intensity of these peaks also decreased significantly when the temperature exceeded 400°C; (2) with the increase of coal rank, the P_2 value of coal tends to increase to a certain extent. However, for the same coal sample, with increasing temperature, the P_2 value of coal gradually decreases. This indicates that temperature changes the molecular structure of coal and weakens the absorption of hydrogen on aliphatic and aromatic hydrocarbon groups; (3) with the increase of coal rank, P_3 value shows a certain increasing trend. For low-rank coal, there is a strong absorption peak at 1600 cm^{-1}, while for high-rank coal, the intensity of the absorption peak decreases. This absorption peak may be attributed to the overlap of hydrogen-bonded carbonyl group and aromatic ring C=C double bond absorption, which gradually weakens with the deepening of coalification. By analyzing the infrared spectrum of coal, it can be seen that the attenuation of 1600 cm^{-1} absorption peak is higher than that of 1700 cm^{-1}, so P3 value increases. Moreover, the reduction of P_3 value during coal pyrolysis shows that, with the increase of polymerization and decrease in the atomic O/C ratio, the amount of oxygen-containing functional groups in coal starts to decrease progressively; (4) P_4 and P_5 are both ratios reflecting the intensity of hydrogen absorption peaks on aliphatic and aromatic hydrocarbon groups. With the increase of coal rank, they all show different degrees of increase. For the same coal sample, as the temperature increases, they also gradually increase. This reflects that temperature increases the vibration intensity of aromatic and aliphatic CH_x groups, resulting in an increase in the number of some oxygen-containing functional groups in coal.

3.3. XRD Investigation

3.3.1. XRD Characteristics of Coal before Pyrolysis. More generally, coal is amorphous, but its aromatic structure tends to graphitize gradually during coalification. Thus, the XRD

TABLE 2: Statistic results of the infrared spectral structural parameters of five coal samples before and after pyrolysis.

Sample number	Temperature (°C)	Structural parameters				
		P_1	P_2	P_3	P_4	P_5
GZ	Raw coal	–	1.965	0.885	1.062	0.585
	100°C	–	1.941	0.880	1.063	0.592
	200°C	–	1.894	0.867	1.066	0.599
	300°C	–	1.817	0.842	1.072	0.606
	400°C	–	1.693	0.794	1.087	0.612
	500°C	–	1.492	0.699	1.127	0.619
HB	Raw coal	–	1.843	0.830	0.997	0.546
	100°C	–	1.821	0.825	0.997	0.552
	200°C	–	1.777	0.813	1.000	0.559
	300°C	–	1.705	0.790	1.005	0.565
	400°C	–	1.588	0.745	1.020	0.572
	500°C	–	1.399	0.656	1.057	0.578
HN	Raw coal	–	1.766	0.812	0.954	0.541
	100°C	–	1.721	0.776	0.931	0.555
	200°C	–	1.701	0.771	0.932	0.562
	300°C	–	1.660	0.760	0.934	0.568
	400°C	–	1.592	0.738	0.939	0.575
	500°C	–	1.483	0.695	0.953	0.581
KL	Raw coal	2.947	1.687	0.760	0.912	0.502
	100°C	2.916	1.667	0.756	0.913	0.508
	200°C	2.852	1.626	0.745	0.915	0.514
	300°C	2.743	1.561	0.723	0.920	0.520
	400°C	2.559	1.454	0.682	0.934	0.526
	500°C	–	1.281	0.600	0.968	0.532
SX	Raw coal	2.668	1.528	0.688	0.826	0.455
	100°C	2.641	1.510	0.684	0.827	0.461
	200°C	2.583	1.473	0.674	0.829	0.466
	300°C	2.484	1.413	0.655	0.833	0.472
	400°C	2.317	1.316	0.617	0.845	0.477
	500°C	–	1.160	0.543	0.877	0.483

The parameter calculation method: P_1 = area$_{2920+2860\,cm^{-1}}$/area$_{1600\,cm^{-1}}$; P_2 = area$_{1380+1460\,cm^{-1}}$/area$_{1600\,cm^{-1}}$; P_3 = area$_{1700\,cm^{-1}}$/area$_{1600\,cm^{-1}}$; P_4 = area$_{1380\,cm^{-1}}$/area$_{1460\,cm^{-1}}$; P_5 = area$_{1380\,cm^{-1}}$/area$_{2920\,cm^{-1}}$.

pattern of coal presents certain regularity with the increase of the degree of coalification.

Figure 6(a) is an XRD pattern of five coal samples before pyrolysis using the Debye–Scherrer method [28], with coal rank increasing sequentially from top to bottom. It can be seen from the diffraction pattern that anthracite (GZ) has the uppermost diffraction peak (002) and the narrowest width (002). The diffraction width and height of bituminous coal (HB and HN) and subbituminous coal (KL) are similar, but both are lower and wider than those of anthracite. Compared with other coals, the diffraction height and width of lignite (SX) are the lowest and widest. The shape change of diffraction curves of five coal samples has certain regularity. With increasing coal rank, the diffraction height on the diffraction curves also increases correspondingly, while the diffraction width becomes narrow progressively, and the diffraction position shifts gradually to the high angle ($2\sin\theta/\lambda$) region. Taking the 2θ angle values corresponding to (002) crystal mesh of five coal samples as an example, this change can be clearly observed (Table 3) that the position of (002) band shifts gradually to a high angle value ($2\sin\theta/\lambda$) from 0.27 to 0.29.

As shown in Figure 6(a), the asymmetric phenomenon of the (002) band suggests the existence of another band (λ) at its left-hand side, which makes the left side of the (002) band broad and diffused [25]. The diffraction height of the (λ) band at low angle becomes relatively higher due to strong scattering between low-angle particles. As coal rank increases, the height of the (002) band becomes higher but the height of the (λ) band becomes lower.

Taking the SX coal sample as an example, its XRD pattern was fitted using Gaussian function (Figure 6(b)). In conjunction with Scherrer equations (1)–(3), the XRD structural parameters of five coal samples were calculated and are presented in Table 3.

Table 3 shows that the interlayer spacing of carbon hexagons (d_{002}) is 0.3567 nm in lignite (SX), 0.3571 nm in subbituminous coal (KL), 0.3561 nm in bituminous gas coal (HN), 0.3518 nm in bituminous coal (HB), and 0.3498 nm in anthracite (GZ). The results show with the increase of coal rank, the d_{002} values decreases accordingly. Compared with the d_{002} value (0.335 nm) of graphite, the d_{002} values of five coal samples were significantly greater, which suggests a lower degree of the crystalline order in these coals. From Table 3, the crystallite height (L_c) value increases from 1.379 nm for lignite (SX) to 1.984 nm for anthracite (GZ), but the crystallite diameter (L_a) value shows the different trend relative to L_c. In all coal samples, the L_a remains

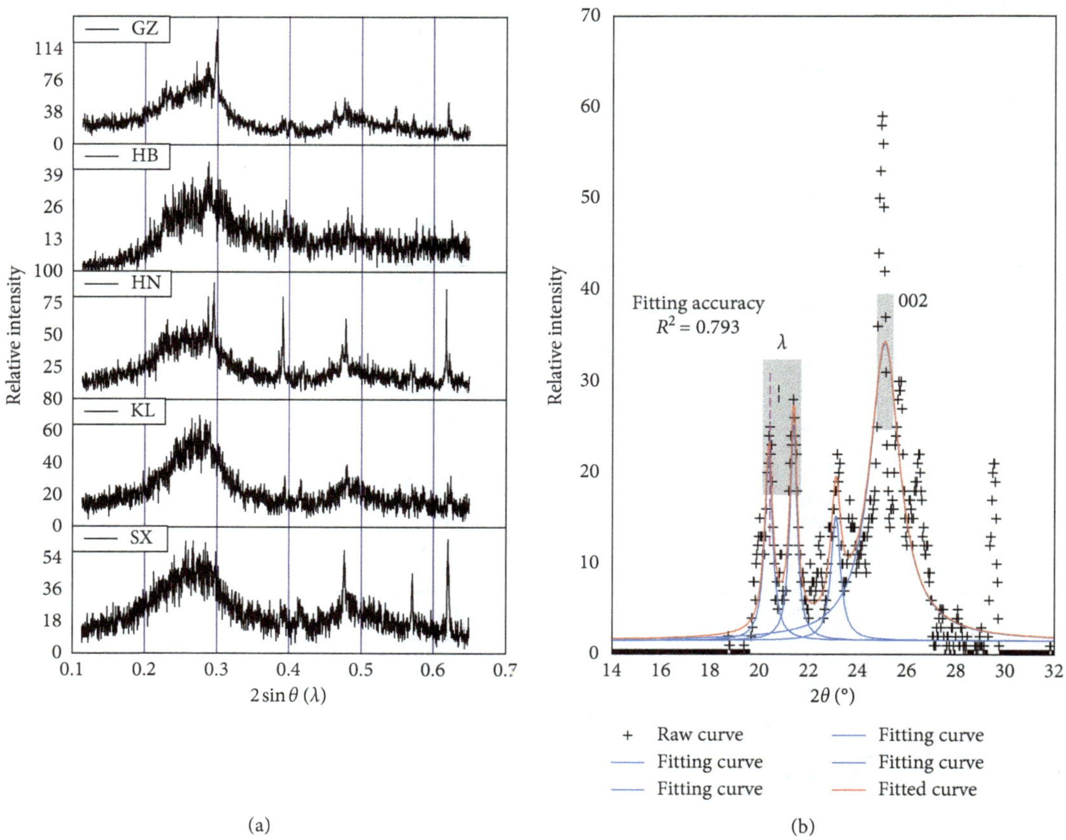

FIGURE 6: XRD pattern of five coal samples before pyrolysis (a) and the curve fitting result of the SX coal sample before pyrolysis as a function of Gaussian function (b).

constant at nearly 2.979 nm. The same conclusion was also documented in the literature [36]. It can be deemed that with increasing coal rank, the degree of polymerization between adjacent carbon crystals was restricted by molecular forces between side chains [36]. Combining L_a and L_c changes, it can be speculated that the natural orientation under pressure and geothermal gradient can be direction-dependent during coalification [8].

3.3.2. In Situ XRD Evolution Characteristics of Different Coals during Pyrolysis. The in situ XRD patterns of five coal samples during pyrolysis are shown in Figure 7.

As can be seen from Figure 7, the shape of the XRD pattern of five coal samples did not change significantly before 500°C. In comparison to coal before pyrolysis, the XRD patterns of all coals present a more sharp band (λ) at approximately 0.23 ($2\sin\theta/\lambda$) in the range of 25 to 500°C. When temperature is up to 900°C, the intensity of the (λ) band decreases significantly. On the other hand, the position of the (002) band remains unchanged firstly between 25°C and 500°C, but above 500°C, the shape of the (002) band becomes sharper and shifts to higher 2θ angle values. In addition, this band tends to be more asymmetric above 500°C than at lower temperatures (Figures 7(c)–7(e)).

3.3.3. Effect of Temperature on Structural Parameters. Taking the HB coal sample as example, Origin 8.0 software was used to fit its XRD patterns in the 2θ range of 14–32° to obtain (002) and (λ) bands and their structural parameters (d_{002}, L_a, and L_c). The in situ XRD curve fitting results of the HB coal sample at different temperatures (300, 600, and 900°C) are shown in Figure 8.

Table 3 shows in situ XRD structural parameters of five coal samples at different temperatures. Figure 9 plots the correlation between temperature and the above structural parameters. Some conclusions were obtained as follows:

(1) L_a. Figure 9(a) shows that, with increasing temperature, the variation range of L_a value is extremely small (approximately 0.09 nm), suggesting that the relationship between the L_a value of coal and temperature becomes quite inappropriate. According to Hirsch [29] and Lu et al. [36], it can be deemed that, with increasing temperature, the bridge structure in coal becomes unstable as a reaction to increasing forces in the aromatic nucleus but this structure may differ in molecular weight and may contain several types of geometric and positional isomers. As a result, the change of L_a for aromatic nucleus is complexity and heterogeneity. In this study, the variation of L_a value is in accordance with the reported in the literature [8, 36].

TABLE 3: Statistic results of the XRD parameters of five coal samples before and after pyrolysis.

Sample number	Temperature (°C)	▲ (°)		▼ (nm)		
		$2\theta_{002}$	$2\theta_{100}$	d_{002}	L_c	L_a
GZ	Raw coal	26.15	46.16	0.3498	1.9841	3.0507
	25°C	26.17	44.64	0.3496	1.9848	2.9172
	100°C	26.26	45.93	0.3485	1.9842	3.0073
	200°C	26.21	45.09	0.3491	1.9842	2.9479
	300°C	26.27	45.28	0.3483	1.9842	2.9611
	400°C	26.26	44.38	0.3485	1.9837	2.8998
	500°C	26.24	44.14	0.3487	1.9833	2.8839
	600°C	26.35	44.61	0.3474	1.9868	2.9152
	700°C	26.40	45.30	0.3468	1.9876	2.9625
	800°C	26.54	45.02	0.3451	1.9897	2.9431
	900°C	26.59	45.51	0.3444	1.9901	2.9773
HB	Raw coal	25.99	45.38	0.3518	1.8797	2.9681
	25°C	26.04	45.26	0.3512	1.8801	2.9598
	100°C	26.00	45.53	0.3516	1.8796	2.9787
	200°C	26.00	45.25	0.3516	1.8792	2.9591
	300°C	26.00	45.26	0.3517	1.8805	2.9598
	400°C	25.96	45.16	0.3521	1.8781	2.9528
	500°C	25.93	45.60	0.3525	1.8793	2.9837
	600°C	26.19	45.53	0.3493	1.8826	2.9787
	700°C	26.25	44.98	0.3486	1.8828	2.9404
	800°C	26.40	45.12	0.3468	1.8866	2.9500
	900°C	26.43	44.96	0.3464	1.8868	2.9390
HN	Raw coal	25.66	45.72	0.3561	1.6786	2.9922
	25°C	25.69	45.87	0.3557	1.6796	3.0030
	100°C	25.65	45.66	0.3562	1.6786	2.9879
	200°C	25.61	45.76	0.3567	1.6772	2.9951
	300°C	25.72	45.01	0.3553	1.6781	2.9424
	400°C	25.54	45.57	0.3576	1.6816	2.9815
	500°C	25.63	46.72	0.3565	1.6806	3.0656
	600°C	25.88	46.64	0.3532	1.6852	3.0596
	700°C	25.89	45.87	0.3531	1.6847	3.0030
	800°C	26.17	45.60	0.3496	1.6885	2.9837
	900°C	26.19	45.66	0.3493	1.6882	2.9879
KL	Raw coal	25.58	45.56	0.3571	1.5786	2.9808
	25°C	25.65	45.64	0.3562	1.5796	2.9865
	100°C	25.58	46.20	0.3571	1.5786	3.0270
	200°C	25.47	45.31	0.3585	1.5772	2.9632
	300°C	25.54	46.55	0.3576	1.5781	3.0529
	400°C	25.80	45.36	0.3543	1.5816	2.9667
	500°C	25.73	45.36	0.3552	1.5806	2.9667
	600°C	26.07	46.60	0.3509	1.5852	3.0566
	700°C	26.03	46.20	0.3513	1.5847	3.0270
	800°C	26.31	45.59	0.3479	1.5885	2.9830
	900°C	26.29	45.70	0.3481	1.5882	2.9908
SX	Raw coal	25.61	45.43	0.3567	1.3791	2.9717
	25°C	25.61	44.61	0.3567	1.3791	2.9152
	100°C	25.54	46.54	0.3576	1.3781	3.0521
	200°C	25.54	45.31	0.3576	1.3781	2.9632
	300°C	25.54	45.66	0.3576	1.3781	2.9879
	400°C	25.47	45.30	0.3585	1.3772	2.9625
	500°C	25.42	44.78	0.3592	1.3765	2.9267
	600°C	25.64	46.01	0.3563	1.3794	3.0131
	700°C	25.76	45.71	0.3548	1.3812	2.9915
	800°C	25.88	43.53	0.3532	1.3826	2.8444
	900°C	25.92	44.67	0.3527	1.3832	2.9192

▲: characteristic parameters of diffraction peak. ▼: crystallite structure parameters.

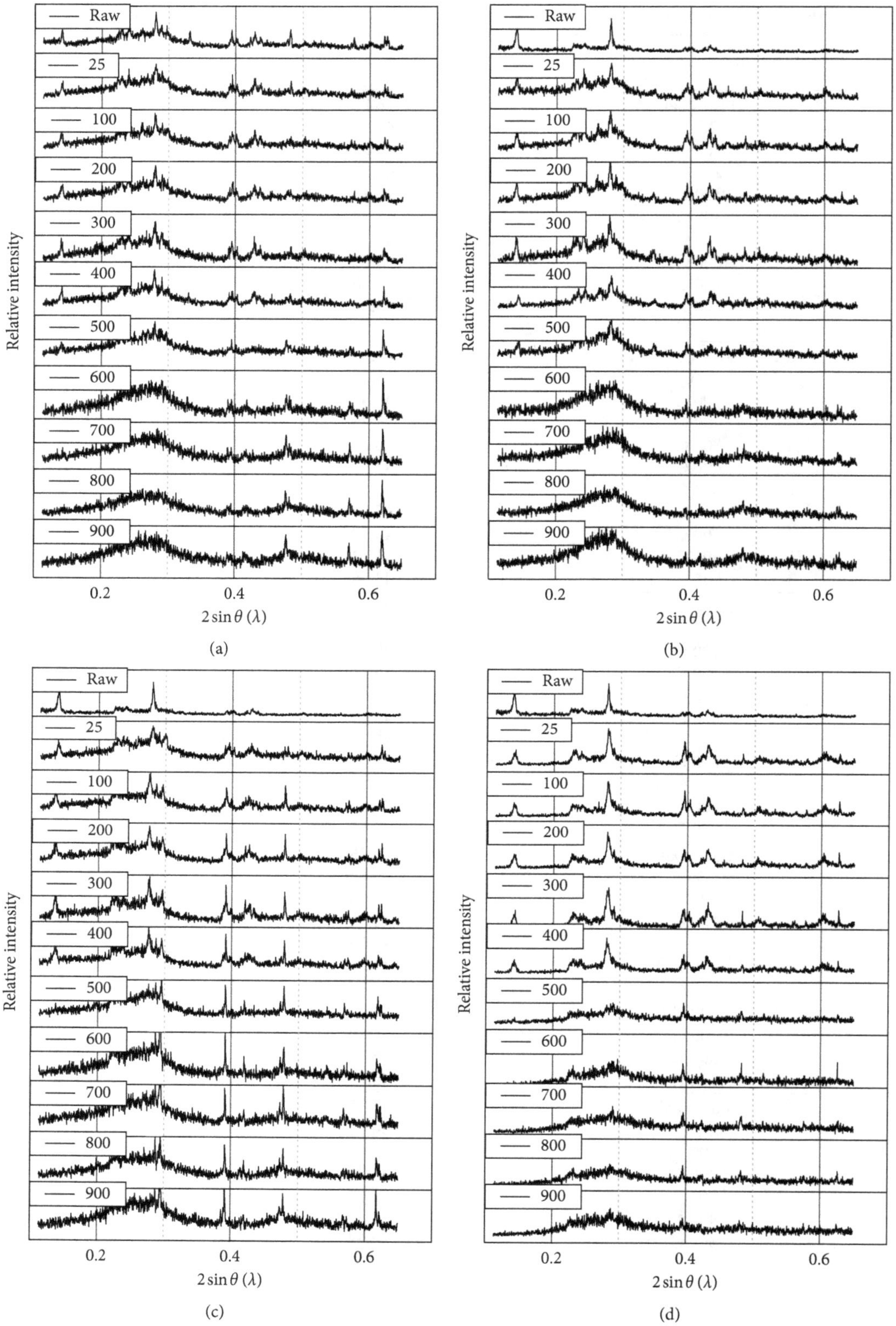

Figure 7: Continued.

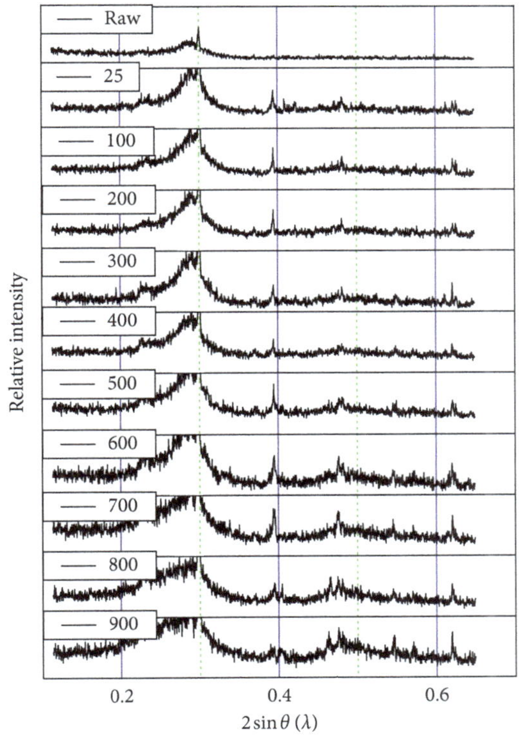

(e)

FIGURE 7: In situ XRD pattern of five coal samples at different temperatures: (a) SX; (b) KL; (c) HN; (d) HB; (e) GZ.

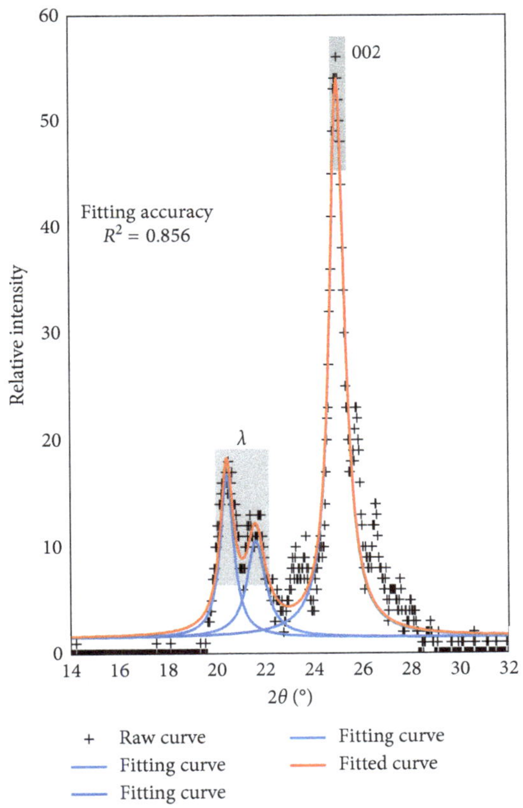

(a)

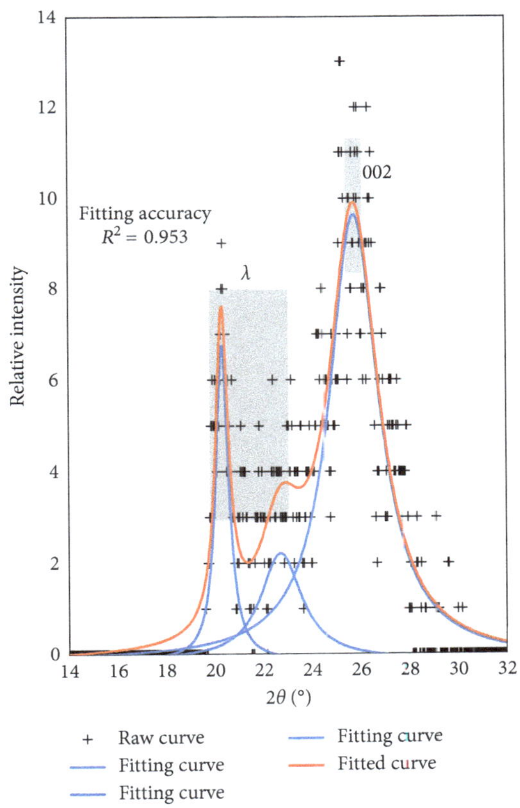

(b)

FIGURE 8: Continued.

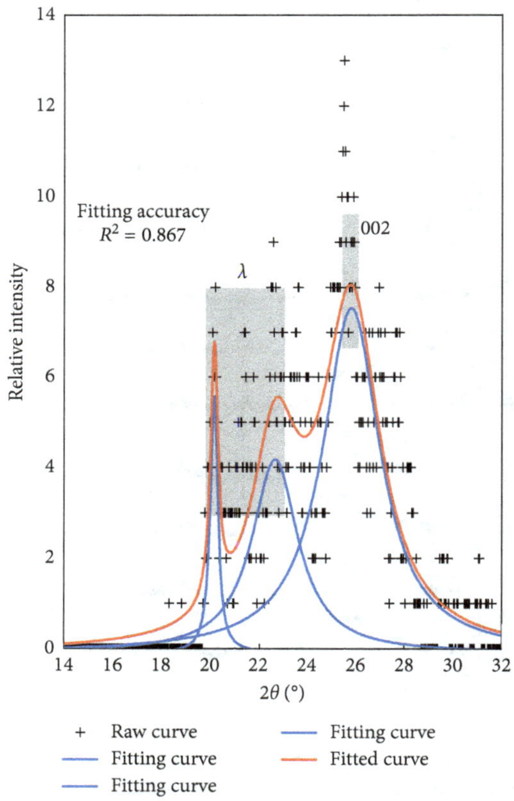

(c)

Figure 8: In situ XRD curve fitting results of the HB coal sample at different temperatures: (a) 300°C; (b) 600°C; (c) 900°C.

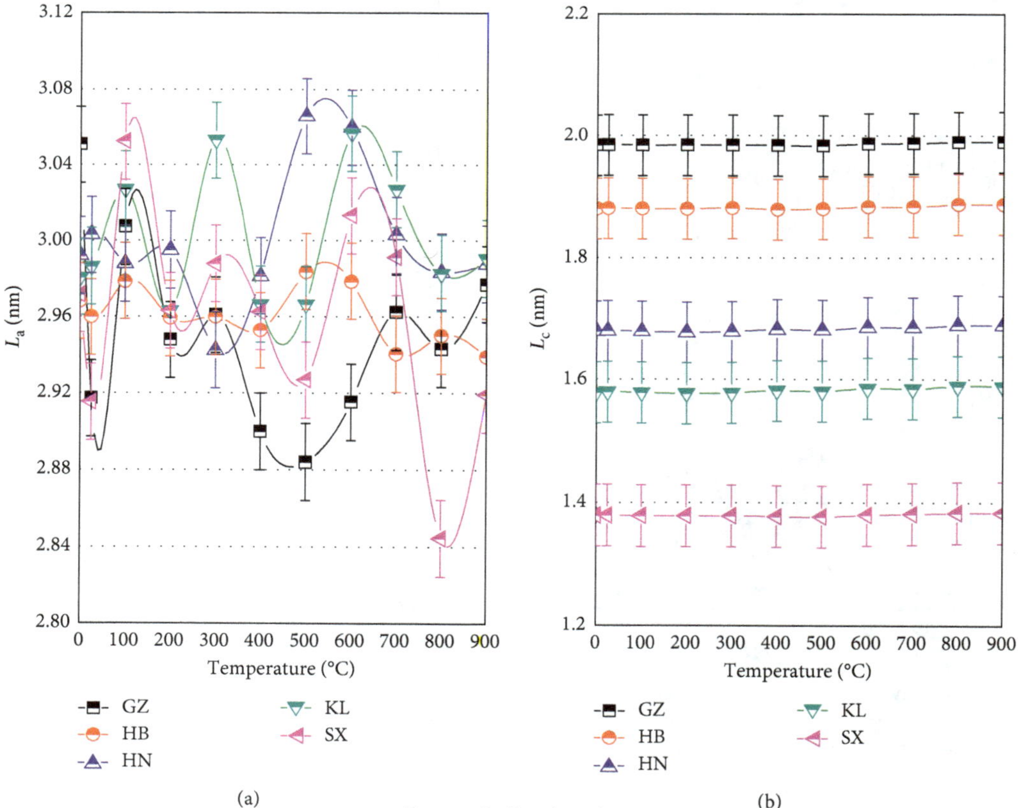

Figure 9: Continued.

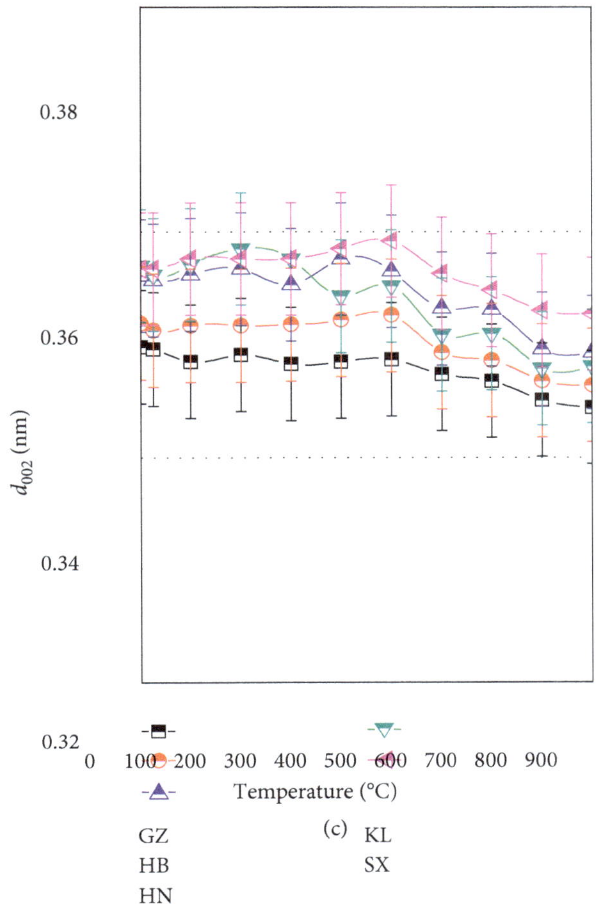

FIGURE 9: Variation of XRD structural parameters with temperature.

(2) L_c. Figure 9(b) shows that, with increasing temperature, the change trend of L_c value can be divided into two main stages. There is a slight decline in the range of 25 to 500°C. After 500°C, the L_c shows a degree of increase. This phenomenon is likely to be related to an evident change in the structure caused by the expulsion of hydrocarbons and other gases physically trapped in the inertinite network [30].

(3) d_{002}. The d_{002} that is considered to be a measure of the perfection in the two-dimensional turbostratic system is characteristic of the substances with two-dimensional periodicity [29]. Figure 9(c) shows that, with increasing temperature, the change of d_{002} values reflects two different changing processes. One stage is in the range of 25–500°C. In this stage, the aggregate d_{002} value increases obviously in lignite (SX) and bituminous coal (HN and HB), whereas it decreases slightly in anthracite (GZ). The other stage is between 500°C and 900°C, where the d_{002} value decreases considerably. This is a good evidence to demonstrate positively such change dependence on the increase of temperature rather than coal rank.

4. Conclusions

The structural evolution signatures of different ranked coal during pyrolysis were deeply investigated using in situ XRD and FTIR analytical methods. The main conclusions are as follows: (1) the FTIR spectra analysis shows that when temperature reaches between 400 and 500°C, coal structure occurs in abrupt change in the intensity of absorption peak; (2) the X-ray diffraction analysis illustrates that with the increase of temperature, two parameters (d_{002} and L_c) change accordingly, but the value of L_a remains unchanged; (3) combining the results of the two experiments, it can be seen that, in the temperature range of less than 500°C, the size of aromatic lamellae of coal decreases, the distance between lamellae decreases, and the degree of bridge bond breaking, aliphatic side chain cracking, and oxygen-containing functional group cracking rises, suggesting that the degree of aromatization of coal increase.

Conflicts of Interest

The authors declare that there are no conflicts of interest regarding the publication of this paper.

Acknowledgments

We acknowledge support from the Research Projects of Department of Land and Resources of Anhui Province, China (2016-K-2), National Natural Science Foundation of China (No. 41502152), Open Projects of Research Center of Coal Resources Safe Mining and Clean Utilization, Liaoning (LNTU17KF10), Natural Science Foundation General Program of Anhui Program (1808085MA07), and Natural Science Foundation Youth Program of Anhui Program (1808085QE150).

References

[1] B. Feng, S. K. Bhatia, and J. C. Barry, "Structural ordering of coal char during heat treatment and its impact on reactivity," *Carbon*, vol. 40, no. 4, pp. 481–496, 2001.

[2] K. Wang, J. Deng, Y.-n. Zhang, and C.-p. Wang, "Kinetics and mechanisms of coal oxidation mass gain phenomenon by TG-FTIR and in situ IR analysis," *Journal of Thermal Analysis and Calorimetry*, vol. 132, no. 1, pp. 591–598, 2018.

[3] Q. Lin and J. M. Guet, "Characterization of coals and macerais by X-ray diffraction," *Fuel*, vol. 69, no. 7, pp. 821–825, 1990.

[4] S. Wang, S. Liu, Y. Sun, D. Jiang, and X. Zhang, "Investigation of coal components of late permian different ranks bark coal using AFM and Micro-FTIR," *Fuel*, vol. 187, pp. 51–57, 2017.

[5] J. Ibarra, R. Moliner, and A. J. Bonet, "FT-I.R. investigation on char formation during the early stages of coal pyrolysis," *Fuel*, vol. 73, no. 6, pp. 918–924, 1994.

[6] T. Das, "Evolution characteristics of gases during pyrolysis of maceral concentrates of Russian coking coals," *Fuel*, vol. 80, no. 4, pp. 489–500, 2001.

[7] Q. Shi, B. Qin, Q. Bi, and B. Qu, "An experimental study on the effect of igneous intrusions on chemical structure and combustion characteristics of coal in Daxing Mine, China," *Fuel*, vol. 226, pp. 307–315, 2018.

[8] B. Tian, Y. y. Qiao, Y. y. Tian, and Q. Liu, "Investigation on the effect of particle size and heating rate on pyrolysis characteristics of a bituminous coal by TG-FTIR," *Journal of Analytical and Applied Pyrolysis*, vol. 121, pp. 376–386, 2016.

[9] J. Yu, Y. He, J. Hao et al., "A mathematical model to characterize the degree of coalification based on the low-angle region of the X-ray diffractogram," *Journal of X-Ray Science and Technology*, vol. 26, no. 1, pp. 71–81, 2018.

[10] K. Mae, T. Maki, H. Okutsu, and K. Miura, "Examination of relationship between coal structure and pyrolysis yields using oxidized brown coals having different macromolecular networks," *Fuel*, vol. 79, no. 3-4, pp. 417–425, 2000.

[11] S. Wang, Y. Tang, H. H. Schobert, G. D. Mitchell, F. Liao, and Z. Liu, "A thermal behavior study of Chinese coals with high hydrogen content," *International Journal of Coal Geology*, vol. 81, no. 1, pp. 37–44, 2010.

[12] Q. Sun, W. Li, H. Chen, and B. Li, "The variation of structural characteristics of macerals during pyrolysis⋆," *Fuel*, vol. 82, no. 6, pp. 669–676, 2003.

[13] K. Angoni, "Remarks on the structure of carbon materials on the basis of Raman spectra," *Carbon*, vol. 31, no. 4, pp. 537–547, 1993.

[14] V. Mennelia, G. Monaco, I. Colangeli, and E. Bussoletti, "Raman spectra of carbon based materials excited at 1064 nm," *Carbon*, vol. 33, no. 2, pp. 115–121, 1995.

[15] P. B. Hirsch, "Recent results on the structure of dislocations in tetrahedrally coordinated semiconductors," *Le Journal de Physique Colloques*, vol. 40, no. 6, pp. 6–27, 1979.

[16] I. Watanabe, K. Sakanishi, I. Mochida, and A. Akimoto, "Changes in coal aggregate structure by heat treatment and their coal rank dependency," *Energy & Fuels*, vol. 16, no. 1, pp. 18–22, 2002.

[17] O. O. Sonibare, T. Haeger, and S. F. Foley, "Structural characterization of Nigerian coals by X-ray diffraction, Raman and FTIR spectroscopy," *Energy*, vol. 35, no. 12, pp. 5347–5353, 2010.

[18] W. Zhang, S. Chen, F. Han, and D. Wu, "An experimental study on the evolution of aggregate structure in coals of different ranks by in situ X-ray diffractometry," *Analytical Methods*, vol. 7, no. 20, pp. 8720–8726, 2015.

[19] M. Li, F. Zeng, H. Chang, B. Xu, and W. Wang, "Aggregate structure evolution of low-rank coals during pyrolysis by in-situ X-ray diffraction," *International Journal of Coal Geology*, vol. 116-117, pp. 262–269, 2013.

[20] J. Kister, M. Guiliano, C. Largeau, S. Derenne, and E. Casadevall, "Characterization of chemical structure, degree of maturation and oil potential of Torbanites (type I kerogens) by quantitative FT-I.R. spectroscopy," *Fuel*, vol. 69, no. 11, pp. 1356–1361, 1990.

[21] A. A. Christy, A. L. Hopland, T. Barth, and O. M. Kvalheim, "Quantitative determination of thermal maturity in sedimentary organic matter by diffuse reflectance infrared spectroscopy of asphaltenes," *Organic Geochemistry*, vol. 14, no. 1, pp. 77–81, 1989.

[22] J. Wang, J. Du, L. Chang, and K. Xie, "Study on the structure and pyrolysis characteristics of Chinese western coals," *Fuel Processing Technology*, vol. 91, no. 4, pp. 430–433, 2010.

[23] D. Wu, W. Y. Zhang, B. Fu, and G. Q. Hu, "Chemical structure and gas products of different rank coals during pyrolysis," *Journal of Thermal Analysis and Calorimetry*, vol. 1, pp. 1–15, 2018.

[24] R. M. Bustin, J.-N. Rouzaud, and J. V. Ross, "Natural graphitization of anthracite: experimental considerations," *Carbon*, vol. 33, no. 5, pp. 679–691, 1995.

[25] D. Wu, G. J. Liu, R. Y. Sun, and X. Fan, "Investigation of structural characteristics of thermally metamorphosed coal by FTIR spectroscopy and X-ray diffraction," *Energy Fuels*, vol. 27, no. 10, pp. 5823–5830, 2013.

[26] V. I. José, M. Edgar, and M. Rafael, "FTIR study of the evolution of coal structure during the coalification process," *Organic Geochemistry*, vol. 24, no. 6-7, pp. 725–735, 1996.

[27] S. Yao, K. Zhang, K. Jiao, and W. Hu, "Evolution of coal structures: FTIR analyses of experimental simulations and naturally matured coals in the Ordos basin, China," *Energy Exploration & Exploitation*, vol. 29, no. 1, pp. 1–19, 2011.

[28] A. Bishop and G. D. Abbott, "Vitrinite reflectance and molecular geochemistry of Jurassic sediments: the influence of heating by Tertiary dykes (northwest Scotland)," *Organic Geochemistry*, vol. 22, no. 1, pp. 165–177, 1995.

[29] P. B. Hirsch, "X-ray scattering from coals," *Proceedings of the Royal Society of London. Series A. Mathematical and Physical Sciences*, vol. 226, no. 1165, pp. 143–169, 1954.

[30] A. Arenillas, F. Rubiera, and J. J. Pis, "Simultaneous thermogravimetric-mass spectrometric study on the pyrolysis behaviour of different rank coals," *Journal of Analytical and Applied Pyrolysis*, vol. 50, no. 1, pp. 31–46, 1999.

[31] A. Arenillas, F. Rubiera, C. Pevida, and J. J. Pis, "A comparison of different methods for predicting coal devolatilisation

kinetics," *Journal of Analytical and Applied Pyrolysis*, vol. 58-59, pp. 685–701, 2001.

[32] A. Arenillas, F. Rubiera, J. J. Pis et al., "Thermal behaviour during the pyrolysis of low rank perhydrous coals," *Journal of Analytical and Applied Pyrolysis*, vol. 68-69, pp. 371–385, 2003.

[33] L. Shi, Q. Liu, X. Guo, W. Wu, and Z. Liu, "Pyrolysis behavior and bonding information of coal - a TGA study," *Fuel Processing Technology*, vol. 108, pp. 125–132, 2013.

[34] D. W. Van Krevelen, "Graphical-statistical method for the study of structure and reaction process of coal," *Fuel*, vol. 29, pp. 269–15, 1950.

[35] P. C. Painter, R. W. Snyder, M. Starsinic, M. M. Coleman, D. W. Kuehn, and A. Davis, "Concerning the application of FT-IR to the study of coal: a critical assessment of band assignments and the application of spectral analysis programs," *Applied Spectroscopy*, vol. 35, no. 5, pp. 475–485, 1981.

[36] L. Lu, V. Sahajwalla, C. Kong, and D. Harris, "Quantitative X-ray diffraction analysis and its application to various coals," *Carbon*, vol. 39, no. 12, pp. 1821–1833, 2001.

16

Multivariate Method Based on Raman Spectroscopy for Quantification of Dipyrone in Oral Solutions

Luciana Lopes Guimarães,[1,2] Letícia Parada Moreira,[1,2] Bárbara Faria Lourenço,[1,2] Walber Toma,[1,2] Renato Amaro Zângaro,[3] Marcos Tadeu Tavares Pacheco,[2,3] and Landulfo Silveira Jr.[2,3]

[1]*Laboratory of Natural Products, Department of Pharmaceutical Sciences, Universidade Santa Cecília (UNISANTA), Rua Cesário Mota, 8, F 83A, 11045-040 Santos, SP, Brazil*
[2]*Department of Ecology, Universidade Santa Cecília (UNISANTA), Rua Cesário Mota, 8, 11045-040 Santos, SP, Brazil*
[3]*Center for Innovation, Technology and Education-CITE, Universidade Anhembi Morumbi (UAM), Estr. Dr. Altino Bondensan, 500, 12247-016 São José dos Campos, SP, Brazil*

Correspondence should be addressed to Luciana Lopes Guimarães; lubioq@gmail.com

Academic Editor: Simona C. Pinzaru

This work employed a quantitative model based on Raman spectroscopy and principal component regression (RS/PCR) to quantify the active ingredient dipyrone (metamizole) in commercially available formulations as an analytical methodology for quality control in the pharmaceutical industry. Raman spectra were collected using a dispersive Raman spectrometer (830 nm, 250 mW excitation, and 20 s exposure time) coupled to a Raman probe. Solutions of dipyrone diluted in water in the range of 80 to 120% of the concentration of commercial formulations (500 mg/mL) were used to develop a calibration model based on PCR to obtain the figures of merit for class I validation from the Brazilian Sanitary Surveillance Agency (ANVISA, RE no. 899/2003). This spectral model was then used to predict the concentration of dipyrone in commercial formulations from distinct brands with 500 mg/mL. A prediction error of 6.5 mg/mL (1.3%) was found for this PCR model using the diluted samples. Commercial formulations had predicted concentrations with a difference below 5.0% compared to the label concentration, indicating the applicability of Raman spectroscopy for quality control in the final product.

1. Introduction

The dipyrone metamizole is a popular analgesic, nonopioid drug, which is commonly used in human and veterinary medicine for pain management in patients where nonsteroidal anti-inflammatory drugs (NSAIDs) are contraindicated [1]. Apart from its analgesic effect, the medication is an antipyretic and spasmolytic agent. Metamizole is predominantly applied in the therapy of pain of different etiologies, of spastic conditions, especially affecting the digestive tract, and of fever refractory to other treatments. It is a relatively safe pharmaceutical preparation, although it is not completely free from undesirable effects. Among these side effects, the most serious one that raises the most controversy is the myelotoxic effect. Despite the evidence showing no risk of teratogenic and embryotoxic effects, the drug must not be administered to pregnant women [2].

Quality control is responsible for verifying that a pharmaceutical product complies with the specifications described in the pharmacopoeias [3]. Routine techniques for measuring dipyrone in industry's quality control include titration in oral metamizole solutions [4]. The process analytical technology (PAT) aims at the development of strategies to the primary and secondary control of the production processes in the pharmaceutical industry. By definition of the Food and Drug Administration (FDA), PAT includes systems for the analysis and control of production processes, based on opportune measures of the quality parameters and critical attributes of the material performance, optimizing the production time, energy, and material, thus contributing to sustainability and reducing environmental impact [5]. The PAT Guide announcement by FDA in 2004 was aimed at research into new and faster analytical methodologies to be used on-line and at-line. Optical techniques such as absorption spectroscopy and Raman spectroscopy have been claimed to be consistent, reliable, rapid, and low-cost analyses of chemical species in the industrial field for analytical instrumentation [5–7], which could successfully meet the PAT requirements.

Due to the increased interest in new equipment and techniques for the quality control of drugs, the Brazilian Sanitary Surveillance Agency (ANVISA) created resolutions to regulate such procedures and equipment to be considered during the validation of analytical methods, referring to the qualitative, semiquantitative and/or quantitative validation of drugs and other substances in pharmaceutical products. The Brazilian resolution RE no. 899/2003 states that the validation should guarantee, through experimental studies, that the method meets the requirements of analytical applications, ensuring the reliability of results [8]. It should provide specificity, linearity, range, precision (repeatability), sensitivity, robustness, accuracy, and limit of detection and quantification appropriate to the analysis. RE 899/2003, which is mostly based on the guidelines developed by the ICH Expert Working Group under the International Conference on Harmonisation of Technical Requirements for Registration of Pharmaceuticals for Human Use [9, 10], classifies the procedures into four distinct classes according to their purpose; for each class, a set of tests is required. The validation method for the quantification of commercial pharmaceutical drugs can be accomplished by class I, in which the figures of merit required are specificity, linearity, range, precision, robustness, and accuracy [8].

Classical techniques for the evaluation of pharmaceutical drugs are based on liquid chromatography followed by mass spectrometry (LC/MS). Although sensitive and very specific, LC/MS techniques are time-consuming and present excessive costs, requiring sample preprocessing prior to analysis, which may destroy the sample. To address this problem, optical techniques have been proposed for the quantitative determination of active pharmaceutical ingredients (API) in drugs and medicines, most of which are based on absorption and reflectance spectrometric techniques in the ultraviolet-visible and infrared regions [11–13], with advantages such as rapid API evaluation with minimal or even no sample preparation [11].

Vibrational spectroscopy, particularly Raman spectroscopy, has been used in the quantitative analysis of API, with advantages such as the need for small amounts of sample, no sample preparation, no sample destruction, and a very short time required to perform the analysis [11]. Raman spectroscopy can be applied for qualitative and quantitative drug analysis, being one of the techniques of greatest interest in pharmaceutical applications [14]. Raman spectra can be obtained from solids, liquids, and gases, either directly, in glass containers or tubes or even through blisters [15]. In this sense, several pharmaceutical presentations have been evaluated by Raman spectroscopy, such as pills, tablets, capsules, solutions, suspensions, and injectable solutions, among others, aiming at the quantification of the API and other formulation constituents [15–17]. Raman spectral information provides information about the sample's chemical structure, molecular composition of the study material, and quantification [16, 18]. Dispersive Raman spectrometers make fast scanning/imaging possible using an imaging spectrograph and multichannel detector such as a Peltier-cooled charge-coupled device (CCD) camera [18]; the use of near-infrared excitation (830 and 1064 nm wavelength) would diminish or eliminate the sample's intrinsic fluorescence [18, 19].

The prediction of the amount of a specific compound in a sample using the spectral information can be reached with suitable chemometric techniques [20]. These techniques have a great impact in the pharmaceutical area, since it deals with the analysis of chemical data of multivariate nature, processed in multichannel equipment, where the intensities of the absorption, scattering, or emission in different wavelengths are recorded in a single spectrum [21]. Chemometrics uses mathematical and statistical methods to design experiments in an optimized way and to provide as much chemical information as possible from the obtained data [20–22].

Chemometric tools commonly used in analytical chemistry include partial least squares regression (PLSR) [23] and principal component regression (PCR) [24]. Both techniques are efficient quantitative methods for dealing with experimental noise, collinearities, and nonlinearities and are sufficiently robust to allow the inclusion of new samples to the model with minimum or no changes in the input parameters. Another advantage of the multivariate methods is the possibility to develop a multivariate calibration model, even with the presence of interferences, whether these are of known or unknown nature [21]. PLSR and PCR are methods intended to model response variables (concentrations) when there are a large number of predictor variables (spectra), and those predictors are highly correlated or even collinear. These methods extract new predictor variables from the dataset, known as components, as linear combinations of the original predictor variables, but each technique extracts those components differently. PCR extracts the components with high variations in the predictor variables, without considering the response variable (concentration), whereas PLSR creates these components taking the response variable into account and therefore can model "complicated" data and is often able to fit the response variable with fewer components [25]. PCR uses the principal component scores (the PCs), which are calculated by applying principal component analysis (PCA) in the dataset composed of spectra taken from the samples in known concentrations, and uses these scores (which can be correlated to the concentrations) and the concentrations to develop a linear regression model to estimate the samples' concentration [23–25]. The number of scores that will be used in the linear model depends on the amount of variation presented in the data set, which is determined by the spectral features present in the principal component loading vectors. The first principal component vectors describe the directions of the maximum variance and are normally related to the spectral features of the API in the drug formulation; therefore, these PCs are normally used in the model.

Optical spectroscopy and multivariate calibration, mainly PLSR and PCR, have been applied in the pharmaceutical field to calculate the amount of API in prepared solutions or directly in the commercial medicines, such as quantitative determination of acetylsalicylic acid and acetaminophen in tablets by FT-Raman using both PCR and PLSR [26]; identification of caffeine, acetaminophen, and acetylsalicylic acid in a mixed solution using ultraviolet absorption and PLSR [22]; the simultaneous quantification of samples containing mixtures of amiloride hydrochloride, atenolol, hydrochlorothiazide, and timolol maleate using classical least squares, PCR, and PLS applied on UV spectral data, evidencing similar results for the PCR and PLS models [27]; the simultaneous determination of sodium dipyrone and papaverine hydrochloride based on PLSR and ultraviolet spectrophotometric measurements in commercial formulation (Melpaz®) [28]; the quantification of diclofenac sodium in tablets and capsules by means of FT-Raman and PLSR, PCR, and counterpropagation artificial neural networks [29]; the quantification of nimesulide in excipient and commercial samples using FT-NIR absorption compared to ultraviolet absorption and the PLSR model [30]; and the novel application of inkjet-printed pharmaceuticals such as haloperidol, used as a model API, and lactic acid and ethanol, used as printable ink bases, applied to three different inorganic and organic substrates and with the quantification of haloperidol performed by the PLS model [31].

Although dipyrone has great relevance in Brazilian clinical practice [32], commercialization of this drug is prohibited in most developed countries in the face of severe adverse reactions that may occur, such as the depressive effect of the bone marrow, which may result in aplastic anemia and, mainly, agranulocytes [33, 34]. There are few studies in the literature that have addressed the quality control of dipyrone production. Therefore, the purpose of this study was to evaluate the feasibility of Raman spectroscopy to assay the API concentration in water solutions containing dipyrone using a spectral model based on principal component regression (PCR) in attending the figures of merit of the Brazilian Agency ANVISA for class I validation (specificity, linearity, range, precision, robustness, and accuracy), in order to assess the ability of Raman spectra to be used as an analytical technique according to the ANVISA and PAT requirements. The developed quantitative model was then employed to predict the concentration of dipyrone in samples of commercial formulations (drop's presentation) containing this API with 500 mg/mL, aiming its use as an analytical technique of choice for quantitative validation.

2. Materials and Methods

2.1. Solution Preparation. Solutions with seven different concentrations of dipyrone were prepared in distilled water (400, 450, 475, 500, 525, 550, and 600 mg/mL), representing the range from 80 to 120% of the concentration found in the commercially available drug containing dipyrone. These samples were prepared in triplicate. Also, 10 flasks of medicines in drops containing dipyrone with 500 mg/mL of API produced by several laboratories (4 generics, 4 similar, and 2 reference brands) were obtained from a local drugstore.

2.2. Raman Spectroscopy. The dipyrone-prepared samples and commercial medicines were submitted to Raman spectroscopy using a benchtop Raman spectrometer (model Dimension P-1, Lambda Solutions Inc., Waltham, MA, USA). The laser is coupled to a fiber-optic-based Raman probe (model Vector Probe, Lambda Solutions Inc., Waltham, MA, USA), as shown in Figure 1. The spectrometer is composed of a diode laser (830 nm), an imaging spectrograph, and a high-sensitivity charge-coupled device (CCD) camera. The system is controlled by a computer. The laser power at the Raman probe tip was adjusted to 250 mW. The dilutions of dipyrone and solutions of commercial medicines were placed in an aluminum sample holder with holes of about 80 μL for Raman measurements, and the Raman signal was collected in 10 scans of 2 s each. All spectra were collected at the same day and experimental conditions in triplicate.

The procedures of spectral calibration and preprocessing (fluorescence background subtraction) were performed using a routine developed under Matlab 7.4 (The MathWorks Inc., Natick, MA, USA). The Raman shift calibration was performed by collecting the spectrum of naphthalene and correlating the spectrum band positions (pixel) with the known band shift (cm^{-1}) with a 3rd-order polynomial fitting. The intensity response of all optical components as a function of the wavelength was obtained using a tungsten calibration lamp. To remove the fluorescence background, the polynomial baseline correction proposed by Lieber and Mahadevan-Jansen [35] was performed, in which a 5th-order polynomial is fitted in the minima of the spectra, and this polynomial is subtracted from the gross spectrum, maintaining the high-frequency Raman bands.

2.3. PCR Quantitative Model. In order to develop a model for the quantitative analysis of API (dipyrone) in commercial solutions, a quantitative model using principal component regression (PCR) was developed using the spectra of the API dilutions in the concentration range from 400 to 600 mg/mL and the nominal concentrations. Then, the linear equation from this model was applied to estimate the concentrations of dipyrone in commercial solutions.

For the PCR model, the principal component analysis (PCA) was initially employed in the spectra of dilutions of 80–120% of dipyrone in water, in order to decompose the spectra onto the principal component loading vectors (LVs) and scores (PCs). In the PCA decomposition, the LVs indicate the spectral features that are important in the diluted samples, and the PCs indicate the intensities of these LVs in each spectrum. PCA was calculated using the function *princomp.m* through the software Matlab 7.4/Statistics Toolbox (The MathWorks Inc., Waltham, MA, USA).

The quantitative PCR model was built using PC1, since it has been demonstrated that LV1 represents 99% of the

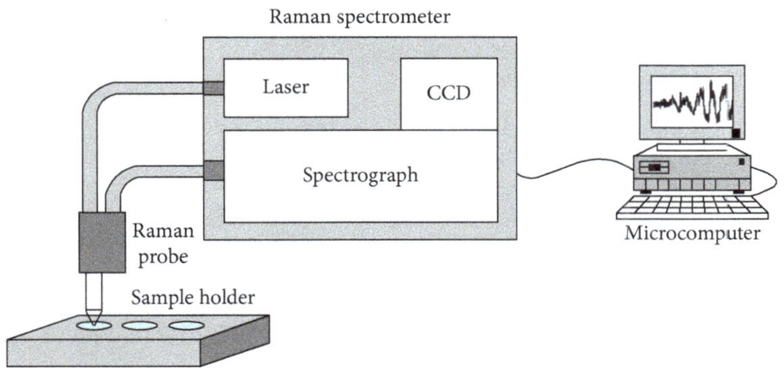

FIGURE 1: Schematic diagram of the dispersive Raman spectrometer used in the experiment. Laser power measured at the Raman probe tip: 250 mW, wavelength: 830 nm, spectrometer resolution in the 400–1800 cm^{-1} range: 2 cm^{-1}.

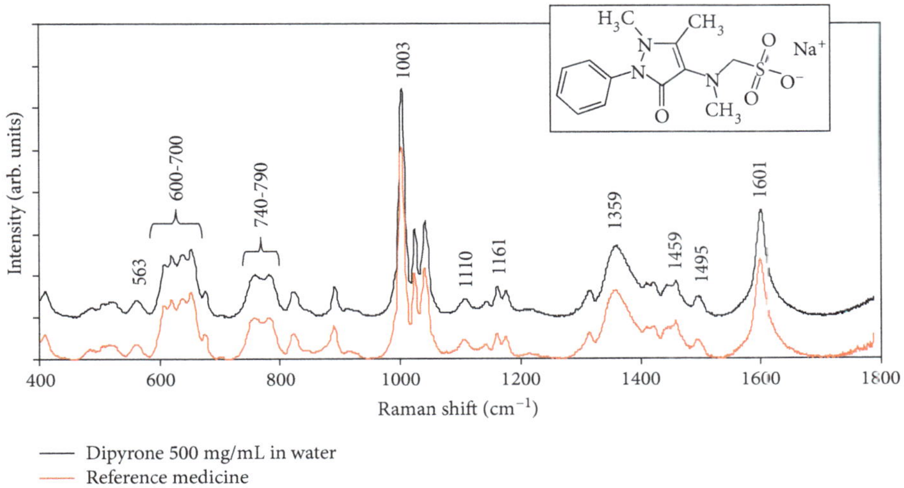

FIGURE 2: Plot of the spectrum of dipyrone diluted in water in the concentration of 500 mg/mL (100% dilution) and the first three principal component loading vectors (LVs), indicating that LV1 is related to the active ingredient dipyrone. LV2 and LV3 do not carry relevant spectral features.

spectral variation in the diluted sample dataset (plot of the spectral information present in the LV1 is seen in Figure 2, Results). For the development of the model, PC1 of the triplicates was plotted versus the concentration of API in the dilution, and the linear equation (PC ordinate, API concentration abscissa) was obtained. Thus, the concentrations of diluted samples were estimated through this equation. These triplicates were used to obtain the figures of merit according to the ANVISA, class I validation and the IUPAC recommendations [36].

Finally, the linear equation was used to estimate the concentration of API in the solutions of commercial medicines. For this, the first step is to obtain the PCs that are present in each sample spectrum of commercial medicine and needed for the ordinate in the regression equation. The PCs were estimated by least-square fitting the first LVs and the sample spectrum, where the fitting coefficients were referred to these "new PCs" of each LV present in each Raman spectrum of the commercial medicines. The fitting was performed using *lsqlin.m* under Matlab 7.4/Optimization Toolbox. PC1 estimated by least-square fitting was used in the linear equation, and the concentration of the API was then obtained.

2.4. Figures of Merit according to ANVISA. The predicted concentrations in the replicate samples were submitted to ANVISA criteria [8], with the figures of merit for class I validation:

(i) Range: in this study, a range of 400 to 600 mg/mL was determined.

(ii) Specificity/selectivity: in order to verify whether the proposed technique is able to identify and quantify dipyrone, even in different excipients, a solution of dipyrone (500 mg/mL) in pure water and dipyrone (500 mg/mL) from the reference commercial medicine (3 replicates of each) was considered.

(iii) Linearity: the curve slope (m), which determines the proportionality of the analytical function, the linear coefficient (h), which is used to represent the correlation, and the correlation coefficient (r) were calculated. The criterion for acceptance is $r = 0.99$.

(iv) Precision: for the study, the concentrations 400, 500, and 600 mg/mL were used in triplicate, totaling nine solutions. The precision was expressed as relative standard deviation (RSD), according to

$$\text{RSD} = \frac{\text{SD}}{\text{DAC}} \times 100, \quad (1)$$

where SD = standard deviation and DAC = determined average concentration of the model's measures. SD is calculated as

$$\text{SD} = \sqrt{\frac{\sum_{i=1}^{n}(\text{VO} - \text{DAC})^2}{n-1}}, \quad (2)$$

where VO = value obtained in the test, DAC = determined average concentration, and n = number of samples. For the ANVISA, the maximum RSD for the method should not exceed 5%.

(v) Accuracy (recovery): this was determined from nine determinations in the range of the procedure, that is, concentrations of 400, 500, and 600 mg/mL, in triplicate. The equation used to calculate the accuracy (AC) is

$$\text{AC} = \frac{\text{DAC}}{\text{VA}} \times 100, \quad (3)$$

where VA is value added.

3. Results

3.1. Raman Spectra of Diluted Solutions and Development of the Spectral Model. The Raman spectrum of dipyrone diluted in water at a concentration of 500 mg/mL and the spectrum of a reference medicine are shown in Figure 3. Both spectra show the vibrational bands of dipyrone (labeled peaks), and band assignments are shown in Table 1. No observable or relevant excipient peak was observed.

The data set of the diluted samples was submitted to PCA, and the output variables (principal component loading vectors—LVs—and scores—PCs) were obtained. Figure 2 presents the first 3 LVs and the spectra of dipyrone diluted in water, showing that vector 1 is strongly related to the active ingredient dipyrone. Therefore, PC1 can be used to estimate the concentration of dipyrone in a linear regression model. This model, presented in Figure 4(a), shows the correlation between the concentration of dipyrone in each dilution and the intensity of PC1 in each concentration (in triplicate). This linear equation ($y = 0.0003322x - 0.01034$), a regression curve of PC1 versus concentration, was then used to estimate the concentration in the same samples (Figure 4(b)). An excellent agreement of the data points can be seen with the zero-error diagonal, revealing an excellent model fit.

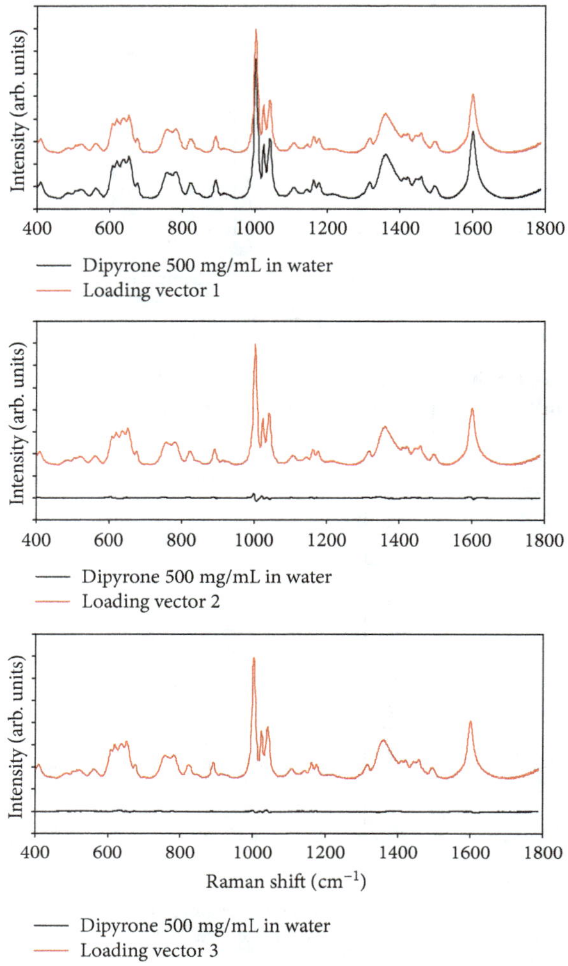

Figure 3: Raman spectra of a solution of 500 mg/mL of dipyrone diluted in distilled water (100% dilution) and the reference commercial solution of dipyrone 500 mg/mL. Table 1 presents the labeled peaks in the reference solution that are related to dipyrone and their assignments. Inner plot: metamizole chemical formula {[(1,5-dimethyl-3-oxo-2-phenylpyrazol-4-yl)-methylamino] methanesulfonic acid sodium salt}.

By applying the PCR model in the Raman spectra of triplicate samples, it is possible to calculate the predicted concentration and provide a root mean squared error (RMSE) of prediction [41], to which the model is subject. The RMSE obtained in this correlation curve was 6.5 mg/mL (1.3%). The resulting prediction error is below that preconized by the Brazilian Sanitary Surveillance Agency [8], which is 5%.

3.2. Figures of Merit for the PCR Model. The results of ANVISA figures of merit for the ANVISA class I parameters are presented.

(i) Interval: The range of 400 to 600 mg/mL (80 to 120%) was established.

TABLE 1: Assignments of the main Raman bands of dipyrone found in Figure 3 based on the literature of vibrational spectroscopy.

Peak position (cm^{-1})	Assignment
563	SO_2 scissor deformation (sulfone) [37]
600–700	Ring in-plane and out-of-plane deformations (phenyl group)
740–790	C-H out-of-plane bending (phenyl group) [38]
1003	C=C stretching and angular CCC in-plane ring deformation (symmetric stretching—"breathing" of the phenyl ring) [37, 39]
1100	SO_2 symmetric stretch (sulfone) [37]
1359	SO_2 antisymmetric stretch (sulfone) [37]
1459 1495	C=C ring stretching (degenerate vibrations pair of the phenyl ring, associated to the symmetric out-of-phase and antisymmetric stretching of the two triatomic systems); N-CH_3 deformation (amine) [38, 40]
1601	C=O and C=C stretching (phenyl group); HCH deformation (CH_3 an CH_2 groups) [39]

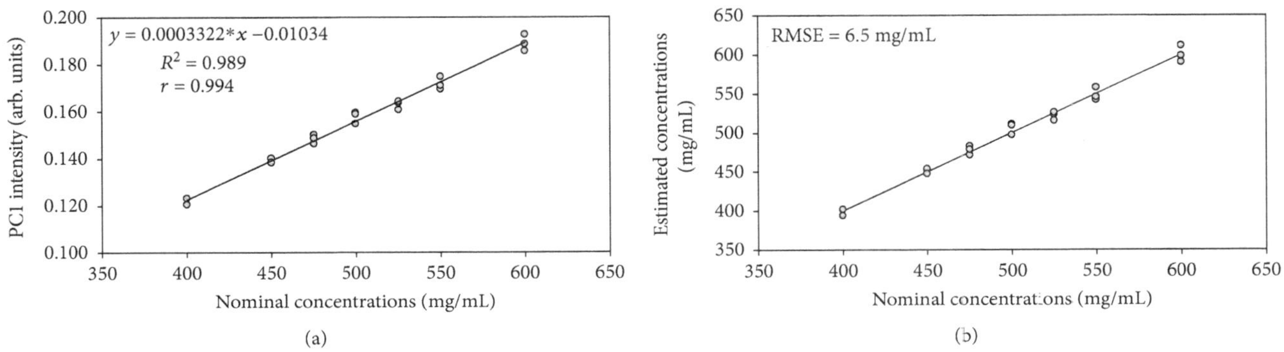

FIGURE 4: (a) Plot of the nominal concentrations of the seven dilutions of dipyrone, in the range from 400 to 600 mg/mL (80 to 120%) versus the intensities of PC1, showing high correlation of the PCR model ($r = 0.994$). (b) Plot of the model-estimated concentrations versus the nominal concentrations and estimated RMSE = 6.5 mg/mL (1.3%).

(ii) Specificity/selectivity: The largest relative error of dipyrone in full excipient compared to dipyrone in water was 3.3%, showing that the technique allows the detection of the major component of the sample in different excipients (Table 2).

(iii) Linearity: The concentrations of five solutions were determined at different concentrations of the drug in water (400, 450, 475, 500, 525, 550, and 600 mg/mL) and with the concentrations of the real dilutions. These data are shown in Figure 4(a). The curve obtained to verify the linearity was $y = 0.0003322x - 0.01034$ ($y = mx + b$, where m is the slope and b is the intercept of the line with the vertical axis). The linear regression coefficient obtained was $r = 0.994$, indicating a high correlation between the dilution effectively performed and the predicted concentration.

(iv) Precision (repeatability): Three concentrations (low, intermediate, and high) were used with three replicate samples of the following concentrations: 400, 500, and 600 mg/mL. RSD was calculated for each concentration obtained by the PCR model in comparison with the real dilution values. ANVISA criteria accept values of precision (RSD) $\leq 5\%$. Table 3 presents the data used to calculate the RSD for the solutions of low, intermediate, and high concentrations (400, 500, and 600 mg/mL). It was found that all three concentrations presented an RSD below the maximum limit specified by ANVISA, ranging from 0.95 to 1.76%.

(v) Accuracy: this was calculated by using the predicted concentrations of 400, 500, and 600 mg/mL, in triplicate, being determined by means of the relative deviation (RD) and recovery, as shown in Table 3. The acceptance criteria established were RD $\leq 2\%$, and satisfactory recovery values are found within the range from 98 and 102% [8]. The solutions showed a relative deviation lower than 1.5%, and the recovery showed values ranging from 99.2% to 101.2%, being within the limits of acceptance.

(vi) Robustness: Due to the lack of evaluation of the influence of hydrogen potential and temperature on the results of the present study, the robustness of the method was not estimated.

Table 2: Predicted concentrations of dipyrone 500 mg/mL in water and in the reference commercial pharmaceutical preparation using the PCR model and the relative errors used to calculate the selectivity.

Sample	Predicted concentration of dipyrone diluted in water (mg/mL)	Predicted concentration of dipyrone in the reference commercial pharmaceutical preparation (mg/mL)	Relative error (%)
1	511	505	1.2%
2	497	485	2.4%
3	509	492	3.3%

Table 3: Solutions of different concentrations in triplicate (400, 500, and 600 mg/mL) used to calculate the precision (RSD) and accuracy (AC).

Real concentration (mg/mL)	VO RS/PCR (mg/mL)	DAC (mg/mL)	SD (mg/mL)	RSD (%)	RD (%)	AC (%)
	395					
400	394	397	3.81	0.95	0.75	99.2
	401					
	511					
500	497	506	7.58	1.49	1.2	101.2
	509					
	611					
600	598	600	10.61	1.76	0	100.0
	590					

VO: value obtained in the assay test of RS/PCR; RS/PCR: Raman spectroscopy combined with chemometric method PCR; DAC: determined average concentration; RD: relative deviation.

3.3. Quantitative Model Applied to the Commercial Formulations. The PCR model has been applied to estimate the concentration of API in the commercial products. The PCs of each medicine sample spectrum were estimated by least-square fitting the first three LVs and the medicine spectra, fitting the coefficient of PCs. Since PC1 was used for the PCR model applied to the samples of dipyrone diluted in water, the concentration of dipyrone in each medicine sample was obtained by estimating x in the linear equation of Figure 4(a) [$x = (y + 0.01034)/0.003322$]. The predicted concentrations of the 10 samples of commercial medicines through the PCR model were correlated to the concentrations presented in the label (Figure 5), and all samples fell below the ±5% error line, indicating excellent results in determining the API concentration.

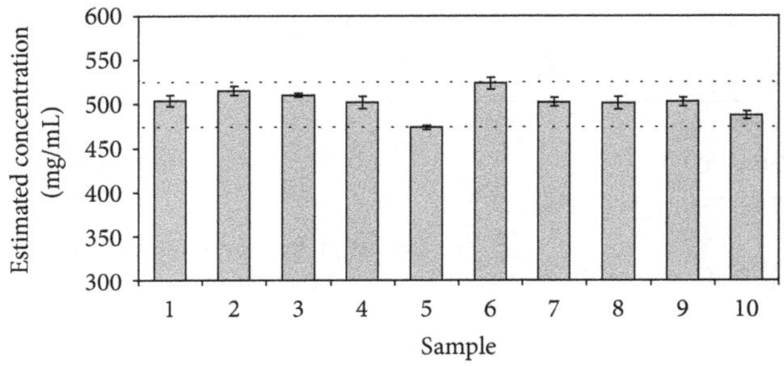

Figure 5: Estimated concentration values using the PCR model applied to the Raman spectra of the commercial solutions containing dipyrone 500 mg/mL. The error bars are based on the triplicate spectra. The ±5% precision limit of ANVISA is shown.

4. Discussion

The pharmaceutical industry is seeking methods to better track drugs throughout the production process, since most of the high-performance techniques currently used require sample preparation using solvents, are destructive and time-consuming, and mostly need expensive inputs [8, 16, 42]. The use of PAT in the pharmaceutical industry and in biotechnological processes led to advances regarding the ability to analyze/monitor key process and quality attributes in the biotech industry, to utilize the collected data for the subsequent control of the process, in order to achieve the optimum yield and product quality [43]. The literature indicates that Raman spectroscopy, combined with chemometric methods such as PLSR and PCR, would meet the challenges for a reliable quantitative determination of API in pharmaceutical drugs [44, 45].

The present study was based on the spectral model to assay dipyrone in commercial medicines, in order to evaluate this optical technique as a method for on-line and at-line analysis, in accordance with the PAT requirements for an industrial analysis system. PCR has been employed for quantitative measurements in chemical analysis when applied to spectroscopic methods [24, 25]. In complex and multivariate systems, such as API in medicines and formulations, the classical univariate regression method may become unreliable because of possible band interactions with the solvents and excipients and band superimpositions. The PCR models all variables that are correlated to the concentration of the API, and thus, the calibration can be performed efficiently, even in the presence of interferences using the whole of the spectral information. With the PCR model, new samples can be modeled at any time. The PCR model showed a smaller prediction error when using the first principal component score. In this way, it was verified that only one spectral variation was found in the samples, which is related to changes in the dipyrone concentration.

Since the intensity of Raman peaks is directly proportional to the concentration of the components on the scattering sample [46], Raman could be used to develop linear models for the evaluation of a medicine's composition. Since the intensity of a Raman band is dependent on the changes in the molecule's polarizability presented in the sample, this has a relevant consequence in investigating pharmaceutical materials because active ingredients are organic molecules that contain covalent bonds and usually aromatic conjugated systems, which show strong Raman transitions, in contrast to the excipients in most formulations, which are generally weaker Raman scatterers due to a lack of C bonds.

In this work, PCR was validated by evaluating the ANVISA RE no. 899/2003 figures of merit such as specificity, linearity, precision, and accuracy in the same way as Orkoula et al. [47], Rossignoli et al. [48], and Borio et al. [49]. The PCR model showed excellent linearity with a correlation coefficient $r = 0.994$, high precision, and RSD as low as 1.76%. Corroborating the method validation employed by Orkoula et al. [47], commercial samples of the reference drug containing 500 mg/mL of dipyrone were used to validate the model. These samples showed very low errors, not exceeding 2.5%, suggesting that dispersive Raman spectroscopy using an optical probe can detect the amount of API present in the commercial samples, and all samples conformed, which is different to the study of Knappmann and Melo [32].

Although dipyrone has been used for more than 90 years, research about quality control, harm, and health risk of this active ingredient is scarce [50]. Dipyrone concentration has been evaluated in several commercial drugs containing dipyrone by FT-Raman spectroscopy [39]; the results showed agreement with the percentages of dipyrone present in tablets. In Brazil, the quality control of medicines is restricted to certain government laboratories [39]. Despite this, Pereira et al. [51] performed the spectrophotometric determination of dipyrone in commercial samples by flow injection triiodide generation, with sixty results obtained per hour. The results found were within the 3% error range at the 95% confidence level. The quality control of distinct brands of dipyrone in oral solutions, marketed in Brazilian drugstores, was also determined by Knappmann and Melo [32], and the results indicated that the quality control of similar medicines needs to be improved, since similar drugs were the ones that presented the most deviations of quality, being disapproved in the study trial.

Dispersive Raman spectroscopy has been considered the most suitable technique to measure API in different pharmaceutical preparations, with advantages such as the ability to obtain a spectrum with a high signal-to-noise ratio using a low-power near-infrared laser with reduced exposure time. In addition, the entire spectrum is obtained at once through a multichannel detector and imaging spectrograph.

According to the PAT guide, the gains in processes quality, safety, and/or efficiency can be increased depending on the product; these changes come from factors such as the prevention of rejections, the disposal and reprocessing of products, and the possibility of carrying out analysis in real time with reduced or no sample preparation [5, 52], leading to a reduction of human error and increasing the safety of the operator, with no risk to the environment [9, 10, 53]. The PCR model could address most of these factors, because each sample can be irradiated rapidly and examined directly with the use of an optical probe, without prior preparation, to obtain a single spectrum, which minimizes human error. Furthermore, the use of Raman spectroscopy as an analytical technique offers the possibility of measurements in real time at-line, avoiding the storage time required until the medicine's stock release, preventing possible rejections/disposal by furnishing actual concentration values, and even measuring the concentration through containers by standard Raman and offset spatially Raman spectroscopy [54–56].

5. Conclusions

This study demonstrated the potential of dispersive Raman spectroscopy to attend ANVISA figures of merit for class I validation (specificity, linearity, range, precision, and accuracy) for evaluation of the concentration of dipyrone in commercial solutions. An effectiveness of the quantitative method was found, with high linearity, specificity of 3.3%, precision < 2%, and relative deviation (accuracy) < 1.5%, with recovery values in the range of 99.2% to 101.2% and RMSE of 6.5 mg/mL (1.3%) for quantifying dipyrone diluted in water. For the quantification of API in the commercial formulation, the model showed concentration values closer to 500 mg/mL, with an error of less than 2.5% compared to the labeled concentration. It has been demonstrated that the PCR model could support the industry's applications of monitoring and quality control, rapidly and without any sample preparation, attending the PAT requirements of the pharmaceutical industry for quality assurance of the final product due to the security and agility provided.

Conflicts of Interest

The authors declare that there is no conflict of interests regarding the publication of this article.

Acknowledgments

Landulfo Silveira Jr. thanks FAPESP (São Paulo Research Foundation) for the grant for the Raman spectrometer (Grant no. 2009/01788-5) and CNPq (National Council for Scientific and Technological Development) (Grant no. 305680/2014-5) for the productivity fellowship. Letícia Parada Moreira and Bárbara Faria Lourenço would like to acknowledge CAPES (Coordenação de Aperfeiçoamento de Pessoal de Nível Superior) for the scholarships.

References

[1] J. Konijnenbelt-Peters, C. Van der Heijden, C. Ekhart, J. Bos, J. Bruhn, and C. Kramers, "Metamizole (dpyrone) as an alternative agent in postoperative analgesia in patients with contraindications for nonsteroidal anti-inflammatory drugs," *Pain Practice*, vol. 17, no. 3, pp. 402–408, 2017.

[2] A. Jasiecka, T. Maślanka, and J. J. Jaroszewski, "Pharmacological characteristics of metamizole," *Polish Journal of Veterinary Sciences*, vol. 17, no. 1, pp. 207–214, 2014.

[3] L. Lachman, H. A. Lieberman, and J. L. Kanig, *The Theory and Practice of Industrial Pharmacy*, Lea & Febiger, Philadelphia, PA, USA, 3rd edition, 1986.

[4] Brazil, *Brazilian Pharmacopoeia*, Brazilian Sanitary Surveillance Agency, Brazil, 5th edition, 2010.

[5] D. C. Hinz, "Process analytical technologies in the pharmaceutical industry: the FDA's PAT initiative," *Analytical and Bioanalytical Chemistry*, vol. 384, no. 5, pp. 1036–1042, 2006.

[6] F. Adar, R. Geiger, and J. Noonan, "Raman spectroscopy for process/quality control," *Applied Spectroscopy Reviews*, vol. 32, no. 1-2, pp. 45–101, 1997.

[7] G. J. Fini, "Applications of Raman spectroscopy to pharmacy," *Journal of Raman Spectroscopy*, vol. 35, no. 5, pp. 335–337, 2004.

[8] Brazil, Resolution RE No. 899/2003, "Ministry of Health. Brazilian Sanitary Surveillance Agency (ANVISA), 2003," October 2017, https://anvisa.gov.br/legis/resol/2003/re/899_03re.htm.

[9] Food and Drug Administration, International Conference on Harmonization, "Guideline on Validation of Analytical Procedures: Definitions and Terminology; Availability," *Federal Register*, vol. 60, no. 40, pp. 11260–11262, 1995.

[10] Food and Drug Administration, International Conference on Harmonization, "Guideline on the Validation of Analytical Procedures: Methodology; Availability," *Federal Register*, vol. 62, no. 96, pp. 27463–27467, 1997.

[11] A. W. Newman and S. R. Byrn, "Solid-state analysis of the active pharmaceutical ingredient in drug products," *Drug Discovery Today*, vol. 8, no. 19, pp. 898–905, 2003.

[12] M. Blanco, M. Bautista, and M. Alcalà, "API determination by NIR spectroscopy across pharmaceutical production process," *AAPS PharmSciTech*, vol. 9, no. 4, pp. 1130–1135, 2008.

[13] O. Kolomiets, U. Hoffmann, P. Geladi, and H. W. Siesler, "Quantitative determination of pharmaceutical drug formulations by near-infrared spectroscopic imaging," *Applied Spectroscopy*, vol. 62, no. 11, pp. 1200–1208, 2008.

[14] C. J. Strachan, T. Rades, K. C. Gordon, and J. Rantanen, "Raman spectroscopy for quantitative analysis of pharmaceutical solids," *Journal of Pharmacy and Pharmacology*, vol. 59, no. 2, pp. 179–192, 2007.

[15] T. Vankeirsbilck, A. Vercauteren, W. Baeyens et al., "Applications of Raman spectroscopy in pharmaceutical analysis," *TrAC Trends in Analytical Chemistry*, vol. 21, no. 12, pp. 869–9877, 2002.

[16] C. Eliasson, N. A. Macleod, L. C. Jayes et al., "Non-invasive quantitative assessment of the content of pharmaceutical capsules using transmission Raman spectroscopy," *Journal of Pharmaceutical and Biomedical Analysis*, vol. 47, no. 2, pp. 221–229, 2008.

[17] B. B. Koleva, T. M. Kolev, D. L. Tsalev, and M. Spiteller, "Determination of phenacetin and salophen analgesics in solid binary mixtures with caffeine by infrared linear dichroic and Raman spectroscopy," *Journal of Pharmaceutical and Biomedical Analysis*, vol. 46, no. 2, pp. 267–273, 2008.

[18] E. B. Hanlon, R. Manoharan, T. W. Koo et al., "Prospects for *in vivo* Raman spectroscopy," *Physics in Medicine & Biology*, vol. 45, no. 2, pp. R1–R59, 2000.

[19] M. S. Hwang, S. Cho, H. Chung, and Y. A. Woo, "Nondestructive determination of the ambroxol content in tablets by Raman spectroscopy," *Journal of Pharmaceutical and Biomedical Analysis*, vol. 38, no. 2, pp. 210–215, 2005.

[20] P. R. Mobley, B. R. Kowalski, J. J. Workman, and R. Bro, "Review of chemometrics applied to spectroscopy: 1985-95, Part 2," *Applied Spectroscopy Reviews*, vol. 31, no. 4, pp. 347–368, 1996.

[21] M. M. C. Ferreira, A. M. Antunes, M. S. Melgo, and P. L. O. Volpe, "Chemometrics I: multivariate calibration, a tutorial," *Química Nova*, vol. 22, no. 5, pp. 724–731, 1999.

[22] M. M. Sena and R. J. Poppi, "N-way PLS applied to simultaneous spectrophotometric determination of acetylsalicylic acid, paracetamol and caffeine," *Journal of Pharmaceutical and Biomedical Analysis*, vol. 34, no. 1, pp. 27–34, 2004.

[23] P. Geladi and B. R. Kowalski, "Partial least-squares regression: a tutorial," *Analytica Chimica Acta*, vol. 185, pp. 1–17, 1986.

[24] S. Boneh and G. R. Mendieta, "Regression modeling using principal components," in *4th Annual Conference on Applied Statistics in Agriculture*, Manhattan, KS, USA, September 19922017, http://newprairiepress.org/agstatconference/1992/proceedings/20.

[25] "The Mathworks Documentation, Partial Least Squares Regression and Principal Components Regression," September 2017, https://www.mathworks.com/help/stats/examples/partial-least-squares-regression-and-principal-components-regression.html.

[26] R. Szostak and S. Mazurek, "Quantitative determination of acetylsalicylic acid and acetaminophen in tablets by FT-Raman spectroscopy," *Analyst*, vol. 127, no. 1, pp. 144–148, 2002.

[27] M. C. Ferraro, P. M. Castellano, and T. S. Kaufman, "Chemometric determination of amiloride hydrochloride, atenolol, hydrochlorothiazide and timolol maleate in synthetic mixtures and pharmaceutical formulations," *Journal of Pharmaceutical and Biomedical Analysis*, vol. 34, no. 2, pp. 305–314, 2004.

[28] A. P. do Nascimento, M. G. Trevisan, E. R. M. Kedor-Hackmann, and R. J. Poppi, "Simultaneous determination of dipyrone and papaverine in pharmaceutical formulation using PLS regression and UV spectrophotometry," *Analytical Letters*, vol. 40, no. 5, pp. 975–986, 2007.

[29] S. Mazurek and R. Szostak, "Quantitative determination of diclofenac sodium in solid dosage forms by FT-Raman spectroscopy," *Journal of Pharmaceutical and Biomedical Analysis*, vol. 48, no. 3, pp. 814–821, 2008.

[30] W. F. C. Rocha, A. L. Rosa, J. A. Martins, and R. J. Poppi, "Determination and validation of nimesulide in pharmaceutical formulation by near infrared spectroscopy," *Journal of the Brazilian Chemical Society*, vol. 21, no. 10, pp. 1929–2010, 2010.

[31] M. Edinger, D. Bar-Shalom, J. Rantanen, and N. Genina, "Visualization and non-destructive quantification of inkjet-printed pharmaceuticals on different substrates using Raman spectroscopy and Raman chemical imaging," *Pharmaceutical Research*, vol. 34, no. 5, pp. 1023–1036, 2017.

[32] A. L. Knappmann and E. B. D. Melo, "Quality of over-the-counter medicines: a study with dipyrone brands commercialized in a drugstore in Cascavel city (Paraná, Brazil)," *Ciência & Saúde Coletiva*, vol. 15, suppl 3, pp. 3467–3476, 2010.

[33] F. Arellano and J. A. Sacristán, "Metamizole: reassessment of its therapeutic role," *European Journal of Clinical Pharmacology*, vol. 38, no. 6, pp. 617–619, 1991.

[34] B. Baro-Oz, M. Clementi, E. Di Giantonio et al., "Metamizol (dipyrone, optalgin) in pregnancy, is it safe? A prospective comparative study," *European Journal of Obstetrics & Gynecology and Reproductive Biology*, vol. 119, no. 2, pp. 176–179, 2005.

[35] C. A. Lieber and A. Mahadevan-Jansen, "Automated method for subtraction of fluorescence from biological Raman spectra," *Applied Spectroscopy*, vol. 57, no. 11, pp. 1363–1367, 2003.

[36] A. C. Olivieri, N. M. Faber, J. Ferré, R. Boqué, J. H. Kalivas, and H. Mark, "Uncertainty estimation and figures of merit for multivariate calibration (IUPAC Technical Report)," *Pure and Applied Chemistry*, vol. 78, no. 3, pp. 633–661, 2006.

[37] N. Colthup, L. Daly, and S. Wiberley, *Introduction to Infrared and Raman Spectroscopy*, Academic Press, London, 1990.

[38] D. W. Mayo, F. A. Miller, and R. W. Hannah, *Course Notes on the Interpretation of Infrared and Raman Spectra*, John Wiley & Sons, Hoboken, NJ, USA, 2004.

[39] A. O. Izolani, M. T. de Moraes, S. Téllez, and C. A. Claudio, "Fourier transform Raman spectroscopy of drugs: quantitative analysis of 1-phenyl-2,3-dimethyl-5-pyrazolone-4-methylaminomethane sodium sulfonate: (dipyrone)," *Journal of Raman Spectroscopy*, vol. 34, no. 10, pp. 837–843, 2003.

[40] D. Lin-Vien, N. B. Colthup, W. G. Fateley, and J. G. Grasselli, *The Handbook of Infrared and Raman Characteristic Frequencies of Organic Molecules*, Academic Press, San Diego, CA, USA, 1991.

[41] H. Martens and M. Martens, *Multivariate Analysis of Quality: An Introduction*, John Wiley & Sons, West Sussex, 2001.

[42] B. A. Olsen, "Developing and using analytical methods to achieve quality by design and efficiency in drug development," *Pharmaceutical Technology*, vol. 1, pp. S14–S25, 2005.

[43] A. S. Rathore, R. Bhambure, and V. Ghare, "Process analytical technology (PAT) for biopharmaceutical products," *Analytical Bioanalytical Chemistry*, vol. 398, no. 1, pp. 137–154, 2010.

[44] M. Kim, H. Chung, Y. Woo, and M. S. Kemper, "A new non-invasive, quantitative Raman technique for the determination of an active ingredient in pharmaceutical liquids by direct measurement through a plastic bottle," *Analytica Chimica Acta*, vol. 587, no. 2, pp. 200–207, 2007.

[45] E. Katainen, M. Elomaa, U. M. Laakkonen et al., "Quantification of the amphetamine content in seized street samples by Raman spectroscopy," *Journal of Forensic Sciences*, vol. 52, no. 1, pp. 88–92, 2007.

[46] European Pharmacopoeia 8.7, *Chapter 2.2.48: Raman spectroscopy*, European Directorate for the Quality of Medicines & HealthCare, Council of Europe, Strasbourg, 2016, p. 8.7-5464.

[47] M. G. Orkoula, C. G. Kontoyannis, C. K. Markopoulou, and J. E. Koundourellis, "Quantitative analysis of liquid formulations using FT-Raman spectroscopy and HPLC: The case of diphenhydramine hydrochloride in Benadryl," *Journal of Pharmaceutical and Biomedical Analysis*, vol. 41, no. 4, pp. 1406–1411, 2006.

[48] P. Rossignoli, R. Pontarolo, C. J. Correr, G. Cordeiro, and P. Peralta-Zamora, "Multivariate spectroscopic determination of glucosamine and chondroitin in pharmaceutical formulations," *Química Nova*, vol. 31, no. 6, pp. 1285–1289, 2008.

[49] V. G. Borio, R. Vinha, R. A. Nicolau, H. P. M. de Oliveira, C. J. de Lima, and L. Silveira, "Quantitative evaluation of acetaminophen in oral solutions by dispersive Raman spectroscopy for quality control," *Spectroscopy: An International Journal*, vol. 27, no. 4, pp. 215–228, 2012.

[50] T. Kötter, B. R. da Costa, M. Fässler et al., "Metamizole-associated adverse events: a systematic review and meta-analysis," *PLoS One*, vol. 10, no. 4, article e0122918, 2015.

[51] A. V. Pereira, L. Penckowski, M. Vosgerau, M. F. Sassá, and O. Fatibello Filho, "Spectrophotometric determination of metamizol in pharmaceutical formulations by flow injection triiodide generation," *Química Nova*, vol. 25, no. 4, pp. 553–557, 2002.

[52] Food and Drug Administration, "Guidance for industry PAT - a framework for innovative pharmaceutical development, manufacturing, and quality assurance, 2004," October 2017 https://www.fda.gov/downloads/drugs/guidances/ucm070305.pdf.

[53] R. P. Cogdill, C. A. Anderson, M. Delgado et al., "Process analytical technology case study: part II. Development and validation of quantitative near-infrared calibrations in support of a process analytical technology application for real-time release," *AAPS PharmSciTech*, vol. 6, no. 2, pp. E273–E283, 2005.

[54] I. Malik, M. Poonacha, J. Moses, and R. A. Lodder, "Multispectral imaging of tablets in blister packaging," *AAPS PharmSciTech*, vol. 2, no. 2, pp. 38–44, 2001.

[55] P. De Peinder, M. J. Vredenbregt, T. Visser, and D. J. De Kaste, "Detection of Lipitor® counterfeits: a comparison of NIR and Raman spectroscopy in combination with chemometrics," *Journal of Pharmaceutical and Biomedical Analysis*, vol. 47, no. 4-5, pp. 688–694, 2008.

[56] K. Dégardin, A. Guillemain, and Y. Roggo, "Comprehensive study of a handheld Raman spectrometer for the analysis of counterfeits of solid-dosage form medicines," *Journal of Spectroscopy*, vol. 2017, Article ID 3154035, 13 pages, 2017.

17

Near-Infrared Spectroscopy Based on Supervised Pattern Recognition Methods for Rapid Identification of Adulterated Edible Gelatin

Hao Zhang,[1] Haifeng Sun,[1] Ling Wang,[1] Shun Wang,[1] Wei Zhang,[1] and Jiandong Hu[1,2]

[1]College of Mechanical and Electrical Engineering, Henan Agricultural University, Zhengzhou 450002, China
[2]State Key Laboratory of Wheat and Maize Crop Science, Zhengzhou 45002, China

Correspondence should be addressed to Hao Zhang; hao.zhang2016@hotmail.com

Academic Editor: Arnaud Cuisset

The aim of this work is to identify the adulteration of edible gelatin using near-infrared (NIR) spectroscopy combined with supervised pattern recognition methods. The spectral data obtained from a total of 144 samples consisting of six kinds of adulterated gelatin gels with different mixture ratios were processed with multiplicative scatter correction (MSC), Savitzky–Golay (SG) smoothing, and min-max normalization. Principal component analysis (PCA) was first carried out for spectral analysis, while the six gelatin categories could not be clearly distinguished. Further, linear discriminant analysis (LDA), soft independent modelling of class analogy (SIMCA), backpropagation neural network (BPNN), and support vector machine (SVM) were introduced to establish discrimination models for identifying the adulterated gelatin gels, which gave a total correct recognition rate of 97.44%, 100%, 97.44%, and 100%, respectively. For the SIMCA model with significant level $\alpha = 0.05$, sample overlapping clustering appeared; thus, the SVM model presents the best recognition ability among these four discrimination models for the classification of edible gelatin adulteration. The results demonstrate that NIR spectroscopy combined with unsupervised pattern recognition methods can quickly and accurately identify edible gelatin with different adulteration levels, providing a new possibility for the detection of industrial gelatin illegally added into food products.

1. Introduction

Edible gelatin is produced by hydrolyzed collagen extracted from fresh skin and bones of animals (such as pig, cattle, or fish) through a series of complicated procedures, which is rich in 18 kinds of amino acids needed by the human body. Due to its high protein, no fat, and no cholesterol, edible gelatin is widely used as a food thickener in the food industry (such as jelly, yogurt, and galantine) [1]. Industrial gelatin is commonly made from waste leather by decoloring, bleaching, and washing, which requires a large amount of chrome-containing tanning agents, so the heavy metal chromium content in industrial gelatin seriously exceeds the standard. Once the human body ingests a large amount of heavy metal chromium, it will cause kidney damage, and in serious cases, may lead to cancer [2]. Therefore, the Chinese government prohibits industrial gelatin from being used in food and pharmaceutical products, and the national standards for "edible gelatin" clearly stipulate that the maximum allowable limit of the chromium content in edible gelatin is 2.0 mg/kg. However, due to the low price and simple production of industrial gelatin, some unscrupulous merchants added industrial gelatin to food products instead of edible gelatin [3]. In China, food safety accident is in high incidence, such as the "poison capsule" and "old yogurt" incidents reported in 2012, as well as the "poison bean jelly" incident reported in 2014. At present, there is no standard detection method for industrial gelatin illegally added in food products, so a fast and nondestructive method is needed to identify adulteration of edible gelatin.

Currently, some analytical methods for the identification of gelatin in products include the electrophoretic method

[4, 5], enzyme linked immune sorbent assay (ELISA) [6–8], high performance liquid chromatography (HPLC) [9–11], and polymerase chain reaction (PCR) [12–15]. However, most of these methods are expensive and time consuming although the sensitivity and accuracy are high. In recent years, Fourier transmission infrared spectroscopy (FTIRS) has been used as a nondestructive spectroscopic analytical method to differentiate different gelatin origins, for example, Hashim et al. [16] combined FTIR with attenuated total reflectance (ATR) to obtain a clear distinction between gelatin samples of bovine and porcine origins, and Cebi et al. [17] used FTIRS combined with principal component-cluster analysis to achieve a better distinction between pig gelatin, bovine gelatin, and fish gelatin.

However, most of these methods were used to differentiate gelatin origins by combining with chemometric methods; few studies were performed to explore if the edible gelatin was adulterated with the toxic industrial gelatin. As a fast and nondestructive method, NIR spectroscopy has been widely used in the detection of adulteration identification of agricultural products and foods [18–22]. In this paper, NIR spectroscopy combined with four kinds of supervised pattern recognition methods was used to study the feasibility of identifying edible gelatin adulterated with industrial gelatin. For this purpose, NIR spectra expressed as absorbance were measured on six kinds of gelatin gels with different adulteration ratios, principle component analysis (PCA) was first used for exploratory analysis, and then discrimination models based on linear discriminant analysis (LDA), soft independent modelling of class analogy (SIMCA), back-propagation neural network (BPNN), and support vector machine (SVM) methods were developed, where two parameters of recognition rate and rejection rate were used to assess the predictive performance of discrimination models.

2. Materials and Methods

2.1. Sample Preparation. The edible gelatin and industrial gelatin used in the experiment were purchased from Henan Boyang Biotechnology Co., Ltd (see Table S1). Each sample was weighed 10 g to prepare for gelatin gels with six different mixture ratios, i.e., 10 g edible gelatin, 8 g edible gelatin + 2 g industrial gelatin, 6 g edible gelatin + 4 g industrial gelatin, 4 g edible gelatin + 6 g industrial gelatin, 2 g edible gelatin + 8 g industrial gelatin, and 10 g industrial gelatin, which were denoted by pure edible gelatin (Class 1), 4 : 1 adulterated gelatin (Class 2), 3 : 2 adulterated gelatin (Class 3), 2 : 3 adulterated gelatin (Class 4), 1 : 4 adulterated gelatin (Class 5), and pure industrial gelatin (Class 6), respectively, as shown in Table S2 and Figure S1.

The preparation process of gelatin gels is as follows: (1) the mixed 10 g gelatin sample was placed in a 250 mL beaker and added with 100 mL purified water to soak for about half an hour; (2) the beaker was placed in a water bath pan for heating by setting the temperature at 50°C, and the solution was stirred while heating until the gelatin was completely dissolved; (3) after cooling to room temperature, the gelatin solution was poured into a small square mold and then placed in a refrigerator to cool about 5 hours for shaping.

2.2. Spectral Acquisition. The NIR measuring setup used in this experiment was constructed with a visible near-infrared halogen lamp light source, an optical fiber adjustable attenuator, two multimode optical fiber with core diameter of 600 μm, a transflection mount, an NIR spectrometer (NIR Quest 256-2.5, Ocean Optics, USA), and a computer. The wavelength range of the light source is 350–2400 nm, and the spectral range of the NIR spectrometer is 900–2500 nm. Data points were collected every 3 nm, with 512 data points for each spectral curve. In the experiment, the NIR absorption spectrum was measured in a transmission geometry and expressed as absorbance for spectral analysis. Each sample was measured at 5 different positions and its average spectrum was taken as the NIR spectrum of the sample.

2.3. Spectral Analysis. To establish a feasible NIRS identification model, some preprocessing methods, including multiplicative scatter correction (MSC), Savitzky–Golay (SG) smoothing, and min-max normalization, were applied to reduce the background noise and increase the signal-noise ratio. In this study, as one of the most commonly used unsupervised machine learning method, PCA was first employed for identification analysis, by reducing the high dimension of the original spectral matrix to a new low-dimensional spectral matrix with the minimal loss of information [23]. Afterwards, four kinds of supervised pattern recognition methods, LDA, SIMCA, BPNN, and SVM, were introduced to establish the discrimination models, respectively.

LDA is to find out a so-called discriminant function which best separates the classes by minimizing the distance of within-class samples and maximizing the distance of between-class samples [24]. SIMCA is based on establishing a PCA model for each class, where two critical values of the Euclidean distances and Mahalanobis distances assessed separately by the residual Q statistic and the Hotelling's T^2 statistic are used to determine boundaries around the samples belonging to one particular class [25, 26]. BPNN is an artificial neural network (ANN) based on the back-propagation method; to obtain the output of a well-trained network, the errors of each layer are calculated layer by layer from the beginning of the output layer and then used to optimize the network structure by adjusting the weights of each layer [27]. SVM achieves the goal of classification by using a kernel function to solve a linear or nonlinear problem, and the commonly used kernel functions include linear kernel, polynomial kernel, Gaussian kernel, and radial basis function; the generalization performance of SVM is highly related to the kernel function and the parameters in kernel function [28].

2.4. Discrimination Model Evaluation. In order to evaluate the performance of the established discrimination models, two parameters of recognition rate and rejection rate were proposed to reflect the reliability of clustering among different classes. The recognition rate refers to the ratio of the number of samples identified by one certain class to the total number of samples from the class, while the rejection rate refers to the ratio of the number of samples rejected by one certain class to the total number of samples from other classes. When both values of recognition rate and rejection rate are 100%, there is no overlap between the different classes of samples, meaning that they could be fully distinguished.

2.5. Software. The preprocessing of NIR spectra was carried out by the self-designed MATLAB programs. LDA and SIMCA models were built using the Unscrambler X 10.4 software (Camo AS, Oslo, Norway), the BPNN model was built using MATLAB programs based on ANN toolbox, and SVM models were built using LibSVM toolbox developed by professor Lin Zhiren from Taiwan University [29]; all MATLAB programs were run under MATLAB R 2016A software (Mathworks Inc., Natick, MA, USA).

3. Results and Discussions

3.1. Spectral Characteristics of Gelatin Gels. After the preprocessing of MSC, SG smoothing, and min-max normalization, the spectra improved a lot, as shown in Figure 1, displaying the mean values and standard deviations for each group of gelatin, where the standard deviations of spectra are very small. As can be seen clearly, in the wavelength range of 800–2200 nm, the normalized spectra have a similar spectral shape but different intensities. In addition, gelatin spectra show some prominent characteristic absorption peaks. The peaks approximating to 940 nm, 1490 nm, and 1930 nm are mainly caused by the overtones and combination bands of O–H group in water molecules. The peaks at about 1200 nm and 1730 nm should be generated by the overtones and combination bands of C–H group in gelatin samples, while the multiple peaks at 2000–2200 nm are supposed to primarily stem from the combination bands of C–H and N–H [30–32]. Due to the similarity of the spectral shapes, it is very difficult to straightforward determine if edible gelatin was added with industrial gelatin only from NIR spectrum.

3.2. Discrimination Models. In order to establish an appropriate discrimination model and evaluate its validity, the original data were divided into a training set and a validation set, where the training set was used to establish discrimination models and the validation set was used to verify the validity of the models. To improve the prediction ability of the models, according to a ratio of 2 : 1, 75 samples were randomly selected from a total of 114 samples of different gelatin varieties as a training set, and the remaining 39 samples were utilized as validation set, as shown in Tables 1 and S2.

3.2.1. LDA Model. In this work, PCA was first performed on the normalized spectral data; the first three principal components PC1, PC2, and PC3 accounted for 93.24%, 5.62%, and 0.51% of the total variance, respectively, making a 99.37% of the cumulative total variance, which indicates that the first three PCs can represent most of the NIR spectral information of gelatin gels. The loading plot and score plot of the first three PCs are shown in Figure 2. In the loading plot (Figure 2(a)), the spectral shape of the pure PC1 loading vector exhibits most of the characteristic absorption peaks shown in Figure 1, although both positive and negative contributions are included. PC2 and PC3 also give the positive and negative contributions, shifting the intensity maximum of the pure PC1 spectrum toward to the real spectrum. Figure 2(b) shows the score scatter plot of the first two and three PCs, respectively; it can be seen that some samples belonging to different classes are clustering, while others belonging to the same class are dispersing into two clusters, resulting in that the six classes of gelatin gels cannot be clearly distinguished due to the vague boundaries.

After principal component analysis, the optimal principal component number of 8 was determined by leave-one-out cross validation (LOOCV) and selected as the input variables of the LDA model, and the training set was selected to establish the LDA recognition model for gelatin adulteration. Then, the model was applied to the training set and the validation set, as shown in Table 2. The results show that both the recognition rate and rejection rate in the training set are 100%, while in the validation set, the recognition rate and rejection rate of other kinds of gelatin except pure edible gelatin and 4 : 1 adulterated gelatin are 100%, which means that these four kinds of gelatin can be well recognized. In the validation set, as shown in Table S2, a sample belonging to pure edible gelatin was incorrectly identified as 4 : 1 adulterated gelatin, resulting in a recognition rate of 90.91% for pure edible gelatin and a rejection rate of 96.88% for 4 : 1 adulterated gelatin. Since one sample in the validation set was wrongly classified, it led to a total correct recognition rate of 97.44%. The results show that the LDA discrimination model has a good classification ability on gelatin adulterated samples.

3.2.2. SIMCA Model. The SIMCA model was built based on the PCA model for each class of adulterated gelatin gels in the training set, and then the PCA models were individually performed on each class in the validation set for classification. For the PCA models, the optimal number of PCs was determined by the cross validation method (Table 3). Table 3 shows the predicting results of the SIMCA model in the training set and validation set when significance level $\alpha = 0.05$ and $\alpha = 0.1$, respectively.

As can be seen clearly, when the significance level is 0.05, except for the pure edible gelatin, the recognition rate and rejection rate for each class were 100% both in the training set and validation set. For the pure edible gelatin, the recognition rate was also 100%, while the rejection rate was 91% and 82%, separately for the training set and validation set. As listed in Tables S4 and S5, six samples in the training set belonging to 4 : 1 adulterated gelatin were incorrectly identified as pure edible gelatin, while five samples in the validation set were misclassified. When the significance level is 0.1, the recognition rate of 100% and rejection rate of 98% were obtained for the training set, indicating that one sample was overlapping

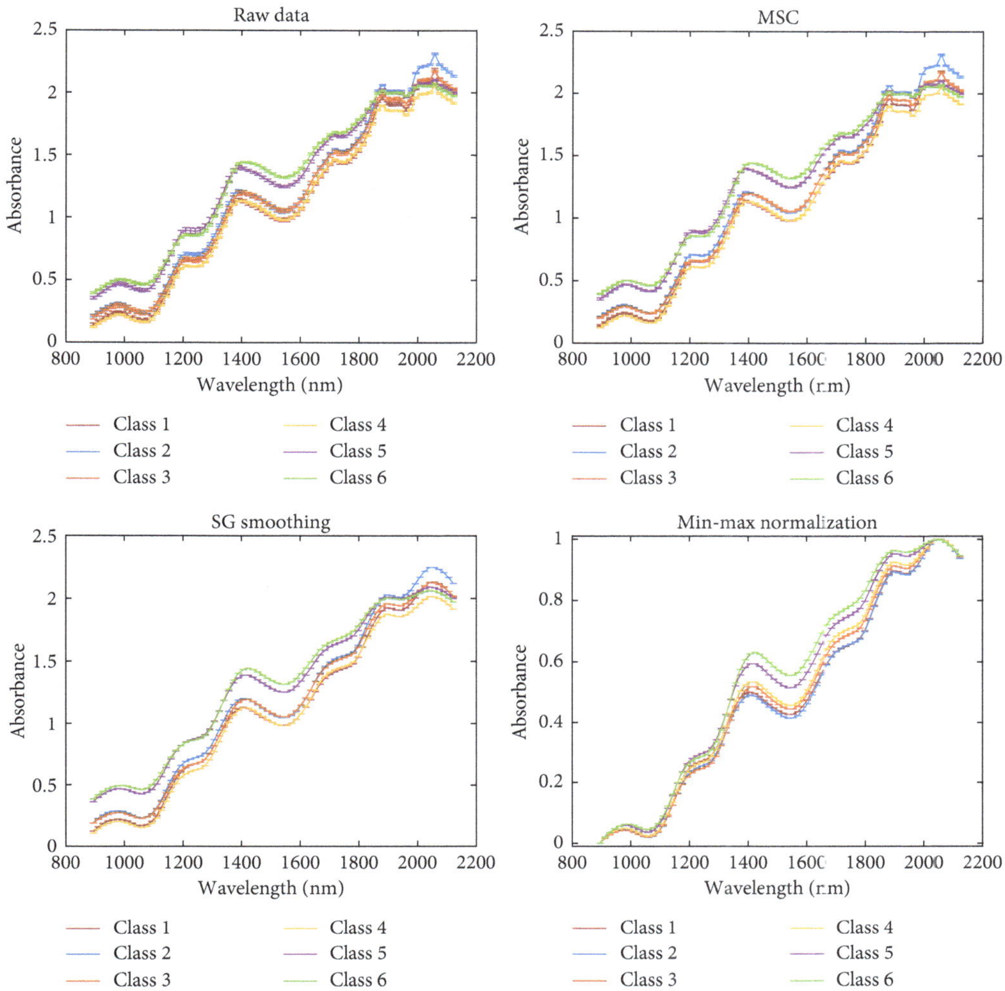

FIGURE 1: Mean values and standard deviations of NIR spectra for each group of gelatin before and after preprocessing.

TABLE 1: Training set and validation set of gelatin samples with six different adulteration levels.

Category	Training set	Validation set	Total number
Pure edible gelatin	9	11	20
4 : 1 adulterated gelatin	13	7	20
3 : 2 adulterated gelatin	13	5	18
2 : 3 adulterated gelatin	13	7	20
1 : 4 adulterated gelatin	12	6	18
Pure industrial gelatin	15	3	18
Total number	75	39	114

which appeared in two classes (Table S6). While in the validation set, the recognition rate and rejection rate for each class were 100% (Table S7), meaning that the SIMCA model can fully recognize the six classes of gelatin gels. Therefore, although 100% of the recognition rate can be achieved for the SIMCA model, the rejection rate less than 100% in particular for the prediction set indicates the possible presence of clustering overlap between the first two classes.

3.2.3. BPNN Model. Since the whole spectral band from 894 nm to 2302 nm consists of 432 data points, the number of input nodes is 432 when using the whole band data for modelling, leading to that not only the calculation becomes very slow but also the classification accuracy would be affected. Therefore, in this work, PCA was first used to reduce the dimensions of the whole spectrum data. According to the cross validation results, the first 10 PCs were selected as the input node m of the BP neural network, the output node n was set to 6 (i.e., 6 classes of gelatin gels), and the hidden layer node l was determined by the empirical formula of $l = \sqrt{m+n} + a$, where a is a constant between 0 and 10. During the training process, the network learning rate was set to 0.1, the number of training iterations was 1000, and the target error was 0.0001. The model was optimized by adjusting the hidden layer nodes, and an optimal three-layer BPNN model of 10 (input node)-12 (hidden layer node)-6 (output node) was obtained.

As described by Vitale et al. [18], in the case of classification involving m classes, a dummy binary-coded m-dimensional matrix can be used to describe the class that one

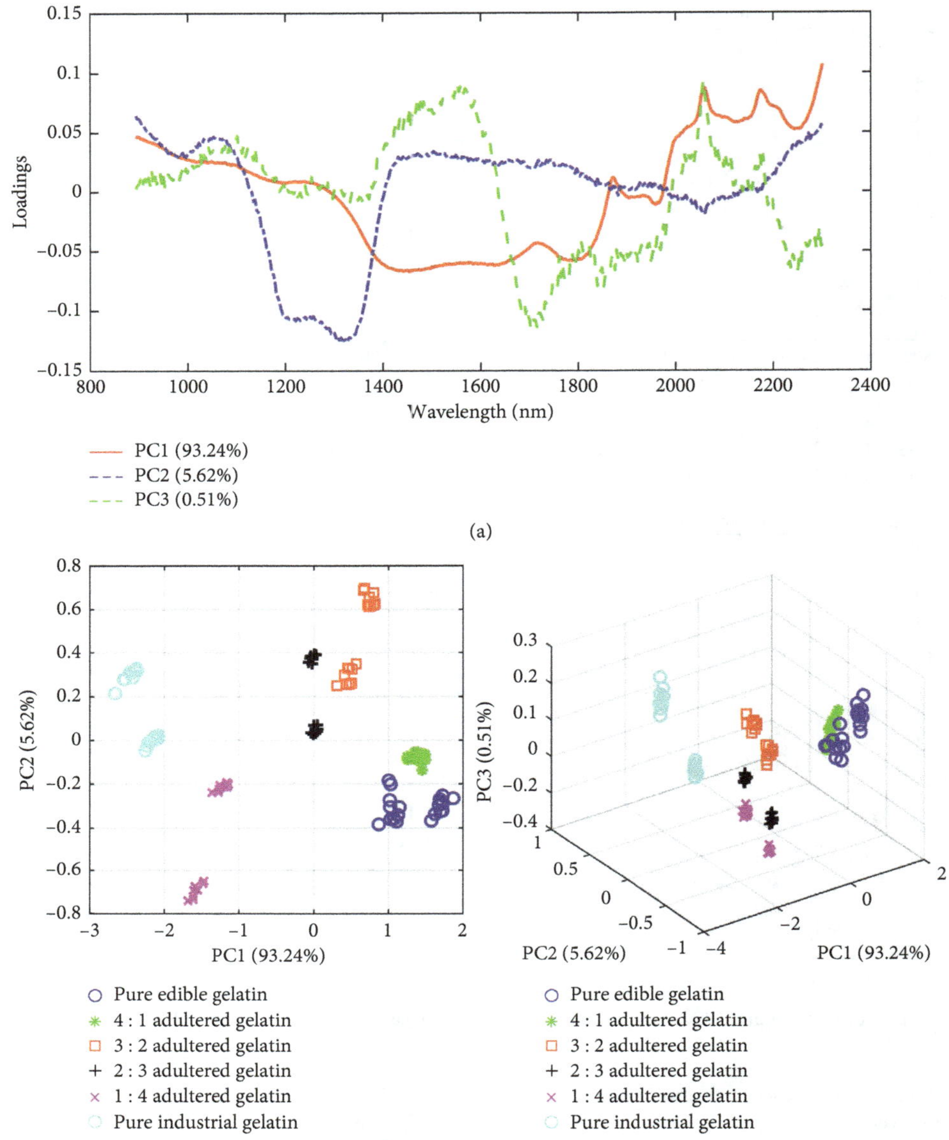

FIGURE 2: PCA loading plot: 2D and 3D score scatter plot of PC1, PC2, and PC3 for NIR spectra of gelatin samples.

TABLE 2: Predicting results of the LDA model in the training set and validation set.

Category	Training set		Validation set	
	Recognition rate	Rejection rate	Recognition rate	Rejection rate
Pure edible gelatin	100% (11/11)	100% (28/28)	90.91% (10/11)	100% (28/28)
4:1 adulterated gelatin	100% (7/7)	100% (32/32)	100% (7/7)	96.88% (31/32)
3:2 adulterated gelatin	100% (5/5)	100% (34/34)	100% (5/5)	100% (34/34)
2:3 adulterated gelatin	100% (7/7)	100% (32/32)	100% (7/7)	100% (32/32)
1:4 adulterated gelatin	100% (6/6)	100% (33/33)	100% (6/6)	100% (33/33)
Pure industrial gelatin	100% (3/3)	100% (36/36)	100% (3/3)	100% (36/36)

sample belongs to. In this study, samples belonging to pure edible gelatin (denoted by Class 1) were described by the vector [1 0 0 0 0 0], samples belonging to 4 : 1 adulterated gelatin (denoted by Class 2) by the vector [0 1 0 0 0 0], samples belonging to 3 : 2 adulterated gelatin (denoted by Class 3) by the vector [0 0 1 0 0 0], samples belonging to 2 : 3 adulterated gelatin (denoted by Class 4) by the vector [0 0 0 1 0 0], samples belonging to 1 : 4 adulterated gelatin (denoted

TABLE 3: Predicting results of SIMCA model in the training set and validation set when the significance level is 0.05 and 0.1, respectively.

Category	PCs	$\alpha = 0.05$				$\alpha = 0.1$			
		Training set		Validation set		Training set		Validation set	
		Recognition rate (%)	Rejection rate (%)	Recognition rate (%)	Rejection rate (%)	Recognition rate (%)	Rejection rate (%)	Recognition rate (%)	Rejection rate (%)
Pure edible gelatin	2	100	91	100	82	100	98	100	100
4 : 1 adulterated gelatin	7	100	100	100	100	100	100	100	100
3 : 2 adulterated gelatin	2	100	100	100	100	100	100	100	100
2 : 3 adulterated gelatin	4	100	100	100	100	100	100	100	100
1 : 4 adulterated gelatin	2	100	100	100	100	100	100	100	100
Pure industrial gelatin	4	100	100	100	100	100	100	100	100

by Class 5) by the vector [0 0 0 0 1 0], and samples belonging to pure industrial gelatin (denoted by Class 6) by the vector [0 0 0 0 0 1].

Under these assumptions, when the BPNN model was used for predicting 6 varieties of gelatin gels in the training set, all 75 samples were correctly identified without clustering overlap, producing a total recognition rate of 100%. As shown in Table 4, the predicting results of the BPNN model in the validation set show that only one sample (No. 23) was misclassified, which should be belonging to Class 4 and now incorrectly classified as Class 3, resulting in a recognition rate of 85.71% (6/7) for Class 4 and a rejection rate of 97.06% (33/34) for Class 3. Except for Class 4, the recognition rate of 100% for other classes was obtained, resulting in a recognition rate of 97.44% (38/39) for the validation set.

3.2.4. SVM Model. As we know, SVM is initially designed for binary classification. However, in recent years, SVM has been widely used for multiclass classification. To construct a SVM multiclass classifier, there are two common methods: one-versus-one (OVO) and one-versus-rest (OVR). In this work, we utilized the LIBSVM tool developed by Lin et al. to perform the SVM analysis, where LIBSVM tool adopted OVO method, thus 15 SVM classifiers were required for 6 groups of gelatin samples (i.e., k classes need $k(k-1)/2$ SVM classifiers). Radial basis function (RBF) was chosen as the kernel function for SVM classification. To obtain an optimal SVM model, the penalty parameter C and kernel parameter γ for the RBF function were optimized by using grid search (GS), genetic algorithm (GA), and particle swarm optimization (PSO) methods, the best C and γ were determined by the accuracy of cross validation (CV). The best optimization parameters and the predicting results of the SVM models are shown in Table 5, where the classified results of validation set (39 spectra) are summarized in a confusion matrix presented in Table 6, where each row represents actual class of samples and each column represents the classification by SVM models. The results show that the CV accuracy of these three optimization methods was 100%, and both the recognition rate and rejection rate in the training set and validation set were 100%, indicating that GS, GA, and PSO methods have the same optimization ability to the SVM discrimination model. Moreover, the modelling results also demonstrate that the SVM model has a preferable discrimination ability to identify adulterated edible gelatin.

3.2.5. Performance Comparison of the Four Discrimination Models. Table 7 summarizes the predicting results of adulterated gelatin gels in the training set and validation set using LDA, SIMCA, BPNN, and SVM discrimination models, respectively. The results show that these four pattern recognition methods can well identify the adulterated gelatin gels, where the recognition rate in the training set was as high as 100%, and the recognition rate in the validation set was higher than 97%. For SIMCA and SVM models, the recognition rate both in the training set and validation set are 100%, which demonstrates that these two methods could completely recognize the 114 gelatin samples with 6 different adulteration levels. However, as displayed in Table 3, for the SIMCA model, when the significance level $\alpha = 0.05$, although the recognition rate for each class was 100%, five samples were overlapping which appeared in two classes. Only when the significance level $\alpha = 0.1$, all kinds of samples can be completely recognized. For the SVM model, the recognition rate and rejection rate both in the training set and verification set for each class were 100%, no matter which parameter optimization method was used. Therefore, among the four supervised pattern recognition methods, SVM is superior to the other three methods, which can fully identify the six classes of gelatin gels with different adulteration levels without clustering overlap.

4. Conclusion

Aiming at the problem of edible gelatin adulterated with industrial gelatin in the food industry, NIR absorption spectra of 114 gelatin gels from 6 kinds of adulteration gelatin were obtained with a compact NIR measuring setup; MSC, SG smoothing, and min-max normalization methods were first used for spectral preprocessing, and then PCA was utilized for the exploratory investigation, while PCA could

TABLE 4: Predicting results of BPNN model in the validation set.

No.	Real class	Predicted class	Result	No.	Actual species	Predicted class	Result	No.	Real class	Predicted class	Result
1	0 0 0 1 0 0	4	Correct	14	0 0 0 0 0 1	6	Correct	27	1 0 0 0 0 0	1	Correct
2	0 1 0 0 0 0	2	Correct	15	0 0 1 0 0 0	3	Correct	28	0 0 0 0 1 0	5	Correct
3	0 0 1 0 0 0	3	Correct	16	1 0 0 0 0 0	1	Correct	29	0 0 0 1 0 0	4	Correct
4	0 1 0 0 0 0	2	Correct	17	0 0 0 1 0 0	4	Correct	30	0 1 0 0 0 0	2	Correct
5	0 0 0 0 0 1	6	Correct	18	1 0 0 0 0 0	1	Correct	31	0 0 0 0 1 0	5	Correct
6	0 0 0 1 0 0	4	Correct	19	0 1 0 0 0 0	2	Correct	32	0 1 0 0 0 0	2	Correct
7	0 0 0 0 1 0	5	Correct	20	0 0 0 0 1 0	5	Correct	33	1 0 0 0 0 0	1	Correct
8	0 0 1 0 0 0	3	Correct	21	0 0 0 1 0 0	4	Correct	34	1 0 0 0 0 0	1	Correct
9	0 0 0 0 1 0	5	Correct	22	0 1 0 0 0 0	2	Correct	35	0 0 0 1 0 0	4	Correct
10	0 1 0 0 0 0	2	Correct	23	0 0 0 1 0 0	3	Error	36	1 0 0 0 0 0	1	Correct
11	0 0 1 0 0 0	3	Correct	24	1 0 0 0 0 0	1	Correct	37	0 0 1 0 0 0	3	Correct
12	0 0 0 0 1 0	5	Correct	25	0 0 0 0 0 1	6	Correct	38	1 0 0 0 0 0	1	Correct
13	1 0 0 0 0 0	1	Correct	26	1 0 0 0 0 0	1	Correct	39	1 0 0 0 0 0	1	Correct

Note. [1 0 0 0 0 0] represents class 1, i.e., pure edible gelatin ; [0 1 0 0 0 0] represents class 2, i.e., 4 : 1 adulterated gelatin ; [0 0 1 0 0 0] represents class 3, i.e., 3 : 2 adulterated gelatin ; [0 0 0 1 0 0] represents class 4, i.e., 2 : 3 adulterated gelatin ; [0 0 0 0 1 0] represents class 5, i.e., 1 : 4 adulterated gelatin; [0 0 0 0 0 1] represents class 6, i.e., pure industrial gelatin.

TABLE 5: Optimization parameters and predicting results of SVM models based on three parameter optimization algorithms.

Parameter optimized methods	Best C	Best γ	CV accuracy	Recognition rate	
				Training set	Validation set
GS-SVM	0.3299	5.2780	100%	100% (75/75)	100% (39/39)
GA-SVM	1.1135	11.4377	100%	100% (75/75)	100% (39/39)
PSO-SVM	0.5283	15.7922	100%	100% (75/75)	100% (39/39)

TABLE 6: The classification results of SVM models in the validation set.

Number	Class 1	Class 2	Class 3	Class 4	Class 5	Class 6
Class 1	11	0	0	0	0	0
Class 2	0	7	0	0	0	0
Class 3	0	0	5	0	0	0
Class 4	0	0	0	7	0	0
Class 5	0	0	0	0	6	0
Class 6	0	0	0	0	0	3

TABLE 7: Performance comparison of 4 kinds of discrimination models.

Discrimination model	Recognition rate	
	Training set	Validation set
LDA	100% (75/75)	97.44% (38/39)
SIMCA	100% (75/75)	100% (39/39)
BPNN	100% (75/75)	97.44% (38/39)
SVM	100% (75/75)	100% (39/39)

not realize the obvious distinction of adulterated gelatin. Subsequently, the feasibility of adulterated gelatin identification was studied by building the LDA, SIMCA, BPNN, and SVM models, respectively. The modelling results show that all these four discrimination models present good recognition ability for the adulterated gelatin, the recognition rate of which in the validation set is higher than 97%. Especially for the SIMCA and SVM models, the correct recognition rate reaches up to 100%. However, due to the possible clustering overlap between different classes when using the SIMCA model, the SVM model shows the preferable ability for the recognition of edible gelatin adulteration. Therefore, it is feasible and effective to use NIR spectroscopy combined with supervised pattern recognition methods for identifying edible gelatin adulteration rapidly and accurately.

Conflicts of Interest

The authors declare that there are no conflicts of interest.

Acknowledgments

This work was financially supported by the China Postdoctoral Science Foundation (grant no. 2017M612399), the Science and Technology Project of Henan Province (grant nos. 182102110427 and 182102110250), the Science and Technology Innovation Project of Henan Agricultural University (grant no. KJCX2018A09), the National Natural Science Foundation of China (grant no. 31671581), and the Natural Science Foundation of Henan Province (grant no. 162300410143).

Supplementary Materials

In the supplementary files, Table S1 gives the list of used gelatin samples, Table S2 describes a total of 114 gelatin gelatin samples in the training set and validation set, Figure S1 displays the photographs of gelatin samples prepared for NIR spectral measurements, Table S3 gives the prediction results of LDA model in the validation set, the prediction results of the SIMCA model in the training set and validation set when the significance level $\alpha = 0.05$ are separately displayed in Tables S4 and S5, and the prediction results of the SIMCA model in the training set and validation set when the significance level $\alpha = 0.1$ are shown in Tables S6 and S7, respectively. (*Supplementary Materials*)

References

[1] H. Shah and F. Yusof, "Gelatin as an ingredient in food and pharmaceutical products: an Islamic perspective," *Advances in Environmental Biology*, vol. 8, pp. 774–780, 2014.

[2] A. Jan, M. Azam, K. Siddiqui, A. Ali, I. Choi, and Q. Haq, "Heavy metals and human health: mechanistic insight into toxicity and counter defense system of antioxidants," *International Journal of Molecular Sciences*, vol. 16, no. 12, pp. 29592–29630, 2015.

[3] X. Wang and Y. Wu, "Industrial gelatin poisons food; police crack ring," 2016, http://www.chinadaily.com.cn/china/2016-07/30/content_26278365.htm.

[4] T. N. Azira, I. Amin, and Y. B. C. Man, "Differentiation of bovine and porcine gelatins in processed products via Sodium Dodecyl Sulphate-Polyacrylamide Gel Electrophoresis (SDS-PAGE) and principal component analysis (PCA) techniques," *International Food Research Journal*, vol. 19, pp. 1175–1180, 2012.

[5] T. N. Azira, Y. B. C. Man, R. N. R. M. Hafidz, M. A. Aina, and I. Amin, "Use of principal component analysis for differentiation of gelatine sources based on polypeptide molecular weights," *Food Chemistry*, vol. 151, pp. 286–292, 2014.

[6] H. Doi, E. Watanabe, H. Shibata, and S. Tanabe, "A reliable enzyme linked immunosorbent assay for the determination of bovine and porcine gelatin in processed foods," *Journal of Agricultural and Food Chemistry*, vol. 57, no. 5, pp. 1721–1726, 2009.

[7] N. A. Tukiran, A. Ismail, S. Mustafa, and M. Hamid, "Determination of porcine gelatin in edible bird's nest by competitive indirect ELISA based on anti-peptide polyclonal antibody," *Food Control*, vol. 59, pp. 561–566, 2016.

[8] N. A. Tukiran, A. Ismail, S. Mustafa, and M. Hamid, "Development of antipeptide enzyme-linked immunosorbent assay for determination of gelatin in confectionery products," *International Journal of Food Science & Technology*, vol. 51, no. 1, pp. 54–60, 2015.

[9] G. Zhang, T. Liu, Q. Wang et al., "Mass spectrometric detection of marker peptides in tryptic digests of gelatin: a new method to differentiate between bovine and porcine gelatin," *Food Hydrocolloids*, vol. 23, no. 7, pp. 2001–2007, 2009.

[10] M. T. Yilmaz, Z. Kesmen, B. Baykal et al., "A novel method to differentiate bovine and porcine gelatins in food products: NanoUPLC-ESI-Q-TOF-MSE based data independent acquisition technique to detect marker peptides in gelatin," *Food Chemistry*, vol. 141, no. 3, pp. 2450–2458, 2013.

[11] M. I. Azilawati, D. M. Hashim, B. Jamilah, and I. Amin, "RP-HPLC method using 6-aminoquinolyl-N-hydroxysuccinimidyl carbamate incorporated with normalization technique in principal component analysis to differentiate the bovine, porcine and fish gelatins," *Food Chemistry*, vol. 172, pp. 368–376, 2015.

[12] H. Cai, X. Gu, M. S. Scanlan, D. H. Ramatlapeng, and C. R. Lively, "Real-time PCR assays for detection and quantitation of porcine and bovine DNA in gelatin mixtures and gelatin capsules," *Journal of Food Composition and Analysis*, vol. 25, no. 1, pp. 83–87, 2012.

[13] S. A. Mutalib, N. M. Muin, A. Abdullah et al., "Sensitivity of polymerase chain reaction (PCR)-southern hybridization and conventional PCR analysis for Halal authentication of gelatin capsules," *LWT-Food Science and Technology*, vol. 63, no. 1, pp. 714–719, 2015.

[14] H. I. A. Amqizal, H. A. Al-Kahtani, E. A. Ismail, K. Hayat, and I. Jaswir, "Identification and verification of porcine DNA in commercial gelatin and gelatin containing processed foods," *Food Control*, vol. 78, pp. 297–303, 2017.

[15] B. Jannat, K. Ghorbani, H. Shafieyan et al., "Gelatin speciation using real-time PCR and analysis of mass spectrometry-based proteomics datasets," *Food Control*, vol. 87, pp. 79–87, 2018.

[16] D. M. Hashim, Y. B. C. Man, R. Norakasha, M. Shuhaimi, Y. Salmah, and Z. A. Syahariza, "Potential use of Fourier transform infrared spectroscopy for differentiation of bovine and porcine gelatins," *Food Chemistry*, vol. 118, no. 3, pp. 856–860, 2010.

[17] N. Cebi, M. Z. Durak, O. S. Toker, O. Sagdic, and M. Arici, "An evaluation of Fourier transforms infrared spectroscopy method for the classification and discrimination of bovine, porcine and fish gelatins," *Food Chemistry*, vol. 190, pp. 1109–1115, 2016.

[18] R. Vitale, M. Bevilacqua, R. Bucci, A. D. Magrì, A. L. Magrì, and F. Marini, "A rapid and non-invasive method for authenticating the origin of pistachio samples by NIR spectroscopy and chemometrics," *Chemometrics and Intelligent Laboratory Systems*, vol. 121, pp. 90–99, 2013.

[19] X. Ding, Y. Ni, and S. Kokot, "NIR spectroscopy and chemometrics for the discrimination of pure, powdered, purple sweet potatoes and their samples adulterated with the white sweet potato flour," *Chemometrics and Intelligent Laboratory Systems*, vol. 144, pp. 17–23, 2015.

[20] F. Mabood, F. Jabeen, M. Ahmed et al., "Development of new NIR-spectroscopy method combined with multivariate analysis for detection of adulteration in camel milk with goat milk," *Food Chemistry*, vol. 221, pp. 746–750, 2017.

[21] L.-Q. Hu, C.-L. Yin, and Z.-P. Zeng, "Detection of adulteration in acetonitrile using near infrared spectroscopy coupled with pattern recognition techniques," *Spectrochimica Acta Part A: Molecular and Biomolecular Spectroscopy*, vol. 151, pp. 34–39, 2015.

[22] P. Liu, J. Wang, Q. Li, J. Gao, X. Tan, and X. Bian, "Rapid identification and quantification of Panax notoginseng with its adulterants by near infrared spectroscopy combined with chemometrics," *Spectrochimica Acta Part A: Molecular and Biomolecular Spectroscopy*, vol. 206, pp. 23–30, 2019.

[23] H. Zhang, S. Wang, K. Chang et al., "Optical characterization of paper aging based on laser-induced fluorescence (LIF) spectroscopy," *Applied Spectroscopy*, vol. 72, no. 6, pp. 913–920, 2018.

[24] G. McLachlan, *Discriminant Analysis and Statistical Pattern Recognition*, John Wiley & Sons Inc., New York, NY, USA, 2005.

[25] S. Wold and M. Sjöström, "SIMCA: a method for analysing chemical data in terms of similarity and analogy," in *Chemometrics, Theory and Application*, American Chemical Society, Washington, DC, USA, pp. 243–282, 52nd edition, 1977.

[26] B. Mertens, M. Thompson, and T. Fearn, "Principal component outlier detection and SIMCA: a synthesis," *The Analyst*, vol. 119, pp. 2777–2784, 1994.

[27] C. M. Bishop, *Neural Networks for Pattern Recognition*, Clarendon Press, Oxford, UK, 1995.

[28] C. Cortes and V. Vapnik, "Support-vector Networks," *Machine Learning*, vol. 20, no. 3, pp. 273–297, 1995.

[29] C.-C. Chang and C.-J. Lin, "Libsvm," *ACM Transactions on Intelligent Systems and Technology*, vol. 2, no. 3, pp. 1–27, 2011.

[30] L. Cséfalvayová, M. Pelikan, I. Kralj Cigić, J. Kolar, and M. Strlič, "Use of genetic algorithms with multivariate regression for determination of gelatine in historic papers based on FT-IR and NIR spectral data," *Talanta*, vol. 82, no. 5, pp. 1784–1790, 2010.

[31] A. Duconseille, D. Andueza, F. Picard, V. Santé-Lhoutellier, and T. Astruc, "Molecular changes in gelatin aging observed by NIR and fluorescence spectroscopy," *Food Hydrocolloids*, vol. 61, pp. 496–503, 2016.

[32] A. Duconseille, D. Andueza, F. Picard, V. Santé-Lhoutellier, and T. Astruc, "Variability in pig skin gelatin properties related to production site: a near infrared and fluorescence spectroscopy study," *Food hydrocolloids*, vol. 63, pp. 108–119, 2017.

18

Rapid Identification and Quality Evaluation of Medicinal Centipedes in China Using Near-Infrared Spectroscopy Integrated with Support Vector Machine Algorithm

Sihe Kang[1,2] **Haiying Deng**[3] **Long Chen**[1,4] **Xiaoxuan Zeng**[1] **Yimei Liu**[1] **and Keli Chen**[1]

[1]*Key Laboratory of Ministry of Education on Traditional Chinese Medicine Resource and Compound Prescription, Hubei University of Chinese Medicine, Wuhan 430065, China*
[2]*NMPA Key Laboratory of Quality Control of Chinese Medicine, Hubei Institute for Drug Control, Wuhan 430075, China*
[3]*Medical College of Wuhan University of Science and Technology, Hubei Province Key Laboratory of Occupational Hazard Identification and Control, Wuhan 430065, China*
[4]*Xiangyang Central Hospital, Xiangyang 441021, China*

Correspondence should be addressed to Yimei Liu; liuyimei1971@126.com and Keli Chen; kelichen@126.com

Academic Editor: Eugen Culea

To investigate the feasibility of rapid identification and quality evaluation of Chinese medicinal centipedes using NIR spectroscopy, the qualitative and quantitative analysis models were explored. A PCA-SVC model was optimized to differentiate five species of the genus *Scolopendra*. When the model was validated with the calibration and prediction sets, the prediction accuracy was 100% and 81.82%, respectively; it can meet the requirement for rapid and preliminary identification. Based on nitrogen content detected by the chemical method, and the dimensionality of spectral data reduced with PLS, the quantitative analysis models were successfully built by PLSR and SVR algorithms. After spectra were pretreated and parameters were optimized, the performance, rationality, and prediction ability of the models were validated and evaluated with RMSECV, RMSEP, RMSEE, R^2, and RPD. Compared with the features and advantages of these two models, the PLS-SVR model had better performance and stronger prediction capacity, and it was finally regarded as the optimal quantitative analysis model to predict nitrogen content. The relative deviation between the predictive value and the reference was 2.69%, and the average recovery was 99.02%, which indicated it has potential for rapid prediction and evaluation of the quality of medicinal centipedes. This research suggested that NIR spectroscopy can be used as a rapid detection method to identify species and evaluate the quality of medicinal centipedes in China.

1. Introduction

Animals of the genus *Scolopendra* are widely distributed in the world, especially in tropical and subtropical areas [1]. In China, there are 14 species which are mainly distributed in the southern region [2]. Recently, the medicinal value of centipedes had become a research hotspot; the venom was reported to be used for relieving pain and anticoagulation [3, 4]. As an important source of Chinese medicinal materials, five species of *Scolopendra* are commonly used in China [2], which were reported to possess analgesic [3], anti-inflammatory [5], and antitumor [6, 7] activities and to improve blood rheology [8]. However, *Scolopendra mutilans* is the only species recorded in Chinese Pharmacopoeia 2015 (ChP 2015), and the other four species are just used in local regions; for instance, *S. multidens* is used in Guangxi and *S. mojiangica* in Yunnan [2]. As the animals of *Scolopendra* are poisonous, the venom and toxic ingredients of some species will bring high risk to humans [9], and the activity and quality of species are different. Therefore, in order to ensure safety and clinical efficacy, a simple, rapid, and accurate method is needed to identify the species and control the quality of medicinal centipedes.

Previously, medicinal centipedes were mostly identified using morphological description, but some similar characteristics were probably shown among closely related species.

If samples were damaged or powdered, they were difficult to be identified, and confusion and misuse would be unavoidable. Presently, molecular methods are gradually applied to identify *Scolopendra* species [10]. However, the complexity of operation and high technical requirements make it difficult to obtain rapid and accurate results, especially in mixed samples. Proteins or amino acids are recognized as the main active ingredient of medicinal centipedes [11–13], and their content is usually used for quality evaluation. Because of the diversity of ingredients, and the complexity of chemical determination methods, this measurement is usually cumbersome [14].

Near-infrared (NIR) spectroscopy combined with chemometrics is a fast, nondestructive, and environmentally friendly analysis technique that can realize multicomponent analysis. Nowadays, it is widely used in agriculture and medicine [15–17]. NIR spectroscopy mainly reflects the absorption of overtone and combination peaks containing hydrogen bonds of C-H, O-H, and N-H [16]. Lipids and proteins are considered to be important medicinal components, which are rich in centipedes and show characteristic absorption in the near-infrared region. However, the application of NIR spectroscopy on medicinal centipedes has not yet been reported. In this study, the NIR spectroscopy analysis methods were investigated, a PCA-SVC model was explored to identify the species of *Scolopendra*, and in light of nitrogen content determined by the chemical method, a quantitative model was established for quality prediction using regression algorithms.

2. Materials and Methods

2.1. Instruments and Software. The nitrogen content of samples was determined with the DK 20 Heating Digester (VELP, Italy) and UDK 149 Automatic Distillation Unit (VELP, Italy). Spectra were collected with an MPA FT-NIR spectrometer (Bruker Optics Co., Ltd., Germany) and analyzed using the OPUS 7.5 spectrum analysis software (Bruker), MATLAB R2014a data analysis software (MathWorks, Inc., USA), and Unscrambler 9.7 data analysis software (CAMO Software AS, Norway).

2.2. Samples and Identification. A total of 64 samples from 28 batches have been collected from field surveys or market commodity in China since 2015. All samples were identified into five nominal species according to characteristics recorded by Siriwut et al. [1], Kang et al. [2], Song et al. [18], and Zhang and Wang [19]. The sample information is summarized in Table 1. All samples were kept below −20°C and housed in Hubei University of Chinese Medicine, Wuhan, China.

2.3. Content Determination. After being scanned with a near-infrared spectrometer, the nitrogen content of 50 mg powder of each sample was determined with the semimicro quantitative nitrogen determination method referring to the guideline of ChP 2015. The samples were digested using the DK 20 Heating Digester with a program as follows: 200°C for 5 min, then up to 260°C sustaining for 5 min, 340°C for 5 min, and 420°C for 40 min, and at last cooled down to 200°C. The sample solution was measured using the UDK 149 Automatic Distillation Unit with a program as follows: 50 ml H_2O and 20 ml 40% NaOH were added to the digested solution, 20 ml 2% H_3PO_4 was used for receiving, the steam quantity was 50%, the distillation time was 4 min, and then titration was done with 0.025 mol/L H_2SO_4 standard solution (Metrological Testing Technology Research Institute of Shanghai; Batch number 150901).

2.4. Spectra Acquisition. After samples were smashed and dried at 55°C for 24 h, the powder of 2 g of individuals was scanned using the MPA FT-NIR spectrometer with a diffuse reflection integral sphere. The spectra were obtained in a range of 12000~4000 cm^{-1} by the coaddition of 32 scans at a resolution of 8 cm^{-1}. Each sample was scanned three times, and the average of three spectra was used for analysis. The spectra diagram is shown in Figure 1.

2.5. Spectral Pretreatment Method. Usually, the raw spectrum includes a lot of irrelevant information or noise, which would lead to baseline drift and instability. Therefore, spectrum pretreatment is a critical step in spectral analysis. There are many pretreatment methods, and each has advantages to improve model performance. For instance, vector normalization (VN) can be used to eliminate influences of the optical path change on the spectrum. The derivative methods including the first derivative (FD) and second derivative (SD) are always employed to eliminate spectral difference from baseline [20], while multiple scattering correction (MSC) is commonly performed to process diffuse reflection spectra [21]. In this study, methods such as VN, FD, SD, and MSC or combined pretreatments were employed by OPUS to optimize model performance.

2.6. Spectral Data Compression Method

2.6.1. Principal Component Analysis (PCA) Method. PCA is a commonly used method for data compression. It performs dimensionality reduction of a high-dimensional dataset, while retaining its variation as much as possible. This method can transform a number of possibly correlated variables (the original data matrix) into one or a few important variables (principal components (PCs)) to reveal the internal structure. Each PC is a linear combination of the original data. The new variables are not related to each other, which can eliminate the overlapped part of information. Moreover, these new variables include the most informative dimensions of the original variables without losing too much information. Commonly, the number of PCs is determined by the contribution rate to original variables. When the cumulative contribution rate is more than 85%, the main components can represent most of the information provided by the original variable [22]. In our identification research, the PCA was used to reduce the dimension of original spectral data.

Table 1: Sample information of medicinal centipedes.

Number	Species	Batch no.	Nitrogen content (%)	Origin
1	*S. mutilans* L. Koch	WG 002-1	10.09	Suizhou, Hubei
2	*S. mutilans* L. Koch	WG 003-1	11.36	Suizhou, Hubei
3	*S. mutilans* L. Koch	WG 004-1	11.82	Jingmen, Hubei
4	*S. mutilans* L. Koch	WG 004-2	10.60	Jingmen, Hubei
5	*S. mutilans* L. Koch	WG 005-1	10.77	Xiangyang, Hubei
6	*S. mutilans* L. Koch	WG 005-2	10.49	Xiangyang, Hubei
7	*S. mutilans* L. Koch	WG 006-1	9.47	Yichang, Hubei
8	*S. mutilans* L. Koch	WG 013-1	10.43	Suizhou, Hubei
9	*S. mutilans* L. Koch	WG 014-1	9.25	Jinshan, Hubei
10	*S. mutilans* L. Koch	WG 014-2	10.22	Jinshan, Hubei
11	*S. mutilans* L. Koch	WG 016-1	10.10	Suizhou, Hubei
12	*S. mutilans* L. Koch	WG 016-2	9.83	Suizhou, Hubei
13	*S. mutilans* L. Koch	WG 017-1	8.20	Anlu, Hubei
14	*S. mutilans* L. Koch	WG 017-2	10.15	Anlu, Hubei
15	*S. mutilans* L. Koch	WG 018-1	11.02	Yichang, Hubei
16	*S. mutilans* L. Koch	WG 019-1	10.05	Nanzhang, Hubei
17	*S. mutilans* L. Koch	WG 019-2	10.16	Nanzhang, Hubei
18	*S. mutilans* L. Koch	WG 020-1	9.06	Anhui
19	*S. mutilans* L. Koch	WG 020-2	11.74	Anhui
20	*S. mutilans* L. Koch	WG 027-1	10.55	Henan
21	*S. mutilans* L. Koch	WG 027-2	11.10	Henan
22	*S. mutilans* L. Koch	WG 032 -1	9.96	Machang, Hubei
23	*S. mutilans* L. Koch	WG 032-2	10.01	Machang, Hubei
24	*S. mutilans* L. Koch	WG 045-1	9.68	Machang, Hubei
25	*S. mutilans* L. Koch	WG 045-2	9.68	Machang, Hubei
26	*S. multidens* Newport	WG 012-1	11.27	Yulin, Guangxi
27	*S. multidens* Newport	WG 012-2	10.83	Yulin, Guangxi
28	*S. multidens* Newport	WG 012-3	10.54	Yulin, Guangxi
29	*S. multidens* Newport	WG 012-4	11.74	Yulin, Guangxi
30	*S. multidens* Newport	WG 012-5	11.33	Yulin, Guangxi
31	*S. multidens* Newport	WG 012-6	9.77	Yulin, Guangxi
32	*S. multidens* Newport	WG 012-7	11.27	Yulin, Guangxi
33	*S. multidens* Newport	WG 012-8	11.94	Yulin, Guangxi
34	*S. multidens* Newport	WG 021-1	9.85	Guangxi
35	*S. multidens* Newport	Wg039-1	11.92	Guangxi
36	*S. multidens* Newport	Wg039-2	10.98	Guangxi
37	*S. multidens* Newport	Wg040-1	10.25	Mengzi, Yunnan
38	*S. multidens* Newport	Wg040-2	9.81	Mengzi, Yunnan
39	*S. dehaani* Brandt	WG 028-1	12.31	Yunnan
40	*S. dehaani* Brandt	WG 038-1	12.62	Yunnan
41	*S. dehaani* Brandt	WG 038-2	12.17	Yunnan
42	*S. dehaani* Brandt	WG 038-3	12.58	Yunnan
43	*S. dehaani* Brandt	WG 038-4	12.30	Yunnan
44	*S. dehaani* Brandt	WG 038-5	12.36	Yunnan
45	*S. dehaani* Brandt	WG 038-6	11.67	Yunnan
46	*S. dehaani* Brandt	WG 038-7	12.46	Yunnan
47	*S. dehaani* Brandt	WG 038-8	11.93	Yunnan
48	*S. mojiangica* Zhang et Chi	WG 007-1	8.37	Mojiang, Yunnan
49	*S. mojiangica* Zhang et Chi	WG 007-2	8.04	Mojiang, Yunnan
50	*S. mojiangica* Zhang et Chi	WG 007-3	8.16	Mojiang, Yunnan
51	*S. mojiangica* Zhang et Chi	WG 007-4	7.47	Mojiang, Yunnan
52	*S. mojiangica* Zhang et Chi	WG 007-5	8.18	Mojiang, Yunnan
53	*S. mojiangica* Zhang et Chi	WG 008-1	8.92	Mojiang, Yunnan
54	*S. mojiangica* Zhang et Chi	WG 008-2	9.06	Mojiang, Yunnan
55	*S. mojiangica* Zhang et Chi	WG 041-2	8.54	Bixi, Yunnan
56	*S. mojiangica* Zhang et Chi	WG 041-3	8.37	Bixi, Yunnan
57	*S. negrocapitis* Zhang et Wang	WG 022-1	10.48	Suizhou, Hubei
58	*S. negrocapitis* Zhang et Wang	WG 022-2	10.07	Suizhou, Hubei
59	*S. negrocapitis* Zhang et Wang	WG 022-3	10.52	Suizhou, Hubei
60	*S. negrocapitis* Zhang et Wang	WG 022-4	10.65	Suizhou, Hubei

Table 1: Continued.

Number	Species	Batch no.	Nitrogen content (%)	Origin
61	*S. negrocapitis* Zhang et Wang	WG 022-5	11.45	Suizhou, Hubei
62	*S. negrocapitis* Zhang et Wang	WG 015-1	10.86	Chaohu, Anhui
63	*S. negrocapitis* Zhang et Wang	WG 015-2	11.71	Chaohu, Anhui
64	*S. negrocapitis* Zhang et Wang	WG 015-3	10.82	Chaohu, Anhui

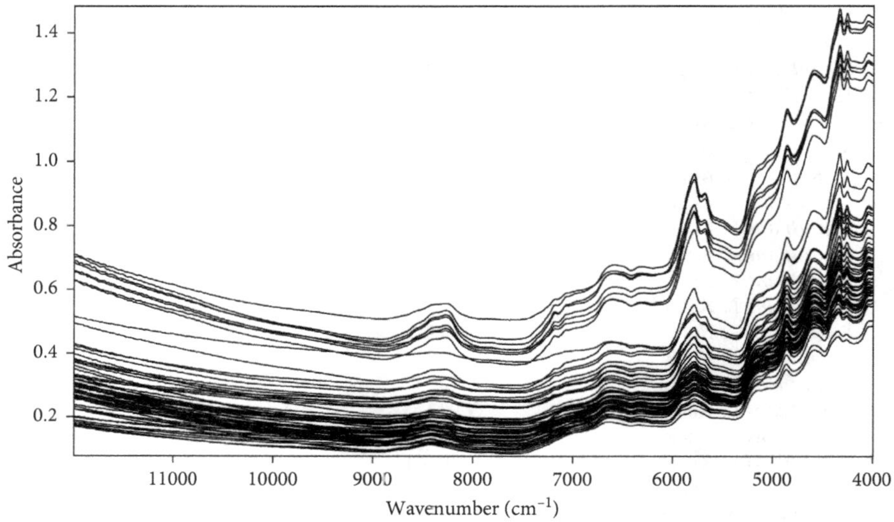

Figure 1: NIR spectra diagram of samples.

2.6.2. Partial Least-Squares (PLS) Method. The PLS is a new multivariate statistical analysis method. It attempts to recombine the original variables (mainly continuous variables) into a group of new independent comprehensive variables and extracts a few comprehensive variables to reflect the information on the original variables as much as possible. The extracted new variables have good interpretation ability for the dependent variables. During modeling, it not only considers factors of the independent variable matrix (spectral matrix) but also takes the "response" matrix (content matrix) into account. The principal component scores extracted by dimension reduction are used as input variables to avoid multicollinearity, improve stability, and simplify the model. Therefore, the PLS has the ability to simplify the model and characteristics of quick calculation and strong prediction ability, and as one of the most classical data processing tools in multiple correlation regression, it is widely applied in NIR spectroscopy quantitative analysis [23–25].

2.7. Support Vector Machine (SVM) Algorithm. SVM is a powerful supervised learning algorithm that was first proposed by Vapnik [26] and successfully extended by researchers in recent years. It is based on the principle of minimization of structural risk in constructing an optimally separating hyperplane that separates different classes of data. In the process, input vectors are mapped to a newly constructed high-dimensional space, and then parallel hyperplanes are constructed to maximize the interplane distance which separates the data. The SVM can solve nonlinear problems in a higher-dimensional space based on radical basis function (RBF) to construct a linear function. Usually, the SVM algorithm includes both support vector machine for classification (SVC) and support vector machine for regression (SVR); the former is used to solve problems of classification, while the latter is used for regression analysis.

RBF is a commonly used kernel function in the SVM algorithm. It has a strong ability to deal with nonlinear problems. It can be expressed as follows:

$$K(x, y) = \exp\left(-g|x - y|^2\right), \quad g > 0. \tag{1}$$

RBF has two important parameters in the SVM algorithm, i.e., penalty factor "C" and kernel function parameter "g," which have a great influence on model prediction ability, and the values should be determined during the model optimization process. Commonly, the optimization methods include the grid search (GS), particle swarm optimization (PSO), and genetic algorithm (GA). The PSO algorithm simulates the flight foraging behaviors of bird clusters through collaboration among birds to achieve the best objective. The GA is an operation based on biological natural selection and genetic mechanism to realize the optimal result, while the GS iterates through every intersection in the grid to find the best combination of parameters (C, g) and makes cross-validation most accurate. The RMSECV was used to guide the optimization of internal parameters.

During modeling of the SVM algorithm, the input data need to be mapped to a higher-dimensional space to realize dimension reduction and regression fitting. So, the data should be firstly pretreated and compressed.

2.8. Model Validation and Evaluation

2.8.1. Qualitative Model. In the PCA-SVC qualitative model, the model performance was evaluated by 3-fold cross-validation (3-CV) of the calibration set. The internal parameters C and g were optimized with the GS method. When the accuracy of 3-CV reached maximum values, the optimal C and g were determined. After models were established, the calibration set and prediction set were input to the model, and the prediction accuracies were used as indexes to evaluate the prediction ability.

2.8.2. Quantitative Model. During the process of modeling, the calibration set was used for internal cross-validation to validate model performance, the internal cross-validation adopted 6-fold cross-validation, and the root mean square error of internal cross-validation (RMSECV), coefficient of determination (R^2), and ratio of performance to deviation (RPD) were taken to guide the model optimization process. The predication set was used for external validation to evaluate the model, with the root mean square error of prediction (RMSEP), R^2, and RPD taken as indexes to evaluate prediction ability. Generally, the smaller the RMSECV and the larger the R^2 are, the better the model performance would be; the smaller the RMSEP and the greater the R^2 are, the stronger the model prediction ability is. Moreover, when RPD >2, it indicates that the model has excellent reliability. After the model was established, the calibration set used for full cross-validation was input to the model again, and the root mean square error of evaluation (RMSEE) was used as an index to further evaluate the reasonability of the model. Theoretically, the RMSECV value was higher than the RMSEE value, which indicated that the modeling process was reasonable and feasible.

3. Results and Discussion

3.1. Determination of Nitrogen Content. The content of 64 samples was measured. Samples used in the analysis are as follows: 25 specimens of *S. mutilans*, 13 of *S. multidens*, 9 of *S. mojiangica*, 8 of *S. negrocapitis*, and 9 of *S. dehaani*. The nitrogen content of species was between 8.19% and 12.27%, the total mean was 10.0%, and the mean value of each species was 10.25%, 11.04%, 8.19%, 10.58%, and 12.27%, respectively. The results are shown in Table 1.

3.2. Analysis of NIR Spectra. The NIR spectra of samples were scanned in the range of 12000–4000 cm^{-1}; the spectra diagram is shown in Figure 1. It indicated that the characteristic wavenumber was mainly in the range of 9000 to 4000 cm^{-1}, while the spectral characteristics showed high similarity among samples that it was difficult to distinguish species from peak data. Hence, the chemometric method was needed for spectral pretreatment and characteristic information extraction in qualitative and quantitative analysis.

3.3. Qualitative Model Based on PCA-SVC Algorithm

3.3.1. Sample Classification. The spectra of 64 samples were randomly classified into calibration and prediction sets in a proportion of approximately 2 : 1. Finally, 42 samples of the calibration set were used for model establishment, and 22 samples of the prediction set were used for model evaluation. The species were represented with category label numbers 1 to 5. The classification information is shown in Table 2.

3.3.2. Optimization of Pretreatment. In this qualitative analysis, the three methods VN, FD, and SD were used to pretreat the raw spectra. The PCA method was used to reduce dimensions of raw and three pretreated spectra. The accumulative contribution rates of PCs were calculated. The result showed that the contribution rates of the first two PCs (PC1 and PC2) were more than 85%, which can represent most of the spectrum information [22]. Hence, the PC1-PC2 correlation diagram was attempted for a preliminary investigation to differentiate the samples. However, most species overlapped together in space and cannot be discriminated, except *S. mojiangica* in the FD and SD.

To further investigate the influence of different pretreatments, a group of PCA-SVC models was established using the scores of the first 2 PCs as input variables and category labels as output variables. The model performance was evaluated by 3-fold cross-validation (3-CV) of the calibration set. The internal parameters of the SVC algorithm were optimized with the GS method. The values of best C and g were obtained based on the initial optimization in a range of $\log 2 c \in [0, 50]$ and $\log 2 g \in [0, 50]$ and in steps of 5 and then the second fine optimization in an adjusted narrow scope. When the 3-CV accuracy reached the maximum, the optimal values of C and g were determined. After models were established, the calibration set and prediction set were input to models and predicted, and the prediction accuracies were used to evaluate the prediction ability. As shown in Table 3, it can be seen that "overfitting" and mismatching existed in the raw spectrum model for high accuracy (90.48%) in the calibration set and low accuracy (59.09%) in the prediction set. In contrast, the other three models VN, FD, and SD had nearer accuracies between calibration and prediction sets to possess a relatively rational structure, although the values were not even high. Also, the model of the SD had the highest accuracy among all pretreatments, whether in the calibration or prediction set. Therefore, the SD was regarded as the best pretreatment for its better prediction ability.

3.3.3. Optimization of the Number of Principal Components (NPC). Although the SD was determined as the optimal pretreatment in a preliminary investigation, the accuracy in the model with scores of the first 2 PCs as input variables was just about 70%, which did not meet the requirement of discrimination. Hence, the best NPC still needs to be optimized. In light of the modeling and SD pretreatment method mentioned above, 10 PCA-SVC models (SVC-5 to SVC-14) were established using the scores of the first 1, 2, 3, . . ., 10 PCs of the calibration set as input variables. As shown in Table 4,

TABLE 2: Classified information of the qualitative model of medicinal centipedes.

Sample set	*S. multidens*	*S. dehaani*	*S. negrocapitis*	*S. mojiangica*	*S. mutilans*	Total
Calibration set	8	6	5	6	17	42
Prediction set	5	3	3	3	8	22
Label value	1	2	3	4	5	—

TABLE 3: Different spectral pretreatments of PCA-SVC models.

Model number	Pretreatment	NPC	C	g	Accuracy rate (%)		
					3-fold cross-validation	Calibration set	Prediction set
SVC-1	Raw	2	524288	0.03125	54.7619	90.4762	59.0909
SVC-2	VN	2	$6.71089 * 10^7$	0.0078125	64.2857	66.6667	63.6364
SVC-3	FD	2	16	32768	59.5238	64.2857	63.6364
SVC-4	SD	2	$3.35544 * 10^7$	32768	66.6667	71.4286	68.1818

TABLE 4: Comparison on PCA-SVC models established with different NPCs.

Model number	NPC	C	g	Accuracy rate (%)		
				3-fold cross-validation	Calibration set	Prediction set
SVC-5	1	64	$4.29497 * 10^9$	59.5238	80.9524	63.6364
SVC-6	2	$3.35544 * 10^7$	32768	66.6667	71.4286	68.1818
SVC-7	3	$4.1943 * 10^6$	262144	66.6667	85.7143	81.8182
SVC-8	4	2521.38	$1.27148 * 10^7$	71.4286	90.4762	77.2727
SVC-9	5	23170.5	$3.65135 * 10^6$	73.8095	97.6190	72.7273
SVC-10	6	26615.9	794672	73.8095	90.4762	77.2727
SVC-11	7	26615.9	$1.2045 * 10^6$	78.5714	97.6190	81.8182
SVC-12	8	$5.93164 * 10^6$	5792.62	83.3333	100	81.8182
SVC-13	9	131072	262144	80.9524	100	81.8182
SVC-14	10	$1.04858 * 10^6$	65536	80.9524	100	77.2727

the accuracy improved with the increase of the NPC. When the NPC was 8, the accuracy in the calibration set was 100% and in the prediction set was 81.82%. When the NPC was higher than 8, the accuracy in the calibration set was 100%, while the accuracy in the prediction set did not increase or even decreased. Therefore, number 8 was considered the best NPC, and model SVC-12 was the optimal qualitative model.

3.3.4. Validation and Evaluation of PCA-SVC Model. According to the research above, SVC-12 was determined as the best qualitative analysis model. After the full spectrum was pretreated with the SD and the dimension was reduced with PCA, the model was established using the scores of the first 8 PCs as input variables and category labels as output variables. The internal parameters of best C and g were optimized with the GS. The initial search was in a range of $\log 2\, c \in [0, 50]$ and $\log 2\, g \in [0, 50]$ and in steps of 5, and then the fine search was in a range of $\log 2\, c \in [20, 23]$ and $\log 2\, g \in [12, 16]$ and in steps of 0.5; when $C = 5.93164 * 10^6$ and $g = 5792.62$, the accuracy of 3-CV was 83.33%. When the model was predicted with the calibration set and prediction set, the accuracy was 100% (42/42) and 81.82% (18/22), respectively, which might be accepted for rapid identification. The optimizations of internal parameters are shown in Figure 2, and the predictive results are shown in Figure 3.

3.4. Quantitative Model Based on PLSR and PLS-SVR Algorithms

3.4.1. Partition of Sample Set. In this quantitative analysis, the Kennard–Stone (K-S) algorithm was used to divide 64-sample spectra into the calibration set and prediction set in a proportion of 2 : 1 in the MATLAB R2014a software; 42 samples of the calibration set were used for validation, while 22 samples of the prediction set were used for prediction.

3.4.2. PLSR Model. The partial least-squares regression (PLSR) model is one of the multiple linear regression (MLR) models; it can easily realize the ideal linear relationship between input variables (spectral information) and output variables (ingredient contents) after high dimensions are compressed by PLS. PLSR has the desirable property to analyze data that are strongly collinear (correlated), noisy, and independent variables and also simultaneously model several response variables; now, it has been developed as a standard tool in chemometrics [27].

The full spectral data (12000~4000 cm^{-1}) were used for modeling. To eliminate noise and other factors, they need to be firstly pretreated. The pretreatments including Raw, VN, FD, FD + VN, MSC, and FD + MSC were applied. After the dimensions were reduced with PLS, the treated spectral data were used as input variables and nitrogen content was the

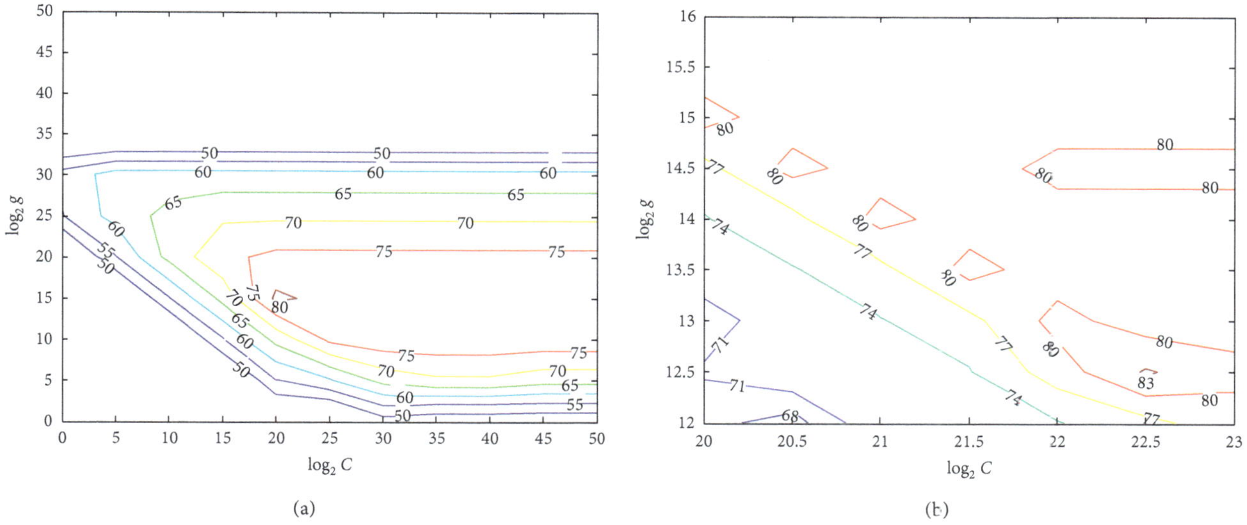

FIGURE 2: Optimization of internal parameters with the grid search of the PCA-SVC model. (a) Initial grid search. (b) Fine search.

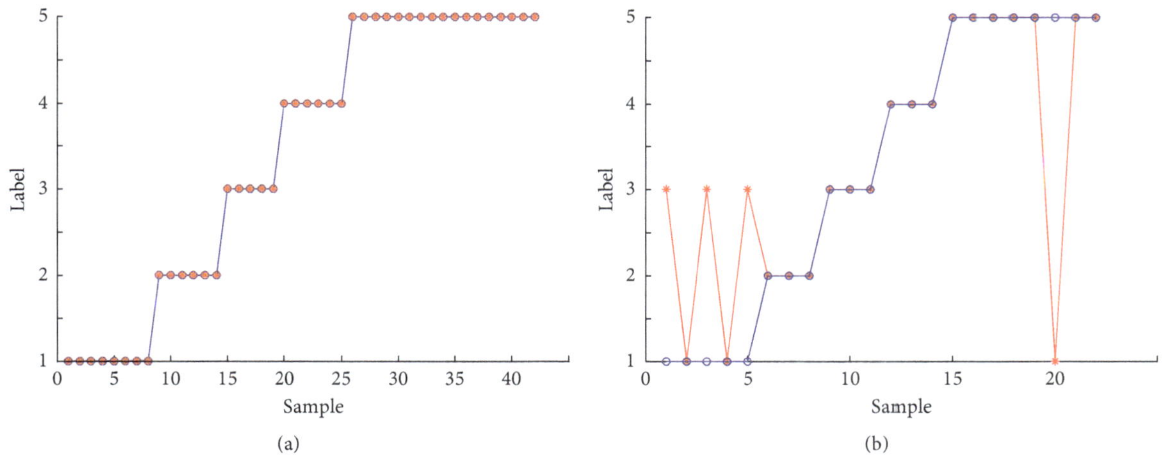

FIGURE 3: Validation results of the PCA-SVC model for medicinal centipedes: (a) calibration set; (b) prediction set. The red points represent validation results, and the blue points represent reference values.

output reference, and a series of PLSR models were established with the Unscrambler 9.7 software.

During the process, the model was validated and evaluated. As shown in Table 5, the RPDs were all over 2 to possess model reliability. All the values of RMSECV were higher than those of RMSEE, which indicated the feasibility of the models. Models PLSR-1, PLSR-4, and PLSR-6 had lower RMSECV and higher R^2 to present good performance, while PLSR-1 had great RMSEP, minimum R^2, in external validation, and the largest NPC, and its structure is unreasonable. There was no significant difference in performance and prediction ability between PLSR-4 and PLSR-6, but considering the rationality of pretreatment, RMSECV, and RMSEE, the PLSR-6 model was considered the best model.

The optimization of the NPC is an important step during modeling. It can be obtained from the RMSECV-NPC diagram. For instance, in the PLSR-6 model, with the change of the NPC, the RMSECV had different values; when the NPC was 5, the RMSECV had a minimum value, and the model had the best performance. Therefore, the optimal NPC was determined as 5. The optimization is shown in Figure 4. The regression equation of the PLSR-6 model between the principal component scores (SPL_i, $i = 1, 2, \ldots, 5$) and the nitrogen content is expressed as follows:

$$\begin{aligned} Y &= 803.21\,SPL_1 + 661.59\,SPL_2 + 589.71\,SPL_3 \\ &\quad + 595.33\,SPL_4 + 476.86\,SPL_5 + 10.26, R^2 = 90.95\%. \end{aligned} \tag{2}$$

As described above, the best PLSR model was finally determined, the optimized pretreatment was determined as FD + MSC, and the NPC was 5. During modeling, 6-fold cross-validation was used as internal validation to validate

Table 5: Validation and predictive results of PLSR models with different pretreatment methods.

Model number	Pretreatment	6-fold cross-validation			External validation			RMSEE (%)	NPC
		RMSECV (%)	R^2 (%)	RPD	RMSEP (%)	R^2 (%)	RPD		
PLSR-1	Raw	0.42	90.50	2.85	0.51	80.78	2.30	0.27	9
PLSR-2	VN	0.47	87.22	2.51	0.43	84.3	2.51	0.34	5
PLSR-3	FD	0.46	88.14	2.53	0.46	84.41	2.39	0.31	6
PLSR-4	FD + VN	0.41	90.71	3.04	0.43	85.84	2.53	0.32	5
PLSR-5	MSC	0.44	89.63	2.69	0.50	81.72	2.39	0.26	8
PLSR-6	FD + MSC	0.40	90.95	2.96	0.44	85.61	2.51	0.31	5

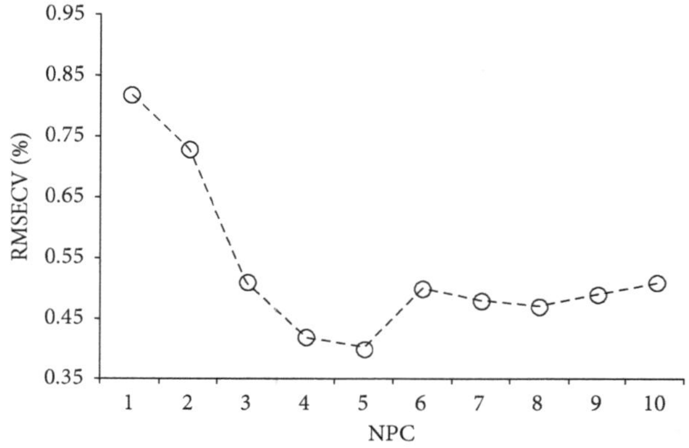

Figure 4: RMSECV-NPC diagram of the PLSR-6 model.

the performance, and the predictive ability was evaluated with external validation using the prediction set. The predictive results are shown in Figure 5. The average relative deviation between the predictive value and the reference was 2.71%, and the average recovery was 98.77%.

3.4.3. PLS-SVR Model. Besides ingredient information, the NIR spectroscopy also contains much other information, such as physical and chemical information, which often causes spectral bands seriously overlapped. Actually, in most cases, it shows nonlinear relationship between sample spectra and content. With the development of application of chemometrics, modern intelligent algorithms have attached more attention to NIR spectroscopy analysis for its strong nonlinear fitting ability and obtained preliminary exploration and application. The SVM algorithm is based on statistics to allow obtain a good fitting effect and stable structure. As a result, it becomes a commonly used nonlinear regression algorithm. Compared with the ANN algorithm which is suitable for solving problems of complex mapping and large sample size [28], the SVR model has undergone much application to become a relatively mature model, and it is suitable for small sample size.

In this study, an SVR algorithm combined with dimensions reduced by the PLS was used to establish a nonlinear regression model. When the parameters determined in the PLSR model (the pretreatment was FD + MSC, and dimensions reduced with PLS and NPC were 5) were introduced into the SVM algorithm, the SVR models were performed in the MATLAB R2014a software. The GS and GA were adopted to optimize the internal parameters (C, g). The model performance was validated with 6-fold cross-validation using the calibration set, and the prediction ability was evaluated with external validation using the prediction set. As shown in Table 6, the RMSECV in PLS-SVR-2 was 0.34, which is less than 0.4 in PLS-SVR-1, and the R^2 in PLS-SVR-2 was 93.29, which is larger than 91.54 in PLS-SVR-1. Therefore, the PLS-SVR-2 model with internal parameters (C, g) optimized with the GS had relatively excellent performance, and it was regarded as the suitable SVR model. The optimization of internal parameters (C, g) and predictive results are shown in Figure 6.

3.4.4. Analysis and Evaluation of Quantitative Models. In this study, the linear regression model of PLSR and nonlinear regression model of PLS-SVR were successfully established. As shown in Tables 5 and 6, the prediction ability had no significant difference between the two models. The relative deviations between the predictive value and the reference were 2.71% and 2.69%, respectively, and the average recoveries were 98.77% and 99.02%. Totally, the two optimized models had a reasonable structure and good prediction ability. Both of them could meet the requirements of accuracy and precision of quantitative analysis and could be used for nitrogen content analysis and quality evaluation of medicinal centipedes.

However, the PLSR model was built based on a linear regression algorithm to have characteristics of fast fitting and simple calculation, when the analysis requirements were not too high, and it would be widely used. In contrast, the SVR model was established based on the nonlinear regression algorithm, and it had the strong nonlinear fitting ability. It

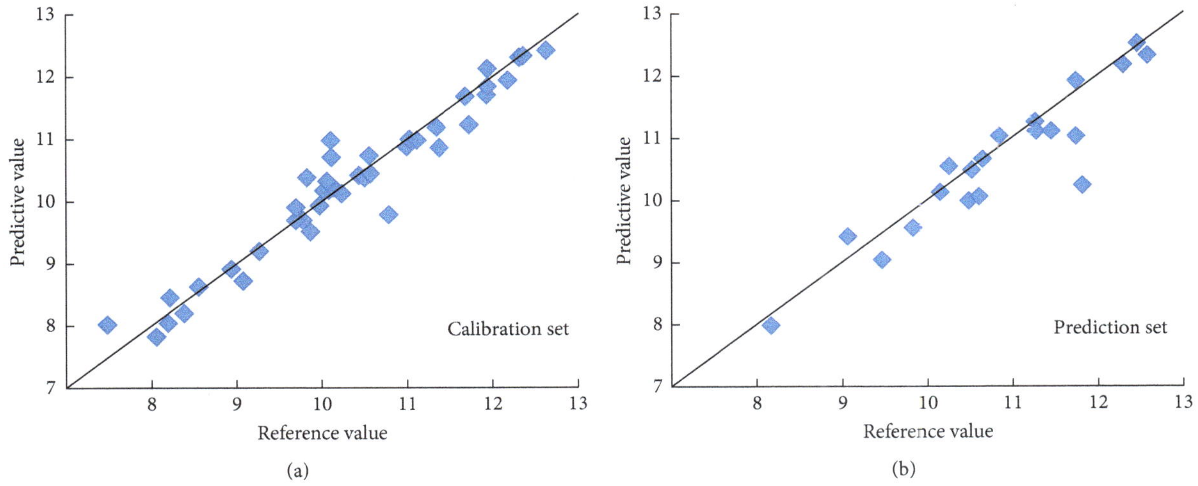

FIGURE 5: Predictive results in the calibration set (a) and prediction set (b) of the PLSR-6 model.

TABLE 6: Validation and evaluation results of SVR models.

Model number	Optimization method	C	g	6-fold cross-validation			External validation			RMSEE (%)
				RMSECV (%)	R^2 (%)	RPD	RMSEP (%)	R^2 (%)	RPD	
PLS-SVR-1	GA	99.99	997.03	0.4	91.54	2.91	0.41	85.89	2.55	0.34
PLS-SVR-2	GS	512	1024	0.34	93.29	3.72	0.43	85.5	2.54	0.32

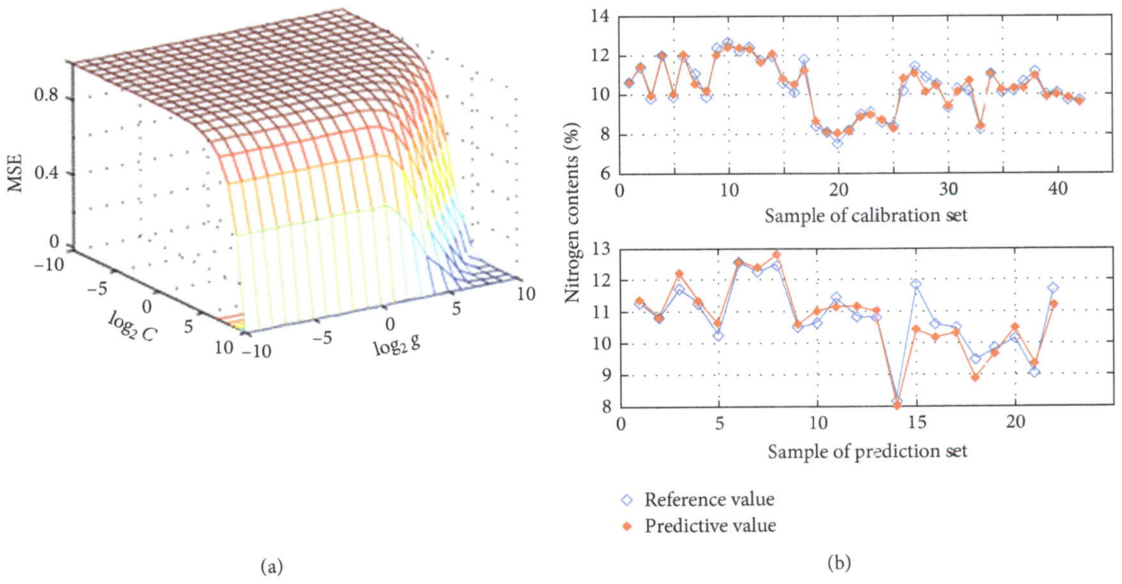

FIGURE 6: Parameter optimization (a) and predictive results (b) of the PLS-SVR-2 model.

was shown from Tables 5 and 6 that the values of RMSECV and RMSEP of SVR models were generally less than those of PLSR models, and the values of R^2 of SVR models were commonly higher than those of PLSR models. It was indicated that the PLS-SVR model had perfect performance and better prediction effect than the PLSR model. For this reason, the PLS-SVR model now becomes the most widely used regression model in NIR spectroscopy analysis. Therefore, the PLS-SVR model with internal parameters (C, g) optimized with the GS was considered the most suitable NIR spectroscopy quantitative model for nitrogen content analysis of medicinal centipedes, and the PLSR model can act as supplement and verification for the analysis.

4. Conclusions

This study was carried out to explore the feasibility of using the NIR spectroscopy method to rapidly differentiate species and evaluate the quality of Chinese medicinal centipedes. In the qualitative analysis, after spectra were pretreated with the SD, dimensions were reduced with PCA, and internal parameters were optimized with the GS algorithm, a PCA-SVC model was set up using the scores of the first 8 PCs as input variables and category labels as output variables. The optimal model (SVC-12) was validated and evaluated, which could identify five species of medicinal centipedes with an accuracy of 100% (42/42) in the calibration set and 81.82% (18/22) in the prediction set. It could be accepted as an objective, rapid, and auxiliary method for identifying the species of medicinal centipedes. Through the spectra pretreated with FD + MSC, data dimension reduced with PLS, and NPC determined as 5, two best quantitative models of PLSR and PLS-SVR were also successfully determined. During the process of modeling, the RMSECV, R^2, and RPD of 6-fold internal cross-validation in the calibration set indicated the better performance and stronger modeling capacity. The RMSEP, R^2, and RPD of external validation in the prediction set proved stronger prediction ability. In addition, to investigate the reasonability of the model, the calibration set used for full cross-validation was input to the models again, and the RMSEE was used as an index. Comparing the characteristics and advantages of two different regression algorithms, the PLS-SVR-2 model had excellent performance and strong prediction capacity, and it was finally considered the most suitable quantitative model of NIR spectroscopy for nitrogen content analysis of medicinal centipedes.

Meanwhile, the pretreatment methods were also optimized in this paper; although the SD was determined in the qualitative model, MSC or its combined methods were applied to pretreat the spectra in quantitative models. The MSC had advantages of weakening or eliminating interference caused by the uneven grain size of solid powder in the diffuse reflection spectrum [29]. In this research, all samples were smashed into powder, and the NIR spectra were obtained with a diffuse reflection spectrum, so the application of FD + MSC was proved to be reasonable.

This study indicated that NIR spectroscopy combined with chemometric algorithms could be successfully used to differentiate species and evaluate the quality of medicinal centipedes in China, which was characterized with rapid, nondestructive, and environmentally friendly properties. However, this study just represented preliminary exploratory research; although 28 batch samples and 64 individuals were conducted, the sample size was still limited. In the future, more samples will be used to improve the prediction ability, and other algorithms will also be considered to simplify the model and improve performance. This study also provided a reference for rapid identification and quality analysis of other animal medicinal materials using NIR spectroscopy.

Conflicts of Interest

The authors declare that there are no conflicts of interest regarding the publication of this paper.

Acknowledgments

This work was supported by a grant from the major drug discovery projects of the National Ministry of Science and Technology of China (no. 2014ZX09304307001).

References

[1] W. Siriwut, G. D. Edgecombe, C. Sutcharit, and S. Panha, "The centipede genus Scolopendra in mainland southeast asia: molecular phylogenetics, geometric morphometrics and external morphology as tools for species delimitation," *PLoS One*, vol. 10, no. 8, Article ID e0135355, 2015.

[2] S. H. Kang, H. Y. Deng, Z. Y. Jiang, Y. M. Liu, J. Li, and K. L. Chen, "Taxonomy and distribution of Chinese medicinal centipedes," *Journal of Chinese Medicinal Materials*, vol. 39, pp. 727–731, 2016.

[3] S. Yang, Y. Xiao, D. Kang et al., "Discovery of a selective NaV1.7 inhibitor from centipede venom with analgesic efficacy exceeding morphine in rodent pain models," *Proceedings of the National Academy of Sciences*, vol. 110, no. 43, pp. 17534–17539, 2013.

[4] Y. Kong, "Cytotoxic and anticoagulant peptide from Scolopendra subspinipes mutilans venom," *African Journal of Pharmacy and Pharmacology*, vol. 7, no. 31, pp. 2238–2245, 2013.

[5] I.-J. Jo, G. S. Bae, K. C. Park et al., "Scolopendra subspinipes mutilans protected the cerulein-induced acute pancreatitis by inhibiting high-mobility group box protein-1," *World Journal of Gastroenterology*, vol. 19, no. 10, pp. 1551–1562, 2013.

[6] W. Ma, D. Zhang, L. Zheng, Y. Zhan, and Y. Zhang, "Potential roles of Centipede Scolopendra extracts as a strategy against EGFR-dependent cancers," *American Journal of Translational Research*, vol. 7, no. 1, pp. 39–52, 2015.

[7] W. Ma, R. Liu, J. Qi, and Y. Zhang, "Extracts of centipede Scolopendra subspinipes mutilans induce cell cycle arrest and apoptosis in A375 human melanoma cells," *Oncology Letters*, vol. 8, no. 1, pp. 414–420, 2014.

[8] Y. Kong, S.-L. Huang, Y. Shao, S. Li, and J.-F. Wei, "Purification and characterization of a novel antithrombotic peptide from Scolopendra subspinipes mutilans," *Journal of Ethnopharmacology*, vol. 145, no. 1, pp. 182–186, 2013.

[9] M. Stankiewicz, A. Hamon, R. Benkhalifa et al., "Effects of a centipede venom fraction on insect nervous system, a native Xenopus oocyte receptor and on an expressed Drosophila muscarinic receptor," *Toxicon*, vol. 37, no. 10, pp. 1431–1445, 1999.

[10] H. Y. Zhang, J. Chen, J. Jia et al., "Identification of Scolopendra subspinipes mutilans and its adulterants using DNA barcode," *China Journal of Chinese Materia Medica*, vol. 39, p. 2208, 2014.

[11] H. Fang, F. Deng, and K. Q. Wang, "Chemical analysis of Scolopendra multidens newport," *Chinese Pharmaceutical Journal*, vol. 32, pp. 202–204, 1997.

[12] H. Fang, F. Deng, Y. C. Yan, and K. Q. Wang, "Chemical constituents of Scolopendra negrocapitis," *Journal of Chinese Medicinal Materials*, vol. 22, no. 5, pp. 226–228, 1999.

[13] X. Chen, H. M. Wen, R. Liu et al., "Analysis of extracted proteins of Scolopendra by nanoflow reversed phase liquid chromatography-tandem mass spectrometry," *Chinese Journal of Analytical Chemistry*, vol. 42, pp. 239–243, 2014.

[14] J. Wang, Y. P. Chu, S. S Yang et al., "Influences of sterilization method on the nitrogen content of Scolopendra," *Asia-Pacific Traditional Medicine*, vol. 12, pp. 27-28, 2016.

[15] Y. He, X. L. Li, and Y. N. Shao, "Discrimination of varieties of apple using near infrared spectra based on principal component analysis and artificial neural network model," *Spectroscopy and Spectral Analysis*, vol. 26, no. 5, pp. 850–853, 2006.
[16] C. Xie, N. Xu, Y. Shao, and Y. He, "Using FT-NIR spectroscopy technique to determine arginine content in fermented Cordyceps sinensis mycelium," *Spectrochimica Acta Part A: Molecular and Biomolecular Spectroscopy*, vol. 149, pp. 971–977, 2015.
[17] Y. Sun, L. Chen, B. Huang, and K. Chen, "A rapid identification method for calamine using near-infrared spectroscopy based on multi-reference correlation coefficient method and back propagation artificial neural network," *Applied Spectroscopy*, vol. 71, no. 7, pp. 1447–1456, 2017.
[18] Z. S. Song, D. X. Song, and M. S. Zhu, "Systematic classification of chilopoda and the order scolopendromorpha (myriapoda)," *Journal of Liaoning Normal University*, vol. 27, no. 1, pp. 69–72, 2004.
[19] C. Z. Zhang and K. Q. Wang, "A new centipede, Scolopendra negrocapitis sp. nov. from Hubei province, China (Chilopoda: Scolopendromorpha: Scolopendridae)," *Acta Zootaxonomica Sinica*, vol. 24, no. 2, pp. 136-137, 1999.
[20] X. L. Chu, H. F. Yuan, and W. Z. Lu, "Progress and application of spectral data pretreatment and wavelength selection methods in NIR analytical technique," *Progress in Chemistry*, vol. 16, no. 4, pp. 528–542, 2004.
[21] Q. Kang, Q. Ru, Y. Liu et al., "On-line monitoring the extract process of Fu-fang Shuanghua oral solution using near infrared spectroscopy and different PLS algorithms," *Spectrochimica Acta Part A: Molecular and Biomolecular Spectroscopy*, vol. 152, pp. 431–437, 2016.
[22] W. Z. Lu, *Near Infrared Spectroscopy Instrument*, Chemical Industry Press, Beijing, China, 2010.
[23] G. Bázár, R. Romvári, A. Szabó, T. Somogyi, V. Éles, and R. Tsenkova, "NIR detection of honey adulteration reveals differences in water spectral pattern," *Food Chemistry*, vol. 194, pp. 873–880, 2016.
[24] Z.-S. Wu, L.-W. Zhou, S.-Y. Dai, X.-Y. Shi, and Y.-J. Qiao, "Evaluation of the value of near infrared (NIR) spectromicroscopy for the analysis of glycyrrhizic acid in licorice," *Chinese Journal of Natural Medicines*, vol. 13, no. 4, pp. 316–320, 2015.
[25] M. Y. Yuan, B. S. Huang, C. Yu, Y. M. Liu, and K. L. Chen, "A NIR qualitative and quantitative model of 8 kinds of carbonate-containing mineral Chinese medicines," *China Journal of Chinese Materia Medica*, vol. 39, pp. 267–272, 2014.
[26] V. Vapnik, *The Nature of Statistical Learning Theory*, Springer Berlin Heidelberg, New York, NY, USA, 1995.
[27] S. Wold, M. Sjöström, and L. Eriksson, "PLS-regression: a basic tool of chemometrics," *Chemometrics and Intelligent Laboratory Systems*, vol. 58, no. 2, pp. 109–130, 2001.
[28] L. Chen, J. Wang, Z. Ye et al., "Classification of Chinese honeys according to their floral origin by near infrared spectroscopy," *Food Chemistry*, vol. 135, no. 2, pp. 338–342, 2012.
[29] W. M. Wang, D. M. Dong, W. G. Zheng, X. D. Zhao, L. Z. Jiao, and M. F. Wang, "Pretteatment method of near-infrared diffuse reflection spectra for sugar content prediction of pears," *Spectroscopy and Spectral Analysis*, vol. 33, no. 2, pp. 359–362, 2013.

19

Calibration Transfer of Soil Total Carbon and Total Nitrogen between Two Different Types of Soils Based on Visible-Near-Infrared Reflectance Spectroscopy

Xue-Ying Li,[1,2,3] **Yan Liu**,[1,2,3] **Mei-Rong Lv**,[1,2,3] **Yan Zou**,[1,2,3] **and Ping-Ping Fan**[1,2,3]

[1] *Institute of Oceanographic Instrumentation, Qilu University of Technology (Shandong Academy of Sciences), Jinan, China*
[2] *Shandong Provincial Key Laboratory of Ocean Environmental Monitoring Technology, Qingdao, China*
[3] *National Engineering and Technological Research Center of Marine Monitoring Equipment, Qingdao, China*

Correspondence should be addressed to Ping-Ping Fan; fanpp_sdioi@126.com

Academic Editor: Khalique Ahmed

Although visible-near-infrared spectroscopy can rapidly and accurately determine soil nutrients without sample destruction, some problems remain unresolved, such as the mismatch of the established spectral model with different types of samples, limiting the wide application of this technology. Here, we took riverside and mountain soils as examples to explore the calibration transfer between two different types of soils by the WMPDS-S/B algorithm (wavelet multiscale piecewise direct standardization combined with Slope/Bias correction method) and by adding new samples. The predicted TN and TC concentrations improved significantly after being transformed. Compared with adding new samples, the WMPDS-S/B algorithm obtained more accurate results. The average relative errors dropped from 440.2% (without transformation) to approximately 6% for TN and from 342.0% to approximately 7% for TC. The maximum relative errors were reduced from 538.1% to less than 20% for TN and from 403.7% to less than 20% for TC. The RMSEP decreased from 2.42 to approximately 0.04 for TN and from 15.74 to approximately 0.4 for TC. The WMPDS-S/B algorithm had advantages in selecting fewer known samples and obtaining better prediction results. In contrast to past studies, which resolved the calibration transfer between different spectrometers and the measurement environment for the same samples, our study resolved the calibration transfer between different types of samples under the same spectrometer and the measurement environment. The former could only be used for correction among instruments, while the latter fundamentally solved the problem of model sharing across different samples.

1. Introduction

Spectral analysis is fast, accurate, and nondestructive, and it can provide rich material information [1–3]; thus, it is widely used in many fields [4–9]. In agriculture, soil nutrients have been successfully predicted by the spectral model, which is established by the known soils sampled in the same sites and under the same environment [10–14]. That is, the spectral model can work only within a specific range. If the unknown samples are different from the model samples due to altering sampling sites, seasons, or other factors, the spectral model will lead to inaccurate results. These questions have hindered the progress of spectroscopy technology; therefore, a resolution for the problem is highly needed.

Currently, studies are concentrated on the similar questions for the same samples derived from different instruments, measurement conditions, or other aspects (called calibration transfer) [15–19]. Many algorithms have successfully resolved these calibration transfer, such as direct standardization (DS) [20, 21], piecewise direct standardization (PDS) [22, 23], orthogonal signal method (OSC) [24, 25], wavelet transform (WT) algorithm [26, 27], and canonical correlation analysis(CCA) [28, 29]. Successful calibration transfer between the same type of samples could keep the accuracy of the same set of spectrometers; however, it could not resolve how to make a spectrometer share a stable spectral model for different types of samples.

Studies on calibration transfer between different types of samples are very scarce. The main solution was to add new samples in the original spectral model. Then, by using the enriched sample set, a new spectral model was rebuilt and was used for predicting the different samples. This method needed to rebuild the spectral model; thus, it was labor-intensive and time-consuming, and the prediction capability of the new spectral model was reduced. Therefore, it was urgent to explore other methods to predict unknown samples of different types quickly and accurately.

The piecewise direct standardization (PDS) algorithm was a calibration transfer method proposed by Wang and Kowalski in 1991 [30]. It required few samples, transferred the correction model faster, and improved the robustness of calibration effectively, so PDS was the most popular method for calibration transfer. However, a fixed transformation matrix was used for the whole spectral data and might not be suitable for every specific spectrum region [31, 32]. The disadvantages of PDS could be avoided by wavelet analysis. Wavelet analysis could disassemble a signal by layers and dimensions; therefore, it had good localization properties and could provide frequency information in the target signal. Its multiresolution analysis combined with PDS had been successfully used in calibration transfer for the same samples derived from different instruments [33–35]. We hypothesized that it could also be used for calibration transfer between different types of samples.

Concentrations of total nitrogen (TN) and total carbon (TC) in soils were better predicted than other soil nutrients by spectral models and were thus selected for the study of calibration transfer [36, 37]. The Slope/Bias algorithm (S/B) was often used to correct the result of the chemical values in calibration transfer, which rebuilt the linear relationships based on abundant data [38]. For instance, Cooper et al. provided a simple slope and bias correction to transfer calibration with different instruments [39]. We would also use S/B to reduce the differences between concentrations of soil nutrients.

Here, we explored the calibration transfer of TN and TC concentrations between riverside and mountain soil samples using wavelet multiscale piecewise direct standardization combined with the Slope/Bias correction method (WMPDS-S/B). Additionally, we compared the results with those by adding new samples. To reduce the influence of other factors on calibration transfer, we sampled spectral data from the same equipment and the same test environment.

2. Experimental

2.1. Samples and Data Acquisition. Two types of upper soils (0–20 cm) were collected from the riverside of Licun River and the foot of Fushan Mountain in Qingdao, China. Licun River soil (riverside soil) was silt loam rich in nutrients, while Fushan Mountain soil (mountain soil) was sandy loam poor in nutrients. We removed foreign bodies such as stones from the soil samples. After drying to a constant weight, all soil samples were filtered through a 0.45 mm nylon screen. Soil TN and TC concentrations were measured with a carbon and nitrogen analyzer (Perkin Elemental Analyzer, USA). The concentrations of TN and TC between two types of soils have a large difference (Table 1).

Soil spectral data were determined by an Ocean Optics QE65000 spectrometer. The sampling interval was 1 nm, and the integral time was 1000 ms. The spectral range was 200–1100 nm, mainly the visible and near-infrared spectrum, including a small amount of the ultraviolet spectrum. The optical fiber probe was inserted into the hole of the probe bracket at a 45° angle, and this made the probe stick to the bracket and just exposed the bracket.

Soil (3–5 g) was gently flattened in a homemade sample box, whose size was the same as the probe bracket and whose cell overlapped the hole of the probe bracket. The schematic diagram of soil spectral data measurements is shown in Figure 1. Each soil sample was measured five times. To reduce the influence of the noise in the reflectance spectrum, the spectral data of 226–975 nm were retained. The reflectance spectrum of the river and mountain soils is shown in Figure 2.

The principal component space was constructed by the scores of the spectral principal component (Figure 3), and the distribution of the principal components in the samples could be used to determine whether the unknown samples were fit to the original model. As shown in Figure 3, the river soil and mountain soil samples were divided into two regions in the principal component space. This outcome indicated that the spectrum of the two types of soil samples was obviously different. Therefore, the model established by the river soil was hardly suitable for predicting the mountain soil directly.

2.2. Data Treatment and Analysis

2.2.1. Preetreatment and Modeling. The reflectance spectral data (226–975 nm) of the river and mountain soils were pretreated using Savitzky–Golay smoothing and differentiation. The calibration set and test set of the two types of soil samples were divided into 3 : 1 proportions by the Kennard–Stone algorithm, with 45 soil samples as the calibration set and the rest of the samples as the test set. The spectral model of the calibration set was built by partial least squares (PLS) [40, 41] (Table 2). Results showed that a better spectral model was built by the river soil (Table 2); therefore, it was set as the First Sample and mountain soil was set as the Second Sample.

2.2.2. Calibration Transfer Algorithm. Here, we use two methods: wavelet multiscale piecewise direct standardization combined with the Slope/Bias correction method (WMPDS-S/B) and the method of adding new samples. WMPDS-S/B is more complicated. Its specific steps are described below.

(1) Wavelet decomposition is used to obtain the coefficients of each wavelet.

First, the wavelet generating function and the number of wavelet decomposition layers N are selected. Then, the spectrum of the First Sample Xm and the spectrum of the

Table 1: Concentrations of TC and TN in the river and mountain soils.

Samples	Number	TN concentrations (mg/kg)		TC concentrations (mg/kg)	
		Range	Mean ± SD	Range	Mean ± SD
River soil	60	4.14–5.66	4.70 ± 0.40	3.8–13.4	10.58 ± 2.24
Mountain soil	58	0.44–0.69	0.55 ± 0.05	0.72–1.74	1.40 ± 0.26

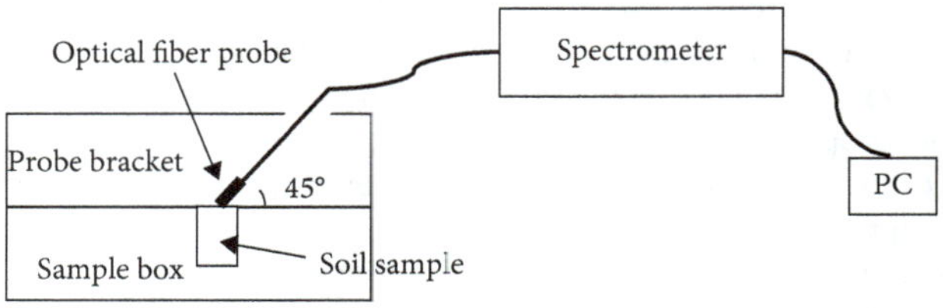

Figure 1: The schematic diagram of soil spectral data measurements.

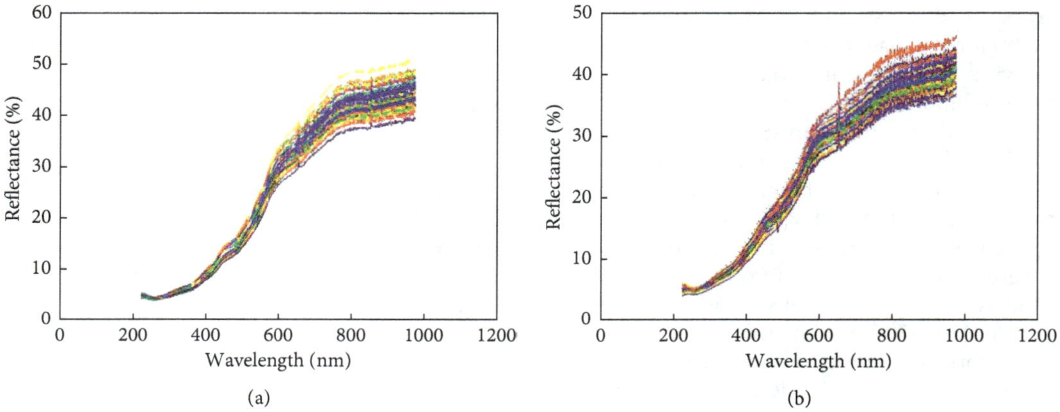

Figure 2: Visible-near-infrared reflectance spectrum of the river soil (a) and mountain soil (b).

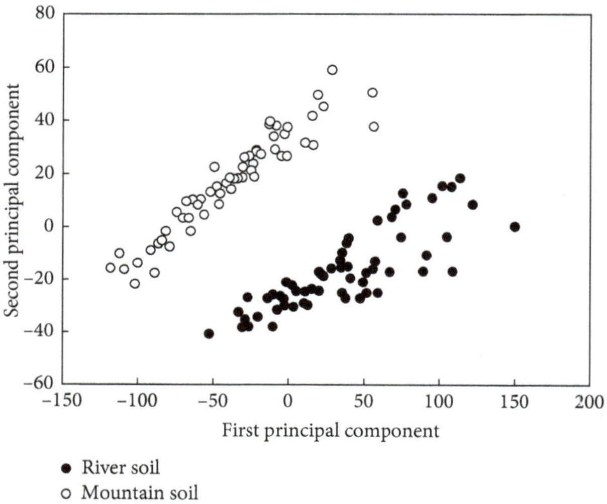

Figure 3: Principal component score diagram of the river and mountain soils.

Second Sample Xs are decomposed by wavelet analysis, and the approximate coefficients Xm_a and Xs_a and detail coefficients Xm_{dn} and Xm_{dn} $(n = 1, 2, \ldots, N)$ of the First Sample and the Second Sample are obtained. The correlation formula is as follows:

$$\begin{aligned} Xm &\xrightarrow{waveletdecomposition} Xm_a + Xm_{d1} + \cdots + Xm_{dN}, \\ Xs &\xrightarrow{waveletdecomposition} Xs_a + Xs_{d1} + \cdots + Xs_{dN}. \end{aligned} \tag{1}$$

(2) The transformation matrices $F_n (n = 0, 1, \ldots, N)$ of each wavelet coefficients are calculated using the PDS algorithm.

A part of the soil samples from the Second Sample are taken as standard samples for calculating the transformation matrices, denoted as standard set Xs_{standard}. The remainders of the Second Sample are unknown samples, denoted as unknown set Xs_{unknown}. If we want to calculate the transformation matrices F_0 of the approximate coefficients between the calibration set of the First Sample and the standard

Table 2: Spectral models built by the river and mountain soils.

Samples	Number	TN concentrations (mg/kg)					TC concentrations (mg/kg)				
		r_c	r_p	RMSEC	RMSEP	RPD	r_c	r_p	RMSEC	RMSEP	RPD
River soil	60	0.97	0.85	0.064	0.142	1.49	0.93	0.85	0.869	1.095	1.64
Mountain soil	58	0.40	0.36	0.046	0.031	1.09	0.86	0.60	0.194	0.323	1.28

set of the Second Sample $Xm_{\text{a,training}}$, $Xs_{\text{a,standard}}$, the average spectral Mm_{a}, Ms_{a} of $Xm_{\text{a,training}}$, $Xs_{\text{a,standard}}$ are first obtained, and the formulas are as follows:

$$Mm_{\text{a}}(j) = \frac{1}{n_1}\sum_{i=1}^{n_1} Xm_{\text{a,training}}(i,j),$$
$$Ms_{\text{a}}(j) = \frac{1}{n_2}\sum_{i=1}^{n_2} Xs_{\text{a,standard}}(i,j), \tag{2}$$

where (i, j) is the jth wavelength point of the ith sample and n_1 and n_2 are the number of samples for $Xm_{\text{a,training}}$ and $Xs_{\text{a,standard}}$.

The spectral band of the window width $(j-k \sim j+k)$ is intercepted near the jth wavelength point. Let $Z_{a,j} = [Ms_{a,j-k}, \ldots, Ms_{a,j}, \ldots, Ms_{a,j+k}]$, make $Mm_a(j) = Z_{a,j} \times f_{a,j}$, and then the regression coefficients $f_{a,j}$ are calculated. Loop j, and find all the $f_{a,j}$. Put $f_{a,j}$ on the main diagonal of the transformation matrix F_0, and set other elements as 0, and finally get the transformation matrix F_0. According to the above algorithms, we can obtain the transformation matrices $F_1, \ldots, F_N$ of the detail coefficients between the calibration set of the First Sample and the standard set of the Second Sample.

(3) The transfer spectrum of the standard set and unknown set in the Second Sample is obtained by using transformation matrices.

According to the transformation matrix F_0, the transfer approximation coefficients $XsF_{\text{a,standard}}$ and $XsF_{\text{a,unknown}}$ are achieved by the transforming approximation coefficients in the standard set and unknown set of the Second Sample $Xs_{\text{a,standard}}$ and $Xs_{\text{a,unknown}}$.

$$XsF_{\text{a,standard}} = Xs_{\text{a,standard}} \cdot F_0,$$
$$XsF_{\text{a,unknown}} = Xs_{\text{a,unknown}} \cdot F_0. \tag{3}$$

Then, the transfer detail coefficients of the standard set and unknown set in the Second Sample $XsF_{\text{dn,standard}}$ and $XsF_{\text{dn,unknown}}$ $(n = 1, 2, \ldots, N)$ are calculated sequentially. Finally, the transfer spectrum $XsF_{\text{all,standard}}$ and $XsF_{\text{all,unknown}}$ are obtained by reconstructing each wavelet coefficient of the Second Sample.

(4) The Slope/Bias correction method is used to calculate the final prediction value.

The least squares solution is obtained by fitting the transfer spectrum $XsF_{\text{all,standard}}$ and the measured nutrient values Y_{standard} in a standard set of the Second Sample with linear regression, providing the slope *slope* and intercept *bias* of the linear model. By substituting $XsF_{\text{all,unknown}}$ into the formula, we can find the prediction values of the unknown set Y_{unknown} by the following equation:

$$Y_{\text{unknown}} = slope \cdot XsF_{\text{all,unknown}} + bias. \tag{4}$$

In this paper, the wavelet generating function chooses "db3," and the decomposition layer number N is 3.

2.2.3. Evaluation Standard. The evaluation standard used the average relative error (error_a), the maximum relative error (error_m), and the prediction root mean square error (RMSEP), which offered a comprehensive analysis. The smaller the average relative error, the maximum relative error and RMSEP were, the better the calibration transfer was.

3. Results and Discussion

3.1. Calibration Transfer of TN Concentrations between Two Different Types of Soils. The calibration transfer of TN concentrations between two different types of soils was realized by WMPDS-S/B. The PDS window width and the number of the standard set in the Second Sample affected the results of calibration transfer (Figure 4). We made the number of the standard set 15 by the Kennard–Stone algorithm, the window width ranged from 3 to 19, and the interval was 2. The influence of different window widths on the results was studied. We studied the effects of the number of the standard set on the results. Setting the window width to 15, the number of the standard set ranged from 11 to 20 by the Kennard–Stone algorithm, and the interval was 1. When the window width was 15 and the number of the standard set was 14, the average relative error was the smallest (Figure 4). Therefore, we chose this parameter for the subsequent calibration transfer.

Using the WMPDS-S/B algorithm, the average relative error (error_a), the maximum relative error (error_m), and RMSEP were obviously decreased (Table 3). The error_a dropped from 440.2% to 5.77%. The error_m were reduced from 538.1% to 16.67%, and the RMSEP was reduced from 2.416 to approximately 0.041 (Table 3).

We used the optimum number of the standard set to divide the standard set and unknown set in the Second Sample by the Kennard–Stone algorithm; the result of the calibration transfer was accidental (Table 3). To avoid the contingency, we used interval sampling by proportion (Table 4) and random sampling (Table 5) to divide the standard set and unknown set. The two methods of sample classification led to similar results of calibration transfer (Tables 4 and 5).

The different proportions between the standard set and unknown set in the Second Sample by the Kennard–Stone

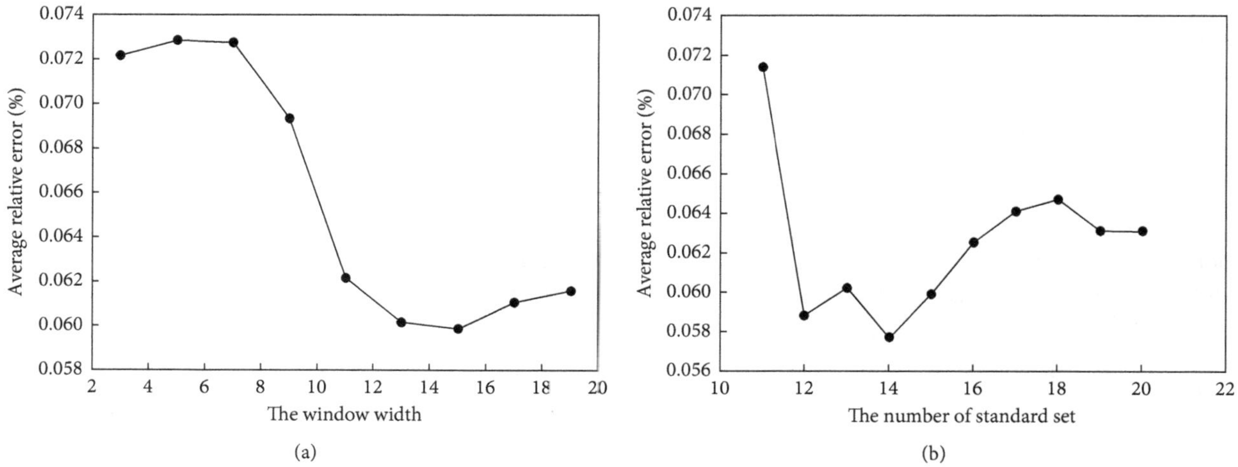

FIGURE 4: Influence of the window width (a) and the number of the standard set (b) on the transfer results of soil TN concentration.

TABLE 3: Comparison of TN concentrations statistical results with and without calibration transfer.

Calibration transfer algorithm	$Error_a$ (%)	$Error_m$ (%)	RMSEP
Without algorithm	440.2	538.1	2.416
WMPDS-S/B	5.77	16.67	0.041

TABLE 4: Comparison of TN concentrations statistical results from dividing standard sets and unknown sets using different proportions.

Proportion	$Error_a$ (%)	$Error_m$ (%)	RMSEP
1:2	5.58	18.93	0.041
1:3	6.05	17.55	0.041
1:4	6.52	15.87	0.044

TABLE 5: Comparison of TN concentrations statistical results from dividing standard sets and unknown sets by using random sample.

Random sampling number	$Error_a$ (%)	$Error_m$ (%)	RMSEP
Random 1	5.34	18.59	0.039
Random 2	5.83	19.76	0.044
Random 3	6.12	17.87	0.044
Random 4	5.93	17.38	0.042
Random 5	5.50	17.79	0.040

algorithm could bring different results of calibration transfer (Table 4). Here, the results showed that. with increased proportion between the standard set and unknown set (from 1 : 2 to 1 : 4), the $error_a$ increased, the $error_m$ decreased, and the RMSEP increased (Table 4). It seemed that the proportion of 1 : 2 was the best, perhaps because the number of the standard set was proper and contained enough information.

Random sampling did not have a big influence on the results of calibration transfer (Table 5). The values of $error_a$, $error_m$, and RMSEP derived from random sampling were similar with those derived from interval sampling by the Kennard–Stone algorithm (Tables 4 and 5).

3.2. Calibration Transfer of TC Concentrations between Two Different Types of Soils. Using the same method, the PDS window width and the number of the standard set in the Second Sample were also found to affect the results of calibration transfer (Figure 5). When the window width was 11 and the number of the standard set was 18, the calibration transfer of the TC concentrations model was the best (Figure 5). Therefore, we chose these parameters for the subsequent calibration transfer of TC concentrations.

Under the best condition, the standard set and unknown set were classified by the Kennard–Stone algorithm, proportion, and random sampling, respectively (Table 6). The accuracy of prediction was significantly improved after the WMPDS-S/B algorithm was used for the calibration transfer (Table 6). The $error_a$ decreased from 341.97 to approximately 6%, the $error_m$ decreased from 403.70% to <20%, and the RMSEP decreased from 15.737 to <0.46 (Table 6).

Among all the prediction results, the best was that from the 1 : 3 proportion between the standard set and unknown set (Table 6), whose $error_a$, $error_m$, and RMSEP were the smallest. The results of the 1 : 2 proportion were relatively poor, maybe because the number of the standard set was too large and some unrelated information was included in the calibration transfer.

Some division approaches could lead to spectral overcorrection and reduce the prediction effect of the calibration transfer. But in this study, when the number of standard set was fixed to 18, the values of $error_a$, $error_m$, and RMSEP were similar across different classification methods (Table 6). This shows that different classification methods produce similar prediction results, suggesting the division approaches in this study were reasonable.

3.3. Comparison between the WMPDS-S/B Algorithm and Adding Samples. Under the condition of selecting the same samples of the standard set by the Kennard–Stone algorithm, we then study the effects of the two methods on the results of calibration transfer in TN and TC concentrations (Tables 7 and 8). The number of the standard set is 10, 15, 20, 25, and

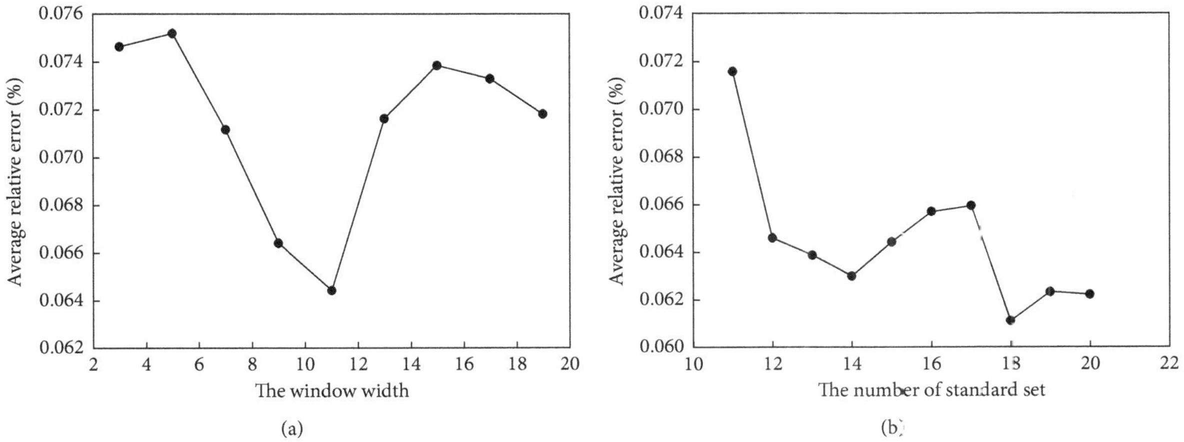

FIGURE 5: Influence of the window width (a) and the number of the standard set (b) on the transfer results of soil TC concentrations.

TABLE 6: Comparison of TC concentrations statistical results with and without calibration transfer.

Calibration transfer algorithm	Standard set division method	$Error_a$ (%)	$Error_m$ (%)	RMSEP
Without algorithm	Kennard-Stone	341.97	403.70	15.737
WMPDS-S/B	Kennard-Stone	6.11	15.37	0.351
WMPDS-S/B	Proportion 1:2	7.17	18.15	0.453
WMPDS-S/B	Proportion 1:3	6.79	15.96	0.380
WMPDS-S/B	Proportion 1:4	6.92	16.27	0.406
WMPDS-S/B	Random 1	5.78	15.75	0.349
WMPDS-S/B	Random 2	6.25	16.54	0.384
WMPDS-S/B	Random 3	6.67	18.57	0.394
WMPDS-S/B	Random 4	6.08	19.06	0.430
WMPDS-S/B	Random 5	6.24	16.50	0.357

TABLE 7: Comparison of different calibration transfer algorithms for the modified result of TN concentrations.

The number of standard set	First Sample			Second Sample		
	Result	WMPDS-S/B	Add samples	Result	WMPDS-S/B	Add samples
10	r_c	0.971	1.000	$Error_a$ (%)	7.52	33.28
	r_p	0.851	0.421	$Error_m$ (%)	20.34	81.10
	RPD	1.49	1.031	RMSEP	0.050	0.214
15	r_c	0.971	0.975	$Error_a$ (%)	5.99	27.81
	r_p	0.851	0.656	$Error_m$ (%)	15.89	76.06
	RPD	1.49	1.069	RMSEP	0.042	0.176
20	r_c	0.971	1.000	$Error_a$ (%)	6.38	25.30
	r_p	0.851	0.594	$Error_m$ (%)	17.18	63.36
	RPD	1.49	1.090	RMSEP	0.043	0.166
25	r_c	0.971	1.000	$Error_a$ (%)	6.21	24.87
	r_p	0.851	0.674	$Error_m$ (%)	15.60	65.58
	RPD	1.49	1.186	RMSEP	0.042	0.168
30	r_c	0.971	1.000	$Error_a$ (%)	6.55	23.68
	r_p	0.851	0.707	$Error_m$ (%)	15.57	64.82
	RPD	1.49	1.195	RMSEP	0.044	0.165

30. The window width of the WMPDS-S/B algorithm in TN concentrations is 15 and in TC concentrations is 11. The results of the First Sample include the relative coefficients of calibration set (r_c), the relative coefficients of test set (r_p), and the RPD. The results of the unknown set of the Second Sample include the $error_a$, $error_m$, and RMSEP.

The prediction performance (the value of the r_c, r_p, and RPD) of the original model did not change by using the

Table 8: Comparison of different calibration transfer algorithms for the modified result of TC concentrations.

The number of standard set	First Sample			Second Sample		
	Result	WMPDS-S/B	Add samples	Result	WMPDS-S/B	Add samples
10	r_c	0.926	0.962	Error_a (%)	6.43	34.69
	r_p	0.853	0.699	Error_m (%)	18.24	86.51
	RPD	1.64	1.016	RMSEP	0.380	1.821
15	r_c	0.926	0.963	Error_a (%)	6.44	31.51
	r_p	0.853	0.733	Error_m (%)	18.80	75.83
	RPD	1.64	1.036	RMSEP	0.378	1.714
20	r_c	0.926	0.966	Error_a (%)	6.22	26.59
	r_p	0.853	0.697	Error_m (%)	15.21	62.72
	RPD	1.64	0.990	RMSEP	0.358	1.403
25	r_c	0.926	0.938	Error_a (%)	6.25	24.98
	r_p	0.853	0.676	Error_m (%)	15.66	59.33
	RPD	1.64	1.096	RMSEP	0.364	1.242
30	r_c	0.926	0.941	Error_a (%)	6.72	26.07
	r_p	0.853	0.679	Error_m (%)	15.95	59.33
	RPD	1.64	1.106	RMSEP	0.391	1.262

WMPDS-S/B in the First Sample model of soil TN and TC concentrations (Tables 7 and 8). With the addition of the number of the standard set, the prediction performance of the model by adding new samples was unstable (Tables 7 and 8), with high r_c but low r_p and RPD. So, WMPDS-S/B could keep the stability of the original model.

Among all calibration transfer algorithms, the best was the WMPDS-S/B algorithm, while the method of adding new samples was the worst. By adding new samples, the error_a, error_m, and RMSEP of TN and TC concentrations were all high. Therefore, it could not be selected for calibration transfer between two different types of soils. The WMPDS-S/B algorithm was based on the S/B algorithm and combined with the WMPDS algorithm to eliminate spectral differences. It could improve the accuracy of nutrient concentration prediction between two different types of soil.

The number of the standard set had a relatively small impact on the results by the WMPDS-S/B algorithm (Tables 7 and 8). With the increase in the number of the standard set in the Second Sample, the error_a, error_m, and RMSEP decreased by adding new samples, showing that the addition of the samples was beneficial to the improvement of the prediction. When the number of the standard set increased to 30, which was the largest samples in both the TN and TC concentrations, the result was the best. However, when the number of the standard set was a certain value in the middle, which was 15 and 20 in the TN and TC concentrations, respectively, the result had the best prediction effect by the WMPDS-S/B algorithm. Compared to the first method, the number of the standard set had a relatively small impact on the results by the WMPDS-S/B algorithm, and this method could get better results in selecting as few a number of standard set as possible. In practical applications, it was impossible to select many standard set samples, which would increase workload and be time-consuming. The new algorithm had advantages in selecting fewer standard samples and obtaining better prediction results.

3.4. Differences from Previous Studies. The calibration transfer studied in this paper was different from previous studies, which were mainly for the calibration transfer between two different types of samples. Previous studies had solved the change of the same sample in different instruments and testing environments, thus ensuring the consistency of the instruments [13–18]. Fernandez et al. used four different calibration transfer techniques (Direct Standardization, Piecewise Direct Standardization, Orthogonal Signal Correction, and Generalized Least Squares Weighting) to offset the effect of temperature change on the prediction of gas sensors [13]. Yahaya et al. compared the calibration transfer effect of the reflectance spectrum of pH for mango among different spectrometers (QE65000, Jaz, and ASD FieldSpec 3) through direct calibration transfer [18]. These studies are all based on the calibration transfer of the same samples under different test conditions or different instruments.

At present, many studies were based on spectroscopy for the prediction of different types of soil nutrients. For example, Kuang et al. used near-infrared spectroscopy to detect the content of soil organic carbon and total nitrogen, and soil samples were collected on five farms. The result for the RPD of organic carbon OC was from 2.66 to 3.39 and of total nitrogen TN was from 2.85 to 3.45 [42]. The soil nutrient concentrations model was established through many soil samples at different locations, and this method still needed a large number of known information about different types of soil nutrient chemical values, so it was not the soil calibration transfer in real sense.

In this study, we solved calibration transfer between two different types of samples in the same instrument and the test environment using one type of soil nutrient concentrations model to predict another type of soil nutrient concentration. If the concentrations model of riverside soil was used to predict the concentrations of mountain soil directly, there was a large error (Tables 3 and 6). After we did the calibration transfer, the concentrations of mountain soil were predicted more accurately. The error significantly declined, and the prediction results could be accepted. The calibration transfer that we studied in this work could fundamentally solve the problem of model sharing between different types of samples and eliminate the technical bottleneck restricting the application of the spectral method. This solution would make it possible for one spectral instrument to rapidly share a spectral model to determine soil nutrients.

3.5. Future Studies on Calibration Transfer with Different Samples. Wavelet analysis has multiscale and multiresolution characteristics. It can process signals and analyze spectral information from coarse to fine. The spectral data can be decomposed into two subspaces by multiscale wavelet processing, that is, low frequency and high frequency. Calibration transfer between low-frequency and high-frequency information is carried out. Different transfer matrices can be obtained according to different frequency domains. The transfer matrices between different regions and different varieties are further refined. However, there are some problems in the WMPDS-SB algorithm.

(1) The PDS window width and the number of standard samples still need to be selected artificially. In the future, we can select the optimal PDS window and the number of samples automatically through algorithms.
(2) The entire band of the spectrum is used in this study. To reduce redundant information and improve prediction performance, we will extract the diagnostic spectral wavelength in the future.
(3) In this study, the method of calibration transfer in soil nutrients is labeled samples, which still require a portion of the chemical value of unknown samples. In the future, we intend to achieve the calibration transfer among different regions or different varieties with smaller labeled samples or without labeled samples.
(4) The WMPDS-S/B algorithm is applied to the calibration transfer in two different types of soil and has good results. In the future, this algorithm can be applied to calibration transfer in different types of soil samples.

4. Conclusions

This paper took mountain and riverside soils as examples to explore the calibration transfer of TC and TN concentrations between two different types of soils using WMPDS-S/B. To reduce the influence of calibration transfer in the same samples, we sampled spectral data with the same equipment and in the same test environment. The riverside soil was set as the First Sample, and the mountain soil was regarded as the Second Sample. First, a calibration model was established by the First Sample. Then, concentrations of total carbon (TC) and total nitrogen (TN) for the Second Sample were predicted by the transferred calibration model. In the calibration transfer of TN concentrations with two different types of soil, when the PDS window width was 15 and the number of standard set in the Second Sample was 14, the prediction accuracy of TN concentrations improved significantly compared to the accuracy without using calibration transfer. The average relative error ($error_a$) dropped from 440.20% to approximately 6%, the maximum relative error ($error_m$) was reduced from 538.1% to less than 20%, and RMSEP was reduced from 2.416 to approximately 0.04. In calibration transfer of TC concentrations between two different types of soils, when the PDS window width was 11 and the number of standard set was 18, the prediction accuracy of TC concentrations in the Second Sample showed significant improvement compared to the accuracy without using calibration transfer. The $error_a$ dropped from 341.97% to approximately 7%, the $error_m$ was reduced from 403.70% to less than 20%, and the RMSEP was reduced from 15.737 to approximately 0.4.

Under the condition of selecting the same samples of standard set, we studied the effects of the three methods on the results of calibration transfer in TN and TC concentrations. Compared to the traditional method of adding new samples, the WMPDS-S/B algorithm obtained better forecasting results. With the increase of the number of standard set in the Second Sample, $error_a$, $error_m$, and RMSEP were decreased by adding new samples. However, the number of the standard set had a relatively small impact on the results from the WMPDS-S/B algorithm. This method had advantages in selecting fewer standard samples and obtaining better prediction results. It was feasible to realize the calibration transfer of soil nutrients between two different types of soils rapidly and accurately, providing new ideas and methods for calibration transfer among different types of samples.

Conflicts of Interest

The authors report there are no conflicts of interest.

Acknowledgments

This work was supported by the National Key Research and Development Program of China (2017YFC1404802), Shandong Provincial Natural Science Foundation, China (ZR2018LD007), the National Key Research and Development Program of China (no. SQ2016YFSF070090), and the National Key Research and Development Program of China (no. SQ 2017YFC1403702).

References

[1] K. S. Hong and H. Santosa, “Decoding four different sound-categories in the auditory cortex using functional near-infrared spectroscopy,” *Hearing Research*, vol. 333, pp. 157–166, 2016.

[2] A. N. Sen, S. P. Gopinath, and C. S. Robertson, "Clinical application of near-infrared spectroscopy in patients with traumatic brain injury: a review of the progress of the field," *Neurophotonics*, vol. 3, no. 3, p. 031409, 2016.

[3] Y. Guo, Y. Ni, and S. Kokot, "Evaluation of chemical components and properties of the jujube fruit using near infrared spectroscopy and chemometrics," *Spectrochimica Acta Part A: Molecular and Biomolecular Spectroscopy*, vol. 153, pp. 79–86, 2016.

[4] M. Schmutzler, A. Beganovic, G. Böhler et al., "Modern safety control for meat products: near infrared spectroscopy utilised for detection of contaminations and adulterations of premium veal products," *NIR News*, vol. 27, no. 4, pp. 11–13, 2016.

[5] I. Marquetti, J. V. Link, and E. Bona, "Partial least square with discriminant analysis and near infrared spectroscopy for evaluation of geographic and genotypic origin of arabica coffee," *Computers and Electronics in Agriculture*, vol. 121, pp. 313–319, 2016.

[6] A. M. Rady, D. E. Guyer, W. Kirk, and I. R. Donis-González, "The potential use of visible/near infrared spectroscopy and hyperspectral imaging to predict processing-related constituents of potatoes," *Journal of Food Engineering*, vol. 135, pp. 11–25, 2014.

[7] H. Li, Q. Chen, J. Zhao, and M. Wu, "Nondestructive detection of total volatile basic nitrogen (TVB-N) content in pork meat by integrating hyperspectral imaging and colorimetric sensor combined with a nonlinear data fusion," *LWT-Food Science and Technology*, vol. 63, no. 1, pp. 268–274, 2015.

[8] U. Khulal, J. Zhao, W. Hu, and Q. Chen, "Nondestructive quantifying total volatile basic nitrogen (TVB-N) content in chicken using hyperspectral imaging (HSI) technique combined with different data dimension reduction algorithms," *Food Chemistry*, vol. 197, pp. 1191–1199, 2016.

[9] F. Kutsanedzie, Q. Chen, M. M. Hassan, M. Yang, H. Sun, and M. H. Rahman, "Near infrared system coupled chemometric algorithms for enumeration of total fungi count in cocoa beans neat solution," *Food Chemistry*, vol. 240, pp. 231–238, 2018.

[10] F. Feyziyev, M. Babayev, S. Priori, and G. L'Abate, "Using visible-near infrared spectroscopy to predict soil properties of Mugan Plain, Azerbaijan," *Open Journal of Soil Science*, vol. 06, no. 3, pp. 52–58, 2016.

[11] L. Zhang and R. B. Zhang, "Measurement of available phosphorus and potassium contents in soil using visible-near-infrared spectroscopy in conjunction with SPA-LS-SVM methods," *Advance Journal of Food Science and Technology*, vol. 10, no. 12, pp. 934–941, 2016.

[12] X. Yu, Q. Liu, Y. Wang, X. Liu, and X. Liu, "Evaluation of MLSR and PLSR for estimating soil element contents using visible/near-infrared spectroscopy in apple orchards on the Jiaodong peninsula," *CATENA*, vol. 137, pp. 340–349, 2016.

[13] N. K. Niazi, B. Singh, and B. Minasny, "Mid-infrared spectroscopy and partial least-squares regression to estimate soil arsenic at a highly variable arsenic-contaminated site," *International Journal of Environmental Science and Technology*, vol. 12, no. 6, pp. 1–10, 2015.

[14] A. M. Mouazen, B. Kuang, J. De Baerdemaeker, and H. Ramon, "Comparison among principal component, partial least squares and back propagation neural network analyses for accuracy of measurement of selected soil properties with visible and near infrared spectroscopy," *Geoderma*, vol. 158, pp. 23–31, 2010.

[15] L. Fernandez, S. Guney, A. Gutierrez-Galvez, and S. Marco, "Calibration transfer in temperature modulated gas sensor arrays," *Sensors and Actuators B Chemical*, vol. 231, pp. 276–284, 2016.

[16] J. H. Kalivas, G. G. Siano, E. Andries, and H. C. Goicoechea, "Calibration maintenance and transfer using Tikhonov regularization approaches," *Applied Spectroscopy*, vol. 63, no. 7, pp. 800–809, 2009.

[17] S. Amat-Tosello, N. Dupuy, and J. Kister, "Contribution of external parameter orthogonalisation for calibration transfer in short waves--near infrared spectroscopy application to gasoline quality," *Analytica Chimica Acta*, vol. 642, no. 1-2, pp. 6–11, 2009.

[18] S. Guo, R. Heinke, S. Stöckel, P. Rösch, T. Bocklitz, and J. Popp, "Towards an improvement of model transferability for Raman spectroscopy in biological applications," *Vibrational Spectroscopy*, vol. 91, pp. 111–118, 2017.

[19] O. K. M. Yahaya, M. Z. Matjafri, A. A. Aziz, and A. F. Omar, "Visible spectroscopy calibration transfer model in determining pH of Sala mangoes," *Journal of Instrumentation*, vol. 10, no. 5, article T05002, 2015.

[20] A. L. B. Brito, A. V. P. Santos, K. D. T. M. Milanez, M. J. C. Pontes, and L. F. B. L. Pontes, "Calibration transfer of flour NIR spectra between benchtop and portable instruments," *Analytical Methods*, vol. 9, no. 21, pp. 3184–3190, 2017.

[21] C. F. Pereira, M. F. Pimentel, R. K. Galvão, F. A. Honorato, L. Stragevitch, and M. N. Martins, "A comparative study of calibration transfer methods for determination of gasoline quality parameters in three different near infrared spectrometers," *Analytica Chimica Acta*, vol. 611, no. 1, pp. 41–47, 2008.

[22] L. S. A. Pereira, M. F. Carneiro, B. G. Botelho, and M. M. Sena, "Calibration transfer from powder mixtures to intact tablets: a new use in pharmaceutical analysis for a known tool," *Talanta*, vol. 147, pp. 351–7, 2016.

[23] Y. Sulub and G. W. Small, "Spectral simulation methodology for calibration transfer of near-infrared spectra," *Applied Spectroscopy*, vol. 61, no. 4, pp. 406–413, 2007.

[24] L. Laghi, A. Versari, G. P. Parpinello, D. Y. Nakaji, and R. B. Boulton, "FTIR spectroscopy and direct orthogonal signal correction preprocessing applied to selected phenolic compounds in red wines," *Food Analytical Methods*, vol. 4, no. 4, pp. 619–625, 2011.

[25] J. Zhang, Z. Zhang, Y. Xiang, Y. Dai, and P. B. Harrington, "An emphatic orthogonal signal correction-support vector machine method for the classification of tissue sections of endometrial carcinoma by near infrared spectroscopy," *Talanta*, vol. 83, no. 5, pp. 1401–1409, 2011.

[26] J. Yoon, B. Lee, and C. Han, "Calibration transfer of near-infrared spectra based on compression of wavelet coefficients," *Chemometrics and Intelligent Laboratory Systems*, vol. 64, no. 1, pp. 1–14, 2002.

[27] C. Tan and M. Li, "Calibration transfer between two near-infrared spectrometers based on a wavelet packet transform," *Analytical Sciences*, vol. 23, no. 2, pp. 201–206, 2007.

[28] J. Bin, X. Li, W. Fan, J.-H. Zhou, and C.-W. Wang, "Calibration transfer of near-infrared spectroscopy by canonical correlation analysis coupled with wavelet transform," *Analyst*, vol. 142, no. 12, pp. 2229–2238, 2017.

[29] W. Fan, Y. Liang, D. Yuan, and J. Wang, "Calibration model transfer for near-infrared spectra based on canonical correlation analysis," *Analytica Chimica Acta*, vol. 623, no. 1, pp. 22–29, 2008.

[30] H. Li, J.-X. Wang, Z.-N. Xing et al., "Influence of improve Kennard/Stone algorithm on the calibration transfer in

near-infrared spectroscopy," *Spectroscopy and Spectral Analysis*, vol. 31, no. 2, pp. 362–365, 2011.

[31] W. Zen, W. Bei, W. Hao et al., "Study on near infrared spectroscopy model transfer by using piecewise direct standardization at different temperature," *Journal of instrumental analysis*, vol. 27, no. 12, pp. 1330–1333, 2008.

[32] X. Chen, F. Yan, and H. Chang, "Evaluation of piecewise direct standardization algorithm for near infrared quantitative model updating," *Chinese Journal of Analytical Chemistry*, no. 09, pp. 1307–1313, 2014.

[33] W. Ju, L. Hua, X. Zhi et al., "Application of wavelet multi-scale piecewise direct standardization on near infrared analysis calibration," *Chinese Journal of Analytical Chemistry*, vol. 39, no. 06, pp. 846–850, 2011.

[34] T. Gao, C. Xiao, Y. Hong et al., "Application of wavelet transform-piecewise direct standardization on the near infrared analysis model transfer Chinese," *Journal of Analytical Chemistry*, vol. 34, no. 07, pp. 927–932, 2006.

[35] C. Da, L. Fan, and L. Qi, "Development of multi-scale modeling methods for calibration transfer in near infrared spectroscopy," *Nanotechnology and Precision Engineering*, vol. 15, no. 2, pp. 121–126, 2017.

[36] R. A. V. Rossel and R. Webster, "Predicting soil properties from the Australian soil visible–near infrared spectroscopic database," *European Journal of Soil Science*, vol. 63, no. 6, pp. 848–860, 2012.

[37] J. Wetterlind, S. Bo, and M. Söderström, "Increased sample point density in farm soil mapping by local calibration of visible and near infrared prediction models," *Geoderma*, vol. 156, no. 3-4, pp. 152–160, 2010.

[38] L. Jiao, L. Xiao, and G. Xiao, "Transfer method among water content detection models for different breeds of pork by hyperspectral imaging technique," *Transactions of the Chinese Society of Agricultural Engineering*, vol. 30, no. 17, pp. 276–284, 2014.

[39] J. Cooper, C. Larkin, and M. Abdelkader, "Virtual standard slope and bias calibration transfer of partial least squares jet fuel property models to multiple near infrared spectroscopy instruments," *Journal of Near Infrared Spectroscopy*, vol. 19, no. 2, pp. 139–150, 2011.

[40] Q. Chen, J. Zhao, M. Liu et al., "Determination of total polyphenols content in green tea using FT-NIR spectroscopy and different PLS algorithms," *Journal of Pharmaceutical and Biomedical Analysis*, vol. 46, no. 3, pp. 568–73, 2008.

[41] L. J. Janik, S. T. Forrester, and A. Rawson, "The prediction of soil chemical and physical properties from mid-infrared spectroscopy and combined partial least-squares regression and neural networks (PLS-NN) analysis," *Chemometrics and Intelligent Laboratory Systems*, vol. 97, no. 2, pp. 179–188, 2009.

[42] B. Kuang and A. M. Mouazen, "Non-biased prediction of soil organic carbon and total nitrogen with vis–NIR spectroscopy, as affected by soil moisture content and texture," *Biosystems Engineering*, vol. 114, no. 3, pp. 249–258, 2013.

Study on the Processing Technology of Calamine Calcination by Near-Infrared Spectroscopy

Xiaodong Zhang,[1] Long Chen,[2] Yu Bai,[3] and Keli Chen[1]

[1]*Key Laboratory of Ministry of Education on Traditional Chinese Medicine Resource and Compound Prescription & Hubei University of Chinese Medicine, Wuhan 430065, China*
[2]*Xiangyang Central Hospital, Xiangyang, Hubei 441500, China*
[3]*Mayinglong Pharmaceutical Group Co., Wuhan 430065, China*

Correspondence should be addressed to Keli Chen; kelichen@126.com

Academic Editor: Feride Severcan

Near-infrared spectroscopy has been widely used in qualitative and quantitative analysis and online monitoring in the production process of traditional Chinese medicines. The aim was to establish a fast determination model of zinc oxide (ZnO) content in calcined calamine and to explore methods through judging the end point of calamine calcination. Eight batches of calamine samples sourced from hydrozincite with different sizes and textures were calcined at different temperatures. During the calcination process, ZnO contents, X-ray diffraction (XRD) patterns, and near-infrared spectra of the samples were used to analyze their changes rules. The model of determining ZnO content of calcined calamine was established to use near-infrared spectroscopy based on the partial least squares (PLS) regression algorithm. In addition, this paper summarized the change rules of calamine in calcination according to XRD patterns, using the "*K* value" quantitative method to define the characteristic *T* value. When the *T* value was equal to 1.00 (100%), that is to say, the calamine sample was completely calcined. Then, matching the near-infrared spectroscopy data with the *T* value and establishing the *T* value analysis model using the PLS algorithm were performed. Through cross and independent validation and evaluation, it was proved that the two models were very effective and had strong predictive abilities. Finally, the purpose of the online monitoring of the calcination process and controlling the quality of the calcined calamine was achieved.

1. Introduction

Calamine, a common mineral used in Chinese Medicine and recorded in China Pharmacopoeia (ChP, the 2015 Edition Volume I), is smithsonite belonging to carbonate minerals in the calcite group and mainly contains zinc carbonate ($ZnCO_3$) [1]. However, according to literature reports and findings in our investigations, calamine is also sourced from hydrozincite [$Zn_5(CO_3)_2(OH)_6$] and the latter is the current mainstream of commercial calamine in the market [2–4]. Calamine must be calcined before being used as a drug. The calcined calamine has the actions to counteract toxicity, to clear the eye off corneal opacity and to relieve itching [1]. It is commonly used in the treatment of eye, skin, anorectal diseases, for example, Musk hemorrhoids ointment for treating hemorrhoids and calamine lotion for relieving itching. The study has shown that the calcination process is the decomposition of $ZnCO_3$ or $Zn_5(CO_3)_2(OH)_6$ into the active ingredient ZnO [5]. The ChP stipulates that the ZnO content of calcined calamine should not be less than 56% [1]. However, in the actual calcination process, if the calamine is not calcined completely, the ZnO content titrated with ethylene diamine tetraacetic acid (EDTA) according to the ChP is actually the total amount of zinc ions in the sample solution, and the obtained result may still comply with the pharmacopoeia. Calamine, if not being completely calcined, used as a drug not only affects the efficacy but also causes waste of resources.

At present, judging whether calamine is completely calcined is mainly through analysis of X-ray diffraction (XRD) patterns before and after calcination [6]. If the phase analysis of the XRD pattern before calcination shows smithsonite or hydrozincite, and after calcination, the smithsonite or hydrozincite phase in the XRD pattern disappeared and produced the ZnO phase, that is to say, the calamine is completely calcined. Although the XRD analysis method is accurate and reliable, the analysis cost is high and online detection cannot be realized in the production process. In addition, since most of the collected calamine samples are sourced from hydrozincite [3, 4], this study aims to explore new methods for judging the calcination end point of the calamine sourced from hydrozincite.

Near-infrared (NIR) spectra are the results of light absorption by organic molecules in the spectral region of 780–2500 nm [7]. Characteristic bands are produced in the NIR spectra of minerals by interaction of electromagnetic radiation with the atoms and molecules that are present in these minerals as a result of electronic and vibrational spectroscopic processes [8, 9]. As a fast, simple, nonpolluting, nondestructive, portable, and on the site analytical technique, NIR spectroscopy has been widely used in the qualitative identification, quantitative analysis, online monitoring, and drug supervision of traditional Chinese medicines [10–13].

In this study, simulated near-infrared spectroscopy online monitoring of calamine calcination process was performed. The change rules of ZnO contents were summarized according to XRD patterns and NIR spectra in the process of calamine calcination. A rapid determination model of ZnO content of calcined calamine was established using near-infrared spectroscopy based on the partial least squares (PLS) algorithm [14]. A method of combining near-infrared spectroscopy with the XRD pattern was also explored to determine whether calamine was completely calcined.

2. Materials and Methods

2.1. Instruments and Software. The instruments and software used for this project consisted of an XPertPro X-ray diffraction instrument (PANalytical Company), a Matrix-F Fourier diffuse reflectance near-infrared spectrometer (Bruker) with sample testing accessories integrated with a fiberoptics probe, an InGaAs detector (Bruker), OPUS 7.5 spectrum analysis software (Bruker), and MDI Jade 6.0 digital signal processing technique software (Materials Data, Inc.).

2.2. Samples. Twenty-eight batches of calamine samples were collected from the Chinese medicine markets, 6 batches of calamine samples were provided by Mayinglong Pharmaceutical Group Co., 18 batches of calamine samples were collected from the Chinese mining areas of Hunan, Guangxi, Guizhou, Yunan, and Sichuan province. A total of 52 batches of calamine samples were obtained [3, 4], and all samples were identified by XRD and the chemical method in the ChP 2015 [1]. Of which, 8 batches of calamine samples were randomly selected, as shown in Table 1.

2.3. Calamine Calcination Process Research

2.3.1. Comparison of XRD Patterns of Calamine before and after Complete Calcination. About 3 g of S1–S6 samples was taken in the crucible, placed within the muffle furnace at 300°C for 100 min; about 3 g of S7-S8 samples were taken in the crucible, placed within the muffle furnace at 400°C for 100 min. Then, all samples were taken out, crashed into powder, and passed through a 200-mesh sieve. Pressed the powder into the groove of the sample plate with a glass plate and collected the XRD pattern according to the instrument operation procedure. XRD patterns were recorded between 3 and 65° (2θ) at a speed of 8 min^{-1}, using a RigakuD/max-1200 diffractometer with CuKα radiation (40 kV, 40 mA).

2.3.2. Variation of ZnO Content and XRD Characteristic T Value. Take out 12 parts from each calamine sample in Table 1 separately, with each part about 3 g. The first part of each sample was taken for calcination for 0 min, and the remaining 11 parts were put into a muffle furnace and then taken out at different calcination time periods. The twentieth parts of the samples were taken out after being calcined for 100 min. For samples S1–S6, calcination temperature was at 300°C. From samples S1, S2, and S4, take out one part every 6 min. From S3 and S5, take out one part every 3 min, and from S6, every 2 min. For sample S7-S8, calcination temperature was at 400°C, and one part was taken out every 2 min. The every part of each sample taken out was cooled and passed through a 200-mesh sieve and then used to determine the ZnO content according to ChP 2015 and collect the XRD patterns and near-infrared spectra.

XRD can be used for quantitative analysis because the intensity of the diffraction peak increases with the content of the phase. However, since the absorption coefficient of different phases is different, the diffraction intensity is not strictly proportional to the content of each phase, so it needs to be corrected [15]. Since 1978, the RIR value, which is the K value, has been added to the PDF card published on the ICDD. It is the ratio of the integrated intensity of the strongest peak of the measured sample to the intensity of the strongest peak of corundum (Al_2O_3) measured by mixing the mass of the sample with Al_2O_3 at a ratio of 1 : 1. It can be written as $K^{A}_{Al_2O_3} = K^A/K^{Al_2O_3} = I_A/I_{Al_2O_3}$, called the K value of the sample in A phase when corundum is the internal standard. If there are A, B, and C phases in a sample, we can use the A phase as the standard sample, find the RIR value of each phase through the PDF card, and calculate the K value of each phase in the sample, namely,

Table 1: Sample information.

Serial number	Source	Characteristics	Chemical precipitation	Phase composition	ZnO (%)	Test result
S1	Yunan	Around 1 cm, incanus, hard	White	Hydrozincite, hemimorphite	66.57	Authentic
S2	Anguo	Around 1 cm, light pink, hard	White	Hydrozincite	69.59	Authentic
S3	Sichuan	Around 1 cm, light pink, soft	White	Hydrozincite, hemimorphite, dolomite	51.57	Authentic
S4	Guangxi	0.3~1.5 cm, incanus, hard	White	Hydrozincite, hemimorphite, dolomite	43.21	Authentic
S5	Guizhou	Around 0.2 cm, light pink, soft	White	Hydrozincite, hemimorphite, dolomite, calcite	38.12	Authentic
S6	Hunan	Around 0.3 cm, light pink, hard	White, blue	Hydrozincite	67.79	Authentic
S7	Guizhou	Around 1 cm, light pink, soft	White	Hydrozincite, hemimorphite, dolomite, calcite	55.92	Authentic
S8	Yunan	Around 1 cm, incanus, hard	White	Hydrozincite, hemimorphite	64.88	Authentic

$$K_A^A = \frac{K_{Al_2O_3}^A}{K_{Al_2O_3}^A},$$
$$K_A^B = \frac{K_{Al_2O_3}^B}{K_{Al_2O_3}^A}, \tag{1}$$
$$K_A^C = \frac{K_{Al_2O_3}^C}{K_{Al_2O_3}^A}.$$

If there are N phases in a system, the mass fraction of the X phase is

$$W_X = \frac{I_X}{K_A^X \Sigma_{i=A}^N I_i / K_A^i}, \tag{2}$$

where A can be any phase of the selected samples, I_i indicates the integrated intensity of the strongest peak of any phase, $i = A \ldots N$ indicates N phases in the sample.

Therefore, the XRD pattern of the calamine calcination process was analyzed by phase analysis to find the RIR value on the PDF card corresponding to hydrozincite and ZnO, the mass fraction of which can be calculated by equation (1). In order to more intuitively describe the trend of hydrozincite changing over time during calamine calcinations, it is defined that

$$T = \frac{W_{ZnO}}{W_{ZnO} + W_{Zn_5(CO_3)_2(OH)_6}}. \tag{3}$$

Substitute equation (1) into equation (2) to get

$$T = \frac{I_{Zno}\mathrm{RIR}_{Zn_5(CO_3)_2(OH)_6}}{I_{Zno}\mathrm{RIR}_{Zn_5(CO_3)_2(OH)_6} + I_{Zn_5(CO_3)_2(OH)_6}\mathrm{RIR}_{Zno}}, \tag{4}$$

where the corresponding PDF card of ZnO is 75–0576 and the RIR value is 5.53; the corresponding PDF card of $Zn_5(CO_3)_2(OH)_6$ is 75–1100, and the RIR value is 2.49.

Obviously, the range of T value is [0, 1]. When T value is 0, $Zn_5(CO_3)_2(OH)_6$ is not decomposed; when T value is at (0,1), part of $Zn_5(CO_3)_2(OH)_6$ is decomposed into ZnO; when T value is 1, $Zn_5(CO_3)_2(OH)_6$ is completely decomposed into ZnO, that is, the hydrozincite is completely calcined.

2.3.3. Near-Infrared Spectra Collection. Spectra were obtained from 12,500–4,000 cm^{-1} (800–2500 nm) by the coaddition of 32 scans at a resolution of 8 cm^{-1}. Each sample was scanned three times, and the average value of three spectra was used as the spectrum information for analyzing.

2.4. Determination of ZnO Content of Calcined Calamine Using NIRS Based on PLS Algorithm

2.4.1. Modeled Samples and Division of Sample Sets. Ninety-six batches of calcined calamine prepared in Section 2.3.2 were used as part of the modeled sample. The other part was randomly selected from the remaining 44 batches of collected calamine samples and placed in a muffle furnace at 400°C [16] for 100 min, and near-infrared spectra and zinc oxide contents were obtained according to the above methods. For the sufficient sample size, a total of 130 batches of calcined calamine samples were randomly divided into training set, validation set, and test set.

2.4.2. Selection of Modeling Spectral Bands and Spectral Pretreatment Methods. Although the PLS regression algorithm is a "full-spectrum method," if the modeling spectrum is too wide, it will be affected by additional components or spectral noise in the sample, which will make the model worse. Therefore, the selection of modeling spectral bands is beneficial to improve the model forecast accuracy [17]. In addition, the collected spectrum contains not only its own information but also other irrelevant information and noise, such as electrical noise, sample background, and stray light [18]. So, it is necessary to select the modeling spectral bands and spectral pretreatment methods during modeling.

This experiment used the OPUS 7.5/QUANT-2 quantitative analysis software to automatically optimize the modeling parameters. The modeling spectral bands and the pretreatment methods were combined and optimized to, respectively, establish quantitative analysis models. Since the near-infrared spectral characteristics of the calamine samples were mainly concentrated in the spectral bands of 7434.3–5781.6, 5445.3–4597, and 4598.2–3998.6 cm^{-1}, the three spectral bands were used as the optimized spectral bands. And, SNV, MSC, FD, SD, FD + SNV, and FD + MSC were selected as the pretreatment methods.

2.4.3. Model Validation and Evaluation. The training set was used to establish the model. The validation set was used to validate the model, of which relative mean square error of validation (RMSEV), coefficient of determination (R^2), and ratio of performance to deviation (RPD) were taken as indexes to guide model optimization process. The test set was used to evaluate the model, of which relative mean square error of prediction (RMSEP), R^2, and RPD were taken as indexes to further evaluate the model prediction ability [19]. The smaller the RMSEV and RMSEP, the bigger the R^2 would be, and thus, the predictive effect of the model would be better. Moreover, three RPD categories were defined to account for the model reliability: excellent models with RPD > 2, fair models with 1.4 < RPD < 2, and unreliable models with RPD < 1.4 [20].

2.4.4. Establishment of Near-Infrared Spectral Analysis Model Using PLS Algorithm to Predict T Value. In Section 2.3.2, it was clear that when the T value of XRD pattern was 1, the calamine was completely calcined. Therefore, match the near-infrared spectroscopy data with the T value, and establish the T-value analysis model using the PLS algorithm. The specific method was as follows.

Sixty-two samples that had been collected for the XRD patterns during the calcination of the above 8 batches of calamine samples were randomly divided into training set and prediction set, including 41 training set samples and 21 predicted set samples. Sample spectrum and corresponding T values were imported into OPUS 7.5/QUANT-2 quantitative analysis software. The modeling spectral bands and pretreatment methods were combined and optimized to, respectively, establish quantitative analysis models. Among them, the three ranges of 7434.3–5781.6, 5445.3–4597, and 4598.2–3998.6 cm^{-1} were selected, and SNV, MSC, FD, SD, FD + SNV, and FD + MSC were selected as the pretreatment methods. For the limited sample size, in the process of establishing the model, internal cross-validation and external validation were used to evaluate the predictive ability of the model. Internal cross-validation adopted the leave-one-out cross-validation, of which relative mean square error of cross-validation (RMSECV), R^2, and RPD were taken as indexes to guide model optimization process. In external validation, RMSEP, R^2, and RPD were taken as indexes to further evaluate the model prediction ability.

3. Results and Discussion

3.1. Comparison of XRD Patterns of Calamine before and after Complete Calcination. The XRD patterns of 8 batches of calamine samples before and after calcination for 100 min are shown in Figure 1. According to Figure 1, the main features of the XRD patterns of the 8 batches of calamine samples before calcination were characterized by all samples and the patterns had the strongest peak at $d = 0.677$ nm, followed by strong peaks at $d = 0.315$, 0.272, and 0.248 nm. The phase retrieval showed that these were characteristic of hydrozincite. In addition, some samples also contained impurities. For example, samples S1, S3, S4, S5, S7, and S8 had strong peaks at $d = 0.654$, 0.541, 0.460, 0.410, 0.330, and 0.306 nm. The phase retrieval showed that these were characteristic of heterogeneous. Samples S3, S4, S5, and S7 had strong peaks at $d = 0.289$, 0.219, and 0.179 nm. The phase retrieval showed that these were characteristic of dolomite. Samples S5 and S7 had strong peaks at $d = 0.286$, 0.304, 0.250, and 0.230 nm. The phase retrieval showed that these were characteristic of calcite. Moreover, phase retrieval of all sample XRD patterns showed no ZnO. After calamine samples were calcined at different temperatures for 100 min, the analysis results of XRD patterns showed that the characteristic peaks of hydrozincite were disappeared while the three strongest peaks were generated at $d = 0.281$, 0.260, and 0.247 nm, which were characteristic of ZnO. It was indicated that the hydrozincite was completely decomposed into ZnO, that is, the above samples were completely calcined. Combined with the characteristics of the sample, it could be obtained that the calamine samples of different sizes, textures, and calcination temperatures could be calcined completely after 100 min. In addition, before and after calcination, the dolomite and calcite peaks still existed in some samples and the relative strength slightly increased, indicating that, under the above calcination conditions, these components did not decompose but the relative content increased. Moreover, if there were heterogeneity in the sample composition, after calcination, the characteristic peaks of the heterogeneous sample were sharper and the dispersion was weakened, indicating that the heterogeneity in the sample after calcination was not decomposed, the relative content was increased, and the crystallinity was enhanced.

3.2. Change of ZnO Content and XRD Characteristic T Value of Calamine Samples during Calcination. The change of ZnO content and XRD characteristic T value of 8 batches of calamine samples during calcination are shown in Figures 2 and 3. From Figure 2, the content of ZnO in the calcination process of calamine increased first with the increase of calcination time and then stabilized at the maximum value. When the ZnO content did not change with time, the calamine was calcined completely, judging that the 8 batches of calamine samples completely calcined at 60, 48, 30, 58, 27, 16, 14, and 18 min. From Figure 3, the T value in the calcination process of calamine was gradually increased from 0 to 1 and then remained unchanged, judging that the 8 batches of calamine samples completely calcined at 60, 48, 30, 60, 27, 25, 14, and 18 min. It was found that the judgment results of S4 and S6 samples had a certain deviation from the actual time according to the change trend of ZnO content. Because the experimental process did not

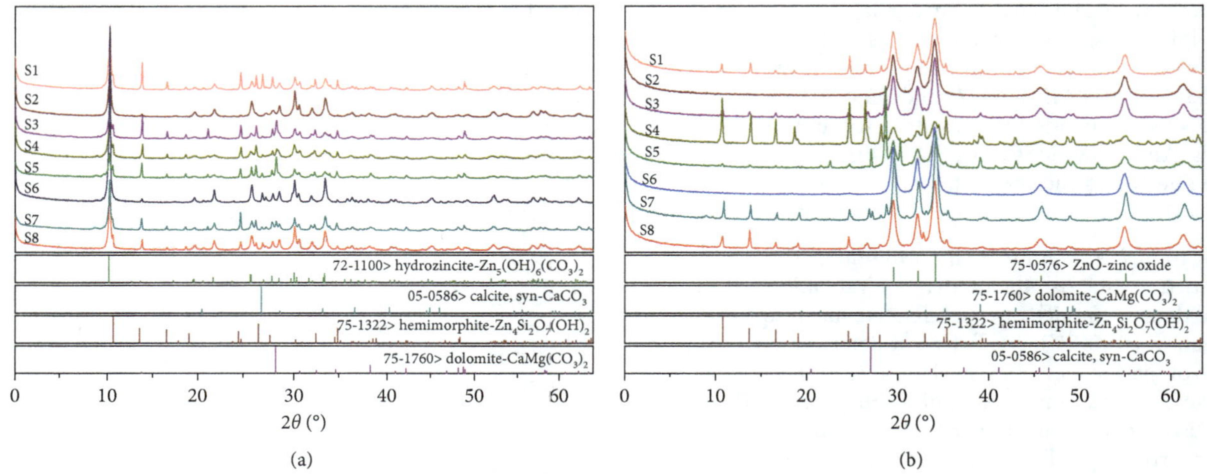

FIGURE 1: XRD patterns of 8 batches of calamine samples before and after calcination for 100 min.

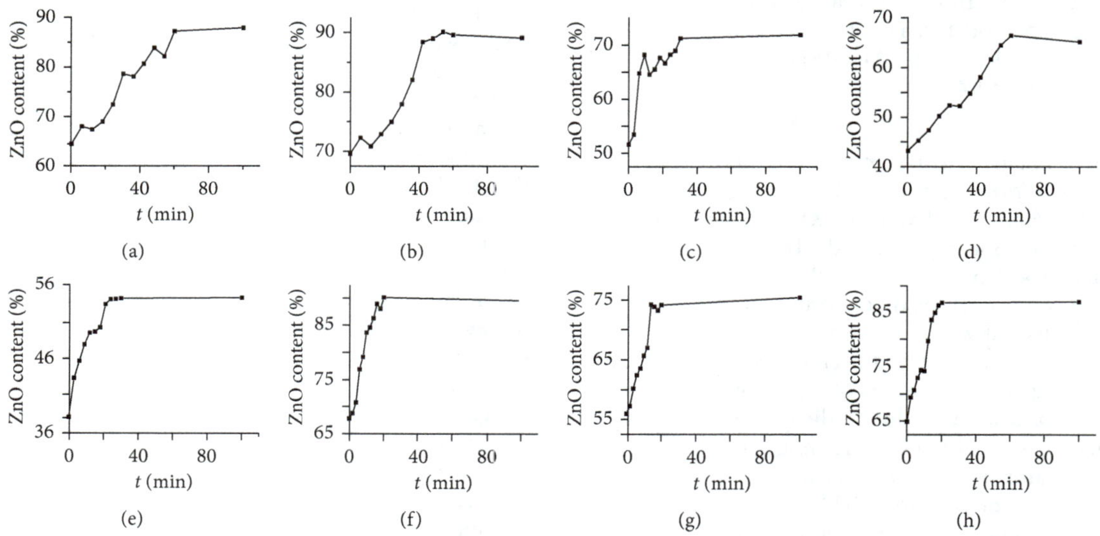

FIGURE 2: Change of ZnO content of 8 batches of calamine samples during calcination. (a) S1. (b) S2. (c) S3. (d) S4. (e) S5. (f) S6. (g) S7. (h) S8.

guarantee that the content of 12 samples in each batch of samples was consistent, there might be a relatively high or low content of individual samples, and then, there would be small fluctuations in the content of ZnO during calcination, eventually leading to the judgment that the end point of calcination was relatively advanced or delayed.

3.3. NIR Spectroscopy during Calamine Calcination. The near-infrared spectrum of calamine calcination could be grouped into three regions:

(1) The spectra in the 12500–7500 cm^{-1} region: The divalent cations of Ca, Fe, Cu, Cd, and Zn played an important role in determining the band positions in this spectral region [7]. These ions also caused shifts for the carbonate bands. Because the sample contained less impurity cations, the absorption characteristics in spectra were not significant.

(2) The spectra in the 7500–5750 cm^{-1} region, which were due to first overtones of the fundamental hydroxyl stretching modes and combination modes of OH stretching and bending modes [21]. All samples had absorption peaks at 7031 cm^{-1} and 6672 cm^{-1}. The absorption peak at 6672 cm^{-1} was related to OH in the structure of hydrozincite, and the hydrozincite was continuously decomposed into ZnO, CO_2, and H_2O during calcination [16]; the intensity of the absorption peak gradually weakened until it disappeared. The absorption peak at 7031 cm^{-1} was related to Zn–OH, while the peak shape was affected

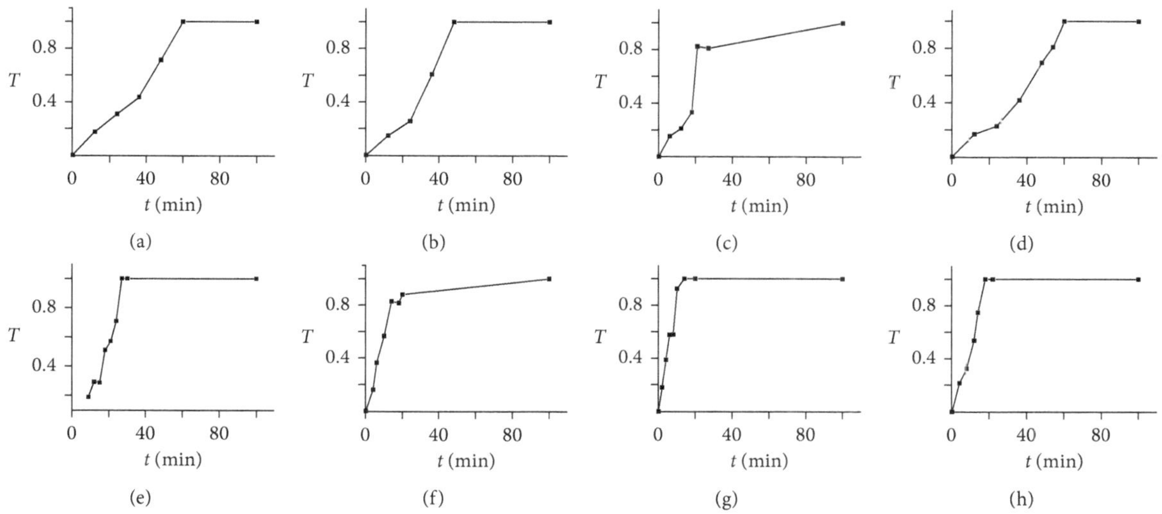

FIGURE 3: Change of XRD characteristic T value of 8 batches of calamine samples during calcination. (a) S1. (b) S2. (c) S3. (d) S4. (e) S5. (f) S6. (g) S7. (h) S8.

by impurities of the hemimorphite $[Zn_4(Si_2O_7)(OH)_2]$. For example, the samples S1, S3, S4, S5, S7, and S8 contained hemimorphite impurities, and the peak shape at 7031 cm^{-1} gradually became sharper with the increase of the relative content and crystallinity of the hemimorphite [22]. In contrast, samples S2 and S6 did not contain hemimorphite impurities, and the peak shape at 7031 cm^{-1} remained unchanged during calcination. The NIR spectrum of 8 batches of calamine samples during calcination in the 7500–5750 cm^{-1} region is shown in Figure 4.

(3) The spectra in the 5750–4000 cm^{-1} region were due to water and the carbonate ions [8]. The absorption feature in the range 5300–4900 cm^{-1} was the contribution of water-OH overtones, and the peak shape was affected by the impurity of hemimorphite. Samples S2 and S6 did not contain hemimorphite, the characteristic absorption of absorbed water was near 5082 cm^{-1} and 4919 cm^{-1}, and the peak shape was slow. Samples S1, S3, S4, S5, S7, and S8 contained hemimorphite, the characteristic absorption of adsorbed water was near 5215 cm^{-1}, 5059 cm^{-1}, and 4934 cm^{-1}, and the peak shape at 5215 cm^{-1} gradually became sharp during calcination. The characteristic of carbonate ion in hydrozincite was near 4349 cm^{-1} and 4170 cm^{-1}; during the calcination process, the intensity of the absorption peak gradually weakened until it disappeared. However, since there were carbonate impurities in some samples, a weak carbonate ion absorption peak would still appear after the hydrozincite was completely decomposed. For example, samples S5 and S7 contained some dolomite and calcite impurities, and after the hydrozincite was completely calcined, the carbonate ion absorption peaks [23] were seen at 4529 cm^{-1}, 4428 cm^{-1} and 4303 cm^{-1}. While the samples S3 and S4 contained dolomite impurities, the amount was very small, so the carbonate ion absorption peak was not obvious. The NIR spectrum of 8 batches of calamine samples during calcination in the 5750–4000 cm^{-1} region is shown in Figure 5.

The above had summarized to the contribution of NIR spectroscopy regions and change rules of absorption peaks during calamine calcination. The characteristic peaks of carbonate ion of hydrozincite were 4349 cm^{-1} and 4170 cm^{-1}; theoretically, both absorption peaks would disappear after the hydrozincite was completely decomposed. However, if the sample contained carbonate impurities, there would still be absorption peak near 4349 cm^{-1} after hydrozincite decomposition. So, the presence or absence of an absorption peak at 4170 cm^{-1} in the spectra could be used for judging the degree of calamine calcination.

3.4. NIR Model of Determining ZnO Content in Calcined Calamine. Information of sample classification is shown in Table 2. The NIR models determining the ZnO content of calcined calamine by different modeling parameters are shown in Table 3. From Table 3, it could be seen that model 1 had the smallest RMSECV value, and from the results of the validation set, there was no significant difference between models 1 and 2, and it was obviously better than other models. Therefore, it was determined that the modeling spectral band was 4601.6–4003.7 cm^{-1}, and the pretreatment method adopted SD. Besides, according to the RMSECV-rank correlation diagram (Figure 6(a)), when the RMSECV was the smallest, the best rank value of the model was determined to be 8. In addition, from the results of the prediction set, the RMSEP value of model 1 was the smallest, indicating the strongest predictability. The prediction results of the model for the training set, validation set, and test set are shown in Figures 6(b)–6(d).

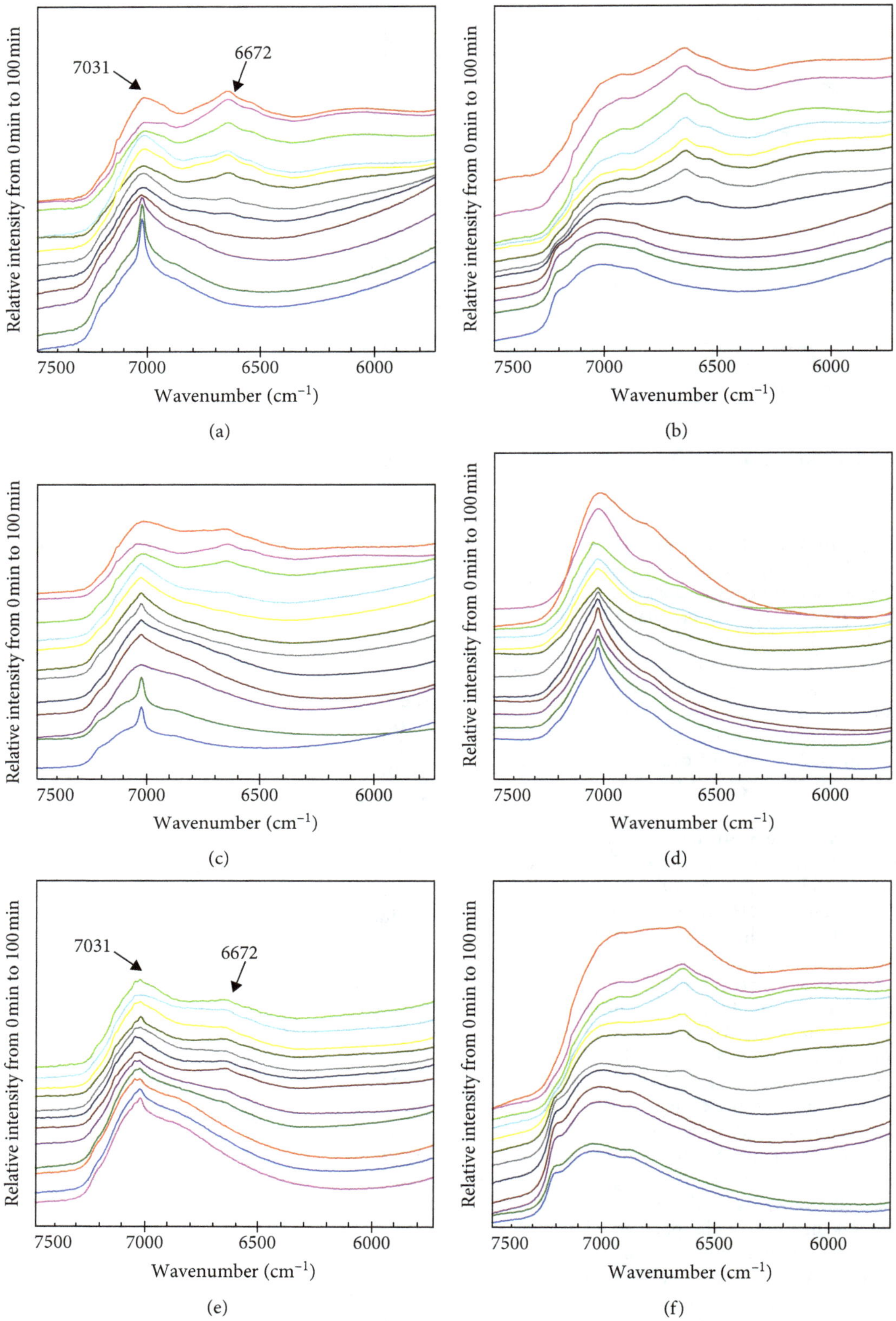

Figure 4: Continued.

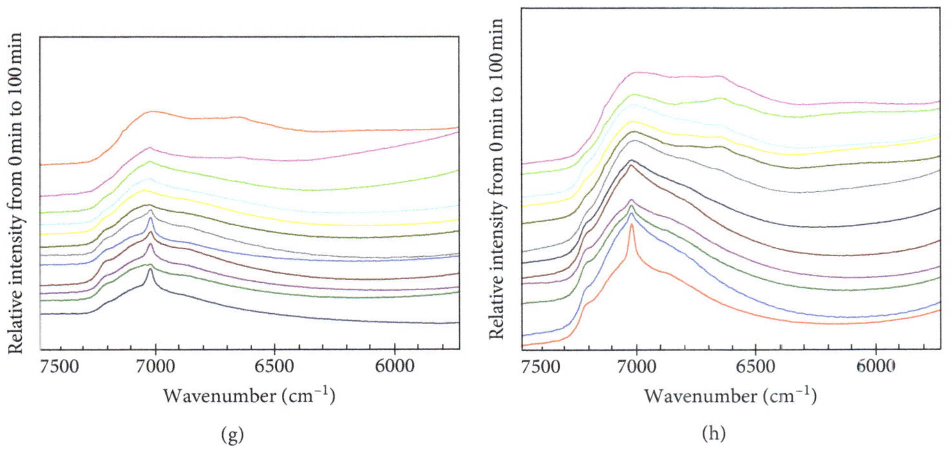

FIGURE 4: NIR spectra of 8 batches of calamine samples during calcination in the 7500–5750 cm^{-1} region. (a) S1 (hydrozincite and hemimorphite). (b) S2 (hydrozincite). (c) S3 (hydrozincite, hemimorphite, and dolomite). (d) S4 (hydrozincite, hemimorphite, and dolomite). (e) S5 (hydrozincite, hemimorphite, dolomite, and calcite). (f) S6 (hydrozincite). (g) S7 (hydrozincite, hemimorphite, dolomite, and calcite). (h) S8 (hydrozincite and hemimorphite).

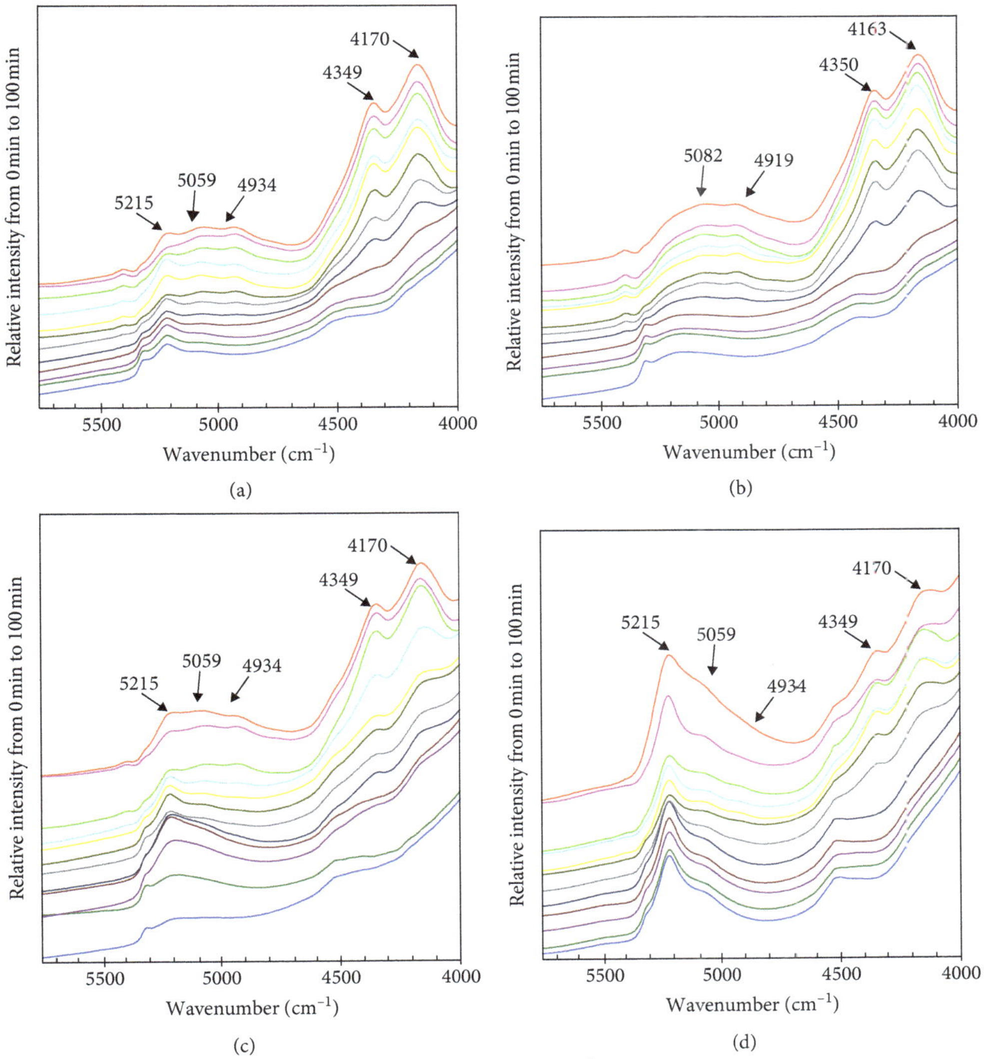

FIGURE 5: Continued.

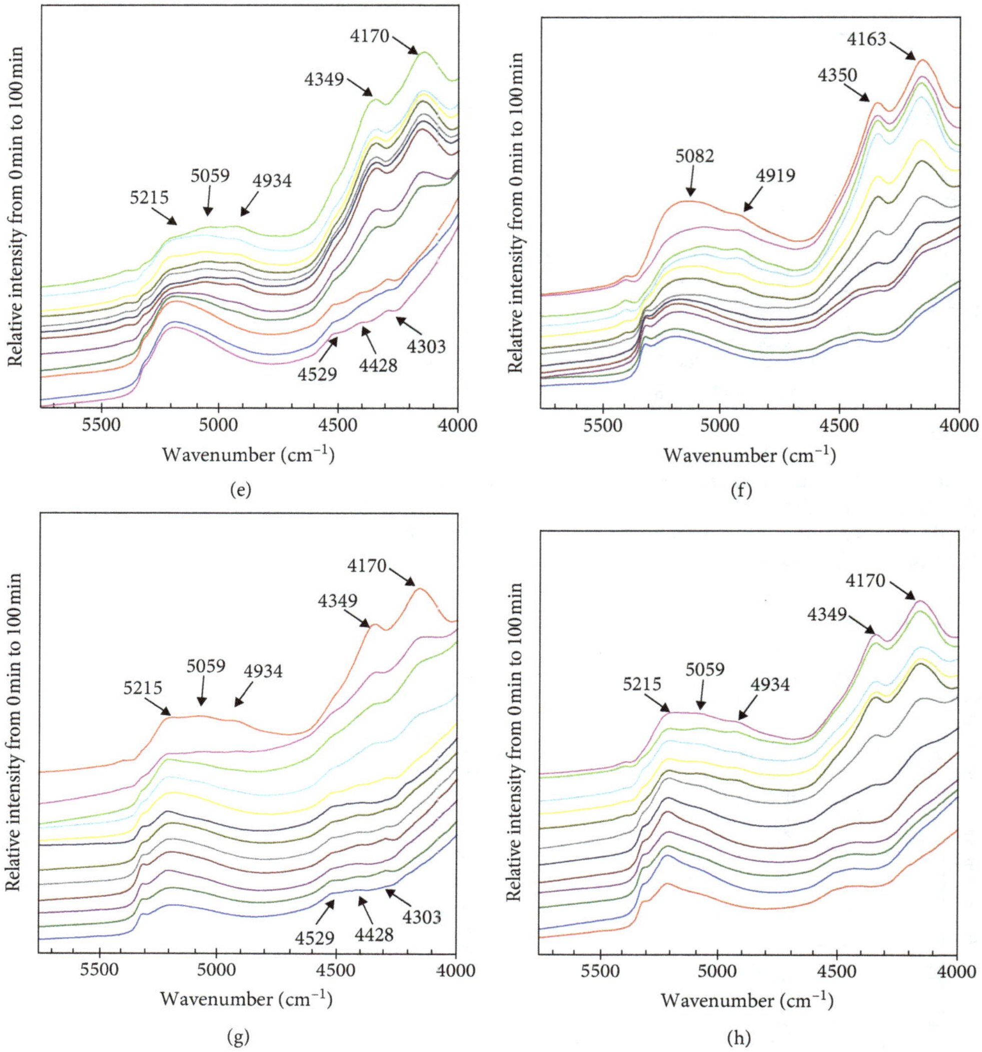

FIGURE 5: NIR spectra of 8 batches of calamine samples during calcination in the 5750–4000 cm^{-1} region. (a) S1 (hydrozincite and hemimorphite). (b) S2 (hydrozincite). (c) S3 (hydrozincite, hemimorphite, and dolomite). (d) S4 (hydrozincite, hemimorphite, and dolomite). (e) S5 (hydrozincite, hemimorphite, dolomite, and calcite). (f) S6 (hydrozincite). (g) S7 (hydrozincite, hemimorphite, dolomite, and calcite). (h) S8 (hydrozincite and hemimorphite).

TABLE 2: Information of sample classification.

Set	Number of samples	Minimum (%)	Maximum (%)	Average value (%)	SD (%)
Training	73	43.43	93.43	73.84	12.36
Validation	32	49.43	93.34	72.47	13.70
Test	25	45.69	93.21	74.51	12.18

TABLE 3: NIR models for determining ZnO content in calcined calamine.

Model	Spectral band (cm^{-1})	Pretreatment	RMSECV (%)	Validation set			Test set			Rank
				R^2 (%)	RMSEV (%)	RPD	R^2 (%)	RMSEP (%)	RPD	
1	4601.6–4003.7	SD	2.81	94.93	3.09	4.52	97.75	1.89	6.44	8
2	5446.3–4003.7	SD	2.9	94.96	3.08	4.62	95.75	2.79	4.36	8
3	7436.5–5778, 4601.6–4003.7	SD	2.97	94.74	3.14	4.41	97.69	1.97	6.18	8
4	7436.5–5778, 5446.3–4003.7	SD	3.01	94.57	3.19	4.42	93.87	3.19	3.82	8
5	7436.5–5778, 5446.3–4003.7	NO	3.09	93.20	3.57	3.89	90.52	4.28	2.85	8
6	7436.5–5778, 5446.3–4003.7	FD	3.14	93.92	3.38	4.14	94.99	2.97	4.11	8

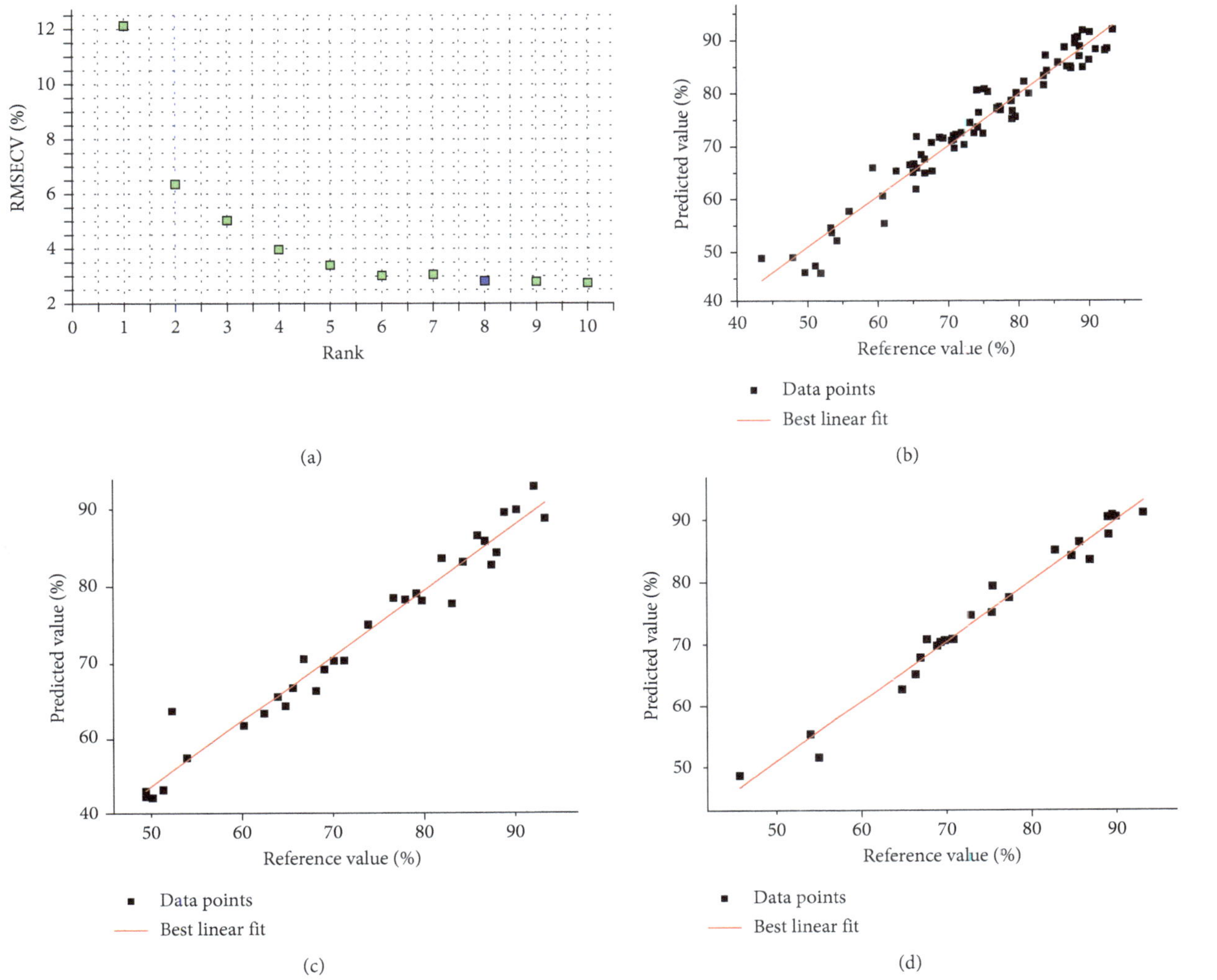

FIGURE 6: RMSECV-rank correlation diagram (a) and the prediction results of the model for the training set (b), validation set (c), and test set (d).

In terms of the internal cross-validation, the RMSECV was 2.81%. In the predicted results of validation set, the RMSEV was 3.09%, the R^2 was 94.93%, and the RPD was 4.52. In the predicted results of test set, the RMSEP was 1.89%, the R^2 was 97.75%, and the RPD was 6.44. It indicated that the prediction accuracy of the model was fine.

3.5. NIR Model for Predicting the XRD Characteristic T Value. The NIR models for predicting the XRD characteristic T value by different modeling parameters are shown in Table 4. From Table 4, it could be seen that model 1 had the smallest RMSECV value. In the external validation, the RMSEP value of model 1 had no significant difference with the minimum of 0.0813%, and model 1 was finally determined to be the best model. According to the RMSECV-rank correlation diagram (Figure 7(a)), when the RMSECV was the minimum, the best rank value of the determined model was 10. The prediction results of the model for the training set and prediction set are shown in Figure 7(c). The model was excellent with RMSECV of 0.0721%, R^2 of 96.01%, and RPD of 5.01 in internal cross-validation and RMSEP of 0.0829%, R^2 of 94.95%, and RPD of 4.64 in external validation. The predicted values of prediction set samples are shown in Table 5.

When the T value was equal to 1, the calamine was completely calcined. However, from the results of Table 5, it could be seen that there was a deviation between the T value predicted by the NIR model and its reference value. Samples whose reference value of T was 1, numbered 1, 10, 28, 37, and 46, were selected and the average of their relative deviations was calculated to be 4.08%. In addition, taking into account some errors in the experimental analysis process, this study stipulated that the calamine calcination was completed when the T value was $1.00 \pm 5\%$.

4. Conclusions

In the earlier stage, our research group established a rapid identification model for calamine and a determination model of ZnO content in calamine by near-infrared spectroscopy [3, 4]. However, the clinical use of calamine

Table 4: NIR models for predicting characteristic T value of XRD pattern.

Model	Spectral band (cm^{-1})	Pretreatment	Internal cross-validation			External validation			Rank
			R^2 (%)	RMSECV (%)	RPD	R^2 (%)	RMSEP (%)	RPD	
1	7436.6–5778, 4601.6–3999.9	SNV	96.01	0.0721	5.01	94.95	0.0829	4.64	10
2	5446.3–3999.9	MSC	94.53	0.0844	4.28	90.26	0.1150	3.51	9
3	7436.6–5778, 5446.3–3999.9	SNV	94.49	0.0848	4.26	95.15	0.0813	5.23	9
4	7436.6–6607.3, 4601.6–3999.9	SNV	93.44	0.0924	3.91	92.36	0.1020	3.9	8
5	6607.3–5778, 4601.6–4300.7	SNV	93.35	0.0931	3.88	93.12	0.0968	3.96	6
6	5446.3–3999.9	SNV	93.28	0.0936	3.89	90.83	0.1120	3.62	9

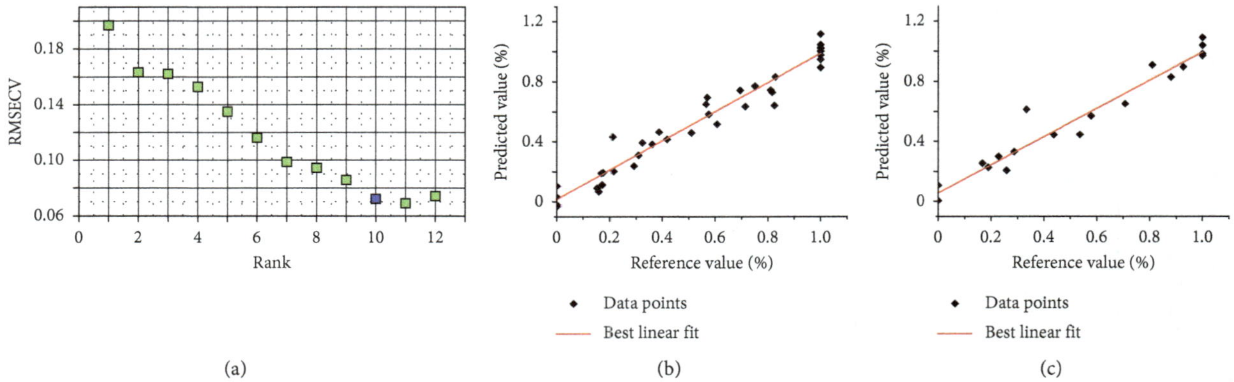

Figure 7: RMSECV-rank correlation diagram (a) and the prediction results of the model for the training set (b) and prediction set (c).

Table 5: Predicted T values of prediction set samples.

Number	Reference value (%)	Predicted value (%)	Deviation (%)	Relative deviation (%)
1	1.0000	1.0887	−0.0887	8.87
4	0	0.0054	−0.0054	∞
7	0.4364	0.4433	−0.0069	1.58
10	1.0000	0.9705	0.0295	2.95
13	0.1667	0.2530	−0.0863	51.77
16	0.8099	0.9086	−0.0987	12.19
19	0.2273	0.2976	−0.0703	30.93
22	0	0.1065	−0.1060	∞
25	0.3331	0.6114	−0.2780	83.46
28	1.0000	1.0370	−0.0370	3.7
31	0.2573	0.2056	0.0517	20.09
34	0.5359	0.4426	0.0933	17.41
37	1.000	0.9816	0.0184	1.84
40	0.5786	0.5687	0.0099	1.71
43	0.9268	0.8948	0.0320	3.45
46	1.0000	0.9696	0.0304	3.04
49	0.8800	0.8275	0.0525	5.97
52	0.1887	0.226	−0.0373	19.77
55	0.2867	0.3292	−0.0425	14.82
58	0.7076	0.6495	0.0581	8.21
61	0	0.0054	−0.0054	∞

required calcination, and currently, there was no good way to monitor the process of calamine calcination online. In this study, the calcination process of 8 batches of calamine samples with different sizes and textures under different calcination temperatures was simulated, to analysis the change rules of ZnO contents, XRD patterns of samples, and the attribution of absorption regions in near-infrared spectra. In the calcination process of calamine, the content of ZnO gradually increased to the maximum and then remained unchanged; at that time, the calamine was completely calcined. The XRD characteristic T value could be used to measure the degree of decomposition of hydrozincite into ZnO. When the T value was 1, the hydrozincite was completely decomposed into ZnO, that is, the calamine was completely calcined. In near-infrared spectroscopy analysis, the absorption of carbonate ion in hydrozincite was $4349\,m^{-1}$ and $4170\,cm^{-1}$, and the absorption peak of $4170\,cm^{-1}$ was not interfered by other carbonic impurities, which would disappear when the calamine was completely calcined.

In addition, this paper established a rapid determination model of ZnO content in calcined calamine using near-infrared spectroscopy based on the PLS algorithm. It was found that the model enjoyed a high accuracy and strong forecasting capacity for analyzing ZnO content ranging within 41.13–93.43%. That is, from the content point of view, the quality of the calcined calamine and the calcination process could be monitored throughout by NIR without titration of ZnO.

This paper also established a NIR model for predicting T value. When applied, the near-infrared spectrum of the sample was brought into the model to obtain the T value. If the T value was $1.00 \pm 5\%$, the sample was completely calcined. This method not only can judge the end point of the calamine calcination process but also can judge the calcination degree of any calamine sample. It had the outstanding advantages of rapidness, environmental protection, and no need to process samples, compared with measuring the change of the ZnO content in the calamine calcination. It was of great significance to optimize the calcination technology of calamine and monitor the quality of calamine processed in actual production.

Conflicts of Interest

The authors declare that there are no conflicts of interest regarding the publication of this paper.

Acknowledgments

This work was supported by the high-technology projects of the National Development and Reform Commission (1399), "Standardized Construction of Musk Hemorrhoids Ointment."

References

[1] National Pharmacopoeia Committee, *China Pharmacopoeia of 2015 Edition*, Vol. 1, China Medical Science and Technology Press, Beijing, China, 2015.

[2] L. J. Yang, Z. J. Zhang, R. R. Li, S. Q. Sun, and R. C. Lin, "Study on the mineralogical origin of medicinal calamine based on phase and composition analysis," *Guang Pu Xue Yu Guang Pu Fen Xi*, vol. 31, no. 11, pp. 3092–3097, 2011.

[3] Y. Sun, L. Chen, B. Huang, and K. Chen, "A rapid identification method for calamine using near-infrared spectroscopy based on multi-reference correlation coefficient method and back propagation artificial neural network," *Applied Spectroscopy*, vol. 71, no. 7, pp. 1447–1456, 2017.

[4] X. Zhang, L. Chen, Y. Sun, Y. Bai, B. Huang, and K. Chen, "Determination of zinc oxide content of mineral medicine calamine using near-infrared spectroscopy based on MIV and BP-ANN algorithm," *Spectrochimica Acta Part A: Molecular and Biomolecular Spectroscopy*, vol. 193, pp. 133–140, 2018.

[5] Y. M. Guo, K. F. Yu, Y. H. Liu, J. Z. Zhao, Z. C. Wang, and H. B. Zhang, "Analysis on processing mechanism of calamine," *Zhongguo Zhongyao Zazhi*, vol. 30, no. 8, pp. 596–599, 2005.

[6] L. Zhou, C. Xu, L. Zhang, and A. Ding, "Processing mechanism of calamine," *Zhongguo Zhongyao Zazhi*, vol. 35, no. 12, pp. 1556–1559, 2010.

[7] F. J. Camino-Sánchez, B. Delgado-Moreno, and P. Navarro-Fuentes, "Quick development of a NIRS method for the analysis of moisture, protein, and fat in nutritional clinical products," *Food Analytical Methods*, vol. 8, no. 7, pp. 1758–1770, 2015.

[8] R. L. Frost, B. J. Reddy, M. C. Hales, and D. L. Wain, "Near infrared spectroscopy of the smithsonite minerals," *Journal of Near Infrared Spectroscopy*, vol. 16, no. 2, pp. 75–82, 2008.

[9] K. L. Chen and L. Chen, "Identification methods on mineral traditional Chinese medicines by near infrared (NIRS) spectroscopy technology," *Zhongnan Minzu Daxue Xuebao*, vol. 33, no. 4, pp. 52–56, 2014.

[10] C. W. Huck, G. Abel, M. Popp, and G. K. Bonn, "Comparative analysis of naphthodianthrone and phloroglucine derivatives in St. John's Wort extracts by near infrared spectroscopy, high-performance liquid chromatography and capillary electrophoresis," *Analytica Chimica Acta*, vol. 580, no. 2, pp. 223–230, 2006.

[11] L. Xiang, G. Fan, J. Li et al., "The application of an artificial neural network in the identification of medicinal rhubarbs by near-infrared spectroscopy," *Phytochemical Analysis*, vol. 13, no. 5, pp. 272–276, 2010.

[12] Y. Chen, M. Y. Xie, Y. Yan et al., "Discrimination of Ganoderma lucidum according to geographical origin with near infrared diffuse reflectance spectroscopy and pattern recognition techniques," *Analytica Chimica Acta*, vol. 618, no. 2, pp. 121–130, 2008.

[13] L. Ye, K. Li, W. Wang, Y. Liu, S. Jiang, and G. Huang, "Near-infrared spectroscopy as a process analytical technology tool for monitoring the parching process of traditional Chinese medicine based on two kinds of chemical indicators," *Pharmacognosy Magazine*, vol. 13, no. 50, pp. 332–337, 2017.

[14] C. B. Luo, L. W. Chen, Y. L. Yan, W. Z. Wang, and Z. Y. Wang, "Study of temperature correction in the PLS near-infrared quantitative analysis of the wheat," *Guang Pu Xue Yu Guang Pu Fen Xi*, vol. 27, no. 10, pp. 1993–1996, 2007.

[15] Z. L. Wang, Y. Bai, J. T. Ding, and F. W. Ning, "Analysis method of mica content of artificial aggregate based on X-ray diffraction," *Shuidian Nengyuan Kexue*, vol. 35, no. 3, pp. 150–153, 2017.

[16] J. H. Zhang, X. J. Shi, Z. Wei, L. Yin, and Y. P. Liu, "Component analysis and thermal stability of calamine before and after processing," *Zhongguo Shiyan Fangjixue Zazhi*, vol. 17, no. 24, pp. 16–18, 2011.

[17] Z. Ni, C. Hu, and F. Feng, "Progress and effect of spectral data pretreatment in NIR analytical technique," *Yaowu Fenxi Zazhi*, vol. 28, no. 5, pp. 824–828, 2008.

[18] H. H. Zhao and Y. L. Yan, "The effects of noise on NIR analysis and related mathematic pretreatments and models," *Guangpuxue Yu Guangpu Fenxi*, vol. 26, no. 5, pp. 842–845, 2006.

[19] B. K. Waruru, K. D. Shepherd, G. M. Ndegwa, P. T. Kamoni, and A. M. Sila, "Rapid estimation of soil engineering properties using diffuse reflectance near infrared spectroscopy," *Biosystems Engineering*, vol. 121, no. 18, pp. 177–185, 2014.
[20] V. Bellon-Maurel, E. Fernandez-Ahumada, B. Palagos, J. M. Roger, and A. McBratney, "Critical review of chemometric indicators commonly used for assessing the quality of the prediction of soil attributes by NIR spectroscopy," *TrAC Trends in Analytical Chemistry*, vol. 29, no. 9, pp. 1073–1081, 2010.
[21] B. J. Reddy and R. L. Frost, "Near infrared spectroscopy of aurichalcite $(Zn,Cu^{2+})_5(CO_3)_2(OH)_6$," *Journal of Near Infrared Spectroscopy*, vol. 15, no. 2, pp. 115–121, 2007.
[22] L. Chen, M. Y. Yuan, and K. L. Chen, "Induction and analysis for NIR features of frequently-used mineral traditional Chinese medicines," *Zhongguo Zhongyao Zazhi*, vol. 41, no. 19, pp. 3528–3536, 2016.
[23] R. L. Frost, B. J. Reddy, D. L. Wain, and M. C. Hales, "An application of near infrared spectroscopy to the study of carbonate minerals-smithsonite, rhodochrosite, sphaerocobaltite and cadmium smithsonite," *Journal of Near Infrared Spectroscopy*, vol. 14, no. 5, pp. 317–324, 2006.

21

Detection of Veterinary Antimicrobial Residues in Milk through Near-Infrared Absorption Spectroscopy

Leandro da Conceição Luiz,[1] **Maria José Valenzuela Bell**,[1] **Roney Alves da Rocha**,[2] **Nayara Lizandra Leal**,[1] **and Virgílio de Carvalho dos Anjos**[1]

[1]*Grupo de Engenharia e Espectroscopia de Materiais, Departamento de Física, Universidade Federal de Juiz de Fora, 36036-900 Juiz de Fora, MG, Brazil*
[2]*Departamento de Ciências de Alimentos, Universidade Federal de Lavras, 37200-000 Lavras, MG, Brazil*

Correspondence should be addressed to Leandro da Conceição Luiz; livroleandro@gmail.com

Academic Editor: Maria Carmen Yebra-Biurrun

This study focuses on detection of antimicrobial residues in milk through Fourier transform near-infrared spectroscopy. Simulated and real samples were considered. The simulated ones take into account veterinary drugs added in milk samples in the following concentrations: enrofloxacin 100 μg/L, terramycin 100 μg/L, and penicillin 4 μg/L. The statistical tool used to discriminate the samples was the principal component analysis (PCA). Our results show that, with this experimental procedure, it is possible to discriminate different types of antimicrobials dissolved in milk. Moreover, the methodology was able to detect real sample milked on different days after the injection of ceftiofur hydrochloride which is in principle a zero grace period antimicrobial. The methodology proved to be fast and accurate within the maximum residue limits allowed by European Agency for Medicinal Products and Ministry of Agriculture Livestock and Food Supply from Brazil.

1. Introduction

According to the Food and Agriculture Organization of the United Nations, milk is one of the most consumed foods in the world. It not only has its importance in the nutritional level, but also plays an important role in the economy. Global consumption of milk and its derivatives exceeds 6 billion of consumers [1].

The milk contains protein, carbohydrates, lipids, minerals, and vitamins which accomplish important biochemical and nutritional functions, particularly to children and elderly people. The bovine milk contains about 87.1% of water, 4.0% of fat, 3.3% of protein, 4.6% of lactose, and 0.7% of ash. The basic protein content in milk is casein in 78.3%, whey protein 19%, and others totaling 2.7% [2, 3]. Concerning carbohydrates, lactose is the main one. Constituted by two monosaccharides, glucose and galactose, they carry nutritional important functions, such as providing 16.8 kJ/g of energy to people [4, 5]. Milk and dairy products are considered a good source of calcium due its high bioavailability. The latter can be understood as the fraction of ingested nutrient and food that is absorbed and used in physiological and normal metabolic functions and storage [6, 7]. Although the potentiality of milk as food is unquestionable, restrictions on its intake exist in people allergic to lactose and casein.

Adulterations in milk have been highly reported in developing countries, such as Pakistan, Brazil, India, and China [8]. Mostly the aim is to increase volume with the addition of water [9]. However, there are other problems, such as the contamination of milk by residues of veterinary drugs that may be present when the cow is milked in the grace period. The most common drugs are antimicrobials and anti-inflammatories. They are widely used in the treatment of dairy cattle diseases, such as mastitis, diarrhea, and lung diseases, also in prevention and control, or to increase the production and growth of animal [10–13]. A study performed by Van Boeckel et al. showed that, between

2000 and 2010, the consumption of antibiotics by the world population has increased by 36% and is related to the appearance of drug-resistant bacteria. From this total, 76% is mainly due to the countries that composes the BRICS (Brazil, Russia, India, China, and South Africa). Fraction of this increase is due to ingestion of animals or their food derivatives contaminated with antibiotics [14].

The overuse of drugs in dairy herd results in detectable traces in the milk. When their concentrations are over the maximum residue limit (MRL), they can cause health damage to the consumer ranging from allergic reactions to bacterial resistance [13] Contaminated milk is harmful to health and should be discarded [11, 15, 16]. In 2014, Van Boeckel et al. related the excessive use of antibiotics in animals to the appearance of super bacteria in humans. In order to monitor the milk content, the regulatory agencies use a variety of analytical methods to detect antibiotics traces in the milk, such as high-performance liquid chromatography (HPLC), gas chromatography coupled to mass spectroscopy (CG-MS), and antimicrobial detection kits. Nevertheless, there are drawbacks in their use such as sample preparation, qualified manpower, complex procedures, time consuming, and high cost. Moreover, normally the tests of antibiotics are specific to a class of antimicrobials.

The search for high sensitive techniques that allow the detection of residues of antibiotics in milk has been carried out for decades. In 1996, Verdon and Couedor used a reverse-phase HPLC technique to determine ampicillin residues which is able to detect 3 μg/L and to quantify up to 10 μg/L of the drug [17]. In 2002, Sivakesava and Irudayaraj carried out a study showing the feasibility of measuring tetracycline at μg/L levels with Fourier transform near-infrared (FT-NIR) spectroscopy and Fourier transform medium-infrared (FT-MIR) spectroscopy. Nevertheless, they reported high prediction errors and suggested that the methodology should be confirmed with naturally contaminated samples and other drug residues [18]. In 2003, Jankovská and Sustová used FT-NIR to analyse cow milks. However, the technique was used to describe physical-chemical characteristics of milk. In addition, partial least squares (PLS) regression was used to develop calibration models for the examined milk components. Through these results, they suggest that the NIR spectroscopy is applicable for a rapid analysis of milk composition [19]. In 2010, Brandão et al. studied fat concentration in milk samples by means of noninvasive techniques, FT-IR and FT-NIR absorption. They concluded that the wavelength of 2308 nm can be used to determine the fat concentration of milk without other components' influence [20]. In 2014, Zhang et al. [21] examined UHT and pasteurized milks to verify the presence of residues of tetracyclines, sulfonamides, sulfamethazine, and quinolone. The milk samples were collected in highly populated cities of China and analysed by the enzyme-linked immunosorbent assay (ELISA). No residue of veterinary drugs has exceeded MRL established by China, European Union, and CAC (Codex Alimentarius Commission). Nevertheless, because of the high number of residues present in milk, they recommended that the control mechanisms should be rigorously applied in order to keep these residues at safe levels. In 2015, Moharana et al. [22] analysed the veterinary drug enrofloxacin in cow milk samples obtained from two cities of India. According to the authors, the enrofloxacin is the most rampantly used drug in veterinary practice. To analyse the samples, they used reverse-phase HPLC. With a limit of detection of 100 μg/L, they have verified that 8% of the samples had values above the MRL. From the brief historical review described above, it is clear that there is a need for more in-depth studies exploring detection limits, embracing bigger classes of antibiotics, and analysing real samples.

Table 1: Result of physical-chemical characterization of milk samples. Average values and their standard deviations.

Analysis	Values	Values reference[a,b]
Cryoscopy	(0.536 ± 0.001) °H	(−0.550 to −0.530) °H
Acidity	(17.3 ± 0.6) °Dornic	(14 to 18) °Dornic
Density	(1.031 ± 0.000) g/mL	(1.029 to 1.040) g/mL
pH to °C	(6.72 ± 0.01)	(6.60 to 6.80)
Fat	(3.65 ± 0.01) %	≥3.00
Protein	(3.14 ± 0.01) %	≥2.90
Lactose	(4.50 ± 0.01) %	≥4.30
Solids	(11.29 ± 0.01) %	≥8.40

[a]IN 62; [b]FAO/TCP/KEN/6611.

This work deals with Fourier transform near-infrared (FT-NIR) spectroscopy associated with principal components analysis (PCA) to detect traces of veterinary antimicrobials (enrofloxacin, terramycin, penicillin, and ceftiofur hydrochloride) bellow the MRL allowed by the legislation of European Agency for Medicinal Products (EMEA) which is adopted in Brazil by Ministry of Agriculture Livestock and Food Supply (MAPA) [23]. The detection of the ceftiofur hydrochloride in milk was performed for two consecutive days after the drug has been administered to the animal.

2. Materials and Methods

The analyses were performed in the Process and Products Laboratory (LPP) and in the Materials Spectroscopy Laboratory (LEM), located in the Physics Department of Federal University of Juiz de Fora, Brazil.

2.1. Milk Samples. Some samples used in this work were raw milk from nonmedicated cows. They were collected at a farm located in the city of Rio Pomba, MG, Brazil, near the laboratory. After milking, the raw milk samples were immediately stored and kept refrigerated at 5°C until analysis which was performed after approximately one hour. The samples were submitted to chemical-physical analysis to verify their conformity with the established standards [24], that is, cryoscopy, density, pH, acidity (Dornic and Alizarol tests), fat, protein, lactose, and solids [25–27]. Each measurement was performed in triplicate. The results are shown in Table 1.

2.2. Contamination Simulation. Initially, two portions were separated, one as a control sample and the other as

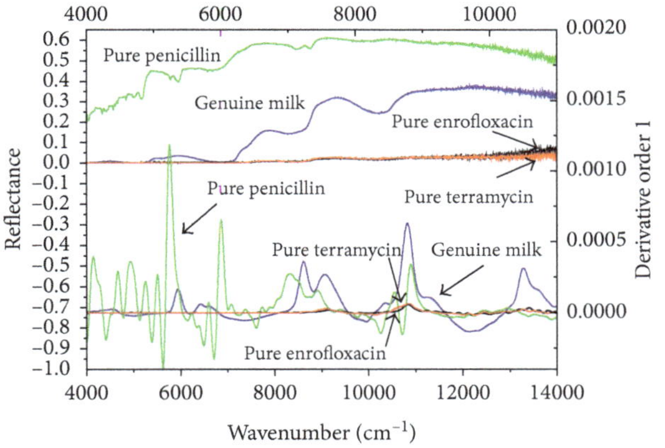

FIGURE 1: Reflectance spectra of pure samples of genuine milk and veterinary drugs with their respective first derivatives.

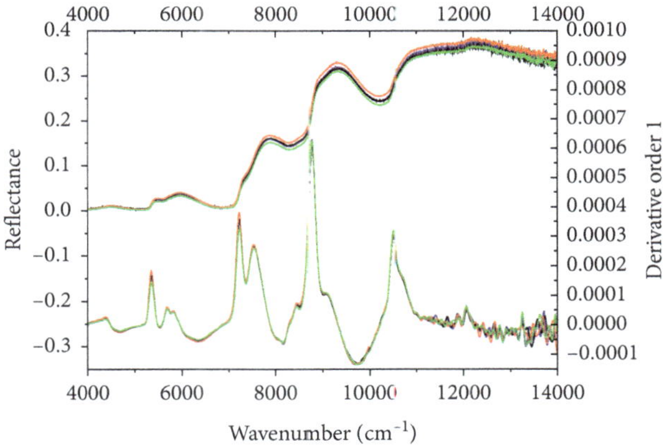

FIGURE 2: Reflectance spectra and their first derivatives of genuine and contaminated milk samples within the MRL.

a self-controlled one with a veterinary drug. The drugs used were the antimicrobials: enrofloxacin Baytril® injective 10% that has 10 g of enrofloxacin plus 100 ml of vehicle in its composition, terramycin/LA Zoetis/Pfizer® injective that has 20 g of oxytetracycline plus 100 g of vehicle in its composition, and reinforced pentabiotic Pfizer with penicillin.

The simulation was done according to the active principle of each drug and not in its volume. More explicitly, in the analysis with PCA, the discrimination of contaminated milk in μg/L is given through the active principle, MRL as provided by legislation, and not through the drug (excipients + active principle). For this purpose, each veterinary drug was first diluted in distilled water, and finally, part of this dilution was added in the genuine milk, in order to reach the concentration of the active principle in the milk. The values equivalent to the MRL allowed by EMEA and MAPA for the antimicrobials and its metabolites are as follows: 100 μg/L to enrofloxacin, 100 μg/L to terramycin, and 4 μg/L to penicillin [22, 28, 29]. This methodology was applied previously to the anti-inflammatory sodium diclofenac [30].

2.3. Real Contaminated Samples. First, the milk free from drugs was collected from a cow used as a control. The CeF-50 Ceftiofur Agner Union injective drug, which has 50 g of ceftiofur hydrochloride plus 1 mL of vehicle in its composition, was then administered to the cow. This drug does not have a grace period. Milk is then collected for two consecutive days. All the samples were immediately stored and kept refrigerated at 5°C until analysis which was performed after approximately one hour.

2.4. Analysis Using FT-NIR Method. Analyses of the samples were carried out with the Multi Purpose FT-NIR Analyser from Bruker operating in the reflectance mode in the range of 13.500 to 3.700 cm^{-1} wavenumbers with a Te-InGaAs detector and 4 cm^{-1} of resolution. The OPUS® software version 5.5 was used for data acquisition. The samples were placed in borosilicate cuvettes with 8 mm thickness. Each analysis was performed in triplicate with 32 scans for simulated and control samples.

2.5. Statistical Analysis. The reflectance spectra and their first derivatives were made and analysed with the software OriginPro®8 SR2 v.8.0891(B891). The eigenvalues were calculated with the software BioStat version 5.3. The principal component analysis was conducted with the software The Unscrambler® X version 10.3.

3. Results

Table 1 shows the average results obtained for the milk quality parameters of the samples and the reference value established by the Brazilian Normative Instruction 62 from MAPA and the FAO/TCP/KEN/6611 [24]. The results are in the range accepted by the legislation.

Figure 1 shows the reflectance spectra (top) and their first derivatives (bottom) of genuine milk and pure antimicrobials. From the figure, one can see that the spectra are very different from each other above all, in the first derivatives.

Figure 2 shows the reflectance spectra (top) and their first derivatives (bottom) of genuine and contaminated milk samples within the MRL. On the contrary of Figure 1, now the differences between the spectral profiles are not apparent due to the very low concentration of the antibiotics. Therefore, we have to rely on a statistical model. This will be accomplished by means of PCA to discriminate control and tampered milk samples.

The principal component of a set of data (dimensionality) is obtained by means of an analysis that consists in finding the eigenvalues of the covariance matrix [31]. Each eigenvector has a corresponding eigenvalue. The eigenvectors with higher eigenvalues are the principal components and are ordered from the higher to the smaller ones furnishing the components in significance degree [32]. Table 2 shows the explained and cumulative variances of the principal components of genuine and contaminated milk samples within the MRL.

Table 2: Explained and cumulative variances of genuine and contaminated milk samples within the MRL.

PC	Explained variance (%)	Cumulative variance (%)
PC1	99.9159	99.9159
PC2	0.0156	99.9315
PC3	0.0105	99.9420
PC4	0.0102	99.9522
PC5	0.0093	99.9615
PC6	0.0084	99.9699
PC7	0.0080	99.9779
PC8	0.0077	99.9857
PC9	0.0069	99.9926
PC10	0.0060	99.9986
PC11	0.0008	99.9993
PC12	0.0007	100.0000

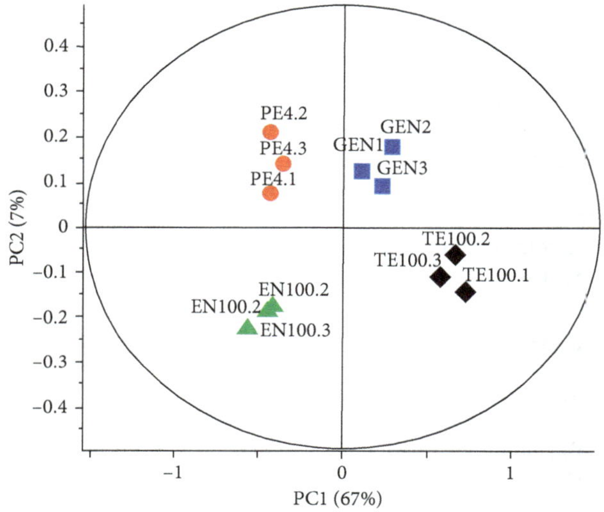

Figure 3: Hotelling score plot of statistical analysis (PCA) showing clustering data for control milk samples GEN1, GEN2, and GEN3 (blue square); milk with 100 μg/L of enrofloxacin, EN100.1, EN100.2, and EN100.3 (green triangles); 100 μg/L of terramycin, TE100.1, TE100.2, and TE100.3 (black rhombuses); and milk with 4 μg/L of penicillin, PE4.1, PE4.2, and PE4.3 (red circles).

Figure 3 shows the score plot with the clustering of milk samples: control (GEN1, GEN2, GEN3), contaminated with 100 μg/L of enrofloxacin (EN100.1, EN100.2, EN100.3), contaminated with 100 μg/L of terramycin (TE100.1, TE100.2, TE100.3), and contaminated with 4 μg/L of penicillin (PE4.1, PE4.2, PE4.3).

From Figure 3, it is clear the formation of clusters resulting from the high degree of similarity between groups of samples. Four groups are present, one in each quadrant. Group 1 (squares) refers to genuine milk (control samples) and is located in the first quadrant. Group 2 is located in the second quadrant (circles). This cluster represents a group of contaminated milk samples with 4 μg/L of penicillin. The third quadrant contains elements of group 3 which is related to contaminated milk samples with 100 μg/L of enrofloxacin

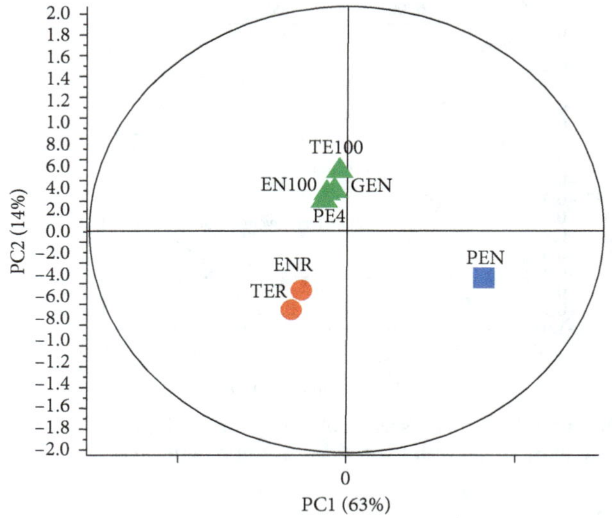

Figure 4: Score plot of statistical analysis (PCA) for average values of samples: pure enrofloxacin (ENR), pure terramycin (TER), pure penicillin (PEN), genuine milk (GEN), milk-enrofloxacin and milk-terramycin in 100 μg/L each (EN100 and TE100), and milk-penicillin in 4 μg/L.

Table 3: Explained and cumulative variances of average values of samples: pure enrofloxacin, pure terramycin, pure penicillin, genuine milk, milk-enrofloxacin and milk-terramycin in 100 μg/L each, and milk-penicillin in 4 μg/L.

PC	Explained variance (%)	Cumulative variance (%)
PC1	81.9197	81.9197
PC2	13.6470	95.5667
PC3	3.6040	99.1707
PC4	0.8143	99.9850
PC5	0.0065	99.9915
PC6	0.0045	99.9660
PC7	0.0040	100.0000

(triangles), while the fourth quadrant is occupied by group 4 (rhombuses) containing contaminated milk samples with 100 μg/L of terramycin. Among the elements of the groups, none is far apart from each other which discard the presence of outliers. The PCA accurately discriminated the samples in groups despite the very low concentration of the antimicrobials. From PC1, one can observe that the penicillin and enrofloxacin have the same score in contraposition to terramycin and genuine milk. This trend may be depicted in the absorption spectra (Figure 2). Therefore, PC1 represents the degree of milk contamination with antibiotics. Note that the group of data containing genuine milk is close to the centre of the axis. The contaminated clusters have different distances and different positions from the centre. This is related to the fact that the concentrations of the drugs are different. For example, samples with 100 μg/L of medication (groups 3 and 4) have similar position. From the above reasoning, we are led to infer that PC2 is related to the milk similitude.

All samples analysed until now were raw milk. Concerning pure antimicrobial data one may verify that the PCA

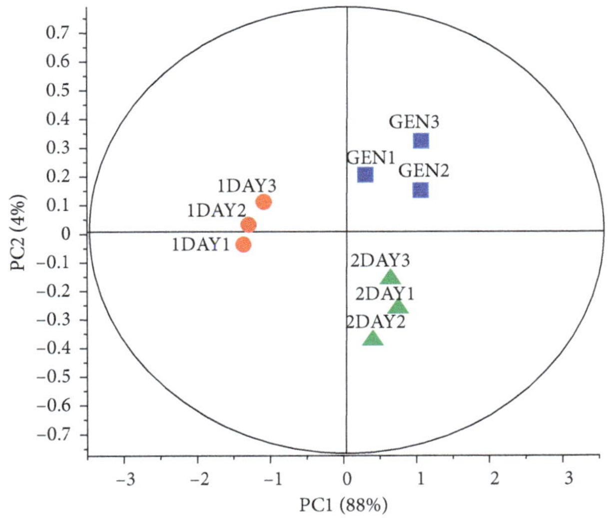

FIGURE 5: Score plot of statistical analysis (PCA) for samples of genuine milk (GEN) and milk collected for 2 days after the cow took the ceftiofur hydrochloride.

TABLE 4: Explained and cumulative variances of samples of genuine milk and milk collected for 2 days after the cow took the ceftiofur hydrochloride.

PC	Explained variance (%)	Cumulative variance (%)
PC1	99.7444	99.7444
PC2	0.1629	99.9073
PC3	0.0170	99.9243
PC4	0.0153	99.9538
PC5	0.0143	99.9538
PC6	0.0137	99.9675
PC7	0.0114	99.9790
PC8	0.0109	99.9898
PC9	0.0102	100.0000

is also able to discriminate between them. Figure 4 deals with such fact. It shows the PCA of average values of reflectance spectra of genuine milk, pure enrofloxacin, pure terramycin, and pure penicillin, and milk-enrofloxacin in 100 μg/L, milk-terramycin in 100 μg/L, and milk-penicillin in 4 μg/L, respectively. It can be seen that the milk-like samples (GEN, EN100, TE100, and PE4) are located very close in relation to PC1 which indicates the degree of similarity between them (cluster formation). For this reason, the samples of enrofloxacin antimicrobial (ENR), penicillin antimicrobial (PEN), and terramycin antimicrobial (TER) are so far apart from the group. Pure penicillin sample is detached from the other clusters because it was the only solid matrix (powder), while the others were liquid.

Table 3 shows the explained and cumulative variances for each of the principal components of the average values of pure enrofloxacin, pure terramycin, pure penicillin, genuine milk, milk-enrofloxacin and milk-terramycin in 100 μg/L each, and milk-penicillin in 4 μg/L.

Figure 5 shows the methodology applied in a real situation. It shows data of a control milk, without medication, and contaminated milk in two consecutive days after the administration of the drug ceftiofur hydrochloride in a cow. Table 4 has the respective explained and cumulative variances from the data of Figure 5.

Cluster formation was observed for the genuine milks (GEN1, GEN 2, and GEN3) and one-day and two-day drug administered milk (1DAY and 2DAY). It turns out that the GEN and 2DAY groups are along the same PC1, showing the similarity between the groups. This is due to the metabolization of the antibiotic in the milked cow after two days of the drug administration. Reinforcement of this supposition is apparent when compared with one day milked sample with genuine milk. The PC2 is connected with time of milking as the spectroscopic measurements were performed after the last day of milking (2DAY).

4. Conclusions

This article dealt with the identification of the antimicrobials enrofloxacin, terramycin, penicillin, and ceftiofur in milk samples in the MRL permitted by regulatory agencies. The methodology developed is based on the combination of Fourier transform near-infrared (FT-NIR) spectroscopy jointly with principal component analysis (PCA). The experimental procedure was able to detect antibiotics traces in a fast and accurate way. The methodology was applied also to detect ceftiofur hydrochloride in milk of a cow where the drug was administered. Beyond trace detection, we are also able to follow the metabolization process in the animal. Our results clearly demonstrate the potentiality of the method for the development of a portable prototype.

Conflicts of Interest

The authors declare that there are no conflicts of interest regarding the publication of this paper.

Acknowledgments

The authors thank the Brazilian funding agencies CAPES (PNPD 2871/2011), CNPq (309100/2016-0), and FAPEMIG (MPR 00004-13 and MPR 01068/16) for financial funding. Leandro da Conceição Luiz would like to thank the Chemical Department of Rural Federal University of Rio de Janeiro, Brazil, for the access to the software The Unscrambler.

References

[1] FAO/WHO, *Dairy Production and Products: Milk and Milk Products*, 2017.

[2] P. Walstra, J. T. M. Wouters, and T. J. Geurts, *Dairy Science and Technology*, CRC Press, Boca Raton, FL, USA, 2nd edition, 2006.

[3] V. K. Tavanti, L. G. Bassi, G. C. C. Ferreira et al., "Composição e capacidade de coagulação de leites de vacas holandesas e girolandas," *Revista do Instituto de Laticínios Cândido Tostes*, vol. 370, no. 64, pp. 5–9, 2009.

[4] Food and Agriculture Organization of United Nations (FAO), *Food Energy—Methods of Analysis and Conversion Factors*, Chapter 3, Food and Agriculture Organization of United Nations, Rome, Italy, 2003.

[5] M. C. S. Pereira, L. P. Brumano, C. M. Kamiyama, J. P. F. Pereira, M. P. Rodarte, and M. A. O. Pinto, "Lácteos com baixo teor de lactose: uma necessidade para portadores de má digestão da lactose e um nicho de mercado," *Revista do Instituto de Laticínios Cândido Tostes*, vol. 67, no. 389, pp. 57–65, 2012.

[6] R. S. Gibson, L. Perlas, and C. Hotz, "Improving the bioavailability of nutrients in plant foods at the household level," *Proceedings of the Nutrition Society*, vol. 65, no. 2, pp. 160–168, 2006.

[7] J. M. Carbonell-Capella, M. Buniowska, F. J. Barba, M. J. Esteve, and A. Frígola, "Analytical methods for determining bioavailability and bioaccessibility of bioactive compounds from fruits and vegetables: a review," *Comprehensive Reviews in Food Science and Food Safety*, vol. 13, no. 2, pp. 155–171, 2014.

[8] C. E. Handford, K. Campbell, and C. T. Elliott, "Impacts of milk fraud on food safety and nutrition with special emphasis on developing countries," *Comprehensive Reviews in Food Science and Food Safety*, vol. 15, no. 1, pp. 130–142, 2016.

[9] W. W. G. Nascimento, M. P. F. Souza, A. C. M. M. Valente, V. Anjos, M. A. M. Furtado, and M. J. V. Bell, "Results from portable and of low cost equipment developed for detection of milk adulterations," *Food Science and Technology*, vol. 37, pp. 38–41, 2017.

[10] M. A. V. P. Brito and C. C. Lange, *Resíduos de Antibióticos no Leite*, Comunicado Técnico 44, Embrapa, Brasília, Federal District, Brazil, 2005.

[11] D. Pontes Netto, M. O. Lopes, M. C. S. Oliveira et al., "Levantamento dos principais fármacos utilizados no rebanho leiteiro do Estado do Paraná," *Acta Scientiarum. Animal Sciences*, vol. 27, no. 1, pp. 145–151, 2005.

[12] T. F. Landers, B. Cohen, T. E. Wittum, and E. L. Larson, "A review of antibiotic use in food animals: perspective, policy, and potential," *Public Health Reports*, vol. 127, no. 1, pp. 4–22, 2012.

[13] R. W. Han, N. Zheng, Z. N. Yu et al., "Simultaneous determination of 38 veterinary antibiotic residues in raw milk by UPLC-MS/MS," *Food Chemistry*, vol. 181, pp. 119–126, 2015.

[14] T. P. Van Boeckel, S. Gandra, A. Ashok et al., "Global antibiotic consumption 2000 to 2010: an analysis of national pharmaceutical sales data," *The Lancet Infectious Diseases*, vol. 14, no. 8, pp. 742–750, 2014.

[15] S. J. Forsythe, *Microbiologia da Segurança Alimentar*, ArtMed, Porto Alegre, RS, Brazil, 2002.

[16] L. Redding, *Understanding the Use of Antibiotics on Small Dairy Farms in Rural Peru*, Publicly Accessible Penn Dissertations Paper 1414, University of Pennsylvania, Philadelphia, PA, USA, 2014.

[17] E. Verdon and P. Couedor, "Determination of ampicillin residues in milk by ion-pair reversed phase high performance liquid chromatography after precolumn derivatization," *Journal of Pharmaceutical and Biomedical Analysis*, vol. 14, no. 8–10, pp. 1201–1207, 1996.

[18] S. Sivakesava and J. Irudayaraj, "Rapid determination of tetracycline in milk by FT-MIR and FT-NIR spectroscopy," *Journal of Dairy Science*, vol. 85, no. 3, pp. 487–449, 2002.

[19] R. Jankovská and K. Sustová, "Analysis of cow milk by near-infrared spectroscopy," *Czech Journal of Food Sciences*, vol. 21, no. 4, pp. 123–128, 2011.

[20] M. C. M. P. Brandão, A. P. Carmo, M. J. V. Bell, and V. C. Anjos, "Characterization of milk by infrared spectroscopy," *Revista do Instituto de Laticínios Cândido Tostes*, vol. 373, no. 65, pp. 30–33, 2010.

[21] Y. D. Zhang, N. Zheng, R. W. Han et al., "Occurrence of tetracyclines, sulfonamides, sulfamethazine and quinolones in pasteurized milk and UHT milk in China's market," *Food Control*, vol. 36, no. 1, pp. 238–242, 2014.

[22] B. Moharana, P. K. Venkatesh, S. P. Preetha, and S. Selvasubramanian, "Quantification of enrofloxacin residues in milk sample using RP-HPLC," *World Journal of Pharmacy and Pharmaceutical Sciences*, vol. 4, no. 10, pp. 1443–1450, 2015.

[23] EMEA, *The European Agency for the Evaluation of Medicinal Products: Veterinary Medicines and Inspections*, Committee for Veterinary Medicinal Products: Enrofloxacin, EMEA, London, UK, 2002.

[24] FAO/TCP/KEN/6611 Project, *Milk Testing and Quality Control*, Milk Processing Guide Series V2.

[25] Food and Agriculture Organization of United Nations (FAO), *Milk Analysis*.

[26] A. F. Bodenmüller, J. C. Damasceno, I. T. S. Previdelli, R. G. Santana, C. E. C. O. Ramos, and G. T. Santos, "Tipologia de sistemas de produção baseada nas características do leite," *Revista Brasileira de Zootecnia*, vol. 39, no. 8, pp. 1832–1839, 2010.

[27] M. S. Robim, *Avaliação de Diferentes Marcas de Leite UAT Comercializadas no Estado do Rio de Janeiro e o Efeito da Fraude por Aguagem na Fabricação, Composição e Análise Sensorial de Iogurte*, Dissertação, Universidade Federal Fluminense, Niterói, Brazil, 2011.

[28] E. Pacheco-Silva, J. R. Souza, and E. D. Caldas, "Resíduos de medicamentos veterinários em leite e ovos," *Química Nova*, vol. 37, no. 1, pp. 111–122, 2014.

[29] A. M. Mangerona, O. D. Mozzer, R. F. Vieira, C. R. Dias, and D. R. Cappelini, "Comparação do período de carência em tecidos e leite bovino após administração de enrofloxacino e diclofenaco sódico (Quinotril Plus®) em diferentes vias parenterais," *A Hora Veterinária*, vol. 33, no. 196, 2013.

[30] L. C. Luiz, M. J. V. Bell, R. A. Rocha, T. O. Mendes, and V. C. Anjos, "Análise de resíduos de diclofenaco sódico veterinário em leites por espectroscopia no infravermelho próximo," *Revista Brasileira de Ciências da Saúde*, vol. 18, no. 3, pp. 219–224, 2014.

[31] R. E. Santos, "Principal component analysis applied to digital image compression," *Einstein*, vol. 10, no. 2, pp. 135–139, 2012.

[32] S. Sayad, *Principal Component Analysis*, University of Toronto, Toronto, ON, Canada, 2010.

22

Surface-Enhanced Raman Spectroscopy Analysis of Human Breast Cancer via Silver Nanoparticles: An Examination of Fabrication Methods

Beata Brozek-Pluska, Monika Kopec, and Jakub Surmacki

Laboratory of Laser Molecular Spectroscopy, Institute of Applied Radiation Chemistry, Lodz University of Technology, Lodz, Poland

Correspondence should be addressed to Beata Brozek-Pluska; beata.brozek-pluska@p.lodz.pl

Academic Editor: Feride Severcan

Breast cancer in a traditional way is diagnosed using mammography, computer tomography, ultrasounds, biopsy, and finally, histopathological analysis. Histopathological analysis is a gold standard in breast cancer diagnostics; however, it is time consuming and prone to the human interpretations. That is why new methods based on optical properties of analyzed human tissue samples are needed to be introduced to the clinical practice objective, costless and fast diagnostic protocols. Nowadays, Raman spectroscopy-based methods are gaining more and more importance. Raman spectroscopy and imaging allow to characterize human tissue samples using an electromagnetic radiation from a safe range, and simultaneously, a minimal sample preparation is required. During measurements, a natural differentiation in tissues components' scattering cross sections is used to build 2D and 3D maps of the chemical component distribution. The paper presents the application of SERS (surface-enhanced Raman spectroscopy) measurements for analysis of human breast cancer (adenocarcinoma). The advantages of SERS application in cancer diagnostics are also discussed. Moreover, the detailed chemical composition of human breast cancer tissue based on Raman bands of DNA/RNA, amino acids, lipids, and proteins which are significantly enhanced is presented. Three different methods of NP preparation are presented, and the effectiveness of Raman signal enhancement of Ag nanoparticles synthetized by these methods is compared. The enhancement effect of NPs synthetized by reduction of silver nitrate with sodium borohydride (method no. 1) and silver nitrate-hydroxylamine hydrochloride reduction (method no. 2) was stronger when compared with the polyol method (method no. 3). Presented SERS results confirmed that the clearly resolved and high-intensity Raman spectra of cancer human breast tissue can be recorded using integration times of the order of fractional seconds and one milliwatt of the excitation laser power.

1. Introduction

In the twenty-first century, cancer, just after cardiovascular diseases, is one of the leading causes of deaths in people around the world. The magnitude of this problem concerns both the epidemiological and diagnostic aspect. In 2017, 1,688,780 new cancer cases and 600,920 cancer deaths were projected to occur in the United States [1]. Such a statistic confirms that cancer is a global problem and remains an unsolved challenge. Various techniques such as MRI, mammography, ultrasonography, and positron emission tomography (PET) are used for detection of pathological cells at different stages of cancer development.

MRI screening is characterized by significantly higher sensitivity than typical for mammography (92.3% versus 30.8%), but simultaneously lower specificity for MRI is noticed (85.9% versus 96.8%) [2]. Even if the sensitivity and the specificity for the abovementioned methods can reach more than 90%, strong magnetic field or X-ray radiation constitutes undisputed disadvantages. Ultrasonography is completely safe for patients but has a low sensitivity equal to around 71.1% among women with predominantly fatty breast and only 57.0% for heterogeneous dense breasts, and the specificity for this technique is estimated around 88.5% [3]. The ability of PET to detect breast cancer strongly depends on the tumor's size, which should be considered as a serious disadvantage. The sensitivity of PET has been reported to be 68% for small (<2 cm) tumors and 92% for larger (2–5 cm) ones; moreover, the accuracy for detecting carcinomas in situ is very low and equal to 2–25% [4].

However, the final outcome for cancer every time is defined by the pathological study of a biological material taken during the biopsy or surgery. Such material is used for preparation of standard histologically stained specimens. Unfortunately, the histopathological diagnostic methods are highly invasive and costly to the health-care system. Moreover, for mass screening patients, the histopathological diagnostic methods are time consuming and strongly prone to the human interpretations. That is why the need to introduce new methods, safe for patients, fast, and objective to reduce the number of biopsies and medical costs, as well as the need for the determination of many cancer markers simultaneously in one measurement is a motivation for many researchers to introduce spectroscopic techniques in diagnostics of cancer [5–14].

Spectroscopic techniques based on Raman scattering have been proven to be powerful diagnostic tools providing objective biochemical fingerprints to distinguish normal, benign, and cancer tissues of many organs: lungs [5] based on nuclei acids and phospholipids; breast [6–13] based on proteins, lipids, and water; esophageal [14] based on DNA/RNA, proteins, and collagen contents; and gastric based on the relative amounts of nucleic acid, collagen, phospholipids, phenylalanine, and saccharide [15].

The key advantages of Raman spectroscopy (RS) over the traditional diagnostics methods are as follows: the capability of providing objective biochemical information about the components of normal, benign, and cancer cells and possibility to identify different cancer markers in one measurement. Advantages of RS over other spectroscopic techniques such as IR spectroscopy, for example, are a lack of interference from water, which is particularly important for live-cell, human tissue analysis and in vivo applications. RS offers also high spatial resolution [7–13]. The main disadvantage of Raman-based optical methods is that even if Raman spectra are really informative and accurate, the Raman signal is very weak, which can limit its applications for diagnosis of early stages of cancer samples.

Therefore, there is a really need to develop new procedures to generate a simple approach that ensures high-quality spectrum. Surface-enhanced Raman spectroscopy (SERS) opens new possibilities to overcome the limitations of conventional RS. SERS enables much better limits of detection enabling single-molecule detection level. Simultaneously, there are many variables that need to be optimized to obtain the enhanced Raman signal of the analyzed biological sample.

SERS is based on the excited surface plasmons in a rough metallic surface. The amplification factors typical for the Raman signal in SERS technique are equal to 10^6–10^8 [16]; that is why in the fingerprint region of recorded vibrational spectra, intense and sharp bands can be observed [17–19]. This makes SERS a great multiplexing technique characterizing by very high sensitivity even in the picomolar range [20].

Metals that are used in SERS techniques are Ag, Au, and Cu. Nanoparticles (NPs) of these metals are characterized by intense absorption bands in the UV-Vis region, which are not present in the spectrum of the bulk metal.

The strongest enhancement in SERS is observed when the frequency of incident photon is in resonance with the oscillation of conduction electrons; that is why the excitation wavelength in SERS experiments must be coupled with absorption properties of synthetized NPs.

Recent SERS technique was successfully applied in molecular diagnosis, in vivo imaging, and drug delivery using nanoparticles such as nanocrystals, quantum dots, biodegradable nanoparticles, and nanoscale carriers [21–24]. Metal NPs functionalized with sensitive Raman reporter molecules followed by bioconjugation have been applied for labeling biological systems in cells and tissues as a noninvasive imaging technique with picomolar sensitivity and multiplexing capabilities [25, 26]. Chemical inactivity of Au and Ag nanoparticles also makes them most suitable for introduction into living cells as SERS nanoprobes to enhance the Raman signal.

In our previous work, we found that spontaneous Raman scattering technique can be applied to distinguish cancerous from noncancerous human breast tissues [7–13].

In this paper, we applied the SERS technique for analysis of nonfixed human breast cancer slices using Ag colloidal nanoparticles. Presented results confirmed that the SERS technique is a powerful tool with high sensitivity for screening of chemical compounds in a sample, being of great importance for the purposes of biomedical research especially the early diagnosis of any cancer. The comparison of physical properties of nanoparticles synthetized using different methods and the comparison of efficiency of signal enhancement will also be discussed.

2. Experimental Methods

2.1. Tissue Samples. We examined cryosectioned slides (6 μm thin) of adenocarcinoma (results for one patient number, P141, from a database of 250 patients, are presented). We recorded 100 spectra for each sample. The research did not affect the course of the operation or treatment of the patients. The tissue was obtained during a surgical operation, cut into thin sections at −25°C, and mounted on CaF_2 windows. For SERS measurements, tissue slices were covered by 4-5 drops (~200–250 μL) of nanoparticle solutions per each sample. Dried samples were analyzed immediately. All procedures have been conducted under a protocol approved by the Bioethical Committee at the Medical University of Lodz (RNN/323/17/KE/17/10/2017).

2.2. Raman Microspectroscopy. Raman spectra were acquired with a confocal Alpha 300RSA+ Raman microscope (WITec) operating at 532 nm laser excitation with an ultrahigh-throughput spectrometer, an Andor Newton DU970N-UVB-353 CCD Camera, and a 40x dry objective (Nikon, CFI Plan Fluor C ELWD DIC-M, NA of 0.60, and a 3.6–2.8 mm WD). A single SERS spectrum was recorded with an integration time of 0.1 second at 532 nm excitation (average laser power: 1 mW). WITec Project Plus (WITec

GmbH, Germany) and Origin (OriginLab) were used for data acquisition and analysis, respectively.

2.3. Nanoparticle Synthesis

2.3.1. Method No. 1. We have mixed 2.0 mL of 1.25×10^{-2} M sodium citrate, 5.0 mL of 3.75×10^{-4} M silver nitrate, and 5.0 mL of 5.0×10^{-2} M hydrogen peroxide. Then, a 20 mL of 1.0×10^{-3} M potassium bromide was added. For the final silver reduction step, 2.5 mL of freshly prepared 5.0×10^{-3} M sodium borohydride was mixed with the solution. Excess sodium citrate was used as a buffer to preserve neutral or weakly basic pH of the solution to stabilize the synthetized silver nanoprisms. Hydrogen peroxide was employed to facilitate formation of shape-selected particles. In the absence of peroxide, uncontrolled growth of NPs with a large number of structural defects can be observed [27]. Finally, the silver NPs were analyzed with a high-resolution scanning electron microscope.

2.3.2. Method No. 2. We have mixed the silver nitrate in the concentration of 10^{-3} M with hydroxylamine hydrochloride in the concentration of 1.5×10^{-3} M. Hydroxylamine hydrochloride was preserved at an initial alkaline pH of the reaction. The final pH of the colloidal solution was optimized by varying the amount of sodium hydroxide added to the hydroxylamine solution, and pH of 7 for the reduction reaction during the experiment was kept. The reaction was completed within a few seconds [28]. Finally, the silver NPs were analyzed with a high-resolution scanning electron microscope.

2.3.3. Method No. 3. Silica NPs were prepared using ammonium hydroxide, dry ethanol, and tetraethyl orthosilicate (TEOS) by stirring mixture for 24 h at 25°C. The formed silica NPs were purified and functionalized with 3-mercaptopropyltrimethoxysilane (MPTS). The polyol method [29] was used for the preparation of silver nanoparticle-embedded silica spheres. Finally, the NPs were analyzed with a high-resolution scanning electron microscope.

2.4. SEM, UV-Vis, and DLS Measurements. The UV-Vis spectroscopy, SEM, and DLS measurements were performed to characterize the size and morphology of the produced colloids. UV-Vis spectra were recorded with a Varian Cary 5E spectrophotometer and 2 mm quartz cells (Hellma). DLS measurements were performed using Zetasizer Nano Z. SEM measurements were obtained using a microscope SEM S-4700 Hitachi.

3. Results and Discussion

In this section, the results of the confocal Raman microspectroscopy for the cancerous human breast tissues of the patient P141 (patient no. 141 from a database of 250 patients) before and after addition of nanoparticles are presented. The typical results for the patient P141 have been used to illustrate the most important findings of research.

In our previous paper, we have proved that Raman spectroscopy can be successfully implemented to discriminate between noncancerous and cancerous breast tissues [7–13]; that is why in this section, we have focused only on the cancerous tissue analysis to show that using SERS one can identify a single pathology cells.

Before we start the analysis of biological SERS spectra, the characterization of NPs synthetized by different methods is needed.

Historically, metallic electrodes were first used as active enhanced supports, and then, metallic NPs in solutions and finally metallic nanostructures like arrays on a flat substrate were used to improve the sensitivity of RS over last 10 years [30].

NP solutions have several advantages compared to the other enhanced supports including ease of colloid formation and straightforward characterization of the colloid solution by simple UV-Vis absorption measurements. On the other hand, the size of NPs in colloidal solutions cannot be uniform, and not negligible problems can be observed due to instability of such a system [31, 32].

Figure 1 presents UV-Vis spectra and SEM images of all three types of NP colloidal solutions used in SERS human breast tissue measurements.

One can see from Figure 1 that the maximum peak position for the absorbance for all NPs is observed in the spectral range 440–485 nm, but not negligible absorbance, compared to the maximum value of 532 nm, used in our experiments as an excitation laser wavelength, is also observed. This observation confirms that the 532 nm wavelength is suitable for the plasmon generation of Ag nanoparticles.

Simultaneously, the comparison for NPs synthetized in our group using different methods shows that the highest value of the absorbance is noticed for the NPs produced by method no. 1 where borohydride was used as a reduction agent.

The colloids characterized by UV-Vis spectra were also analyzed using SEM technique. Figures 1(b), 1(d), and 1(f) show SEM images of colloidal solutions of NPs prepared to test in SERS measurements of human cancer breast tissues.

The characteristic of NP morphology was completed by DLS measurements. Table 1 presents information about the size of NPs prepared by different methods.

The differences between the results obtained using SEM technique and DLS measurements can be explained by the fact that the aggregation of NPs during DLS and SEM measurements is observed.

The SERS activity of synthetized colloids was tested using human cancerous tissues of the same patient P141. Figure 2 presents the SERS spectra of human cancer slices (laser power 1 mW, integration time 0.1 sec) recorded using synthetized NP colloids and the native Raman spectra recorded for the same tissue sample.

One can see that Figures 2(a)–2(f) show the clearly resolved and high-intensity SERS spectra of cancerous human breast tissue with added Ag solutions. The regular

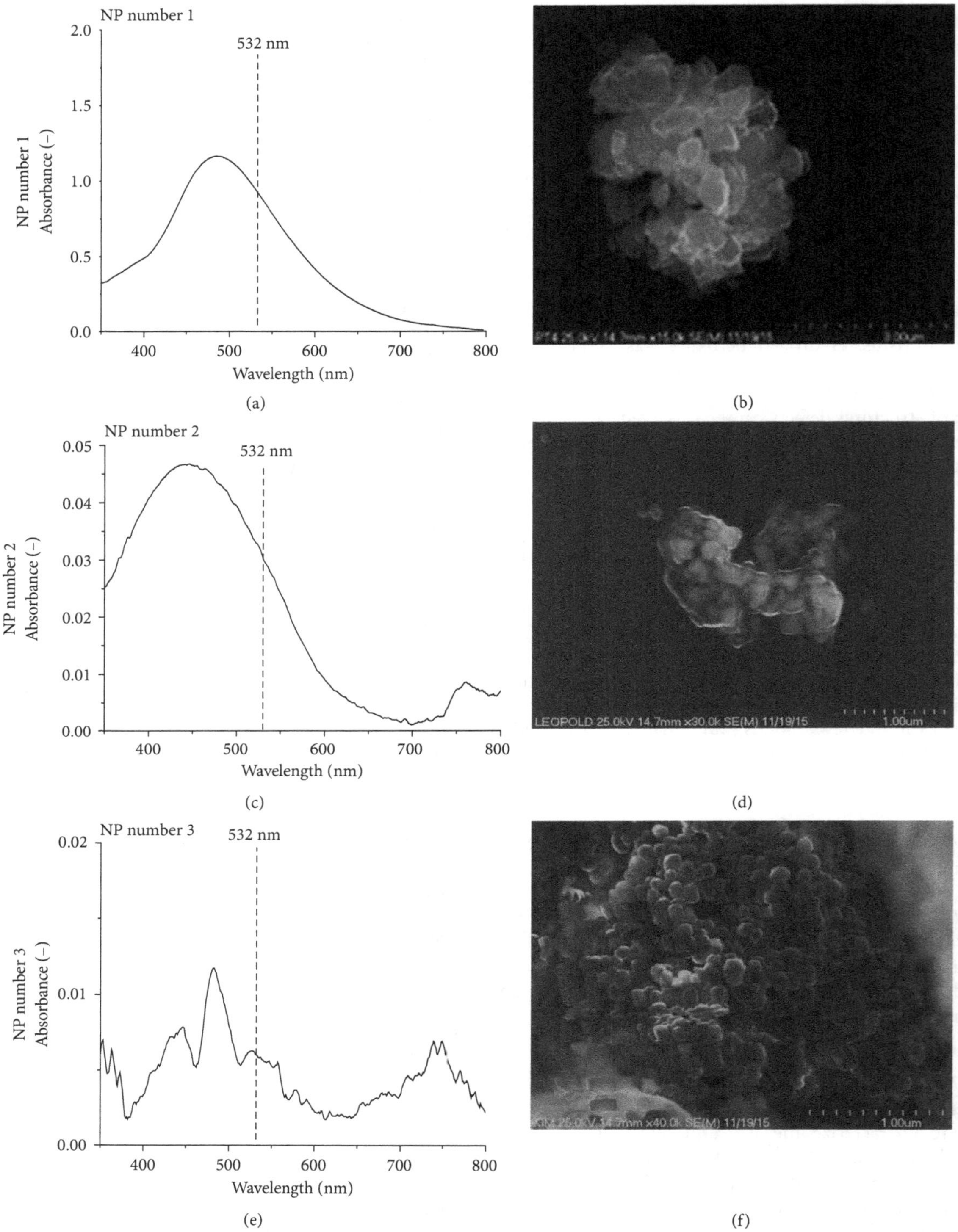

FIGURE 1: UV-Vis spectra (a, c, e) and SEM images (b, d, f) of NPs synthetized by different methods described in detail in the experimental section.

Raman spectrum of the cancerous sample without Ag sol is shown in Figure 2(g). In the native tissue sample without the addition of silver colloid solution for the integration time of 0.1 sec. and the laser power of 1 mW, no Raman peaks were observed. A comparison of data presented in Figures 2(a)–2(g) clearly demonstrates that Ag colloid enhanced dramatically

TABLE 1: The average size of NPs determined using the DLS method.

Method	The average size of NPs determined using the DLS method
No. 1	64 nm
No. 2	118 nm
No. 3	296 nm

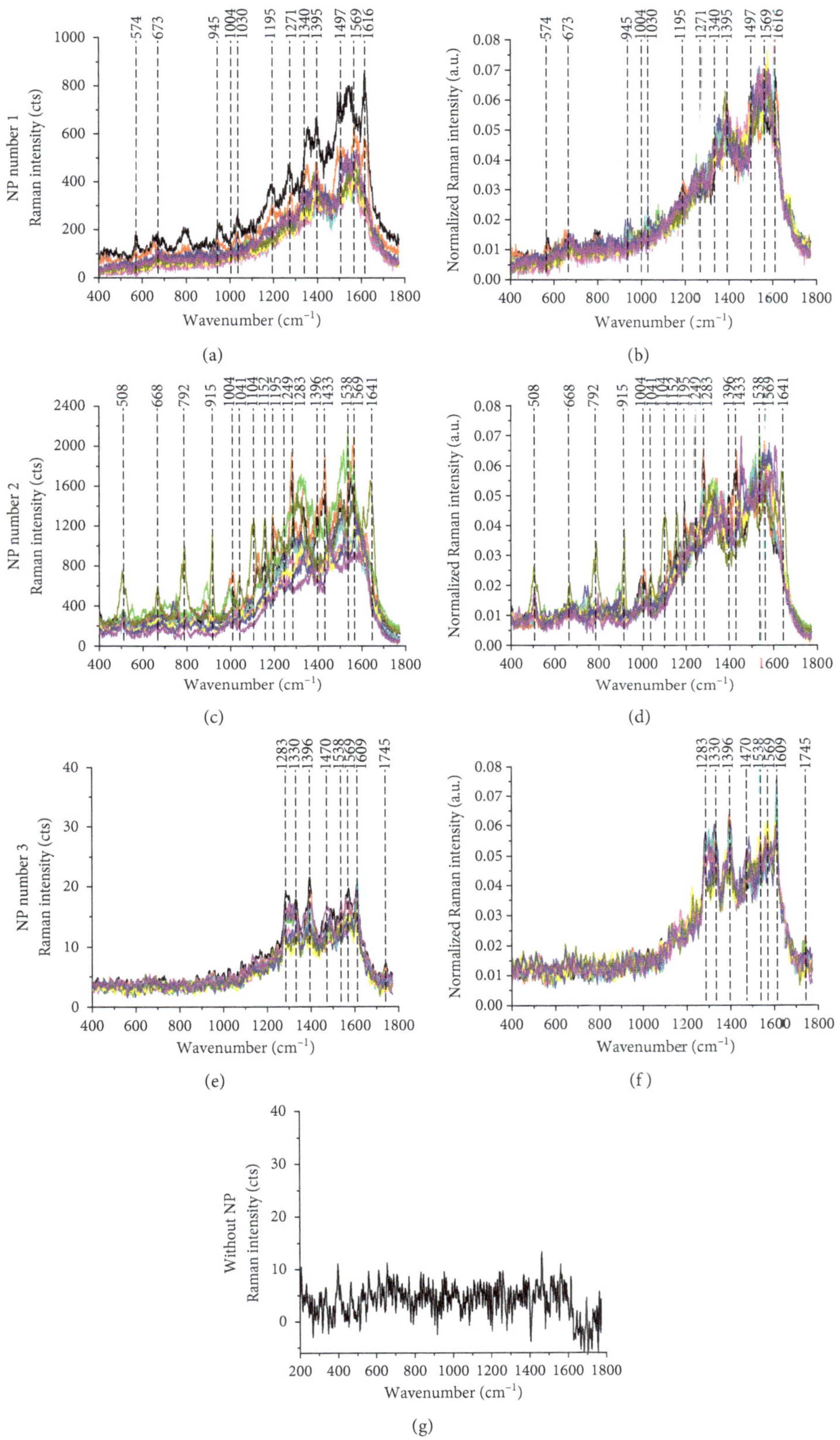

FIGURE 2: Raw (a, c, e) and vector-normalized (b, d, f) SERS spectra of human cancerous breast tissue of the patient P141 and the native Raman spectra (g) recorded for the same tissue sample; excitation wavelength: 532 nm, integration time 0.1 sec, and laser power 1 mW.

Table 2: Maximum peak positions of Raman bands observed for SERS spectra and the tentative assignments of these bands.

Maximum peak position in SERS spectrum (cm^{-1})	Tentative assignment [36]
547	Cholesterol
668, 673	C–S stretching mode of cysteine, T, G (DNA/RNA), C–S stretching mode of cytosine
945	Valine, proline, and polysaccharides
1004	Phenylalanine
1030	Phenylalanine of collagen
1095	Symmetric PO_2^- stretching vibration of the DNA backbone; phosphate backbone vibration as a marker mode for the DNA concentration; C–N of proteins
1152	ν(C–N), proteins
1249	Thymine, adenine, DNA, RNA, phospholipids, Amide III, lipids, fatty acids
1271	Typical phospholipids, Amide III band in proteins, a C–N stretch from alpha helix proteins
1283	Lipids and fatty acids
1340	DNA, RNA
1395	Lipids and fatty acids
1497	Proteins and lipids
1569	Amides, vibrations of C=O, C–N, and N–H groups
1609	Cytosine (NH_2)
1745	ν(C=O), phospholipids, and triglycerides (lipid and fatty acids assignment)

the intensity of vibration bands that characterize the human cancer tissue, indicating that there is a strong interaction between the synthetized silver NPs and the analyzed tissue sample. The analysis of human breast tissue presented in Figures 2(a)–2(f) reveals that the significant SERS spectral signals typical for breast cancer structures are observed in a spectral range 500–1800 cm^{-1} and can be recorded for a split second. One can see from Figures 2(a)–2(f) that the main peaks observed for the tissue samples correspond to DNA/RNA, amino acids, lipids, and proteins. Our analysis confirms preliminary observation by González-Solís et al. [33], Guerrini et al. [34], Karabeber et al. [35], and our previous experiments using spontaneous Raman spectroscopy [7–13]. More detailed analysis of band intensity at ~1396 cm^{-1} observed in all SERS spectra of the three proposed methods shows that the mean ± SD are 527.1 ± 94.9, 906.5 ± 281.9, and 15.2 ± 3.3 for NP nos. 1, 2, and 3, respectively. The weak enhancement effect in the Raman observed in silver NP no. 3 by taking three different synthesis methods into consideration can be explained by smaller efficiency plasmon generation excited at $\lambda = 532$ nm (lower absorbance at 532 nm, see Figures 1(a), 1(c), and 1(e)).

Table 2 presents the maximum peak positions of Raman bands observed for SERS spectra and the tentative assignments of these bands based on [36].

For better comparison of the enhancement efficiency of synthetized NPs, all recorded SERS spectra were vector-normalized (Origin model: divided by Norm (length)).

One can see from Figures 2(b), 2(d), and 2(f) that the efficiency of Raman signal enhancement is comparable for NPs synthetized by the above-described different methods.

The efficiency of Raman signal enhancement for NPs synthetized by method no. 1, which is very simple in preparation compared to method nos. 2 and 3 and offers high Raman peak intensities for not normalized human breast cancerous tissue samples that can be commonly used for screening in medical measurements was also tested using the well-known Raman reporter compound 4-mercaptotoluene (4-MT). The objective of medical screening procedures is to detect cancerous cells in a very short time.

SERS measurements for this model compound, which can be finally used for functionalization of Ag NPs, confirmed that obtained silver nanostructures allow to record Raman spectra of 4-MT for picomolar concentrations. Figure 3 presents SERS spectrum of 4-mercaptotoluene for the concentration equal to 2.5×10^{-3} and 1.0×10^{-15} M (a) and the analysis of SERS Raman signal intensity in the function of 4-MT concentration (in a concentration range of 1.0×10^{-15}–2.5×10^{-3} M for peak recorded at 1079 cm^{-1}) (b).

One can see from Figure 3 that nonlinear correlation in the whole concentration range of 1.0×10^{-15}–2.5×10^{-3} M between SERS signal and the concentration of 4-MT is observed. The high linearity is observed for the concentration below 1×10^{-4} M and for the range 1×10^{-4}–2.5×10^{-3} M.

4. Conclusions

In this work, we have presented the biochemical characteristic of human breast cancer using SERS technique by recording fingerprint vibrations effectively enhanced by silver NPs. The application of NPs allowed to identify the biomolecules as amino acids (phenylalanine, tyrosine, and tryptophan), proteins, and DNA. We have shown that the SERS technique reduces exposure times to a split second for collecting good quality spectra with well-defined and isolated bands, whereas the conventional Raman spectroscopy requires at least one of several to hundred seconds for analysis. Our study confirmed that surface-enhanced Raman scattering maintains great possibilities to acquire hundreds

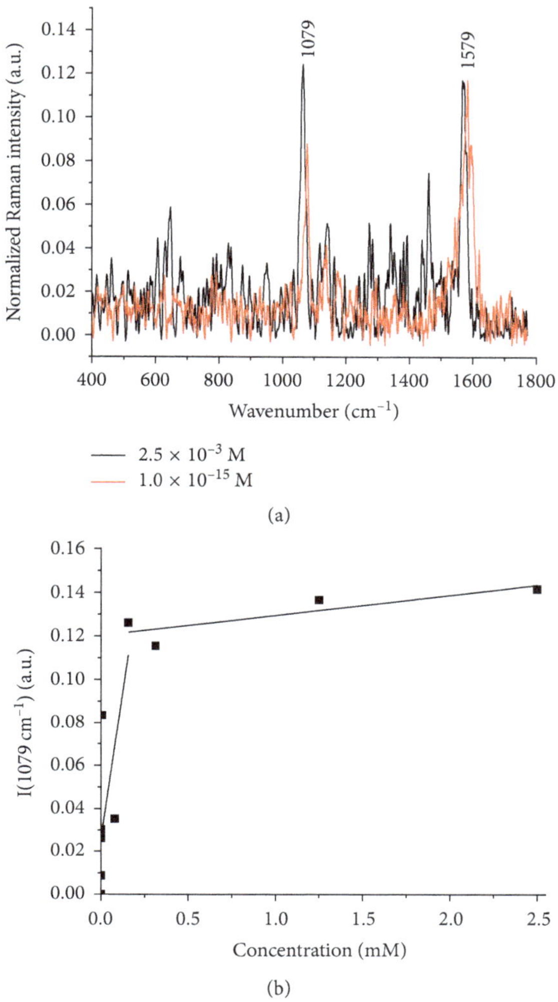

FIGURE 3: SERS spectrum of 4-mercaptotoluene (2.5×10^{-3} and 1.0×10^{-15} M) with NPs synthetized by method no. 1 (a). SERS Raman signal intensity in the function of 4-MT concentration in a concentration range of 1.0×10^{-15}–2.5×10^{-3} M (b).

of spectra of tissues in a very short time which is important for medical patient screening. Moreover, it is shown that the enhancement effect of NPs synthetized by simple reduction method is stronger in comparison with the polyol method. We have shown also that nonlinear correlation in the whole concentration range of 1.0×10^{-15}–2.5×10^{-3} M between SERS signal and the concentration of 4-MT is observed. The high linearity was observed for the concentration below 1×10^{-4} M and for the range 1×10^{-4}–2.5×10^{-3} M.

Conflicts of Interest

The authors declare that there are no conflicts of interest regarding the publication of this paper.

Acknowledgments

The project was funded through the National Science Centre Poland Grant UMO-2015/19/B/ST4/01878 and Dz. St. 2018.

References

[1] R. L. Sieqel, K. D. Miller, and A. Jemal, "Cancer statistics, 2017," *CA: A Cancer Journal for Clinicians*, vol. 67, no. 1, pp. 7–30, 2017.

[2] A. Raikhlin, B. Curpen, E. Warner, C. Betel, B. Wright, and R. Jong, "Breast MRI as an adjunct to mammography for breast cancer screening in high-risk patients: retrospective review," *American Journal of Roentgenology*, vol. 204, no. 4, pp. 889–897, 2015.

[3] E. Devolli-Disha, S. Manxhuka-Kerliu, H. Ymeri, and A. Kutllovci, "Comparative accuracy of mammography and ultrasound in women with breast symptoms according to age and breast density," *Bosnian Journal of Basic Medical Sciences*, vol. 9, no. 2, pp. 131–136, 2009.

[4] S. K. Yang, N. Cho, and W. K. Moon, "The role of PET/CT for evaluating breast cancer," *Korean Journal of Radiology*, vol. 8, no. 5, pp. 429–437, 2007.

[5] Z. Huang, A. McWilliams, H. Lui, D. I. McLean, S. Lam, and H. Zeng, "Near-infrared Raman spectroscopy for optical diagnosis of lung cancer," *International Journal of Cancer*, vol. 107, no. 6, pp. 1047–1052, 2003.

[6] A. S. Haka, K. E. Shafer-Peltier, M. Fitzmaurice, J. Crowe, R. R. Dasari, and M. S. Feld, "Diagnosing breast cancer by using Raman spectroscopy," *Proceedings of the National Academy of Sciences*, vol. 102, no. 35, pp. 12371–12376, 2005.

[7] H. Abramczyk and B. Brozek-Pluska, "New look inside human breast ducts with Raman imaging. Raman candidates as diagnostic markers for breast cancer prognosis: mammaglobin, palmitic acid and sphingomyelin," *Analytica Chimica Acta*, vol. 909, pp. 91–100, 2016.

[8] B. Brozek-Pluska, M. Kopec, J. Surmacki, and H. Abramczyk, "Raman microspectroscopy of the noncancerous and the cancerous human breast tissues. Identification and phase transitions of linoleic and oleic acids by Raman spectroscopy and Raman low-temperature studies," *Analyst*, vol. 140, no. 7, pp. 2134–2143, 2015.

[9] J. Surmacki, B. Brozek-Pluska, R. Kordek, and H. Abramczyk, "The lipid-reactive oxygen species phenotype of breast cancer. Raman spectroscopy and mapping, PCA and PLSDA for invasive ductal carcinoma and invasive lobular carcinoma. Molecular tumorigenic mechanisms beyond Warburg effect," *Analyst*, vol. 140, no. 7, pp. 2121–2133, 2015.

[10] B. Brozek-Pluska, M. Kopec, I. Niedzwiecka, and A. Morawiec-Sztandera, "Label-free determination of lipids composition and secondary proteins structure of human salivary noncancerous and cancerous tissues by Raman microspectroscopy," *Analyst*, vol. 140, no. 7, pp. 2107–2113, 2015.

[11] H. Abramczyk, B. Brozek-Pluska, J. Surmacki, J. Musial, and R. Kordek, "Oncologic photodynamic diagnosis and therapy: Confocal Raman/fluorescence imaging of metal phthalocyanines in human breast cancer tissue in vitro," *Analyst*, vol. 139, no. 21, pp. 5547–5559, 2014.

[12] H. Abramczyk, B. Brozek-Pluska, M. Krzesniak, M. Kopec, and A. Morawiec-Sztandera, "The cellular environment of cancerous human tissue. Interfacial and dangling water as a "hydration fingerprint"," *Spectrochimica Acta Part A: Molecular and Biomolecular Spectroscopy*, vol. 129, pp. 609–623, 2014.

[13] H. Abramczyk and B. Brozek-Pluska, "Raman imaging in biochemical and biomedical applications. Diagnosis and treatment of breast cancer," *Chemical Reviews*, vol. 113, no. 8, pp. 5766–5781, 2013.

[14] S. Feng, J. Lin, Z. Huang et al., "Esophageal cancer detection based on tissue surface-enhanced Raman spectroscopy and

multivariate analysis," *Applied Physics Letters*, vol. 102, pp. 1–5, 2013.

[15] S. Feng, R. Chen, J. Lin et al., "Gastric cancer detection based on blood plasma surface-enhanced Raman spectroscopy excited by polarized laser light," *Biosensors and Bioelectronics*, vol. 26, no. 7, pp. 3167–3174, 2011.

[16] D. Graham, W. E. Smith, A. M. T. Linacre, C. H. Munro, N. D. Watson, and P. C. White, "Selective detection of deoxyribonucleic acid at ultralow concentrations by SERRS," *Analytical Chemistry*, vol. 69, no. 22, pp. 4703–4707, 1997.

[17] K. Faulds, R. P. Barbagallo, J. T. Keer, W. E. Smith, and D. Graham, "SERRS as a more sensitive technique for the detection of labelled oligonucleotides compared to fluorescence," *Analyst*, vol. 129, no. 7, pp. 567-568, 2004.

[18] M. Vendrell, K. K. Maiti, K. Dhaliwal, and Y.-T. Chang, "Surface enhanced Raman scattering in cancer detection and imaging," *Trends in Biotechnology*, vol. 31, no. 4, pp. 249–257, 2013.

[19] T. Troy, D. Jekic-McMullen, L. Sambucetti, and B. Rice, "Quantitative comparison of the sensitivity of detection of fluorescent and bioluminescent reporters in animal models," *Molecular Imaging*, vol. 3, no. 1, pp. 9–23, 2004.

[20] D. Graham and K. Faulds, "Quantitative SERRS for DNA sequence analysis," *Chemical Society Reviews*, vol. 37, no. 5, pp. 1042–1051, 2008.

[21] R. Duncan, "Polymer conjugates as anticancer nanomedicines," *Nature Reviews Cancer*, vol. 6, no. 9, pp. 688–701, 2006.

[22] J. R. Mccarthy, K. A. Kelly, E. Y. Sun, and R. Weissleder, "Targeted delivery of multifunctional magnetic nanoparticles," *Nanomedicine*, vol. 2, no. 2, pp. 153–167, 2007.

[23] M. C. Woodle and P. Y. Lu, "Nanoparticles deliver RNAi therapy," *Materials Today*, vol. 8, no. 8, pp. 34–41, 2005.

[24] Z. Medarova, W. Pham, C. Farrar, V. Petkova, and A. Moore, "In vivo imaging of siRNA delivery and silencing in tumors," *Nature Medicine*, vol. 13, no. 3, pp. 372–377, 2007.

[25] C. L. Zavaleta, B. R. Smith, I. Walton et al., "Multiplexed imaging of surface enhanced Raman scattering nanotags in living mice using noninvasive Raman spectroscopy," *Proceedings of the National Academy of Sciences*, vol. 106, pp. 13511–13516, 2009.

[26] M. Y. Sha, H. Xu, M. J. Natan, and R. Cromer, "Surface-enhanced Raman scattering Tags for rapid and homogeneous detection of circulating tumor cells in the presence of human whole blood," *Journal of the American Chemical Society*, vol. 130, no. 51, pp. 17214-17215, 2008.

[27] A. J. Frank, N. Cathcart, K. E. Maly, and V. Kitaev, "Synthesis of silver nanoprisms with variable size and investigation of their optical properties: a first-year undergraduate experiment exploring plasmonic nanoparticles," *Journal of Chemical Education*, vol. 87, no. 10, pp. 1098–1101, 2010.

[28] N. Leopold and B. Lendl, "A new method for preparation of highly surface-enhanced Raman scattering (SERS) active silver colloids at room temperature by reduction of silver nitrate with hydroxylamine hydrochloride," *Journal of Physical Chemistry B*, vol. 107, no. 24, pp. 5723–5727, 2003.

[29] J.-H. Kim, J-S. Kim, H. Choi et al., "Nanoparticle probes with Surface enhanced Raman spectroscopic tags for cellular cancer targeting," *Analytical Chemistry*, vol. 78, no. 19, pp. 6967–6973, 2006.

[30] S. Duraipandian, M. S. Bergholt, W. Zheng et al., "Real-time Raman spectroscopy for in vivo, online gastric cancer diagnosis during clinical endoscopic examination," *Journal of Biochemical Optics*, vol. 17, no. 8, article 081418, 2012.

[31] Y. Qin, X. Ji, J. Jing, H. Liu, H. Wu, and W. Yang, "Size control over spherical silver nanoparticles by ascorbic acid reduction," *Colloids and Surfaces A: Physicochemical and Engineering Aspects*, vol. 372, no. 1–3, pp. 172–176, 2010.

[32] X. Dong, X. Ji, H. Wu, L. Zhao, J. Li, and W. Yang, "Shape control of silver nanoparticles by stepwise citrate reduction," *Journal of Physical Chemistry C*, vol. 113, no. 16, pp. 6573–6576, 2009.

[33] J. S. González-Solís, G. H. Luévano-Colmenero, and J. Vargas-Mancilla, "Surface enhanced Raman spectroscopy in breast cancer cells," *Laser Therapy*, vol. 22, no. 1, pp. 37–42, 2013.

[34] L. Guerrini, N. Pazos-Perez, E. Garcia-Rico, and R. Alvarez-Puebla, "Cancer characterization and diagnosis with SERS-encoded particles," *Cancer Nanotechnology*, vol. 8, no. 1, 2017.

[35] H. Karabeber, R. Huang, P. Iacono et al., "Guiding brain tumor resection using surface-enhanced Raman scattering nanoparticles and a hand-held Raman scanner," *ACS Nano*, vol. 8, no. 10, pp. 9755–9766, 2014.

[36] Z. Movasaghi, S. Rehman, and J. U. Rehman, "Raman spectroscopy of biological tissues," *Applied Spectroscopy Reviews*, vol. 42, no. 5, pp. 493–541, 2007.

23

Rapid Recognition of Geoherbalism and Authenticity of a Chinese Herb by Data Fusion of Near-Infrared Spectroscopy (NIR) and Mid-Infrared (MIR) Spectroscopy Combined with Chemometrics

Haiyan Fu,[1] Qiong Shi,[1] Liuna Wei,[1] Lu Xu,[2] Xiaoming Guo,[1] Ou Hu,[1] Wei Lan,[1] Shunping Xie,[3] and Tianming Yang[1]

[1]*The Modernization Engineering Technology Research Center of Ethnic Minority Medicine of Hubei Province, College of Pharmacy, South-Central University for Nationalities, Wuhan 430074, China*
[2]*College of Material and Chemical Engineering, Tongren University, Tongren 554300, Guizhou, China*
[3]*State Key Laboratory Breeding Base of Green Chemistry-Synthesis Technology, College of Chemical Engineering, Zhejiang University of Technology, Hangzhou 310032, China*

Correspondence should be addressed to Lu Xu; lxchemo@163.com and Tianming Yang; tmyang@mail.scuec.edu.cn

Academic Editor: Massimo Lucarini

Fourier transform near-infrared (NIR) spectroscopy and mid-infrared (MIR) spectroscopy play important roles in all fingerprint techniques because of their unique characteristics such as reliability, versatility, precision, and ease of measurement. In this paper, a supervised pattern recognition method based on the PLSDA algorithm by NIR and the NIR-MIR fusion spectra has been established to identify geoherbalism of *Angelica dahurica* from different regions and authenticity of *Corydalis yanhusuo* W. T. Wang. Comparing principle component analysis (PCA) cannot successfully identify geographical origins of *Angelica dahurica*. Linear discriminant analysis (LDA) also hardly distinguishes those origins. Furthermore, the PLSDA model based on the data fusion of NIR and IR was more accurate and efficient. But, the identification of authenticity of *Corydalis yanhusuo* W. T. Wang was still inaccurate in the PLSDA model. Consequently, data fusion of NIR-MIR original spectra combined with moving window partial least-squares discriminant analysis was firstly used and showed perfect properties on authenticity and adulteration discrimination of *Corydalis yanhusuo* W. T. Wang. It indicated that data fusion of NIR-MIR spectra combined with MWPLSDA could be considered as the promising tool for rapid discrimination of the geoherbalism and authenticity of more Chinese herbs in the future.

1. Introduction

Herbal medicines are of effective pharmacological functions, low toxicity, and less side effects to human body, so they have been widely used all over the world [1–3]. However, herbal medicines with different geographical origins have different chemical compositions and pharmacological activities [4, 5]. In addition, the processing of herbal medicines often removes morphological properties of species, and some herbal medicines at high cost are often the subject of fraudulent practices by replacing them with ones at low cost [6, 7], which may lead to an unfair competition in the pharmaceutical and harm the interest of consumers. Thus, the quality analysis method of herbal medicines to distinguish the origins is an important concern for consumers [8–10]. Traditional methods such as high-performance liquid chromatography and mass spectroscopy are time-consuming, expensive, and laborious and have to be performed by highly trained technicians [11, 12]. Therefore, a rapid, more accurate, and sensitive identification method is required to determine herbal medicines.

Most studies focused on specific pharmacological ingredients in herbal medicines; however, the pharmacological activity of herbs is the result of the interaction of all ingredients rather than specific ingredients. Therefore, the specific ingredients could not be used as a proper criterion for characterization of the overall quality of the herbs

[13, 14]. Fourier transform mid-infrared (MIR) [15] and near-infrared (NIR) [16, 17] techniques are efficient tools for studying food and pharmaceutical quality control because of their fast and nondestructive analytical characteristics. For example, Zhu et al. used FT-IR and 2DCOS-IR methods to discriminate the cultivated *Codonopsis lanceolata* in different ages [18], and in the research done by Gayo and Hale, near-infrared spectroscopy was applied to detect and quantify the species authenticity in Crabmeat [19]. By studying the characteristic information of the spectra, different types of samples can be accurately distinguished. Nonetheless, the information obtained from by NIR spectra may be difficult to interpret directly because of the highly overlapped spectra. Although MIR spectra provide some significant differences of spectral peaks, they do not give abundant chemical and structural information of samples like NIR spectra. Therefore, establishing effective and robust chemometric methods has been extensively concerned [20, 21]. For example, Woo's team used Mahalanobis distance and discriminant PLS2 combined with NIR spectroscopy to discriminate herbal medicines according to geographical origins, but there are only two different classes from different geographical origins [22]. Frizon et al. used the PLS in determination of total phenolic compounds in yerba mate and predicted total phenolics with associated errors of 12% [23]. Liu et al. studied on the differentiation of the root of various ginseng by FT-IR and two-dimensional correlation IR spectroscopy, and the cluster analysis demonstrated that the three kinds of ginseng can be distinguished clearly from each other but with an exception [24]. PCA is a multivariate statistical technique that reduces the multidimensionality of data while minimizing information loss [25]. LDA can establish linear transformations to find the best boundary and achieve maximum separation between classes by constructing discriminant functions [26]. From another aspect, as a powerful pattern recognition method, PLSDA has successfully been applied to solve classification problems in many scientific fields [27, 28]. Furthermore, a global model with moving window partial least-squares (MWPLS) [29, 30] like other variable selection methods, MWPLSDA was successfully applied to spectra interval selection for calibration problems, and desirable results were obtained [31]. A subset of the whole wavelengths to develop the calibration model, the wavelengths carrying serious heteroscedastic noises, and especially the spectral ranges contaminated by external factors are excluded from the model, and wavelength ranges sensitive only to the chemical compositions of the samples are selected to develop a simplified yet stable calibration model.

Sometimes, it is difficult to discriminate the origins of herbal medicines only through the pattern recognition method by single NIR or MIR spectra [32] combined with chemometrics, and it is necessary to extract from the data fusion of NIR and MIR spectroscopy [33]. There is abundant information related to combinatory MIR and NIR spectroscopy coupled with chemometrics for quality control of herbal medicines.

In this study, different supervised pattern recognition algorithms including principal component analysis (PCA), linear discriminant analysis (LDA), and partial least squares discriminant analysis (PLSDA) with raw NIR spectra were used to discriminate five different geographical origins of *Angelica dahurica*. Moreover, moving window partial least-squares discriminant analysis (MWPLSDA) and the fusion spectra variables evaluate authenticity and adulteration of *Corydalis yanhusuo* W. T. Wang. The result shows that PLSDA model is of great performance than PCA and LDA in identifying geographical origins of herbal medicines. In addition, the full spectra information fused by NIR and MIR combined with MWPLSDA showed the best ability in determination of authenticity of herbal medicines. This method provides pattern recognition models that can be applied in geographical origin discrimination or authenticity and adulteration recognition at the same time and can further be widely used in various herbal medicines.

2. Material and Methods

2.1. Collection of Raw Materials. A total of 50 *Angelica dahurica* samples from five geographical origins (Hebei, Anhui, Yunnan, Zhejiang, and Sichuan) were purchased from the Derentang pharmacy, and each region included 10 batches. Besides, two kinds of authentic *Corydalis yanhusuo* W. T. Wang (Zhejiang) were purchased from the Derentang pharmacy and the Kangderuiqi flagship store, while three kinds of adulterations *Corydalis decumbens* (Thunb.) Pers., *Typhonium flagelliforme* (Lodd.) Blume, and *Dioscorea opposita* (Thunb.) were, respectively, collected from Anhui, Jiangsu, and Fujian, and the aforementioned five samples for identification of adulteration were collected in 10 batches.

2.2. Apparatus. The following apparatuses were used: Nicolet 6700 FT-IR, OMNIC 8.2 spectral collecting software (Thermo Fisher Scientific Inc., USA); Antaris II FT-NIR spectrometer, RESULT 3.0 spectral collecting software (Thermo Electron Co., USA); DZF-6021 vacuum oven (Shanghai YIHENG Technical Co., Ltd); and FW135 herbal grinder (Tianjin Taisite Instrument Co., LTD).

2.3. Methods of Sample Measurement and Data Preprocessing by NIR and MIR. All samples used in NIR were crushed with the grinder, sieved into fine powders by a 200 mesh sieve, then vacuum-dried at 60°C for 24 hours, and stored in a dryer spare. The sample powder was placed directly into the quartz cup, and the air background was subtracted. Spectra were collected by integrating sphere diffuse reflectance with the collecting region at 4000–10000 cm^{-1} and a resolution of 8 cm^{-1}. Data processing was performed using the average of the five measured spectra for each sample. In total, 250 spectra from different geographic origins (5 samples × 10 batched × 5 measurements) were obtained. And 250 spectra were discriminated for the authenticity and adulteration of *Corydalis yanhusuo* W. T. Wang.

2.4. Method of Chemometrics. PCA, LDA, PLSDA, and MWPLSDA methods were written and performed through a Matlab 2010a (MathWorks, Natick, MA. USA). All preprocessing in those chemometrics only used the original spectra. PLSDA is based on the simultaneous decomposition response matrix and the class matrix extraction factor. By arranging the extraction factors in order of their correlation, the virtual vectors are encoded to represent different classes, wherein the virtual vector *fj* encoded for the *j*th element is 1. The other elements are 0 for the *j*th class, and then each column of the response matrix is associated with the class matrix. The principle of moving window partial least-squares discriminant analysis (MWPLSDA) is that a suitable window moves along the full spectral interval according to our past study [34, 35]. In MWPLSDA, a suitable window of width H is constructed and moved along the entire spectrum to select useful wavelength intervals, and then the selected spectral spacing is used to construct the PLSDA model. The principle of MWPLSDA is based on the virtual setting of a window, which contains the number of variables from the first wavelength to the end of $(i+H-1)$ wavelength. A series of submatrices are obtained continuously by moving the window. According to the variables in the moving window, a series of PLS submodels are constructed. Then, according to the principle of least residual square (SSR), the interval of measurement matrix with smaller classification error and latent variable is selected as the final MWPLSDA model.

3. Results and Discussion

3.1. Geographical Origin Discrimination of Angelica dahurica by NIR. In order to analyze the five different samples more effectively, the classical quick data analysis, and non-destructive analytical technique, NIR was used in the measurement. The average NIR spectra of each group are displayed to reflect the overlay in Figure 1. The peaks located at 8319 cm^{-1} might be associated with the second overtone of C-H, O-H, and N-H stretching modes and those around 6780 cm^{-1} were caused by the C-H deformation vibration of CH_3. Due to the second overtone of the C=O stretching vibration, bands at 5164 cm^{-1} emerge and the C-H combination and second overtone can be seen at 4200–4300 cm^{-1}. However, owing to the overlaps and the systematic noise in NIR spectra, chemometric methods were required to extract useful information for the recognition of *Angelica dahurica* samples. Herein, three classical chemical pattern recognition methods using principal component analysis (PCA), linear discriminant analysis (LDA), and partial least squares discriminant analysis (PLSDA) models were associated with virtual coding of original NIR spectral variables of different sample sets. The 250 sample spectra of five different *Angelica dahurica* samples were randomly divided into a training set and a prediction set (Table 1). The model was built using the training set, the number of latent variables (LVs) was determined to be 5 by eightfold cross-validation using the prediction set, and the discrimination results were analyzed for comparison.

Firstly, as a common method in the chemical pattern recognition which is mainly used for classification in the analytical processes of Chinese herbal identification, principal component analysis (PCA) is one of the most classic high-dimensional methods, which reduces the high-dimensional data of FTNIR and converts 1557 raw variables into fewer new principal components. PCA used fewer principal component features to represent the original features of the sample by decomposing the sample matrix in the training set and prediction. Based on the PCA technology, the vector scores of the training and prediction sets of the aforementioned samples are reflected in Figure 1(b), and all samples from five different geographic origins in the training and prediction sets could not be clearly distinguished, but these samples were with same shape. This phenomenon could be attributed to small differences in the chemical properties reflected in its geographical origin. The results demonstrated that the PCA method can effectively reduce and extract fewer new variables from the original high-dimensional data, but the restoration process also leads to loss some information useful for sample differentiation.

Other than looking for the vector space that can best describe the original data like PCA, linear discriminant analysis (LDA) is a linear discriminant function based on input response variables for searching linear transformations and dimensionality reduction. The axes of interest for LDA can maximize the distinction between classes, projecting feature spaces (multidimensional samples in the dataset) into smaller dimensional k-dimensional subspaces while maintaining information that distinguishes categories. Figure 1(c) shows the vector scores of the first two latent variables based on the LDA model for the training and prediction sets of samples. It clearly distinguished samples from different geographical origins in the training set, while those in the prediction set were not clearly distinguished. The result may be due to some special requirements of the LDA model, of which at least one of the needs to be nonsingular. In addition, when the so-called outlier class dominates in estimating the scattering matrix, the LDA model cannot guarantee that the optimal subspace is found [36]. Furthermore, PLSDA can reduce the effects of multicollinearity between variables, and it can simultaneously decompose the extraction factors of the prediction measurement matrix and the class matrix and arrange them according to the correlation between them. Five different geographical sources of *Angelica dahurica* are identified based on the maximum virtual coding position of the NIR spectral data. In order to optimize the predictive power of the PLSDA model and simplify the complexity of the PLSDA model, we selected the number of latent variables (LV) as 5 by 8-fold cross-validation. Figure 1(d) shows the plots of dummy codes of the training and prediction sets for five group samples of different geographic origins. Table 1 shows the virtual code attribution maps for the training and prediction sets of the original spectra in the PLSDA model. We encode five sets of samples into f1 (1, 0, 0, 0, 0), f2 (0, 1, 0, 0, 0), f3 (0, 0, 1, 0, 0), f4 (0, 0, 0, 1, 0), and f5 (0, 0, 0, 0, 1), respectively, according to the position of the largest virtual code. As shown in Figure 1(d), all training and prediction samples belonging to

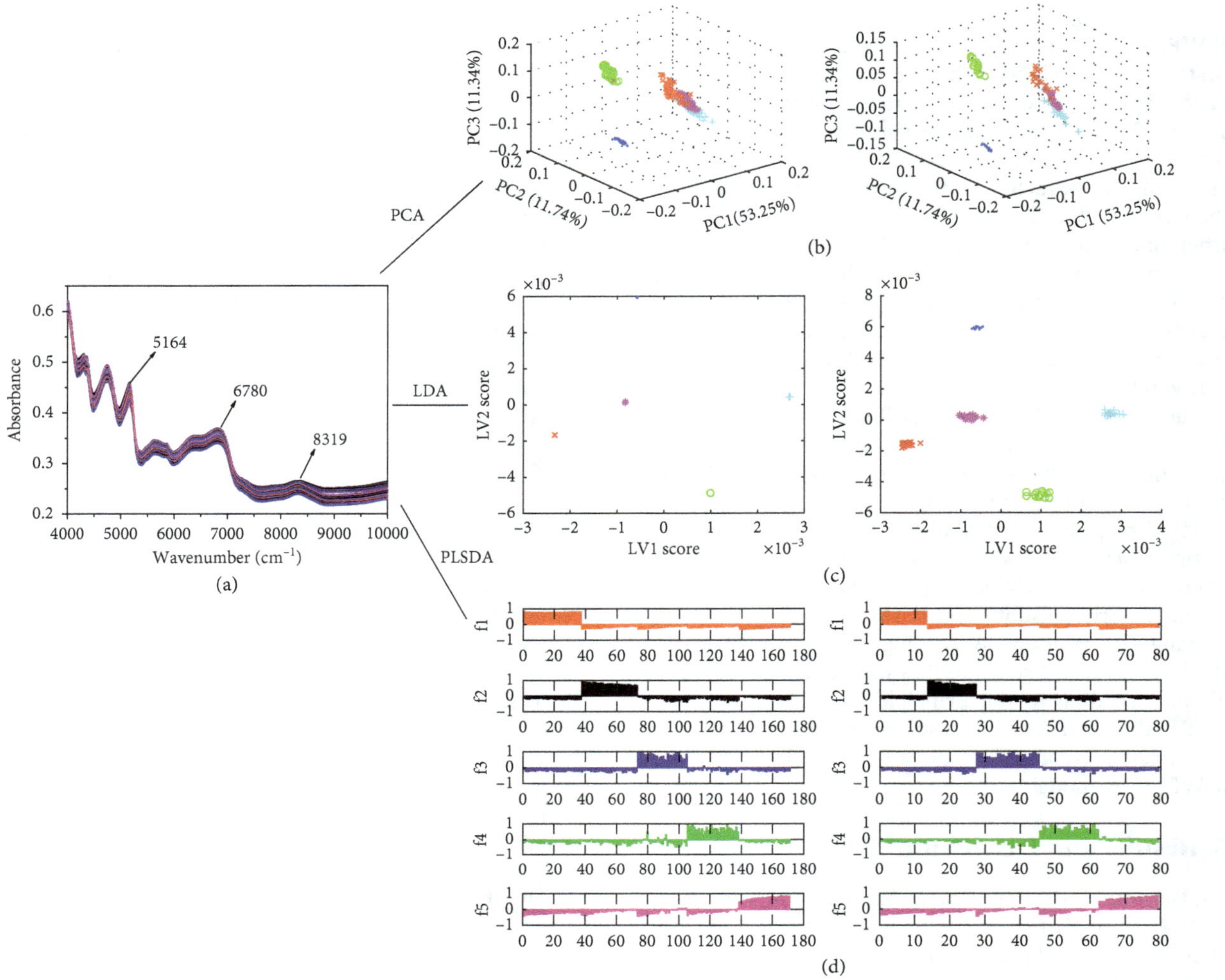

FIGURE 1: The raw NIR spectra of five different origins of *Angelica dahurica* (a) and the results by PCA (b), LDA (c), and PLSDA (d).

TABLE 1: A detailed list for the training set and the prediction set of five different kinds of *Angelica dahurica* samples.

Groups code	f1	f2	f3	f4	f5
Sample group	1	2	3	4	5
Symbol	·	○	×	+	*
PCA, LDA, or PLSDA training set	1st–37th	38th–73rd	74th–105th	106th–138th	139th–171st
PCA, LDA, or PLSDA prediction set	1st–13th	14th–27th	28th–45th	46th–62nd	63rd–79th

all groups of *Angelica dahurica* by original NIR spectra combined with PLSDA were identified accurately with a perfect recognition rate of 100%. This demonstrated that the PLSDA model successfully discriminates *Angelica dahurica* samples of different geographic origins. This further revealed that NIR spectroscopy combined with PLSDA method can be used to identify herbal medicines more rapidly, effectively, and reliably than the traditional ones.

3.2. Authenticity and Adulteration Discrimination of Corydalis yanhusuo W. T. Wang by NIR and Combinatory of NIR. Herbal medicine processing often removes morphological properties of species, which leads to failure of distinguishing one type from another. For this reason, NIR spectra were used to discriminate the authenticity and adulteration of *Corydalis yanhusuo* W. T. Wang. As is shown Figure 2(a), the peaks around 6826 cm^{-1} were due to the C-H deformation vibration of CH_3. Due to the C-H first overtone of $-CH_2-$ groups, bands at 5800 and 5600 cm^{-1} were observed and bands at 5172 cm^{-1} were the second overtone of the C=O stretching vibration. Furthermore, the C-H combination and second overtone can be seen at 4200–4300 cm^{-1}. The seriously overlapped raw spectra hardly reflect the differences between samples. Thus, PCA technology and LDA and PLSDA models were used to relate the dummy code for the full original and preprocessing spectral variables. 250 sample spectra of two kinds of authenticity,

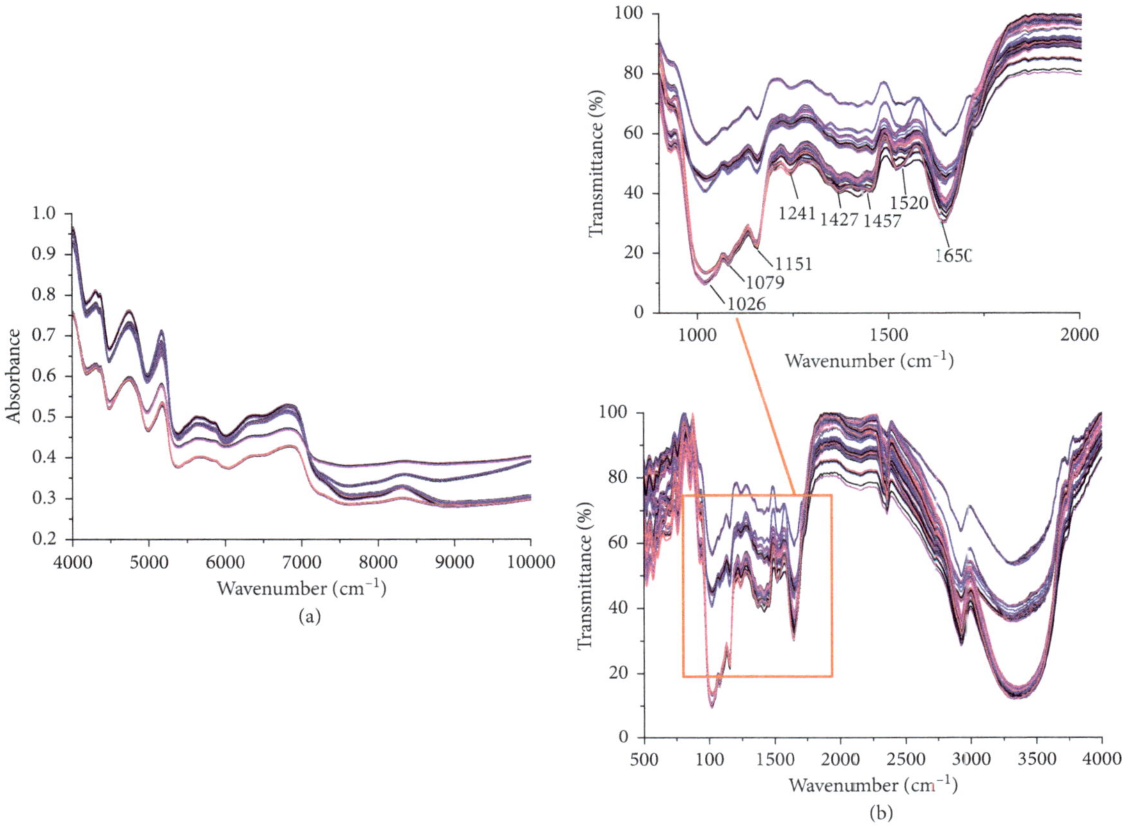

FIGURE 2: Raw NIR (a) and MIR (b) spectra of two kinds of *Corydalis yanhusuo* W. T. Wang, *Corydalis decumbens* (Thunb.) Pers, *Typhonium flagelliforme* (Lodd.) Blume, and *Dioscorea opposita* (Thunb.).

Corydalis yanhusuo W. T. Wang 1 and 2, and three kinds of adulteration, *Corydalis decumbens* (Thunb.) Pers. (3), *Typhonium flagelliforme* (Lodd.) Blume (4), and *Dioscorea opposita* (Thunb.) (5) were randomly divided into the training set and the prediction set (Table 2). However, both PCA technology and LDA model failed to show the correct results in prediction sets for five different groups by NIR (not shown here). Thus, PLSDA was adopted for the identification of authentic *Corydalis yanhusuo* W. T. Wang.

In our work, all training and prediction samples were correctly identified except for the two samples in the training set (34th and 88th) and the two samples in the prediction set (35th and 82nd). The 34th sample in the training set of f2 is incorrectly discriminated as f1, and the 84th sample of f5 is erroneously classified as f2. Furthermore, the 35th sample in the prediction set of f2 is incorrectly assigned as f3, and the 82nd sample in the prediction set belonging to f5 is incorrectly classified as f2. It may account for the useless information of some spectral variables. The total correction rate was 97.94% on the test set in PLSDA models. On the other hand, MIR spectroscopy provides more specific and distinct absorption bands than NIR spectroscopy. As is shown in Figure 2(b), the band centered at 2931 cm^{-1} is due to a stretching vibration of aliphatic C-H in terminal CH_3 groups. The strong single peak of the C=O stretching vibration of ketone groups is observed at about 1635 cm^{-1}, whereas the band centered at 1250 cm^{-1} is due to the antisymmetric stretching vibrations of =C-O-C.

In order to better identify the origin of Chinese herbal medicines, we combined the mid-infrared spectrum with the near-infrared spectrum to obtain fusion spectra with more abundant sample information (Figure 3). The PLSDA was also applied to relate the dummy code for the full fused spectral variables.

As is shown in Figure 3(a), only the 17th sample in the prediction set of f1 was misclassified as f2 in fusion spectra (Table 3). It suggested that fusion spectra of NIR and MIR spectroscopy combined with PLSDA has better use in authenticity and adulteration discrimination of *Corydalis yanhusuo* W. T. Wang. But, it also failed to get 100% predictive accuracy.

In MWPLSDA, the appropriate window with H width is constructed, and the useful wavelength range is selected by moving the whole spectrum. Then, all the selected windows are constructed into the PLSDA model. Finally, according to the minimum SSR principle of the MWPLSDA algorithm, the feature differences among the five samples are extracted. As shown in Figure 3(b), when the window size is 20, the optimum number of potential variables in the MWPLSDA model is 12. Figure 3(b) shows that when the number of variables in the fusion spectrum of NIR and MIR is 140-200, 750-930, and 1250-1380, the SSR is the smallest. At this time, MWPLS-DA

Table 2: A detailed list of two kinds of *Corydalis yanhusuo* W. T. Wang, *Corydalis decumbens* (Thunb.) Pers., *Typhonium flagelliforme* (Lodd.) Blume, and *Dioscorea opposita* (Thunb.)

Groups code	f1	f2	f3	f4	f5
Sample group	1	2	3	4	5
NIR training set	1st–33rd	34th–66th	67th–89th	90th–121st	122nd–154th
NIR prediction set	1st–17th	18th–34th	35th–61st	62nd–79th	80th–96th
Fusion training set	1st–33rd	34th–66th	67th–89th	90th–121st	122nd–154th
Fusion prediction set	1st–17th	18th–34th	35th–61st	62nd–79th	80th–96th

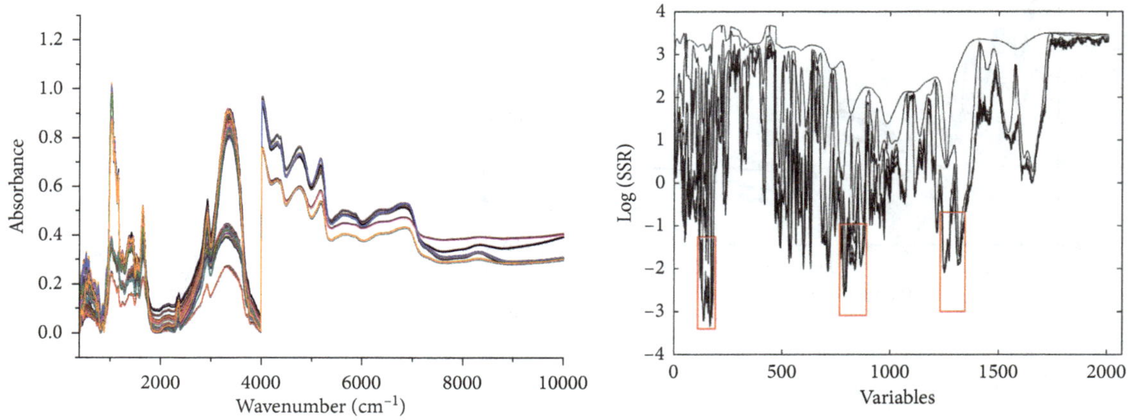

(a) (b)

Figure 3: Data fusion of NIR and MIR spectroscopy for authenticity or adulteration discrimination of *Corydalis yanhusuo* W. T. Wang (a) and the residue line obtained by MWPLSDA for the training sets (b).

Table 3: Forecasted results of NIR-MIR fusion spectra of two kinds of *Corydalis yanhusuo* W. T Wang, *Corydalis decumbens* (Thunb.) Pers., *Typhonium flagelliforme* (Lodd.) Blume, and *Dioscorea opposita* (Thunb.) by PLSDA and MWPLSDA.

Sample sets	Discrimination	Combinatory spectra	
		PLSDA	MWPLSDA
Training	Accuracy (%)	100	100
	Error number	0	0
Prediction	Accuracy (%)	98.97	100
	Error number	1	0

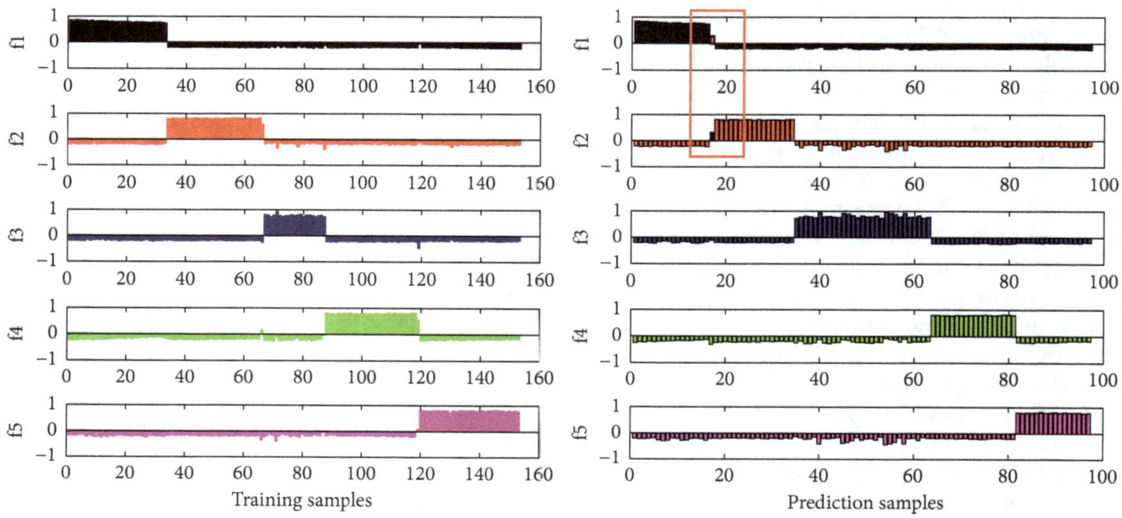

(a)

Figure 4: Continued.

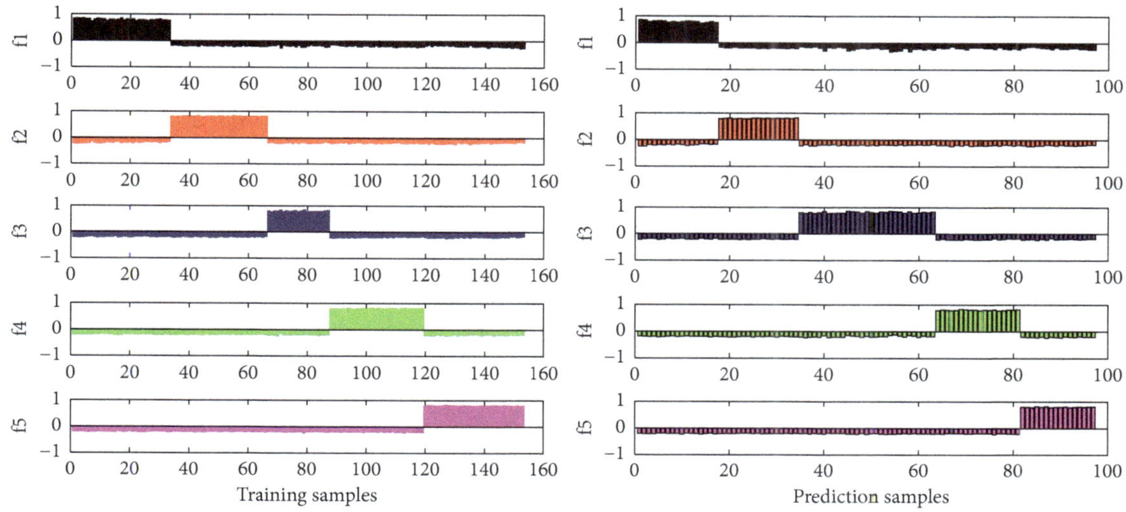

(b)

FIGURE 4: Assigned plots of dummy codes of the training set and prediction set for the raw fusion MIR-NIR spectra in the PLSDA model (a) and the MWPLSDA (b) model.

was benefited in the selection of combined informative fusion spectral regions of 660–950 cm^{-1}, 3550–4400 cm^{-1}, and 5900–6600cm^{-1} for classification modeling of all samples and yielded the results better than that obtained from a partial least squares-discrimination analysis (PLS-DA) model built by using the whole NIR spectral region. As shown in Figure 4(b) and Table 3, all training and prediction set samples were correctly predicted by 100%. MWPLSDA can improve sample classification accuracy by eliminating useless information variables and noncomponent-related factors.

4. Conclusions

Supervised pattern recognition methods based on PLSDA and MWPLSDA algorithms by NIR and the data fusion of both NIR and MIR has been established to study *Angelica dahurica* and to identify the authenticity of *Corydalis yanhusuo* W. T. Wang. In addition, it was clarified from the results that other than PCA and LDA that can merely have well learning performance and do well in the training sets, the PLSDA model shows good performance in the area of identification of *Angelica dahurica* or *Corydalis yanhusuo* W. T. Wang and can be employed in the analysis of the geographical origins of *Angelica dahurica* and the authenticity or adulteration of *Corydalis yanhusuo* W. T. Wang. Furthermore, the full spectrum information of NIR and MIR spectroscopy combined with MWPLSDA performed much better than the single NIR spectra or PLSDA model and demonstrated an unparalleled ability of herbal medicine discrimination. This new recognition method provided a promising approach for the identification of herbal medicines widely.

Conflicts of Interest

The authors declare that they have no conflicts of interest.

Acknowledgments

This work was financially supported by the National Natural Science Foundation of China (21575297, 21776321, and 21665022), the Talented Youth Cultivation Program from the "Fundamental Research Funds for the Central University," South-Central University for Nationalities (No. CRZ18002), and the Key Research Program (Nos. 2015ZD001 and 2015ZD002) from the Modernization Engineering Technology Research Center of Ethnic Minority Medicine of Hubei Province. Lu Xu is financially supported by Guizhou Provincial Department of Science and Technology (Nos. QKHJC[2017]1186 and QKHZC[2019]2816) and the Talented Researcher Program from Guizhou Provincial Department of Education (QJHKYZ[2018]073). We also gratefully acknowledge the help of Yao Fan, Ji Yang, Li Liu, Hanyue Lan, Chuang Ni, and Yuan-Bin She.

References

[1] F.-T. Chau, H.-Y. Chan, C.-Y. Cheung, C.-J. Xu, Y. Liang, and O. M. Kvalheim, "Recipe for uncovering the bioactive components in herbal medicine," *Analytical Chemistry*, vol. 81, no. 17, pp. 7217–7225, 2009.

[2] M.-Y. Lee, C.-S. Seo, J.-A. Lee et al., "Anti-asthmatic effects of *Angelica dahurica* against ovalbumin-induced airway inflammation via upregulation of heme oxygenase-1," *Food and Chemical Toxicology*, vol. 49, no. 4, pp. 829–837, 2011.

[3] K.-M. Ng, Z. Liang, W. Lu et al., "In vivo analysis and spatial profiling of phytochemicals in herbal tissue by matrix-assisted laser desorption/ionization mass spectrometry," *Analytical Chemistry*, vol. 79, no. 7, pp. 2745–2755, 2007.

[4] X. R. Liang, M. L. Ma, and W. K. Su, "Fingerprint analysis of *Hibiscus mutabilis* L. leaves based on ultra performance liquid chromatography with photodiode array detector combined with similarity analysis and hierarchical clustering analysis

methods," *Pharmacognosy Magazine*, vol. 9, no. 35, pp. 238–243, 2013.

[5] H. Zhu, Y. Wang, H. Liang, Q. Chen, P. Zhao, and J. Tao, "Identification of *Portulaca oleracea* L. from different sources using GC-MS and FT-IR spectroscopy," *Talanta*, vol. 81, no. 1-2, pp. 129–135, 2010.

[6] Q. Jiang, W. Gao, X. Li, and J. Zhang, "Characteristics of native and enzymatically hydrolyzed *Zea mays* L., *Fritillaria ussuriensis* Maxim. and *Dioscorea* opposita Thunb. Starches," *Food Hydrocolloids*, vol. 25, no. 3, pp. 521–528, 2011.

[7] M. S. M. Preto, M. I. B. Tavares, P. J. O. Sebastião, and R. B. V. Azeredo, "Determination of herb authenticity by low-field NMR," *Food Chemistry*, vol. 136, no. 3-4, pp. 1272–1276, 2013.

[8] S. Y. K. Fong, M. Liu, H. Wei et al., "Establishing the pharmaceutical quality of Chinese herbal medicine: a provisional BCS classification," *Molecular Pharmaceutics*, vol. 10, no. 5, pp. 1623–1643, 2013.

[9] S. Anubala, R. Sekar, and K. Nagaiah, "Development and validation of an analytical method for the separation and determination of major bioactive curcuminoids in *Curcuma longa* rhizomes and herbal products using non-aqueous capillary electrophoresis," *Talanta*, vol. 123, pp. 10–17, 2014.

[10] J. Kang, L. Zhou, J. Sun, J. Han, and D.-A. Guo, "Chromatographic fingerprint analysis and characterization of furocoumarins in the roots of *Angelica dahurica* by HPLC/DAD/ESI-MSn technique," *Journal of Pharmaceutical and Biomedical Analysis*, vol. 47, no. 4-5, pp. 778–785, 2008.

[11] M. Z. Končić, D. Kremer, W. Schühly, A. Brantner, K. Karlović, and Z. Kalođer, "Chemical differentiation of *Berberis croatica* and *B. Vulgaris* using HPLC fingerprinting," *Croatica Chemica Acta*, vol. 83, pp. 451–456, 2010.

[12] S. Su, W. Cui, W. Zhou, J.-A. Duan, E. Shang, and Y. Tang, "Chemical fingerprinting and quantitative constituent analysis of Siwu decoction categorized formulae by UPLC-QTOF/MS/MS and HPLC-DAD," *Chinese Medicine*, vol. 8, no. 1, pp. 5–20, 2013.

[13] D. Kumar, H. Kumar, J. R. Vedasiromoni, and B. C. Pal, "Bioassay guided isolation of α-glucosidase inhibitory constituents from *Hibiscus mutabilis* leaves," *Phytochemical Analysis*, vol. 23, no. 5, pp. 421–425, 2012.

[14] C. Wang, S. Wang, G. Fan, and H. Zou, "Screening of antinociceptive components in *Corydalis yanhusuo* W. T. Wang by comprehensive two-dimensional liquid chromatography/tandem mass spectrometry," *Analytical and Bioanalytical Chemistry*, vol. 396, no. 5, pp. 1731–1740, 2010.

[15] Y. Wang, P. Wang, C. Xu et al., "Macro-fingerprint analysis-through-separation of licorice based on FT-IR and 2DCOS-IR," *Journal of Molecular Structure*, vol. 1070, pp. 1–9, 2014.

[16] J. Lu, B. Xiang, H. Liu, S. Xiang, S. Xie, and H. Deng, "Application of two-dimensional near-infrared correlation spectroscopy to the discrimination of Chinese herbal medicine of different geographic regions," *Spectrochimica Acta Part A: Molecular and Biomolecular Spectroscopy*, vol. 69, no. 2, pp. 580–586, 2008.

[17] A. P. Craig, A. S. Franca, L. S. Oliveira, J. Irudayaraj, and K. Ileleji, "Application of elastic net and infrared spectroscopy in the discrimination between defective and non-defective roasted coffees," *Talanta*, vol. 128, pp. 393–400, 2014.

[18] Y. Zhu, C.-H. Xu, J. Huang et al., "Rapid discrimination of cultivated *Codonopsis lanceolata* in different ages by FT-IR and 2DCOS-IR," *Journal of Molecular Structure*, vol. 1069, pp. 272–279, 2014.

[19] J. Gayo and S. A. Hale, "Detection and quantification of species authenticity and adulteration in crabmeat using visible and near-infrared spectroscopy," *Journal of Agricultural and Food Chemistry*, vol. 55, no. 3, pp. 585–592, 2007.

[20] L. M. Dale, A. Thewis, C. Boudry et al., "Discrimination of grassland species and their classification in botanical families by laboratory scale NIR hyperspectral imaging: preliminary results," *Talanta*, vol. 116, pp. 149–154, 2013.

[21] H. Yang, J. Irudayaraj, and M. Paradkar, "Discriminant analysis of edible oils and fats by FTIR, FT-NIR and FT-Raman spectroscopy," *Food Chemistry*, vol. 93, no. 1, pp. 25–32, 2005.

[22] Y.-A. Woo, H.-J. Kim, J. Cho, and H. Chung, "Discrimination of herbal medicines according to geographical origin with near infrared reflectance spectroscopy and pattern recognition techniques," *Journal of Pharmaceutical and Biomedical Analysis*, vol. 21, no. 2, pp. 407–413, 1999.

[23] C. N. T. Frizon, G. A. Oliveira, C. A. Perussello et al., "Determination of total phenolic compounds in yerba mate (*Ilex paraguariensis*) combining near infrared spectroscopy (NIR) and multivariate analysis," *LWT—Food Science and Technology*, vol. 60, no. 2, pp. 795–801, 2015.

[24] D. Liu, Y.-G. Li, H. Xu, S.-Q. Sun, and Z.-T. Wang, "Differentiation of the root of cultivated ginseng, mountain cultivated ginseng and mountain wild ginseng using FT-IR and two-dimensional correlation IR spectroscopy," *Journal of Molecular Structure*, vol. 883-884, pp. 228–235, 2008.

[25] E. K. Kemsley, "Discriminant analysis of high-dimensional data: a comparison of principal components analysis and partial least squares data reduction methods," *Chemometrics and intelligent laboratory systems*, vol. 33, no. 1, pp. 47–61, 1996.

[26] R. A. Fisher, "The use of multiple measurements in taxonomic problems," *Annals of Eugenics*, vol. 7, no. 2, pp. 179–188, 1936.

[27] J. Peerapattana, H. Shinzawa, K. Otsuka, Y. Hattori, M. Otsuka, and J. Peerapattana, "Partial least square discriminant analysis of mangosteen pericarp powder by near infrared spectroscopy," *Journal of Near Infrared Spectroscopy*, vol. 21, no. 3, pp. 195–202, 2013.

[28] H.-Y. Fu, D.-C. Huang, T.-M. Yang, Y.-B. She, and H. Zhang, "Rapid recognition of Chinese herbal pieces of *Areca catechu* by different concocted processes using Fourier transform mid-infrared and near-infrared spectroscopy combined with partial least-squares discriminant analysis," *Chinese Chemical Letters*, vol. 24, no. 7, pp. 639–642, 2013.

[29] Y. P. Du, Y. Z. Liang, J. H. Jiang, R. J. Berry, and Y. Ozaki, "Spectral regions selection to improve prediction ability of PLS models by changeable size moving window partial least squares and searching combination moving window partial least squares," *Analytica Chimica Acta*, vol. 501, no. 2, pp. 183–191, 2004.

[30] S. Fang, M. Q. Zhu, and C. H. He, "Moving window as a variable selection method in potentiometric titration multivariate calibration and its application to the simultaneous determination of ions in Raschig synthesis mixtures," *Journal of Chemometrics*, vol. 23, no. 3, pp. 117–123, 2010.

[31] L. Xu, H. Y. Fu, C. B. Cai, and Y. B. She, "Quality degradation of Chinese white lotus seeds caused by dampening during processing and storage: rapid and nondestructive discrimination using near-infrared spectroscopy," *Journal of Analytical Methods in Chemistry*, vol. 2015, no. 12, p. 345352, 2015.

[32] M. Casale, P. Oliveri, C. Casolino et al., "Characterisation of PDO olive oil chianti classico by non-selective (UV-visible, NIR and MIR spectroscopy) and selective (fatty acid composition) analytical techniques," *Analytica Chimica Acta*, vol. 712, pp. 56–63, 2012.

[33] Y. Wang, M. Mei, Y. Ni, and S. Kokot, "Combined NIR/MIR analysis: a novel method for the classification of complex substances such as *Illicium verum* Hook. F. and its adulterants," *Spectrochimica Acta Part A: Molecular and Biomolecular Spectroscopy*, vol. 130, pp. 539–545, 2014.
[34] W. Luo, S. Huan, H. Fu et al., "Preliminary study on the application of near infrared spectroscopy and pattern recognition methods to classify different types of apple samples," *Food Chemistry*, vol. 128, no. 2, pp. 555–561, 2011.
[35] H.-Y. Fu, S.-Y. Huan, L. Xu et al., "Construction of an efficacious model for a nondestructive identification of traditional Chinese medicines liuwei dihuang pills from different manufacturers using near-infrared spectroscopy and moving window partial least-squares discriminant analysis," *Analytical Sciences*, vol. 25, no. 9, pp. 1143–1148, 2009.
[36] H. Zhao, B. Guo, Y. Wei, and B. Zhang, "Near infrared reflectance spectroscopy for determination of the geographical origin of wheat," *Food Chemistry*, vol. 138, no. 2-3, pp. 1902–1907, 2013.

Permissions

All chapters in this book were first published in JS, by Hindawi Publishing Corporation; hereby published with permission under the Creative Commons Attribution License or equivalent. Every chapter published in this book has been scrutinized by our experts. Their significance has been extensively debated. The topics covered herein carry significant findings which will fuel the growth of the discipline. They may even be implemented as practical applications or may be referred to as a beginning point for another development.

The contributors of this book come from diverse backgrounds, making this book a truly international effort. This book will bring forth new frontiers with its revolutionizing research information and detailed analysis of the nascent developments around the world.

We would like to thank all the contributing authors for lending their expertise to make the book truly unique. They have played a crucial role in the development of this book. Without their invaluable contributions this book wouldn't have been possible. They have made vital efforts to compile up to date information on the varied aspects of this subject to make this book a valuable addition to the collection of many professionals and students.

This book was conceptualized with the vision of imparting up-to-date information and advanced data in this field. To ensure the same, a matchless editorial board was set up. Every individual on the board went through rigorous rounds of assessment to prove their worth. After which they invested a large part of their time researching and compiling the most relevant data for our readers.

The editorial board has been involved in producing this book since its inception. They have spent rigorous hours researching and exploring the diverse topics which have resulted in the successful publishing of this book. They have passed on their knowledge of decades through this book. To expedite this challenging task, the publisher supported the team at every step. A small team of assistant editors was also appointed to further simplify the editing procedure and attain best results for the readers.

Apart from the editorial board, the designing team has also invested a significant amount of their time in understanding the subject and creating the most relevant covers. They scrutinized every image to scout for the most suitable representation of the subject and create an appropriate cover for the book.

The publishing team has been an ardent support to the editorial, designing and production team. Their endless efforts to recruit the best for this project, has resulted in the accomplishment of this book. They are a veteran in the field of academics and their pool of knowledge is as vast as their experience in printing. Their expertise and guidance has proved useful at every step. Their uncompromising quality standards have made this book an exceptional effort. Their encouragement from time to time has been an inspiration for everyone.

The publisher and the editorial board hope that this book will prove to be a valuable piece of knowledge for researchers, students, practitioners and scholars across the globe.

List of Contributors

Jin-Fang Ma and Fa-Huan Ge
School of Pharmaceutical Sciences, Sun Yat-sen University, Guangzhou 510006, China
Nansha Research Institute, Sun Yat-sen University, Guangzhou 511458, China

Tian-Ling Chen
Nansha Research Institute, Sun Yat-sen University, Guangzhou 511458, China
Guangdong Pharmaceutical University, Guangzhou 510006, China

Xiang-Dong Zhang
Nansha Research Institute, Sun Yat-sen University, Guangzhou 511458, China

Xue Xiao
Guangdong Pharmaceutical University, Guangzhou 510006, China

Jelena Krneta Nikolić, Milica Rajačić, Dragana Todorović, Marija Janković, Nataša Sarap, Gordana Pantelić and Ivana Vukanac
Institute of Nuclear Sciences "Vinča", University of Belgrade, Mike Petrovica Alasa 12-16, 11001 Belgrade, Serbia

Luke O'Neill
FOCAS Research Institute, Dublin Institute of Technology, Kevin St., Dublin 8, Ireland

Laura Perdisatt
FOCAS Research Institute/School of Chemical and Pharmaceutical Sciences, Dublin Institute of Technology, Kevin St., Dublin 8, Ireland

Christine O'Connor
School of Food Science and Environmental Health, Dublin Institute of Technology, Cathal Brugha St., Dublin 1, Ireland

Alan Mašláni, Viktor Sember and Milan Hrabovský
Institute of Plasma Physics, Za Slovankou 3, 18200 Prague, Czech Republic

Peter Ondáč
Institute of Plasma Physics, Za Slovankou 3, 18200 Prague, Czech Republic
Department of Surface and Plasma Science, Faculty of Mathematics and Physics, Charles University, V Holešovičkách 2, 18200 Prague, Czech Republic

Emily Grisanti
Institute of Analytical Chemistry, TU Bergakademie Freiberg, Leipziger Str. 29, 09599 Freiberg, Germany
Robert Bosch GmbH, Renningen, 70465 Stuttgart, Germany

Matthias Otto
Institute of Analytical Chemistry, TU Bergakademie Freiberg, Leipziger Str. 29, 09599 Freiberg, Germany

Maria Totska, Stefan Huber, Christina Krick Calderon, Monika Hohmann and Dominic Lingenfelser
Robert Bosch GmbH, Renningen, 70465 Stuttgart, Germany

Yasenjiang Kahaer and Nigara Tashpolat
College of Resources and Environment Sciences, Xinjiang University, Urumqi 830046, China
Ministry of Education Key Laboratory of Oasis Ecology, Xinjiang University, Urumqi 830046, China

Marta Petrillo and Paolo Grossi
CNR-IVALSA Trees and Timber Institute, 38010 San Michele all'Adige, Italy

Anna Sandak
CNR-IVALSA Trees and Timber Institute, 38010 San Michele all'Adige, Italy
InnoRenew CoE, 6310 Izola, Slovenia

Jakub Sandak
CNR-IVALSA Trees and Timber Institute, 38010 San Michele all'Adige, Italy
InnoRenew CoE, 6310 Izola, Slovenia
University of Primorska, Faculty of Mathematics, Natural Sciences and Information Technology, 6000 Koper, Slovenia

Dominika Janiszewska
Composite Wood Products Department, Wood Technology Institute, Poznan, Poland

Salim Hiziroglu
Natural Resource Ecology and Management, Oklahoma State University, Stillwater, Oklahoma, USA

L. R. Schimleck
Department of Wood Science and Engineering, Oregon State University, Corvallis, OR 97331, USA

J. L. M. Matos, R. Trianoski and J. G. Prata
Department of Wood Science and Technology, Federal University of Paraná, Curitiba, PR, Brazil

Jing Ming, Yan Cao, Chi Yu, Bi-Sheng Huang and Ke-Li Chen
Key Laboratory of Ministry of Education on Traditional Chinese Medicine Resource and Compound Prescription, Hubei University of Chinese Medicine, Wuhan, Hubei 430065, China

Long Chen
Key Laboratory of Ministry of Education on Traditional Chinese Medicine Resource and Compound Prescription, Hubei University of Chinese Medicine, Wuhan, Hubei 430065, China
Xiangyang Central Hospital, Xiangyang, Hubei 441021, China

Hanlie Hong, Ke Yin, Feng Cheng and Qian Fang
School of Earth Sciences, China University of Geosciences, Wuhan, China

Ming Li
School of Earth Sciences, China University of Geosciences, Wuhan, China
Department of Gemmology, Jinling Institute of Technology, Nanjing, China

Chaowen Wang
Gemmological Institute, China University of Geosciences, Wuhan, China

Bin Li, Shuo-Cheng Zhang and Yao-Dan Chi
Institute of Electrical & Computer Engineering, Jilin Jianzhu University, Changchun 130012, China

Yasmin Roye and Uche Udeochu
Department of Natural Sciences, University of Maryland Eastern Shore, 1 Backbone Rd., Princess Anne, Maryland 21853, USA

Maraizu Ukaegbu and Jonathan Onuegbu
Department of Chemistry, Howard University, 525 College St. NW, Washington, DC 20059, USA

M. Aints, P. Paris, M. Laan, K. Piip and I. Tufail
Institute of Physics, University of Tartu, Tartu, Estonia

H. Riisalu
Virumaa College, Tallinn University of Technology, Kohtla-Järve, Estonia

Iqbal Ahmad, Syed Haider Abbas, Zubair Anwar, Muhammad Ali Sheraz, Sofia Ahmed and Nafeesa Mustaan
Baqai Institute of Pharmaceutical Sciences, Baqai Medical University, Toll Plaza, Super Highway, Gadap Road, Karachi 74600, Pakistan

Muhammad Furqan Mobeen
Pharmatec Pakistan (Pvt.) Ltd., D-86/A S.I.T.E., Karachi, Pakistan

Wajiha Gul
Department of Pharmaceutical Chemistry, Dow College of Pharmacy, Dow University of Health Sciences, Ojha Campus, Karachi 74200, Pakistan

Dun Wu
School of Earth and Space Sciences, University of Science and Technology of China, CAS Key Laboratory of Crust-Mantle Materials and Environment, Hefei, Anhui 230026, China
Research Center of Coal Resources Safe Mining and Clean Utilization, LiaoNing Technical University, Fuxin 125105, Liaoning, China
Exploration Research Institute, Anhui Provincial Bureau of Coal Geology, Hefei, Anhui 23008, China

Wenyong Zhang
Exploration Research Institute, Anhui Provincial Bureau of Coal Geology, Hefei, Anhui 23008, China

Luciana Lopes Guimarães, Letícia Parada Moreira, Bárbara Faria Lourenço and Walber Toma
Laboratory of Natural Products, Department of Pharmaceutical Sciences, Universidade Santa Cecília (UNISANTA), Rua Cesário Mota, 8, F 83A, 11045-040 Santos, SP, Brazil
Department of Ecology, Universidade Santa Cecília (UNISANTA), Rua Cesário Mota, 8, 11045-040 Santos, SP, Brazil

Renato Amaro Zângaro
Center for Innovation, Technology and Education-CITE, Universidade Anhembi Morumbi (UAM), Estr. Dr. Altino Bondensan, 500, 12247-016 São José dos Campos, SP, Brazil

Marcos Tadeu Tavares Pacheco and Landulfo Silveira Jr.
Department of Ecology, Universidade Santa Cecília (UNISANTA), Rua Cesário Mota, 8, 11045-040 Santos, SP, Brazil
Center for Innovation, Technology and Education-CITE, Universidade Anhembi Morumbi (UAM), Estr. Dr. Altino Bondensan, 500, 12247-016 São José dos Campos, SP, Brazil

Hao Zhang, Haifeng Sun, Ling Wang, Shun Wang and Wei Zhang
College of Mechanical and Electrical Engineering, Henan Agricultural University, Zhengzhou 450002, China

Jiandong Hu
College of Mechanical and Electrical Engineering, Henan Agricultural University, Zhengzhou 450002, China
State Key Laboratory of Wheat and Maize Crop Science, Zhengzhou 45002, China

Xiaoxuan Zeng, Yimei Liu and Keli Chen
Key Laboratory of Ministry of Education on Traditional Chinese Medicine Resource and Compound Prescription, Hubei University of Chinese Medicine, Wuhan 430065, China

Sihe Kang
Key Laboratory of Ministry of Education on Traditional Chinese Medicine Resource and Compound Prescription, Hubei University of Chinese Medicine, Wuhan 430065, China
NMPA Key Laboratory of Quality Control of Chinese Medicine, Hubei Institute for Drug Control, Wuhan 430075, China

Haiying Deng
Medical College of Wuhan University of Science and Technology, Hubei Province Key Laboratory of Occupational Hazard Identification and Control, Wuhan 430065, China

Long Chen
Key Laboratory of Ministry of Education on Traditional Chinese Medicine Resource and Compound Prescription, Hubei University of Chinese Medicine, Wuhan 430065, China
Xiangyang Central Hospital, Xiangyang 441021, China

Xue-Ying Li, Yan Liu, Mei-Rong Lv, Yan Zou and Ping-Ping Fan
Institute of Oceanographic Instrumentation, Qilu University of Technology (Shandong Academy of Sciences), Jinan, China
Shandong Provincial Key Laboratory of Ocean Environmental Monitoring Technology, Qingdao, China
National Engineering and Technological Research Center of Marine Monitoring Equipment, Qingdao, China

Xiaodong Zhang and Keli Chen
Key Laboratory of Ministry of Education on Traditional Chinese Medicine Resource and Compound
Prescription & Hubei University of Chinese Medicine, Wuhan 430065, China

Long Chen
Xiangyang Central Hospital, Xiangyang, Hubei 441500, China

Yu Bai
Mayinglong Pharmaceutical Group Co., Wuhan 430065, China

Leandro da Conceição Luiz, Maria José Valenzuela Bell, Nayara Lizandra Leal and Virgílio de Carvalho dos Anjos
Grupo de Engenharia e Espectroscopia de Materiais, Departamento de Física, Universidade Federal de Juiz de Fora, 36036-900 Juiz de Fora, MG, Brazil

Roney Alves da Rocha
Departamento de Ciências de Alimentos, Universidade Federal de Lavras, 37200-000 Lavras, MG, Brazil

Beata Brozek-Pluska, Monika Kopec and Jakub Surmacki
Laboratory of Laser Molecular Spectroscopy, Institute of Applied Radiation Chemistry, Lodz University of Technology, Lodz, Poland

Haiyan Fu, Qiong Shi, Liuna Wei, Xiaoming Guo, Ou Hu, Wei Lan and Tianming Yang
The Modernization Engineering Technology Research Center of Ethnic Minority Medicine of Hubei Province, College of Pharmacy, South-Central University for Nationalities, Wuhan 430074, China

Lu Xu
College of Material and Chemical Engineering, Tongren University, Tongren 554300, Guizhou, China

Shunping Xie
State Key Laboratory Breeding Base of Green Chemistry-Synthesis Technology, College of Chemical Engineering, Zhejiang University of Technology, Hangzhou 310032, China

Index

Q

R

S

T

V

W